Foundations of Mathematical and Computational Economics

Second Edition

Kamran Dadkhah

Foundations of Mathematical and Computational Economics

Second Edition

Springer

Kamran Dadkhah
Northeastern University
Department of Economics
Boston
USA
k.dadkhah@neu.edu

First edition published by Thomson South-Western 2007, ISBN 978-0324235838

ISBN 978-3-642-13747-1 e-ISBN 978-3-642-13748-8
DOI 10.1007/978-3-642-13748-8
Springer Heidelberg Dordrecht London New York

Cover design: WMXDesign, Heidelberg

Printed on acid-free paper

Springer is part of Springer Science+Business Media (www.springer.com)

To Karen and my daughter, Lara

Preface

Mathematics is both a language of its own and a way of thinking; applying mathematics to economics reveals that mathematics is indeed inherent to economic life. The objective of this book is to teach mathematical knowledge and computational skills required for macro and microeconomic analysis, as well as econometrics. In addition, I hope it conveys a deeper understanding and appreciation of mathematics.

Examples in the following chapters are chosen from all areas of economics and econometrics. Some have very practical applications, such as determining monthly mortgage payments; others involve more abstract models, such as systems of dynamic equations. Some examples are familiar in the study of micro and macroeconomics; others involve less well-known and more recent models, such as real business cycle theory.

Increasingly, economists need to make complicated calculations. Systems of dynamic equations are used to forecast different economic variables several years into the future. Such systems are used to assess the effects of alternative policies, such as different methods of financing Social Security over a few decades. Also, many theories in microeconomics, industrial organization, and macroeconomics require modeling the behavior and interactions of many decision makers. These types of calculation require computational dexterity. Thus, this book provides an introduction to numerical methods, computation, and programming with Excel and Matlab. In addition, because of the increasing use of computer software such as Maple and Mathematica, sections are included to introduce the student to differentiation, integration, and solving difference and differential equations using Maple and to the concept of computer-aided mathematical proof.

The second edition differs from the first in several respects. Parts of the book are rearranged, some materials are deleted and some new topics and examples are added. In the first edition most computational examples used Matlab and some Excel. In the present edition, Excel and Matlab are given equal weights. These are done in the hope of making the book more reader friendly. Similarly, more use is made of the Maple program for solving non-numerical problems. Finally, many errors had crept into the first edition, which are corrected in the present edition. I am indebted to students in my math and stat classes for pointing out some of them.

I would like to thank Barbara Fess of Springer-Verlag for her support in preparing this new edition. I also would like to thank Saranya Baskar and her colleagues at Integra for their excellent work in producing the book.

As always, my greatest appreciation is to Karen Challberg, who during the entire project gave me support, encouragement, and love.

Contents

Part I
Basic Concepts and Methods

Chapter 1
Mathematics, Computation, and Economics

1.1 Mathematics

Many believe that mathematics is one of the most beautiful creations of humankind, second only to music. The word *creation*, however, may be disputed. Have humans created something called mathematics? Then how is it that we increasingly discover that the world, and indeed the universe around us, obey one or another mathematical law? Perhaps to the faithful, the answer is clear: A higher power is the greatest mathematician of all. A worldly answer may be that humans discovered, rather than created, mathematics. Thus, wherever we look, we see mathematical laws at work. Whether humans created mathematics and it just so happened that the world seems to be mathematical, or the universe is a giant math problem and human beings are discovering it, we cannot deny that mathematics is extremely useful in every branch of science and technology, and even in everyday life. Without math, most likely we would be still living in caves. But what is mathematics? And why are so many afraid of it?

To begin with, for the greatest part mathematics is nothing but logical thinking. Indeed, mathematics consists of three parts: *definitions*, *logical thinking*, and *axioms*. We will take up each of these parts in turn, but let's clear up one confusion right away: What about all those dreaded Greek and other strange symbols? First, we use Greek symbols because Greeks started the systematic study of mathematics and some reward must accrue to pioneers. Second, the Greek alphabet and symbols simply serve as a shorthand that makes understanding mathematics easier. If you don't believe this, consider how one of the great mathematicians of the Middle Ages, Omar Khayyam,[1] explained the solution of a quadratic equation.

[1]Omar Khayyam (1048–1131), the great Iranian mathematician, is better known in the West for his *Rubaiyat*, a book of poetry freely translated by Edward J. Fitzgerald. Khayyam has several treatises on mathematics including *Treatise on Demonstration of Problems of Algebra*, from which the quotation in the text is taken. He worked with a group of learned men on the compilation of astronomical tables and the reform of the calendar, which resulted in the Jalali calendar (after Jalaleddin Malik Shah Seljuq), a quite accurate calendar for his time and for many centuries to come. Khayyam measured the length of a year with amazing accuracy. He also knew the Pascal

K. Dadkhah, *Foundations of Mathematical and Computational Economics*, 2nd ed.,
DOI 10.1007/978-3-642-13748-8_1,

> The second species. *A square and a number equal a root*. It is necessary in this class that the number should not be greater than the square of half the number of roots, otherwise the problem would be impossible to solve. If the number is equal to the square of half the root-number, then half the root-number is equal to the root of the square. If the number is less than the square of half the root, subtract it from the square of half the roots and extract the square root of the remainder, adding the result to half the roots or subtracting it from the latter. The sum, in case of addition, or the remainder, in case of subtraction, is the root of the square. The arithmetical proof of this theorem is known when its geometrical demonstration is known.[2]

Translation, please! Put more clearly, the question is finding the solution of a quadratic equation of the form

$$x^2 + c = bx \qquad \text{or} \qquad x^2 - bx + c = 0$$

The solution to this equation, which we shall discuss in Chap. 2, is

$$x = \frac{b \pm \sqrt{b^2 - 4c}}{2}$$

Khayyam argues that the equation does not have a solution if $(b/2)^2 < c$ or $b^2 - 4c < 0$ because this would result in a complex number and at the time complex numbers were not known. But if $(b/2)^2 = c$, then x = b/2 and if $(b/2)^2 > c$, then the solutions are

$$x_1 = \frac{b + \sqrt{b^2 - 4c}}{2} \quad \text{and} \quad x_2 = \frac{b - \sqrt{b^2 - 4c}}{2}$$

Thus, Khayyam had correctly solved the problem, yet a modern reader would have difficulty understanding it. Nevertheless, if you think that 1000 years after Khayyam you can do better, then please write down a description of the problem and its solution in plain English. Clarity and economy require the use of symbols and conventions.

One other important lesson from the above quotation is that mathematics and symbols are two different things. Khayyam was doing mathematics; the fact that he did not use symbols made no difference. Symbols cannot make a banal statement about mathematics, and not using symbols cannot detract from the essence of a mathematical argument.

Going back to our discussion, the necessity of definitions should be clear. If we are to have a logical discussion, we should all speak the same language. In other

---

triangle and the coefficients of the binomial expansion, but because of the limitation of mathematics in his time, he could not express them in the general form of today. Khayyam was also a philosopher, and his poetry is mostly musings on the meaning of life.

[2] *The Algebra of Omar Khayyam*, translated by Daoud Kasir, New York, Teachers College, Columbia University, 1931, p. 61.

words, an expression or statement should mean the same thing to all of us. A mathematical definition should be such that, in no uncertain terms, it includes all of the items that conform to the definition and excludes all that do not. In this matter, mathematicians are a bit obsessive-compulsive. They go to great lengths to make sure we are speaking about the same items and situations. But there are exceptions. In every branch of mathematics, a number of concepts are deemed too fundamental or basic to be defined. For example, in geometry, concepts such as point and line are at best vaguely defined. We assume that all are well aware of what we are talking about. The same is true about the concept of a set in set theory and a random event in probability theory. The rationale should be clear. Every concept has to be defined in terms of some other concept or object. We cannot start from nothing and expect to create a host of meaningful definitions. We may end up trying to define the meaning of the word *word*.

By far the largest part of mathematics is logical thinking, that is, starting with assumptions and deriving implications from them. In mathematics, every statement flows logically from another. As such, you do not get anything from math unless you have already put it in. It is like the work of a magician, who pulls a rabbit out of the hat as if it materialized out of thin air. But the rabbit was there all along, except that the audience could not see it—perhaps because of distraction or you did not pay enough attention to see something that was in full view. Of course, let us not forget the skill of the magician. Not everyone can perform a magic trick like Houdini or Siegfreid and Roy.

Let us try some such trick. If you have seen it before, humor me. If you haven't, please do not look up the answer until you have answered all my questions.

1. Pick a number between one and nine (the number of movies you would like to see in a week or the number of times you prefer to eat out).
2. Multiply the number by 2 (what the heck?).
3. Add 5.
4. Multiply the result by 50.
5. If you have already had your birthday this calendar year, add 1750; or else add 1749.
6. Now, what year is this? 2011? 2012? 2013? Whatever it is, add the last digit of the year, say, 11 if it is 2011 or 13 if it is 2013.
7. Now, subtract the year you were born (all four digits, e.g., 1989).
8. You should have a three-digit number. The first digit is your original number; the next two numbers are your age!

To solve the mystery, let x be the number you chose. Then the steps taken were

1. x
2. $2x$
3. $2x + 5$
4. $50(2x - 5) = 100x + 250$
5. $100x + 250 + 1750 = 100x + 2000$

6. $100x + 2000 + 11 = 100x + 2011 = 100x + \text{current year}$ (assuming the year is 2011)
7. $100x + \text{current year} - \text{your birth year} = 100x + \text{your age}$

"Elementary, my dear Watson!"

Similarly, the solution of the quadratic equation we mentioned earlier was already inside the equation. We could write

$$\left(x - \frac{b + \sqrt{b^2 - 4c}}{2}\right)\left(x - \frac{b - \sqrt{b^2 - 4c}}{2}\right) = 0$$

If you carry out the multiplication, you will get the original equation.

Two issues remain to be resolved. First, logical thinking on its own is not adequate to build the magnificent edifice of mathematics. Logical thinking derives the implications of assumptions, and from nothing one cannot derive something. This points out the importance of axioms. Every field of mathematics starts with a number of axioms that cannot be proved. We assume such axioms to be self-evident, or at least we shall treat them as such. Consider the axioms of natural numbers as defined by Peano.[3]

1. Zero is a natural number.
2. Every natural number a is followed by another natural number $a + 1$.
2. Zero is not a successor to a natural number.
3. For any two natural numbers $a \neq b$, we have $a + 1 \neq b + 1$.
4. If zero has a property and if a having that property implies the same for $a + 1$, then all natural numbers have that property.

These axioms seem unassailable and one cannot dispute them. Of course, natural numbers are not the only numbers in the world. We can think of real or imaginary numbers that do not correspond to the above axioms. Perhaps the most famous and talked about set of axioms are those of Euclidean geometry. Originally they were formulated about 2300 years ago by Euclid,[4] whose brilliant work commands the admiration of modern-day mathematicians.

1. Only one straight line passes through two distinct points.
2. Given three distinct points on a straight line, only one lies between the other two.
3. If two line segments AB and CD are both congruent to EF, then AB and CD are congruent.
4. Given any real number $d > 0$, there exists a line segment of length d.

[3]Gioseppe Peano (1858–1932), an Italian mathematician, contributed to the study of differential equations and was a founder of mathematical logic. Bertrand Russell wrote in his autobiography that the 1900 International Congress of Philosophy "was the turning point of my intellectual life, because there I met Peano."

[4]Euclid of Alexandria (~325BC–~265BC) was the most famous mathematician of antiquity, whose book *The Elements* brought together the mathematical knowledge of his day with rigor and clarity. Indeed, the geometry we learn and use in high schools today is based on Euclid's axioms and, therefore, referred to as Euclidean geometry. *The Elements* is still available in bookstores and from online bookshops.

5. Given a line and a point outside it, exactly one and only one line passes through the point and is parallel to the line.

These axioms correspond to our experiences in the real world. They seem to be self-evident. Of course, axioms require definitions of many terms such as *congruent* or *between* that are parts of the system of axioms. We did not present them here, as our objective is not a thorough discussion of geometry. Euclid's axioms or postulates were not complete or 100% rigorous. The present-day axioms of geometry are due to Hilbert,[5] who published them in his 1899 book *Grundlagen der Geometrie*.

For many years, mathematicians wondered if the axioms of geometry could be further reduced. In particular, the fifth postulate (or axiom) was the subject of intense scrutiny. If it could be deduced from the other axioms, then it would be redundant. All attempts in this direction failed. Then in the eighteenth and nineteenth centuries, mathematicians tried changing the last axiom to see if it resulted in a contradiction. If that happened, it meant that the fifth postulate followed from the other four and therefore was not needed. This line of work led to surprising discoveries: the non-Euclidean geometries.

Replacing the fifth postulate with the *hyperbolic axiom*—given a line and a point off the line, more than one line can pass through the point and are parallel to the line—Lobachevsky[6] discovered the *hyperbolic* or *Lobachevsky's geometry*. Reimann[7] replaced the fifth postulate with the *elliptic* axiom—that is, there exists no line that passes through a point outside a line and is parallel to the line—that resulted in *elliptic geometry*.

Finding the bare minimum axioms upon which a field of mathematics could be founded requires nothing short of mathematical genius. The mathematicians' names attached to the axioms of different fields of mathematics bear this out. In addition to those mentioned previously, we can add Cantor[8] (set theory) and Kolmogorov (probability theory).

---

[5]David Hilbert (1862–1943), one of the greatest mathematicians of the twentieth century, is best known for his work for infinite dimensional space referred to as Hilbert space. In a speech he challenged mathematicians to solve 23 fundamental problems, a challenge that is still partly open. Hilbert also contributed to mathematical physics. For more on Hilbert's problems, which have been solved, and who has solved them, see *The Honors Class, Hilbert's Problems and Their Solvers* by Ben H. Yendell (2002).

[6]Nikolai Ivanovich Lobachevsky (1792–1856), the great Russian mathematician, presented his results on non-Euclidean geometry in 1826, although not many of his contemporaries understood it. János Bolyai (1802–1860), a Hungarian mathematician, independently discovered hyperbolic geometry.

[7]German mathematician Georg Friedrich Berhard Riemann (1826–1866), who despite his short life made brilliant contributions to mathematics. Of him it is said that "he touched nothing that he did not in some measure revolutionize." [*Men of Mathematics*, by E. T. Bell (1937)].

[8]Set theory was founded in the late nineteenth century by Georg Cantor (1845–1918). He based his analysis on three axioms. But soon these axioms ran into paradoxes, the discovery of the the most famous of them being due to Bertrand Russell. The interested reader is referred to *Mathematics, The Loss of Certainty*, by Morris Kline (1980). To avoid paradoxes and to assure consistent deductions, set theory is based on Zermelo-Frankel axioms plus the axiom of choice. A discussion of these axioms is well beyond this book.

Aside from axioms, to find the appropriate assumptions and work out their implications, that is, to prove a theorem or lemma,[9] requires mathematical intuition, ingenuity, and knowledge. The process is usually the reverse of what appears in math books. The mathematician "sees" the proposition in the same way that a musician "hears" the as-yet-unwritten music or Michelangelo "saw" the sculpture of David in the massive stone he had in front of him. Then in the same way that the musician composes what he has conceived and Michelangelo extracted his masterpiece, the mathematician works out the necessary assumptions and proofs.

Mathematical work follows the strict rules of logic to deduce a proposition from a set of assumptions. Thus, every statement is either true or false. But this does not mean that we can prove the truth of each and every statement. Gödel[10] proved that in any mathematical structure based on a set of consistent axioms, there are statements that are true but cannot be proved or disproved within that system of axioms. Cohen,[11] using a method called forcing, showed that the truth or falsity of some statements cannot be decided. Cantor had proved that the set of all integers and the set of all rational numbers have the size $\aleph_0$, which is the same as the size of the set of all positive integers. The set of all real numbers is bigger; perhaps its size is $\aleph_1, \aleph_2, \aleph_3$, or .... This is Cantor's continuum problem, which he himself was unable to solve. Gödel showed that one cannot prove that the size of the set of all real numbers is not $\aleph_1$. Cohen showed that this issue cannot be decided within the axioms of set theory. Thus, mathematical statements are either true, false, or undecided.

Economists in the course of their research may never come into contact with a mathematical statement that cannot be decided. Nevertheless, the subject is an important one for understanding mathematics. Suppose we add another axiom to the axioms of set theory by which we can decide that the continuum hypothesis is true, or another axiom that renders it false. It seems that, like geometry, we can have not one, but many set theories. It may be tempting to conclude that mathematics is invented; otherwise how could we account for the variety of geometries? Such a conclusion is not 100% warranted, as each geometry could be relevant to one set of circumstances or one part of the universe. For example, as long as we are on Earth and concerned with short distances, Euclidean geometry rules. On the other hand, non-Euclidean geometry may be relevant for the study of outer space and distances measured in millions of light years. We may have discovered math.

---

[9]Usually, a *lemma* is a mini-theorem.

[10]Kurt Gödel (1906–1978) was born in Brno, Czech Republic, but spent more than half of his life in the United States doing research at the Institute for Advanced Study and at Princeton University, although he did not have lecturing duties. Gödel's fame rests on his three theorems in mathematical logic: the completeness theorem for predicate calculus (1930), the incompleteness theorem for arithmetic (1930), and the theorem on consistency of the axiom of choice with continuum hypothesis (1938).

[11]Paul Joseph Cohen (1934-2007) was a professor in the mathematics department at Stanford University. He received a Fields Medal (the equivalent of a Nobel Prize in mathematics) in 1966.

1.2 Philosophies of Mathematics

Philosophies of mathematics are concerned with the meaning of mathematics and the nature of mathematical truth. As such, many working mathematicians never find a need for nor find the time to spend on the subject. Perhaps the same is true for other professionals regarding their own discipline. Economists may have more reason to be aware of the issues raised in philosophies of mathematics because they have a bearing on the modeling activity and the application of mathematics in economics. Nevertheless, we shall be brief; there are many interesting books on the subject that the reader may want to consult[12] and, in allocation of the limited resource of time, the subject cannot be assigned high priority.

The first question to consider is whether mathematical reality has an independent existence or is only a construction of humans and their minds (or cultures). In other words, are there such things as circle, line, number 431, the number π, the Pythagoras theorem, and Taylor formula, or have humans made them up like the king, bishop, rook, and the rules of their movements in the game of chess? Those who believe that such entities really do exist are referred to as *Platonists* or *realists* whereas those who are not of such persuasion are in two camps of *intuitionists* and *formalists*.

The first group do not believe in the existence of Plato's world of ideal forms but believe that abstractions such as those mentioned above are indeed independent of a person's mind and have a real existence. Thus, mathematicians are really discovering relationships between these abstract entities. Among the realists are such great mathematicians as Gödel, Alfred North Whitehead,[13] and Bertrand Russell.[14]

Intuitionists believe that such entities are our own creation because they are the results of our perception of time and space. Thus, they do not exist unless we perceive them.[15] The formalists, on the other hand, say that the whole mathematics is manipulation of symbols. Those symbols have no particular meaning and therefore need not have any connection to anything in the real world.

The second question is with regard to the nature of truth in mathematics. Based on their answers, mathematicians are divided into logicists, formalists, and constructivists. To clarify this issue note that laws of physics say that one thing leads

---

[12]In particular, I recommend the following books: *Philosophy of Mathematics: A Contemporary Introduction to the World of Proofs and Pictures* (2008) by James Robert Brown; *The Mathematical Experience* (1981) by Philip Davis and Reuben Hersh; and *The Road to Reality, A Complete Guide to the Laws of the Universe* (2005) by Roger Penrose. The last book is concerned with the applications of mathematics to physics.

[13]Alfred North Whitehead (1861–1947) was a British mathematician and a collaborator of Russell.

[14]British mathematician and philosopher, Bertrand Arthur William Russell (1872–1970) is considered one of the most important logicians of the last century. In 1950, he won the Nobel Prize in literature "in recognition of his varied and significant writings in which he champions humanitarian ideals and freedom of thought."

[15]A prominent intuitionist is Luitzen Egbertus Jan Brouwer (1881–1966) of the Netherlands. Economists will become familiar with his name and work through his *fixed point theorem* that is crucial for proving the existence of equilibrium.

to another, for example, water at sea level starts boiling when heated to 100 degrees centigrade. The same is true in economics; for example, if income increases, so does consumption. These are observed relationships and there is no logical reason that one thing should follow from the other. We can logically think of a situation that water starts to boil at 135 centigrade or an increase in income would leave consumption unchanged. Mathematical relationships, on the other hand, are logical. Conclusions perforce follow from the assumptions: $2 + 2$ is always 4, given the axioms of Euclidean geometry, the angles of a triangle sum to π radians, and it is always true that $(a - b)^3 = a^3 - 3a^2b + 3ab^2 - b^3$. In other words, mathematical propositions are derived by the application of logic to axioms and nothing else. In this regard *logicists* and formalists do not differ except that the former started from a realist point of view and applied logic to axioms while the latter insisted on defining everything formally as symbols and rules of deduction. The third group are *constructivists*, who start from the intuitionist point of view but accept a proof only if it can be constructed from the ground up. In other words, it is not enough to prove such and such exists; it has to be shown how to find that which exists.

Logicism and formalism were a reaction to some loose ends mathematicians of the nineteenth century found in calculus and other branches of mathematics. On the other hand, both groups went too far in trying to show that every step follows from the previous one by the application of formal logic.[16] The formalist project was championed by David Hilbert, and later on a group of French mathematicians under the pen name of Nicholas Bourbaki published several volumes presenting different areas of mathematics in completely formal fashion. The volumes are self-sufficient in that there are no references to external sources, and everything is formally defined and proved before it is used in later sections or volumes. Moreover, there is no reference to any geometrical figure or form as an aid for imagining or understanding a concept. It is generally believed that the books are not appetizing nor are they good pedagogically. Nevertheless, the Bourbaki approach had great influence on a generation of mathematicians and, perhaps through them, on economics.[17]

It is believed that most mathematicians are realists and while all believe in rigorous proof, they think that pure formalism is cumbersome and wasteful of time and paper. It is also said that most mathematicians are realists but would defend their profession against outsiders and philosophers by resorting to formalism. Constructivists are a tiny minority.

But what can we make of all these? For the most part we can go about doing mathematics without worrying about these issues. In those rare philosophical moments, I am a realist believing in the existence of mathematics independent of our minds. As recounted in this book, mathematics has been an international project: Greeks started it; Iranians, French, Germans, Russians, Poles, British, Americans,

[16] It took 362 pages in Russell's *Principia Mathematica* to show that $1 + 1 = 2$.

[17] A glimpse of the history of the evolution of mathematics in the twentieth century and its relation to economics is captured in E. Roy Weintraub, *How Economics Became a Mathematical Science* (2002).

Norwegians, and others contributed to it. If their work were not based on some external reality, how could we have only one mathematics? If mathematics were completely the result of mental and cultural activity, we should have a Greek, a Russian, an American, and perhaps a Christian, a Judaic, a Buddhist, and a Moslem mathematics. We don't. Perhaps we have only one mathematics, because all mathematicians or all reasonable people would agree on the same set of assumptions and derivation rules. After all, mathematics cannot simply be the product of just one mind. But how did we reach this consensus, and who is eligible to vote in this process? We cannot allow just everyone to express opinion on mathematical issues, from the Pythagoras theorem to tensor analysis. But if we only allow those who know mathematics to participate in the process of building a consensus, we shall be guilty of circuitous argument. In other words, mathematics is what mathematicians agree on and mathematicians are those who do mathematics. If mathematics were simply a game invented by mathematicians, how could we find it so useful? If mathematics is simply a game, it is not a spectators' game nor has it a high entertainment value as do soccer and baseball. Therefore, if it were just a game for a few, why would different societies spend so many resources in operating mathematics departments and research institutes? This last argument should win over the economists.

Finally, whereas mathematics has to be rigorous and has no room for hand-waving arguments, the Bourbaki's and formalists' way of doing mathematics was both extreme and wasteful.

1.3 Women in Mathematics

In this book, we encounter a number of mathematicians. It turns out that all are men. This may give the impression that mathematics is a man's sport. That is a wrong conclusion. From the beginning, women have shown great aptitude for mathematics. The first known woman mathematician is Hypatia (360–415), who wrote treatises on algebra and commentary on Euclid's *Elements*. During the Enlightenment, there were Sophie Germain (1776–1831), who contributed to the solution of Fermat's last theorem and whose work was admired by Gauss. In the nineteenth century, Ada Augusta Byron Lovelace (1815–1852) was a pioneer in the development of the computer; Mary Fairfax Grieg Somerville (1780–1872), the author of *The Mechanism of the Heavens*; Sofya Korvin-Krukovskaya Kovalevskaya (1850–1891), who made original contributions to the study of partial differential equations, astronomy, and mechanics; and Emmy Noether (1882–1935) of whom Albert Einstein said, "the most significant creative mathematical genius thus far produced since the higher education of women began."

The achievements of these women are more remarkable because for a long time women were barred from pursuing mathematics. Perhaps men were afraid of competition. With the changes in the role of women in science and society in the twentieth and twenty-first centuries, a number of prominent women mathematicians have emerged. Lenore Blum (1942), Sylvia Bozeman (1947), Marjorie

Lee Browne (1914–1979), Fan King Chung (1949), Evelyn Boyd Granville (1924), Rhonda Hughes (1947), Nancy Kopell (1942), Cora Sadosky (1940), and Ingrid Daubechies (1954), who is famous for her work on wavelets.[18] It is noteworthy that among the women mathematicians mentioned above are African Americans, Chinese, Europeans, and Latin Americans. So neither sex nor race have anything to do with mathematics.

1.4 Computation

In everyday life experience and in running any small or large business, numbers and computation play significant roles. From deciding where to take a vacation, how much money you need, whether to buy or lease a car, to the amount to contribute to the retirement account, you need to use numbers and calculations. Similarly, in running a business, no matter how small, you need to have hard data and numbers. A restaurateur has to figure out how much it costs to prepare a dish in the restaurant and how much should be charged for it. A farmer needs to know how to allocate his land to different crops. When we get to giants of industry such as General Motors, Microsoft, and the Marriott Hotel chain, questions of where to locate each site of activity, where to procure material, and how much to outsource require quite sophisticated computations.

A common question at the time of refinancing a mortgage is whether it is worth the cost. Suppose you have a $200,000 mortgage with an interest rate of 6% that has to be paid in the next 15 years. Is it advisable to refinance if you could get a rate of 5.5%? Let's calculate. Staying with the old mortgage would give you a monthly payment of $1,687.71, which includes interest and principal but not property taxes. In a year, you will pay $11,769.23 in interest and in 2 years $23,015.23. If you refinance, you pay $1,634.17 per month. Your interest payment in the first year will be $10,779.60 and in 2 years $21,061.10. Thus, in the first year you save $989.63 and in 2 years $1,954.13. If your marginal tax rate is 15%, then your total saving is $841.18 for 1 year and $1,661.01 for 2 years. Now you can compare these numbers to the cost of refinancing and take into account if there is any chance of selling the house within the next 2 years. Without these calculations we can only philosophize.

We can illustrate the importance of computation with many more examples, but nowhere is the result as dramatic as in economic policy analysis. What is the effect of a tax cut? What is the effect of a quarter percentage point increase in the federal funds rate on investment, consumer expenditures, and on the government budget deficit? Without numbers, the discussion can degenerate into ideological rant. A similar situation arises in analyzing economic models. It would be nice if we could simply say that an increase in government expenditures or taxes, or money supply will unequivocally have such and such effects. But especially in an open economy,

[18]For a more comprehensive list, the reader may want to consult *Notable Women in Mathematics, A Biographical Dictionary* (1998), edited by Charlene Morrow and Teri Perl.

any change could have ambiguous effects. The only way to come up with an estimate of the net effects is to assign numbers to different coefficients and elasticities.

The difficulties with models increase tremendously when we analyze dynamic models over time. Suppose we take up the question of Social Security reform. The whole story depends on computation: Is the system viable and can it function in the foreseeable future? Will it break sometime in the future? If so, when? In such an eventuality, what is the amount of deficit? What does it take to fix the system now? Ten years hence? in 2035? What would be the effect of a rise in Social Security tax? A cut in benefits? Partial privatization? The answers to all these questions require computation. Even if we do not agree on the assumptions or the coefficients of the model used, we can pinpoint the areas of dispute and get a sense of the order of magnitude of difference in our estimates of shortfall in the Social Security fund. Thus, the importance of computation cannot be exaggerated, and all economists need to know where their data come from and how the numbers and quantities are calculated. It is imperative to make sure that data are verified and reliable and that computations are correct.

Modern computation requires two complementary skills. First, we must know, or at least be familiar with, the methods of numerical analysis. Second, we must know the art of computer programming, or at least be familiar with a programming language and a couple of software packages. Numerical analysis is that part of applied mathematics concerned with obtaining accurate numerical values. In ordinary life we simply add two numbers or even take the logarithm of a number. In elementary calculus we learn to find the maximum and minimum of a function. It is possible to imagine, for example, that every calculator has a filing system and that, when we punch in 2.374 and then the `ln` function, the file cabinets are ransacked to find the correct logarithm. Or when we use an optimization program to find the maximum or minimum of a function, we might imagine that the computer takes the partial derivatives of the function, sets them equal to zero, and finds the values of the variables that maximize or minimize the function. Neither perception is correct. Computation is an iterative process. To find the maximum of a function, we start with an initial guess, and step-by-step get closer to the maximum point. We stop when the improvement in our approximation from one step to another is so small as to be negligible by a preset standard.

Numerical analysis is a science and an art in its own right that economists need to learn. Computer programming is also an art and a science. The tendency may be to program an algorithm as an exact replica of its mathematical version. Yet this may be an inefficient use of computing power. An efficient program minimizes the number of evaluations to be performed, saving time and computing power. This issue gains tremendous importance when the problem is large and complicated.

On the other hand, a program should be reusable; a program that can compute only the mortgage payment when the rate is 5.5% is of not much use. A program should be designed so that it can easily be used with any set of parameter values. It should also be portable in the sense that its whole or modules could be used in other programs. Finally, a program should be clear and well documented. The document will help others, but mostly the author of the program herself. It is not unusual to look at one's own program after a few months and wonder whoever wrote this program, hence the importance of extensive documentation and comments in programming.

1.5 Mathematics and Economics

Theorizing in economics as in other sciences starts with a set of assumptions and proceeds to logically draw conclusions that will explain the phenomena of interest and make predictions as to future outcomes or the effects of different interventions. Because this is an exercise in logic and given our definition of mathematics, it should come as no surprise that mathematics has become an indispensable tool of economics, nor that we have a specialized field of mathematical economics. If theory is how we organize our thought and mathematics is an efficient and parsimonious process of logical thinking, it seems rational to use mathematics in economics. This is the road that all sciences have followed and economics hasn't been an exception.

Mathematics forces us to be explicit and precise about our assumptions and conclusions. The precision allows us and others to see if our ideas offer anything of substance. It is always easy to talk oneself or others into believing that something of substance has been said or that a model unequivocally predicts one thing or another. Mathematics takes away the means of such chicanery. In the meantime, mathematics enables the theorist to dispense with extraneous assumptions, basing a model on the minimum assumptions required. Furthermore, the theorist is forced to show that the conclusions and predictions of a model mathematically (logically) follow from her assumptions. Mathematics enables us to verify the validity of a logical argument and detect inconsistencies or double-talk. Last but not least, mathematics allows us to formulate models in such a way that they can be estimated using quantitative data. This, in turn, allows researchers to statistically test different hypotheses. The importance of this function will be better appreciated considering the quantitative nature of economics and the role of computation in the theory and application of economics.

Since World War II, mathematics has occupied an increasingly prominent place in economic analysis. The use of mathematics in economics, however, dates back to the early days of this science, and many great economists were either trained as mathematicians or well versed in the subject. Mathematics and statistics have been used in the work of almost all economists: Cournot's work in the early nineteenth century employed mathematics in an essential way, as did Irving Fisher's dissertation published in 1892.

1.6 Computation and Economics

Joseph Schumpeter noted that economics was the only science that is inherently quantitative. He wrote

> There is, however, one sense in which economics is the most quantitative, not only of "social" or "moral" sciences, but of *all* sciences, physics not excluded. For mass, velocity,

> current, and the like *can* undoubtedly be measured, but in order to do so we must always invent a distinct process of measurement. This must be done before we can deal with these phenomena *numerically*. Some of the most fundamental economic facts, on the contrary, already present themselves to our observation as quantities made numerical by life itself. They carry meaning only by virtue of their numerical character. There would be movement even if we were unable to turn it into measurable quantity, but there cannot be prices independent of the numerical expression of every one of them, and of definite numerical relations among all of them [emphases in the original].[19]

Even Karl Marx felt the need for quantitative analysis. In 1873 he wrote to Engels, "You know the diagrams in which changes over time occurring in prices, discount rates, etc., are presented as rising and falling zig-zag lines. When analyzing crises, I have tried, on various occasions, to compute these ups and downs by fitting irregular curves and I believe that the main laws of the crises could be mathematically determined from such curves. I still believe this to be feasible given sufficient data."[20]

Indeed, quantitative analysis and computation play a pivotal role in economics.

Computation is pivotal in three areas of economic practice: verification or falsification of economic theories, policy evaluation, and forecasting.

When we assert that a higher price results in lower quantity demanded or that the cause of inflation is the growth of money supply, we have to show that indeed there exists statistically significant relationship between quantity demanded and price or between the rate of inflation and growth rate of money supply. Furthermore, we have to be able to document that price negatively affects quantity demanded and money supply positively affects inflation. The tool for such verification is econometrics. Throughout this book we use many examples from econometrics.

Decision making in economics involves forecasting even if it is implicit. Not to carry an umbrella means you are not expecting rain. Buying any asset including real estate, stocks, bonds, and others means that the buyer is forecasting a stream of income and perhaps an appreciation in value of the asset. Further, she must be estimating that it is the best among the alternative purchases. The same can be said about the sale of an asset except that now the seller assumes that there will be little or no appreciation. The Federal Reserve has to base its monetary policy on a forecast of the future course of the economy. Without such a forecast the board of governors would resemble a driver who is driving with windshield covered with snow.

On many occasions the question of policy choice revolves on the orders of magnitude of different effects of the policies being considered. Every economic policy has more than one effect and, of course, many policies impose costs on citizens. To what extent we are ready to accept adverse effects and costs of such policies in order to attain the stated goals. Consider an economy that is experiencing high rates of inflation. A restrictive monetary policy is the remedy. But in the short run it will cause unemployment. To what extent are we ready to endure unemployment

---

[19] Joseph A. Schumpeter, "The Common Sense of Econometrics," *Econometrica*, 1, 1933, pp. 5–6.

[20] Quoted in Leon Smolinski, "Karl Marx and Mathematical Economics," *Journal of Political Economy*, 81, 1973, p. 1200.

to bring under control the rate of inflation and to dampen inflationary expectations? Based on verified relationships in the economy (as discussed above) we should estimate the effects of each policy, as there may be more than one option conducive to our goal. It is only with such computations that we can have an intelligent policy assessment.

In general any economic action, change, and policy initiative has two or more opposing effects. Consider a time of recession when many workers may become unemployed. Leaving them in cold would hurt them and their families. Furthermore, their lack of income would reduce the aggregate demand which, in turn, would aggravate the employment situation. Thus, paying unemployment benefits and extending them would be desirable from both a welfare and macroeconomic point of view. On the other hand, such payments may rob the receiving workers and even others of their incentives to seek jobs or work. It is the job of an economist to assess the two effects and pronounce which is the dominant and if the net effect is beneficial or harmful. To say: "on the one hand and then on the other," is a copout. Similarly, in any intervention it is not whether to intervene or not but to what extent.

Chapter 2
Basic Mathematical Concepts and Methods

This chapter and the next two have three objectives. First, to introduce the reader to some basic concepts and formulas that will be needed in later chapters. Second, to serve as an introduction to computation and numerical methods and the use of Excel and Matlab procedures. The present chapter is devoted to mathematics and Chap. 3 is an introduction to computation and Chap. 4 will concentrate on probability theory and statistics. Those who are familiar with the material may want to glance through these chapters and move on. A third function of the chapters is to provide a handy reference for readers who, in reading later chapters, might feel a need to refresh their understanding of a concept or to check a formula.

2.1 Functions of Real Variables

In studying economic phenomena, we frequently come across cases in which variation in one variable induces variation in another. For example, an increase in income increases consumption, and an increase in price of a good or service reduces its demand. In other words, one variable, say y, depends on another, say x. Such dependencies are not confined to economics; they are observed in physical sciences and in everyday life. For example, the area of a circle, denoted by A, depends on its radius R, that is, $A = \pi R^2$. Similarly, the distance traveled by a car depends on the speed and time traveled.

If the relationship is such that every value of x leads to a unique value of y, then we can write $y = f(x)$ and say that y is a function of x. Note that the same y can be attached to more than one x, but that each x should be attached to only one y. Functions of real variables can be written as a mapping from the extended real line to itself. In other words, every real number in the domain corresponds to a unique real number in the range.

$$f : \Re \rightarrow \Re \tag{2.1}$$

Needless to say, a function need not be confined to one argument. We can write y as a function of x and z or as a function of $x_1, \ldots, x_k$. We can write them as

K. Dadkhah, *Foundations of Mathematical and Computational Economics*, 2nd ed.,
DOI 10.1007/978-3-642-13748-8_2, © Springer-Verlag Berlin Heidelberg 2011

$$f : \Re^2 \to \Re \qquad \text{or} \qquad y = f(x, z) \tag{2.2}$$

and

$$f : \Re^k \to \Re \qquad \text{or} \qquad y = f(x_1, \ldots, x_k) \tag{2.3}$$

We will encounter many kinds of functions in this chapter, and more functions yet throughout this book (examples of functions appear in Fig. 2.1a–c). Among them are polynomial functions, which are of the general form

$$y = \sum_{i=0}^{k} a_i x^i \tag{2.4}$$

Letting $k = 0, 1, 2, 3$, we have

$y = a_0$	Constant function
$y = a_0 + a_1 x$	Linear function
$y = a_0 + a_1 x + a_2 x^2$	Quadratic function
$y = a_0 + a_1 x + a_2 x^2 + a_3 x^3$	Cubic function

Example 2.1 (Utility Function). The utility function is an important tool of economic analysis. But as a function, it has a special feature that we would like to emphasize. The function attaches a real number to any bundle of goods and services. For instance, if the amount of each good or service is denoted by $x_i,\ i = 1, \cdots, n$, then

$$U = U(x_1, \cdots, x_n)$$

This function is such that if a particular bundle, say, bundle a, is preferred to another bundle b, then the utility, U^a, attached to a, is a bigger number than the utility attached to b. That is, $U^a > U^b$. But the numbers themselves do not have any significance in the following sense. Suppose $U^a = 10$ and $U^b = 5$. Clearly, the bundle a is preferred to bundle b. But we could also assign $U^b = 9.5$ to the bundle b and it would make no difference, in the sense that it conveys the same information as $U^b = 5$. The only important consideration is that $U^a > U^b$. Because of the property just described, utility is an *ordinal* number and utility function is an ordinal function. An ordinal number is different from a *cardinal* one like the amount of income. If a person makes \$50,000 a year and another person \$25,000, then there is a \$25,000 difference between their incomes, and the first one makes twice as much as the second. But the difference between $U = 10$ and $U = 5$ does not convey any information, nor does it mean that one bundle is preferred twice as much as the other. Another important property of the utility function is that if we keep all x_i's constant and increase only one of them, then the new bundle is preferred to the old one. To put it simply, the utility function is based on the idea that more is preferred to less.

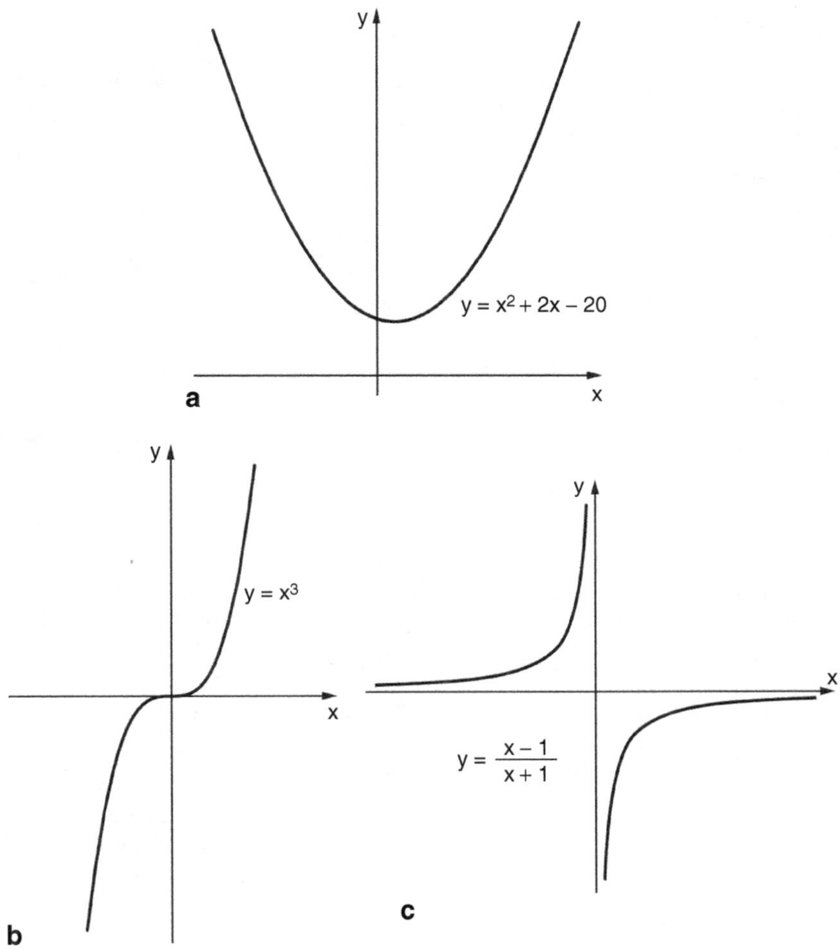

Fig. 2.1 (**a**) Quadratic function or parabola; (**b**) cubic function; (**c**) hyperbola

Example 2.2 (Cobb Douglas Production Function). A production function relates services of labor (L) and capital (K) to the maximum amount of output (Q) attainable from their combination. There are a number of production functions, which we shall discuss in Chap. 9. An important production function is the Cobb Douglas, which has the form

$$Q = AK^{\alpha}L^{\beta}$$

Graphing functions with Matlab is straightforward

Matlab code

```
% Define the domain of the function from -3 to 3 with
% increments of 0.1
x = -3:0.1:3;
% Define the function
y = x.^3;
% Plot the function
plot(x, y)
% You can plot more than one function on the
% same graph
hold on
for k=1:3
      y = x.^3 + 3*k;
      plot(x, y);
end
hold off
```

If you plot the function $y = x^3$, you will get a graph similar to Fig. 2.1b above. The same can be accomplished using Excel. To graph the function $y = x^3 + 3$, create the following on an Excel sheet. Highlight column B and use
>Insert → Line
You can use Select Data and use column A as the horizontal axis.

A	B
−3.0	=A1^3+3
−2.9	=A2^3+3
−2.8	=A3^3+3
...	...
2.9	=A60^3+3
3.0	=A61^3+3

Plotting three-dimensional graphs is slightly different. In Matlab use the following code:

Matlab code

```
% Define the domain of the function
[x, y] = meshgrid(-2:0.1:2,-2:.1:2);
% Define the function
z = x.*exp(-x.^2 - y.^2);
% Plot the function
mesh(x, y, z)
```

You will get the following graph (Fig. 2.2):
Similarly, we can plot the Cobb-Douglas production function.

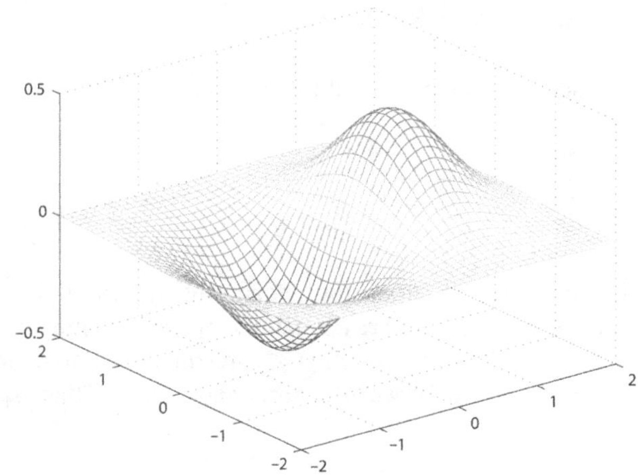

Fig. 2.2 Graph of the function $z = x\exp(-x^2 - y^2)$.

A	B	C	...	AQ
	−2.0	−1.9	...	2.0
−2.0	= A2*exp(− A2^2 − B 2^2)	= A2*exp(− A2^2 − C 2^2)	...	= A2*exp(− A2^2 − AQ 2^2)
−1.9	= A3*exp(− A2^2 − B 2^2)	= A3*exp(− A2^2 − C 2^2)	...	= A3*exp(− A2^2 − AQ 2^2)
−1.8	= A4*exp(− A2^2 − B 2^2)	= A4*exp(− A2^2 − C 2^2)	...	= A4*exp(− A2^2 − AQ 2^2)
...				
1.9	= A41*exp(− A2^2 − B 2^2)	= A41*exp(− A2^2 − C 2^2)	...	= A41*exp(− A2^2 − AQ 2^2)
2.0	= A42*exp(− A2^2 − B 2^2)	= A42*exp(− A2^2 − C 2^2)	...	= A42*exp(− A2^2 − AQ 2^2)

Matlab code

```
% Define the domain of the function
[K, L] = meshgrid(0:0.1:2, 0:0.1:4);
% Define the production function
Q = 5.*(K.^0.4).*(L.^0.6);
% Plot the function
mesh(K, L, Q)
```

Creating a three dimensional graph in Excel is a bit more time consuming and the result not as expressive as that of Matlab. Create the above worksheet in Excel. Highlight the square containing the computed numbers but not the values assigned to x and y. Then click Insert → Other Charts and choose one of the options.

2.1.1 Variety of Economic Relationships

In economics, we encounter three types of relationships:

1. Identities or definitions
2. Causal relationships
3. Equilibrium conditions

Identities or *definitions* are relationships that are true by definition or because we constructed them as such. Examples are the national income identity, in which GDP (Y) is defined as the sum of consumption (C), investment (I), government expenditures (G), and the difference between exports and imports, that is, net exports ($X - M$):

$$Y = C + I + G + X - M$$

Similarly, we define profit as revenues (quantity sold times price) net of cost:

$$\pi = PQ - C$$

Given their nature, such identities are not subject to empirical verification; they are always true. If we estimate the national income identity above using data from any country, we get an $R^2 = 1$ and coefficients that are highly significant and are usually 0.99999 (or -0.99999) and 1.00001 (or -1.00001). More important, since identities do not posit any hypothesis, no amount of algebraic manipulation of them will result in new insights into the workings of an economy.

Causal relationships are the mainstay of economics. They incorporate hypotheses regarding the behavior of economic agents, or technical and legal characteristics of the economy. Therefore, they are subject to empirical testing. Examples of behavioral relationships are consumption function, demand function, demand for imports, production functions, and tax revenues as a function of aggregate income.

By writing one variable as a function of a set of other variables, we implicitly declare that causation runs from the right-hand side (RHS) or explanatory variables to the left-hand side (LHS) or dependent variable. But how do we know this? How could we substantiate such a statement? Unlike physics and chemistry where experiments are the main source for accepting or rejecting a hypothesis, experiments play a very limited role in economics. Economics is an observational science.

Having been denied experiments and knowing well that correlation does not imply causation, econometricians have devised statistical tests of causality. The most widely used test of causality is due to Clive Granger (Nobel Laureate 2003). The test is for the necessary, but not sufficient, condition of the existence of causality in the strict sense. Thus, failing to reject the null hypothesis of no causality via the Granger test shows that x does not cause y in the strict sense. On the other hand, rejecting the null hypothesis establishes the necessary, but not sufficient, condition for causation. Most economic variables mutually affect each other. Money supply affects prices, which in turn affect the demand for money and indirectly the supply of money. Given that many economic variables are measured over arbitrary intervals of a month, quarter, or year, we may observe the mutual causation in the form of simultaneity. Of course, we may also observe simultaneity among economic variables because they are simultaneously determined through interdependent processes.

Equilibrium conditions describe the situation or condition when two or several variables are in such configuration that they need not change. Unlike identities, equilibrium conditions do not always hold. On the other hand, equilibrium conditions differ from causal relations in that a change in one variable does not automatically bring a change in another. Only if equilibrium is restored would a change in one variable bring about a change in the other. An equilibrium condition, if stable, implies that any deviation from equilibrium sets in motion forces that will bring back equilibrium. Therefore, stable equilibrium conditions are subject to statistical testing. Such tests, referred to as tests of cointegration, were proposed by Robert Engle (Noble Laureate 2003) and Granger, and by Søren Johansen. Furthermore, because there must be a force to restore the equilibrium, cointegration implies an error-correction mechanism. Thus, a delayed causation arises through error correction if the equilibrium is stable. We will encounter all these types of economic relationships in this book, and the reader will get a better sense of them after working with several specific examples.

2.1.2 Exercises

E.2.1. Graph the following functions for $-5 < x < 5$.

$$\begin{aligned} &i.\ \ y = 10 + 2x \\ &ii.\ \ y = 5 + 2x + 3x^2 \\ &iii.\ \ y = 7x^3 - 14x + 5 \\ &iv.\ \ y = \frac{1-x}{1+x} \end{aligned}$$

E.2.2 Make a list of economic relationships that you recall from economics courses and classify them as identities, causal relationships, and equilibrium conditions.

2.2 Series

The sequence of numbers

$$x_1 \quad x_2 \quad x_3 \quad \ldots \quad x_{n-1} \quad x_n \tag{2.5}$$

is called a series.

Example 2.3 The following are examples of series

$$\begin{array}{llllll} i. & 1 & 2 & 3 & \dots & n \\ ii. & 2 & 2^2 & 2^3 & \dots & 2^n \\ iii. & \frac{1}{2} & \frac{1}{2^2} & \frac{1}{2^3} & \dots & \frac{1}{2^{n-1}} \quad \frac{1}{2^n} \end{array}$$

Two issues are of importance here. First, could we write a series in a more compact format instead of enumerating its members? This can be done in two ways: by writing a general expression for its n-th term or by writing its recurrence relation. For instance, the n-th terms of series in Example 2.3 can be written as

$$i. \quad x_n = n \qquad ii. \quad x_n = 2^n \qquad iii. \quad x_n = \frac{1}{2^n}$$

Not all series can be written in this format. An alternative is to write their recurrence relation. For the above series the recurrence relations are

$$\begin{array}{ll} i.\ x_1 = 1, & x_n = x_{n-1} + 1 \\ ii.\ x_1 = 2, & x_n = 2x_{n-1} \\ iii.\ x_1 = \frac{1}{2}, & x_n = \frac{1}{2}x_{n-1} \end{array}$$

Example 2.4 Consider the Fibonacci sequence

$$1 \quad 1 \quad 2 \quad 3 \quad 5 \quad 8 \quad 13 \quad 21 \quad \dots$$

It starts with 1 and 1 and then each term is the sum of its two previous numbers. Thus, the recurrence relation is

$$x_0 = x_1 = 1 \qquad \text{and} \qquad x_n = x_{n-1} + x_{n-2} \qquad n = 2, 3, \dots,$$

We cannot always find recurrence relations for a series. For example, if the series is the realization of a random variable, we would not be able to find such a formula.

The second question is whether the sum of a series exists and if so, how we could calculate it. Note that mathematically speaking, when we say something exists, we mean that the entity in question has a finite value. Thus, here the question is whether the sum of a series is finite or tends to infinity. Before discussing these questions, however, we need to learn about the *summation notation* Σ and the concept of *limit*.

2.2.1 Summation Notation Σ

We are all familiar with summing a set of specific numbers. But suppose we would like to talk of the sum of x_1, x_2, x_3, x_4, x_5, x_6, x_7, and x_8. Of course, we can always write it as

$$x_1 + x_2 + x_3 + x_4 + x_5 + x_6 + x_7 + x_8$$

But such a formula is cumbersome and inefficient. It is even more cumbersome when we have 20 or 100 values to add. Even worse is when we want to represent the sum of an infinite series of numbers. We make the following convention:

$$\sum_{i=1}^{n} x_i \tag{2.6}$$

by which we mean the sum of numbers x_1 to x_n inclusive. Note that i is simply a counter and can easily be exchanged with j or k or any other symbol, although by usage, i, j and k are the most commonly used letters for counters. Other examples of summation are

$$\sum_{i=0}^{\infty} x_i, \qquad \sum_{j=-n}^{n} x_j, \qquad \sum_{t=1}^{T} x_{2t}$$

A few properties of sums should be noted:

$$\sum_{i=1}^{n} a = \underbrace{(a+a+\ldots+a)}_{n} = na \qquad \text{where } a \text{ is a constant} \tag{2.7}$$

$$\begin{aligned} \sum_{i=1}^{n} ax_i &= ax_1 + ax_2 + \ldots + ax_n = a(x_1 + x_2 + \ldots + x_n) \\ &= a\sum_{i=1}^{n} x_i \end{aligned} \tag{2.8}$$

$$\begin{aligned} \left(\sum_{i=1}^{n} x_i\right)^2 &= \sum_{i=1}^{n}\sum_{j=1}^{n} x_i x_j = \sum_{i=1}^{n} x_i^2 + \sum_{i\neq}^{n}\sum_{j=1}^{n} x_i x_j \\ &= \sum_{i=1}^{n} x_i^2 + 2\sum_{i<}^{n}\sum_{j=2}^{n} x_i x_j \end{aligned} \tag{2.9}$$

2.2.2 Limit

Consider the series (*ii*) in Example 2.3. As n increases, the last term of the series gets increasingly large. As n tends to ∞, so does the last term of the series. In such cases we say that the series has no limit. Note that ∞ is not a number. On the other hand, as n increases, the last term in (*iii*) in the same example becomes smaller and smaller as depicted in Table 2.1:

It can be seen that as n increases, $1/2^n$ tends to zero and, for all practical purposes, we can take it to be zero. In such cases, we say that the limit of the series exists and as n tends to ∞, $1/2^n$ tends to zero, and we write

$$\lim_{n\to\infty} \frac{1}{2^n} = 0 \tag{2.10}$$

Table 2.1 Approaching limit

N	$1/2^n$
1	0.5
2	0.25
...	...
10	0.0009765
11	0.0004882
...	...
100	7.8886×10^{-31}
101	3.9443×10^{-31}
...	...

Note that the limit needs not always be zero. It can be any number $L < \infty$. Now that we have an intuitive notion of a limit, let us present a formal definition.

Definition 2.1 Let $x_1, x_2, x_3, \ldots$ be a sequence of points on the real line. L is called the limit of this sequence if, for any number $\varepsilon > 0$, we could find a number N such that $|x_n - L| < \varepsilon$ if $n > N$.

If we apply the above definition to series (*ii*) in Example 2.3, we can reason that the series does not have a limit, because no matter what values we choose for L and N and no matter how large or small ϵ is, we cannot have $|2^n - L| < \varepsilon$ for all $n > N$. The reason: As n gets larger, so does 2^{n} and there is no limit to how large it can get. For the series (*iii*) the story is different. Let $L = 0$ and set $\varepsilon = 0.001$, then for all $n > 9$ we have $1/2^n < 0.001$. For example, $1/2^{10} = 0.0009765$. We can set $\varepsilon = 10^{-33}$, that is the decimal point followed by 32 zeros and then one. Now for all $n > 109$, we have $1/2^n < 10^{-33}$. The following properties of limits will prove quite useful.

Property 2.1 Let x_n and y_n represent two series and assume that both

$$\lim_{n\to\infty} x_n \quad \text{and} \quad \lim_{n\to\infty} y_n$$

exist. Then

$$\lim_{n\to\infty} (x_n + y_n) = \lim_{n\to\infty} x_n + \lim_{n\to\infty} y_n \tag{2.11}$$

The proposition is also true for the sum of any finite number of series. If c is a constant, it follows from (2.11) that

$$\lim_{n\to\infty} (c + x_n) = c + \lim_{n\to\infty} x_n \tag{2.12}$$

and

$$\lim_{n\to\infty} c x_n = c \lim_{n\to\infty} x_n \tag{2.13}$$

It is evident that the limit of the series

$$c \quad c \quad c \quad \ldots$$

is c.

Property 2.2 Let the series x_n and y_n be as in Property 2.1, then

$$\lim_{n\to\infty} x_n y_n = \lim_{n\to\infty} x_n \lim_{n\to\infty} y_n \tag{2.14}$$

Also

$$\lim_{n\to\infty} \frac{x_n}{y_n} = \frac{\lim x_n}{\lim y_n} \tag{2.15}$$

provided $\lim_{n\to\infty} y_n \neq 0$.

2.2.3 Convergent and Divergent Series

Consider the sum of the first n terms of a series

$$S_n = \sum_{i=1}^{n} x_i \tag{2.16}$$

Clearly, for every value of n we have a different sum. These sums, referred to as partial sums, form a series themselves. The question is whether the sum S_n exists as $n \to \infty$. In other words, is the following statement true?

$$S = \lim_{n\to\infty} S_n < \infty \tag{2.17}$$

The answer is that the sum exists if

$$\lim_{n\to\infty} x_n = 0 \tag{2.18}$$

If S exists, then the series is called *convergent*, or else it is called *divergent*.

Example 2.5 The sum

$$S = \lim_{n\to\infty} \sum_{i=1}^{n} \frac{1}{2^i}$$

exists because $\lim_{n\to\infty} (1/2^n) = 0$. Later in this chapter we will show how such sums can be calculated.

Example 2.6 Is the sum

$$\frac{1}{3} + \frac{2}{5} + \frac{3}{7} + \ldots + \frac{n}{2n+1} + \cdots$$

convergent or divergent? Because

$$\lim_{n\to\infty} \frac{n}{2n+1} = \lim_{n\to\infty} \frac{1}{2+1/n} = \frac{1}{2} \neq 0$$

we conclude that the series is divergent.

An alternative way of determining if a series is convergent or divergent when all terms are positive is the d'Alembert[1] test.

Property 2.3 (d'Alembert test). The sequence of positive numbers

$$x_1 \quad x_2 \quad x_3 \quad \ldots \quad x_n \quad \ldots \tag{2.19}$$

is convergent and the limit

$$S = \lim_{n\to\infty} S_n = \lim_{n\to\infty} \sum_{i=1}^{n} x_n \tag{2.20}$$

exists, if

$$\lim_{n\to\infty} \frac{x_{n+1}}{x_n} < 1 \tag{2.21}$$

If the above limit is greater than one, then the series is divergent. The case of the limit being equal to one is indeterminate.

Let us apply this test to some of the series we have encountered in this section. Note that all terms in these series are positive.

Example 2.7 The series in *(ii)* in Example 2.3 is divergent because:

$$\lim_{n\to\infty} \frac{2^{n+1}}{2^n} = 2 > 1$$

[1]Probably the most dramatic event in the life of the French mathematician Jean Le Rond d'Alembert (1717–1783) was that as a newborn he was left on the steps of a church. He was found and taken to a home for homeless children. Later, his father found him and provided for his son's living and education. D'Alembert made contributions to mathematics, mechanics, and mathematical physics. The eighteenth century was the age of European enlightenment and nothing represented the spirit of that age better than the *Encyclopédistes*, a group of intellectuals gathered around Diderot including Voltaire, Condorcet, and d'Alembert. They published the 28-volume *Encyclopedia* that contained articles on all areas of human knowledge including political economy.

But, the series in (*iii*) is convergent because

$$\lim_{n\to\infty} \frac{1/2^{n+1}}{1/2^n} = \frac{1}{2} < 1$$

For the series in Example 2.6 we have:

$$\lim_{n\to\infty} \frac{\dfrac{n+1}{2(n+1)+1}}{\dfrac{n}{2n+1}} = 1$$

Thus, in this case the d'Alembert test cannot resolve the issue.

In the following two subsections we will discuss two examples of series: arithmetic and geometric progressions.

2.2.4 Arithmetic Progression

The series

$$a \quad a+d \quad a+2d \quad a+3d \quad \dots \quad a+(n-1)d \tag{2.22}$$

is called *arithmetic progression*. We can write it more compactly as

$$x_n = a+(n-1)d \qquad n = 1, 2, \dots \tag{2.23}$$

or

$$x_1 = a \qquad x_n = x_{n-1} + d \qquad n = 2, 3, \dots \tag{2.24}$$

Thus, every member of the series is equal to its predecessor plus a constant number.

Example 2.8 The following are arithmetic series:

$$\begin{array}{ll} i. & 1 \;\; 2 \;\; 3 \;\; 4 \;\; \dots \;\; 20 \\ ii. & 5 \;\; 8 \;\; 11 \;\; 14 \;\; \dots \end{array}$$

To calculate the sum of arithmetic series in (*i*), above, we can write

$$\begin{array}{l} S \;= 1 \;\;+\;\; 2 + \;\; 3 + \cdots + 20 \\ \underline{S \;= 20 + 19 + 18 + \cdots + \;\; 1} \\ 2S = 21 + 21 + 21 + \cdots + 21 \end{array}$$

Thus,

$$2S = 20 \times 21$$

and

$$S = \frac{20 \times 21}{2} = 210$$

This can be generalized to the sum of any n consecutive integers starting with 1.

$$1 + 2 + 3 + \ldots + n = \frac{n(n+1)}{2} \tag{2.25}$$

Following the same line of reasoning for the general case, the sum of n consecutive terms in an arithmetic progression is,

$$S = n\left[a + \frac{(n-1)d}{2}\right] \tag{2.26}$$

Example 2.9 For the sum of the first 20 integers, we have $a = d = 1$ and $n = 20$. Plugging the numbers into (2.26), we get the sum of 210.

Example 2.10 For the sum of the first 10 integers divisible by 3, we have $a = d = 3$ and $n = 10$. Plugging the numbers into (2.26), we get the sum of 165.

These formulas can be programmed in Matlab in two ways. First, we can simply write a procedure that adds up, one by one, the n terms in a particular series. Alternatively, we can use (2.26) to evaluate the sum of the series.

Matlab code

```
% Initialize n, a, d, and S
n = 20;
a = 1;
d = 3;
S = 0;
% Compute S by adding the 20 terms
for i = 1:n
      S = S + a + (i-1)*d;
end
% Call S
S
% Alternatively you can write
S = n*(a + (n-1)*d/2)
```

Note that you can change n, a, and d to any number and run the procedure again and again.

The same can be accomplished using Excel. By replacing n, a, and d with the desired values, you will get the sum of the arithmetic series in two different ways.

Note that in the illustration below and in subsequent Excel illustrations we make reference to cell numbers. In the next chapter we learn how to name variable in Excel and to refer to them by name.

A	B	C	D	E
=C1		1	3	20
=A1+D$1				
=A2+D$1		=E1*(C1+(E1−1)*D1/2)		
⋮				
=SUM(A1:A20)				

2.2.5 Geometric Progression

The series

$$a \quad aq \quad aq^2 \quad aq^3 \quad \dots \quad aq^{n-1} \quad \dots \tag{2.27}$$

is called a geometric progression. The recurrence relation is

$$x_1 = a \qquad x_n = qx_{n-1} \qquad n = 2, 3, \dots \tag{2.28}$$

We are interested in finding the sum of the first n terms of this series. Let

$$S = \sum_{i=0}^{n-1} aq^i = a + aq + aq^2 + aq^3 + \dots + aq^{n-1} \tag{2.29}$$

Multiplying S by q and subtracting it from S, we have

$$\begin{array}{rl} S = & a + aq + aq^2 + aq^3 + \cdots + aq^{n-1} \\ -Sq = & \quad\; -aq - aq^2 - aq^3 - \cdots - aq^{n-1} - aq^n \\ \hline S - Sq = & a - aq^n \end{array}$$

Thus,

$$S = a\frac{1-q^n}{1-q} \tag{2.30}$$

Example 2.11 Find the following sum:

$$S = 2 + 6 + 18 + 54 + 162 + 486 + 1458$$

Because $a = 2, q = 3$, and $n = 7$, we have

$$S = 2\frac{1-3^7}{1-3} = 2186$$

Geometric progression finds a few applications in macroeconomics including aggregate demand multiplier, money multiplier, and present value.

Example 2.12 (Keynesian multiplier) When discussing the effect of an increase in government expenditures on aggregate demand and income, the following argument is offered. Suppose the government increases its expenditures by $100 billion. These additional expenditures by the government will become the income of individuals who provide the goods and services to the government. Assuming a marginal propensity to consume of 0.92, the additional consumption will be $92 billion. This consumption, in turn, forms the income of those who produce consumer goods and services. But then they will spend 0.92×92 or $84.64 billion on consumption which in turn will be the income of those who produce consumption goods and services. You get the idea. The stream of income generated in different stages is shown in Table 2.2:

The sum of the first 20 terms of the addition to national income can be calculated as

$$S = 100\frac{1 - 0.92^{20}}{1 - 0.92} = 1014.13$$

If we repeat the same calculation for the first 40 terms, we get a total of $1205 billion. The second 20 terms add less than a quarter of the first 20. The sum of the first 100 terms equals 1249.7. The reason: $0.92 < 1$ and when a number whose absolute value is less than one is raised to increasing exponents, it becomes smaller and smaller. The smaller the absolute value of the number, the sooner it reaches zero. For example, if the marginal propensity to consume was 0.5 instead of 0.92, the sum of the first 44 terms would be $250 billion and additional terms would have no effect. Indeed, terms beyond the first 20 would have no practical significance. Thus, if we allow the process to continue indefinitely, that is, letting $n \to \infty$, we will have

$$S = 100\frac{1}{1 - 0.92} = 1250$$

Note 0.92 is marginal propensity to consume, and $1/(1-0.92)$ is our good old multiplier.

Table 2.2 Multiplier effect at work: the effect of government expenditures on income

Steps	Increase in income	
0	100	$= 100$
1	92	$= 100 \times 0.92$
2	84.64	$= 100 \times 0.92^2$
3	77.8688	$= 100 \times 0.92^3$
...	...	
...	...	

We can generalize the results of the last example by noting that

$$\lim_{n\to\infty} q^n = 0 \qquad \text{if} \qquad |q| < 1 \tag{2.31}$$

It follows that

$$\lim_{n\to\infty} S = \lim_{n\to\infty} a\frac{1-q^n}{1-q} = a\frac{1}{1-q} \qquad \text{if} \qquad |q| < 1 \tag{2.32}$$

As in arithmetic progression we can use Matlab to carry out the necessary calculations:

Matlab code

```
% Initialize n, a, q, and S
n = 7;
a = 2;
q = 3;
S = 0;
% Compute S by adding the 20 terms
for i = 1:n
      S = S + a.*q.^(i-1);
end
% Call S
S
% Alternatively you can write
S = a.*(1-q.^n)./(1-q)
```

If you use the second method, you may want to define a function and call it when needed. First you create an M-file in Matlab containing the function.

Matlab code

```
function G = Geoprog(v);
n = v(1);
a = v(2);
q = v(3);
G = a.*(1-q.^n)./(1-q);
```

Then you can call this function for different values of n, a, and q.

Matlab code

```
v = [7 2 3];
S = Geoprog(v);
```

We can perform these computations in Excel as illustrated below:

A	B	C	D	E
=C1		2	3	7
=A1*D$1				
=A2*D$1		=C1*(1−D1^E1)/(1−D1)		
⋮				
=SUM(A1:A7)				

2.2.6 Exercises

E.2.3 Find the sum of all odd numbers from 1 to 451.

E.2.4 Find the sum of all even numbers from 2 to 450.

E.2.5 Find the sum of the following geometric series:

$$1 \quad \frac{1}{2} \quad \frac{1}{4} \quad \frac{1}{8} \quad \ldots$$

$$1 \quad \frac{1}{3} \quad \frac{1}{9} \quad \frac{1}{27} \quad \ldots$$

E.2.6 The *present value* (PV) of a stream of income is $D_t,\ t = 0, 1, \ldots, T$ is defined as

$$\mathrm{PV} = \sum_{t=0}^{T} \frac{D_t}{(1+r)^t}$$

where $t = 0$ is the current year and r is the rate of interest.

i. Compute the present value of a winning lottery ticket that will pay \$200,000 per year for 20 years starting in the present year. Assume an interest rate of 12%. Solve the same problem assuming interest rates of 15% and 20%. [*Hint*: For interest rate of 12%, $r = 0.12$.]
ii. Compute the value of a government bond that pays one dollar every year in perpetuity (i.e., forever) given the interest rate of r.

E.2.7 Show that

$$\sum_{i=0}^{\infty} (i+1)\lambda^i = \frac{1}{(1-\lambda)^2}, \qquad |\lambda| < 1$$

[*Hint*: $\lim_{n\to\infty} n\lambda^n = 0$.]

2.3 Permutations, Factorial, Combinations, and the Binomial Expansion

Counting rules discussed in this section are the elementary building blocks of combinatorics, a branch of mathematics that has applications in many areas including cryptography, computer science, probability theory, statistics, econometrics, and economics. Consider a collection of n items denoted by $\mathbf{A} = \{A_1, A_2, \ldots, A_n\}$. Suppose we choose $r \leq n$ items from $\mathbf{A}$ and arrange them in the order they are chosen. A typical arrangement will look like

$$\overbrace{A_3, A_7, \ldots, A_{r+2}}^{r}$$

How many such collections can we form that are different from each other at least in one item or in the position of one item? We can argue as follows. For the first item, we can choose from n items; for the second place, from among the remaining $n - 1$ items, because one item has already been taken for the first place. Continuing in this way, for the r-th item we can choose from among the remaining $n - (r-1)$ items. Thus, the total possible arrangements are

$$n \times (n-1) \times \ldots \times (n - r + 1) \tag{2.33}$$

For example if we have five objects, we can make $5 \times 4 \times 3 = 60$ different arrangements containing three elements. If we allow $r = n$, then we have

$$1 \times 2 \times 3 \times \ldots \times n = \prod_{i=1}^{n} i = n! \tag{2.34}$$

$n!$ is called "n factorial," and its meaning is quite obvious. Π is similar to the summation notation, except that it stands for the product of a set of numbers or variables. Note that $0!=1$. The reason: We can arrange or permute in only one way the elements of a null set (the set with zero elements). Using the convention of (2.34) we can write (2.33) as

$$\frac{n!}{(n-r)!} \tag{2.35}$$

Now suppose we ask, in how many ways can we pick r elements from the set containing n elements? The number of combinations is

$$\binom{n}{r} = \frac{n!}{r!(n-r)!} \tag{2.36}$$

Example 2.13 Suppose five soccer teams are playing in a tournament. How many games will be played? Let us designate our teams by letters A, B, C, D, and E. Here is the list of games to be played:

$$\text{AB, AC, AD, AE, BC, BD, BE, CD, CE, DE}$$

which makes a total of 10 games. Note that we do not have both AB and BA because when A has played against B, the reverse is also true. The problem is the same as choosing two out of a set of five. Based on (2.36) we have,

$$\binom{5}{2} = \frac{5!}{2!3!} = \frac{5 \times 4}{2} = 10$$

Example 2.14 A mutual fund is a portfolio consisting of a number of equities held in different proportions. For example, it may have 5% of its assets in IBM stock, 6% in Verizon, 10% in Microsoft, and so on. Assume that 1000 stocks in the market are deemed to be appropriate for inclusion in such funds. Further suppose that each fund consists of 30 stocks. How many different portfolios can one form from the 1000 stocks?

$$\binom{1000}{30} = \frac{1000!}{30!970!} = 242960819217375 \times 10^{43}$$

A very large number indeed. As can be seen, the precision of these numbers is 15 digits; that is, the first 15 digits are accurate and the rest give the order of magnitude. What is interesting is that the number of potential mutual funds far exceeds the number of stocks. Note that a mutual fund needs not consist of exactly 30 stocks; it can have 40, 50, 100, 200, or any other number of stocks. For each of those numbers, a large number of funds could be formed. Thus, the total number of potential mutual funds is astronomical.

Two functions in Matlab allow calculations of $n!$ and $\binom{n}{r}$.

Matlab code

```
% for n!
factorial(n)
% for
nchoosek(n,r)
```

Excel has a function for factorial and for combination. In Formulas choose Insert Function and then choose Math & Trig, Finally choose FACT for factorial and COMBIN for combination. Alternatively you could type in:

=FACT(n)	=COMBIN(n,r)

Combinations prove useful in writing the binomial expansion.

$$(a+b)^n = \binom{n}{0} a^n + \binom{n}{1} a^{n-1}b + \binom{n}{2} a^{n-2}b^2 + \cdots + \binom{n}{n-1} ab^{n-1} + \binom{n}{n} b^n = \sum_{i=0}^{n} \binom{n}{i} a^{n-i}b^i \tag{2.37}$$

Example 2.15 We can illustrate the general formula in (2.37) by applying it to $n = 2, 3, 4$.

$$(a+b)^2 = \binom{2}{0} a^2 + \binom{2}{1} ab + \binom{2}{2} b^2 = a^2 + 2ab + b^2$$

$$(a+b)^3 = \binom{3}{0} a^3 + \binom{3}{1} a^2b + \binom{3}{2} ab^2 + \binom{3}{3} b^3 = a^3 + 3a^2b + 3ab^2 + b^3$$

$$(a+b)^4 = \binom{4}{0} a^4 + \binom{4}{1} a^3b + \binom{4}{2} a^2b^2 + \binom{4}{3} ab^3 + \binom{4}{4} b^4$$
$$= a^4 + 4a^3b + 6a^2b^2 + 4ab^3 + b^4$$

2.3.1 Exercises

E.2.8 Evaluate the following expressions using a calculator, the Excel function `COMBIN`, and the Matlab function `nchoosek`.

$$\binom{14}{3}, \binom{9}{6}, \binom{23}{8}, \binom{33}{12}$$

E.2.9 There are 50 delegates at a convention, 32 men and 18 women. In how many ways can we choose a committee of eight equally divided between men and women?

E.2.10 There are 9 potential judges for a contest including 5 women and four men. In how many ways can we choose 5 judges provided that at least two of them are women?

E.2.11 Show that

$$\sum_{j=0}^{n} \binom{n}{j} = \binom{n}{0} + \ldots + \binom{n}{n} = 2^n$$

[*Hint*: Consider the binomial expansion of $(1+1)^n$.]

2.4 Logarithm and Exponential Functions

Most likely you are already familiar with logarithm and exponential functions, because both play important roles in mathematics. In addition, they find many uses in economics, especially in dynamic models and growth theory. Many econometric models involve logarithms of both dependent (endogenous) and explanatory (exogenous) variables.

2.4.1 Logarithm

Suppose

$$y = a^x \qquad y > 0, \quad a > 1 \tag{2.38}$$

then x is the logarithm[2] of y in the base a, which we denote as

$$x = \log_a y \tag{2.39}$$

Note that both a and y are positive real numbers. Logarithms of negative numbers are complex numbers. In this book we confine ourselves to logarithm of positive numbers. Whereas a could be any positive real number, the three important bases are 2, 10, and e. Base 2 is used in information science and communication. Base 10 is convenient for certain calculations; note that the logarithm of 1, 10, 100, 1000, ... in base 10 are 0, 1, 2, 3, ...

The base we will be dealing with in this book is e, an irrational number approximately equal to 2.7182818285. This unusual number somewhat like π, will prove quite useful and will play a significant role in mathematics and computation. In the next section, we have more to say about e, but for the time being consider it a number. The logarithm in base e is referred to as the natural logarithm and sometimes (to avoid confusion) is denoted by ln—a practice we will adopt in this book.

A basic property of logarithm that makes manipulation and calculations easier is that

$$\ln(xy) = \ln x + \ln y \tag{2.40}$$

Let

$$x = e^{\alpha}, \qquad y = e^{\beta}$$

then

[2]John Napier (1550–1617), a Scottish nobleman, conceived the idea of the logarithm. The first tables using base 10 were calculated by Henry Briggs (1561–1631), a professor of geometry at Gresham College.

$$xy = e^{\alpha}e^{\beta} = e^{\alpha+\beta}$$

and

$$\ln(xy) = \alpha + \beta = \ln x + \ln y$$

Repeated application of (2.40) results in

$$\ln\left(x^n\right) = n \ln x \tag{2.41}$$

Combining (2.40) and (2.41), we have

$$\ln\left(\frac{x}{y}\right) = \ln(xy^{-1}) = \ln x + \ln(y^{-1}) = \ln x - \ln y \tag{2.42}$$

Thus, logarithm turns multiplication into addition, division into subtraction, raising to a power into multiplication, and finding the roots of a number into division.

Logarithmic functions are programmed in every calculator and in software such as Excel. In Matlab one can get the logarithm of a positive number in three bases.

Matlab code

```
% Natural logarithm
log(x)
% In base 10
log10(x)
% In base 2
log2(x)
```

Excel has three functions for logarithm: LN for the natural logarithm, LOG10 for the base 10, and LOG for any base the user specifies.

You will hardly ever need the logarithm of a number in any other base, but should such a need arise, the calculation is simple. Suppose you are interested in finding the logarithm of y in the arbitrary base of $b > 1$. Let x, and z be, respectively, logarithms of y in bases e and b. We can write

$$y = e^x = b^z$$

and

$$\ln y = x = z \ln b$$

Therefore,

$$\log_b y = z = \frac{x}{\ln b} = \frac{\ln y}{\ln b} \tag{2.43}$$

Example 2.16

$$\ln 45 = 3.8066625 \quad \ln 10 = 2.3025851$$

$$\log_{10} 45 = \frac{\ln 45}{\ln 10} = \frac{3.8066625}{2.3025851} = 1.6532125$$

Example 2.17

$$\log_2 1024 = \frac{\ln 1024}{\ln 2} = \frac{6.9314718}{0.6931471} = 10$$

2.4.2 Base of Natural Logarithm, e

Next to π, the base of natural logarithm, e, is the most famous irrational number among mathematicians and those who apply mathematics. It is approximately equal to 2.71828182845905 and more precisely

$$e = \lim_{x \to \infty} \left(1 + \frac{1}{x}\right)^x \tag{2.44}$$

We need not dwell on the origin and the logic behind this number. Rather, we can gain an intuitive understanding of it through an example from economics. Suppose you deposit \$1000 in an interest-bearing account with an interest rate of 12%. After a year, your money would be \$1000 (1+0.12) = \$1120. But the underlying assumption in this calculation is that the interest accrues to your money at the end of the year. Why should it be that way? Suppose at the end of 6 months you receive half of the annual interest and increase your account to \$1060. For the next 6 months you earn interest on this new amount and, at the end of the year, your balance would be

$$\$1000 \left(1 + \frac{0.12}{2}\right)^2 = \$1123.60$$

Why should we stop there? Why not ask for the interest to accrue every season, every month, or even instantaneously? Table 2.3 shows the amount of principal plus interest when interest accrues at different frequencies. The last amount is approximately equal to $\$1000 \times e^{0.12} = \$\ 1127.50$. This is the amount you would have had if interest accrued every second. As a matter of notation, sometimes e^x is written as $\exp(x)$. Matlab has a ready-made function for $\exp(x)$.

Matlab code

```
% Exponential function
```

Table 2.3 Effect of the frequency of interest accrual on the total amount of interest

Frequency	Total interest
Annual	$1000(1+0.12) = 1120.00$
Semiannual	$1000\left(1+\frac{0.12}{2}\right)^2 = 1123.60$
Seasonal	$1000\left(1+\frac{0.12}{4}\right)^4 = 1125.51$
Monthly	$1000\left(1+\frac{0.12}{12}\right)^{12} = 1126.83$
Daily	$1000\left(1+\frac{0.12}{365}\right)^{365} = 1127.47$

```
exp(x)
```

The Excel function for exponential is EXP.

2.4.3 Exercises

E.2.12 Graph the following functions for $0.1 < x < 6$:

$$y = \ln(x) \qquad y = e^x \qquad y = e^{-2x}$$

E.2.13 For the annual interest rates 20, 18,15, 12, 10, 8, 5, and 2%,

i. Compute the corresponding daily rates.
ii. Compute the corresponding effective annual rates if the interest is compounded daily.
iii. Compute the corresponding effective annual rates if the interest is compounded instantaneously.
iv. How close are the results in *ii* and *iii*?

E.2.14 Given the following equations, find x and y.

$$y^x = x^y$$
$$y = 2x$$
$$x > 0$$

2.5 Mathematical Proof

In many math books proofs of theorems end with the abbreviation QED that stands for the Latin phrase "Quod Erat Demonstrandum," meaning "which was to be shown." But what is to be shown and what do we mean by a mathematical proof?

2.5.1 *Deduction, Mathematical Induction, and Proof by Contradiction*

Mathematicians prove their propositions in one of three ways: deduction or direct proof, mathematical induction, or by contradiction. As we mentioned in Chap. 1, mathematical propositions are tautologies, although the connection between the assumptions (the starting point) and proposition (the end point) may not be easy to see. The goal of Mathematics is to find and substantiate such connections. The genius of a great mathematician is in discerning an important proposition and in proving how it can be derived from a minimal set of assumptions. On many occasions it is easier to start from a proposition and work backward. Other times, mathematicians must refine the assumptions or add to or subtract from them. In still other cases, the proposition may need adjustment. Once a proposition is proved, others may find easier proofs, discover that the proposition needs less strict assumptions, or that the proposition is simply a special case of a more general theorem. Finding implications of a general proposition, finding interesting applications and special cases for it, and discovering its connections to other propositions provide avenues for further research.

Proof by Deduction. Direct proofs or deductions start with assumptions and lead to the proposition. We have to show that every statement follows logically from the previous one. In other words, we have to show that each step is implied by what we knew in the previous step. In this process we can use any theorem or lemma that has already been proved because by having proved them, we know they are logically correct. Our derivation of the formula for the sums of arithmetic and geometric series, although elementary, are examples of direct proof. Similarly, many propositions you remember from high school geometry are proved by direct reasoning.

Proof by Induction. Another way of proving a proposition is by induction, in which we first prove the validity of a proposition for the case of $n = 1$; then assuming that the proposition is true for the case of $n - 1$, we show that it is also true for n. Since we already know that the theorem or lemma is true for $n = 1$, then it should be true for $n = 2$, and therefore, $n = 3$, and indeed for any n.

Example 2.18 We have already seen that the sum of integers from 1 to n is equal to $n(n+1)/2$. We can verify that this formula is correct for $n = 1$ and indeed for $n = 2$ and $n = 3$. Suppose we know that the formula is true for the sum of $n - 1$ numbers, that is,

$$S_{n-1} = \frac{(n-1)n}{2}$$

Then for the sum of n consecutive integers we have

$$S_n = S_{n-1} + n = \frac{(n-1)n + 2n}{2} = \frac{n(n+1)}{2}$$

Example 2.19 Similarly we showed that the sum of n terms of geometric progression is

$$S = a\frac{1-q^n}{1-q}$$

We can verify that this sum is correct for the first term, and the sum of the first two terms. Now let us assume that the formula is correct for the first $n-1$ terms. Then

$$\begin{aligned} S_n &= S_{n-1} + aq^{n-1} \\ &= a\tfrac{1-q^{n-1}}{1-q} + aq^{n-1} \\ &= a\tfrac{1-q^{n-1}+q^{n-1}-q^n}{1-q} \\ &= a\tfrac{1-q^n}{1-q} \end{aligned}$$

Proof by Contradiction. In proving a proposition by contradiction, we first assume that the proposition is false. Then, deriving the implications of the proposition being false, we show that they contradict some proven theorems or known facts. The conclusion is that the proposition cannot be false.

Example 2.20 One of Euclid's theorems states that the number of primes is infinite. Recall that a prime number is divisible only by one and itself. To prove the theorem, we assume the contrary, that prime numbers are finite. Therefore, we can write them as

$$p_1\, p_2\, p_3 \cdots p_n$$

But now consider

$$p_{n+1} = p_1 \times p_2 \times p_3 \times \ldots \times p_n + 1$$

This number is not divisible by other primes because the division will have a remainder of one, therefore, it is a prime. Thus, no matter how many numbers we have, we can add one more and then another. This contradicts the assumption that there are only a finite number of primes.[3]

[3]Alternatively, the fact that p_{n+1} is not divisible by any prime contradicts the fundamental theorem of arithmetic that states that any integer $k > 1$ has a unique factorization of the form

2.5.2 Computer-Assisted Mathematical Proof

Proofs of mathematical propositions rely on logical steps that are convincing, if not to all ordinary mortals, at least to all who have the proper training. Moreover, such proofs are general in the sense that they apply to a class of problems. For instance, the proof of the solution of the quadratic equation does not rely on any particular values of the coefficients. Rather it is correct for all equations with real coefficients. But suppose we make a statement about all integers less than a particular finite number n. Could we use the computer and check the statement for all such numbers and show that it is true and call it a mathematical proof? Well, it did not happen exactly like that, but in 1976 computers made their first nontrivial appearance in the realm of mathematical proofs. It involved the famous *four-color problem.*

Consider a map of the world on a flat surface or on a globe.[4] We want to color the map with the condition that no two countries with a common border are of the same color. To make the problem more specific, the areas of all countries have to be contiguous, and no common boundary can be only a single point. The first condition rules out countries with two or more pieces; for instance, the United States (because Alaska and Hawaii are not attached to the mainland). The question is, how many colors do we need? Cartographers have dealt with this problem for ages. It was conjectured that the feat could be accomplished with four or fewer colors. But proof of this conjecture seemed to be out of reach.

In 1976 Kenneth Appel and Wolfgang Haken[5] proved the theorem with partial help from a computer. The proof relies on old-fashioned mathematical work, but 1200 hours of computer time were used to check certain difficult cases.

Computer-assisted mathematical proofs are still exceptions and most mathematicians go about their work in the old-fashioned way. It is said that computer proofs are uncertain and cannot be checked and verified. The uncertainty arises because there may be faults in the hardware, problems with the operating system, or bugs in the program. Assuming that these issues have been thoroughly checked, we can be sure with a high probability of the validity of the proof. This is different from traditional proofs that are offered with certainty and there can be no doubt about them, even an infinitesimal one. If computer-assisted proofs become the prevalent mode of work, mathematics would resemble physics, in which laws are tested and either rejected or not rejected, but never 100% accepted. Furthermore, it is said that

---

$$k = p_1 \times p_2 \times \cdots \times p_r$$

Therefore, p_{n+1} must be a prime.

[4]Technically, we are talking about a *planar* map or graph. Suppose we represent every country by a node and connect each pair of the nodes representing adjacent countries by a line. If we are able to draw such a graph without the lines crossing, then the graph is planar.

[5]For a better idea of the problem and its solution you may want to check Appel and Haken's article in *Scientific American* (October 1977) or their book *Every Planar Map is Four Colorable* (1989). A more technical understanding of the subject could be gained from textbooks on graph theory or discrete mathematics.

computer-assisted proofs that involve thousands of lines of codes cannot be verified; no one would spend her energy on the thankless job of checking a complicated computer program.

One may think of these issues as a matter of degree. After all, complicated proofs such as Gödel's theorem or Fermat's last theorem (conjecture)[6] cannot easily be checked even by many mathematicians. On the other hand, if computers make headway in proving mathematical theorems, we can imagine that in the future mathematical proofs will not be checked but confirmed through independent replications and then held to be true with a high probability. There is another way that computers could help in the advancement of mathematics. A computer program could be written to carry out all logical steps necessary for the proof of a theorem. This doesn't mean that the computer is proving the theorem. Rather it is carrying out the instructions of the mathematician. Such a step-by-step operation would be time consuming and too tedious for human beings, but computers don't mind. The procedure would be especially beneficial when the proof runs into tens and perhaps hundreds of pages. Such activities are already under way, but the role of computers in the mathematics of the future is a matter of speculation.

2.5.3 Exercises

E.2.15 Use mathematical induction to show that

$$\text{i.}\quad 1^2 + 2^2 + 3^2 + \ldots + n^2 = \frac{n(n+1)(2n+1)}{6}$$

$$\text{ii.}\quad n! \geq n^2, \qquad \forall n \geq 4$$

$$\text{iii.}\quad \sum_{j=1}^{n} (2j-1) = n^2$$

[6]Consider the equation $x^n + y^n = z^n$. If $n = 2$, we can find integers satisfying the equation $3^2 + 4^2 = 5^2$. But could the same be done for $n \geq 3$? French mathematician Pierre de Fermat (1601–1665) claimed that he could prove that no such solutions could be found. But because he was writing on the margin of a book, he said he could not write it out. In all likelihood he did not have such a proof. Over the years, many contributed to the solution of the problem. In 1993, the British mathematician Andrew John Wiles (1953) (now at Princeton University in the United States) announced that he had proved the theorem. But there was a significant gap in the proof that took Wiles and a co-worker one and a half years to fill. There are two books written for the public on this subject: *Fermat's Last Theorem: Unlocking the Secret of an Ancient Mathematical Problem* (1996) by Amir Aczel, and *Fermat's Enigma: The Epic Quest to Solve the World's Greatest Mathematical Problem* (1997) by Simon Singh. Both are available in paperback.

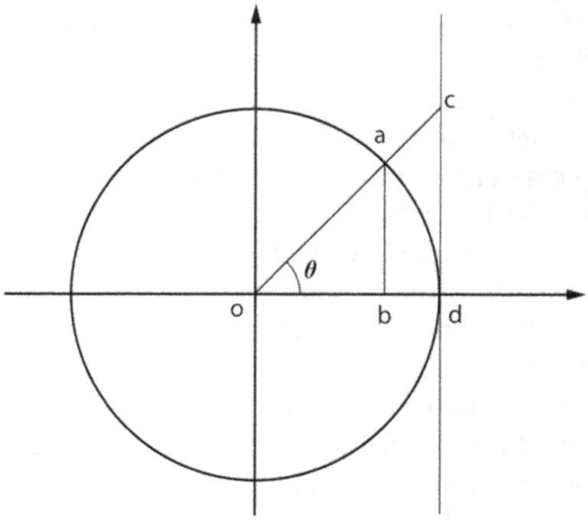

Fig. 2.3. Geometric representation of trigonometric functions

2.6 Trigonometry

Trigonometry, one of the most fun areas of mathematics,[7] has many practical applications in engineering, statistics, and econometrics, as well as in everyday life. What is more, it requires learning only a few basic relationships and the rest is a matter of deduction. Consider the circle in Fig. 2.3. It has a radius of unity. We define the following functions[8] of the angle θ:

$$\sin\theta = \frac{\text{ab}}{\text{oa}} = \text{ab}, \qquad \cos\theta = \frac{\text{ob}}{\text{oa}} = \text{ob}, \qquad \tan\theta = \frac{\text{dc}}{\text{od}} = \text{dc} \tag{2.45}$$

Thus, for the angle θ and the point a on the unit circle, $\cos\theta$ and $\sin\theta$ are, respectively, the coordinates of the point a on the x-and y-axes. If we consider the right-angle triangle oab, then $\sin\theta$ is the ratio of the side opposing the angle to the hypotenuse. Similarly, $\cos\theta$ is the ratio of the side forming the angle to the hypotenuse. This definition applies to all right-angle triangles regardless of the length of the hypotenuse. In the case depicted in Fig. 2.3, the hypotenuse has a length of one and, therefore, we can ignore the denominator of the ratios.

A graph of sin(x) is shown in Fig. 2.4. As Figs. 2.3 and 2.4 show, both sine and cosine functions take values between -1 and 1. The tangent function, however, is bounded neither from below nor from above. If we multiply sine or cosine functions by ρ, the range of the functions is changed from $[-1, 1]$ to $[-\rho, \rho]$ and ρ is referred to as the *amplitude*.

[7] The interested reader is referred to *Trigonometric Delights* by Eli Maor (1998).

[8] Other trigonometric functions exist, but we will not discuss them here because economists rarely if ever come across them and, therefore, we have no reason to clutter the subject with many unfamiliar notations.

Using elementary geometry the following relationships can be deduced:

$$\tan\theta = \frac{\sin\theta}{\cos\theta} \tag{2.46}$$

$$\sin^2\theta + \cos^2\theta = 1 \tag{2.47}$$

The first is based on the Thales theorem[9] and the second on the Pythagoras theorem.[10] Observe the notation for the square of a trigonometric function. It is written $\sin^2\theta$ and not $\sin\theta^2$, as the latter means the angle θ is raised to the power 2. Of course, we could write it as $(\sin\theta)^2$, but we prefer the economy in the universally accepted convention. Recall that angles can be measured in terms of degrees, radians, and grads. A circle spans 360 degrees, 2π radians, and 400 grads. Thus, a right angle would be 90 degrees, $\pi/2$ radians, and 100 grads. In this book and in most mathematics books, angles are measured in radians.

From Fig. 2.3 it is evident that

$$\begin{aligned} &\sin 0 = \sin\pi = \sin 2\pi = 0,\\ &\sin\frac{\pi}{2} = 1,\\ &\sin\frac{3\pi}{2} = -1 \end{aligned} \tag{2.48}$$

Similarly

$$\begin{aligned} &\cos 0 = \cos 2\pi = 1,\\ &\cos\frac{\pi}{2} = \cos\frac{3\pi}{2} = 0,\\ &\cos\pi = -1 \end{aligned} \tag{2.49}$$

In addition, using well-drawn circles and a ruler, the reader should convince herself of the following identities:

$$\begin{aligned} \sin\left(\theta + \frac{\pi}{2}\right) &= \cos\theta, & \cos\left(\theta + \frac{\pi}{2}\right) &= -\sin\theta,\\ \sin(\theta+\pi) &= -\sin\theta & \cos(\theta+\pi) &= -\cos\theta,\\ \sin(\theta+2\pi) &= \sin\theta, & \cos(\theta+2\pi) &= \cos\theta\\ \sin(-\theta) &= -\sin\theta, & \cos(-\theta) &= \cos(\theta) \end{aligned} \tag{2.50}$$

Trigonometric functions are programmed in all scientific calculators. In addition, software such as Excel, Matlab, and Maple also have these functions. Matlab's trigonometric functions follow.

[9] The theorem is named after Thales de Miletos (624 B.C.–547 B.C.) although the germ of the idea dates back to 1650 B.C. and the building of the Pyramids.

[10] This is the famous Pythagoras theorem that the square of hypotenuse is equal to the sum of the squares of the other two sides of a right-angle triangle. Egyptians who built the Pyramids clearly had an empirical understanding of this theorem. Pythagoras (569 B.C.–475 B.C.), for whom the theorem is named, is one of the great mathematicians of antiquity and pioneers of mathematics.

Matlab code

```
% sin, cos and tan of x are obtained using
sin(x)
cos(x)
tan(x)
% Matlab assumes that x is expressed in terms of
% radians. Thus, the sin of π/6 or 30° can be
% calculated in one of the the following ways
sin(pi./6)
% or
x = 30;
sin(x.*pi./180)
% Both return
ans =
        0.5000
```

In Excel the sine, cosine, and tangent of an angel, say $\pi/4$, can be obtained as

=SIN(PI()/4)	=COS(PI()/4)	=TAN(PI()/4)

The fact that $\sin(\theta + \pi/2) = \cos(\theta)$ shows that the sine and cosine functions are out of phase by $\pi/2$ radians. In other words, it takes $\pi/2$ angle rotations for the sine function to catch up with the cosine function. Similarly, the two functions $y_1 = \sin(\theta)$ and $y_2 = \sin(\theta + \pi/4)$ (see Fig. 2.4) are out of phase by $\pi/4$. In general, when we have $\sin(\phi + \theta)$ or $\cos(\phi + \theta)$ with ϕ being a constant, then ϕ is referred to as the *phase*.

In many applications we need to find trigonometric functions of sums or differences of two or more angles. The following relationships exist between trigonometric functions of sums and differences of angles, and the trigonometric functions of the angles themselves.

$$\sin(\theta \pm \phi) = \sin\theta\cos\phi \pm \sin\phi\cos\theta \tag{2.51}$$

$$\cos(\theta \pm \phi) = \cos\theta\cos\phi \mp \sin\theta\sin\phi \tag{2.52}$$

Letting $\phi = \theta$, we have

$$\sin 2\theta = 2\sin\theta\cos\theta \tag{2.53}$$

and

$$\cos 2\theta = \cos^2\theta - \sin^2\theta \tag{2.54}$$

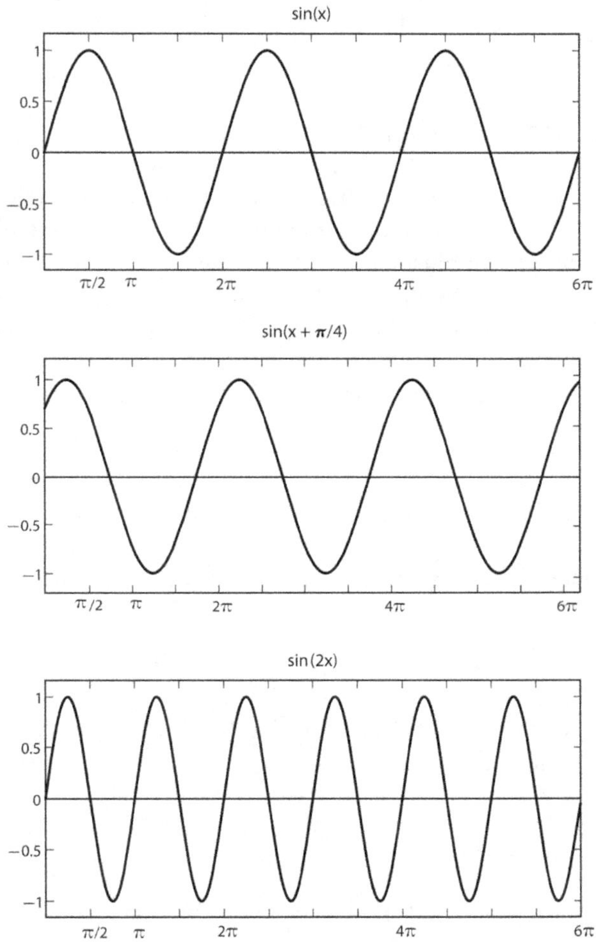

Fig. 2.4 Sin functions with different phases and frequencies

Recalling (2.47), (2.54) can be written as

$$\cos 2\theta = 2\cos^2\theta - 1 = 1 - 2\sin^2\theta \tag{2.55}$$

Example 2.21

$$\sin\left(\frac{\pi}{5}\right) = 0.588 \qquad \cos\left(\frac{\pi}{5}\right) = 0.809$$

$$\sin\left(2\frac{\pi}{5}\right) = 2\sin\left(\frac{\pi}{5}\right)\cos\left(\frac{\pi}{5}\right) = 2 \times 0.588 \times 0.809 = 0.951$$

$$\cos\left(2\frac{\pi}{5}\right) = \cos^2\left(\frac{\pi}{5}\right) - \sin^2\left(\frac{\pi}{5}\right) = 0.809^2 - 0.588^2 = 0.309$$

2.6.1 Cycles and Frequencies

Trigonometric functions are cyclical because as the point on the circle travels counterclockwise, it comes back to the same point again and again (see Fig. 2.4). As a result, the sine and cosine functions assume the same values for angles $\theta, 2\pi + \theta$, and, in general, $2k\pi + \theta$. Similarly, the tan function has the same value for θ and $\theta + \pi$. These functions are called *periodic*.[11] Compare the two functions

$$y_1 = \sin(x) = \sin(x + 2k\pi)$$

and

$$y_2 = \sin(2x) = \sin(2x + 2k\pi) = \sin[2(x + k\pi)]$$

Clearly y_2 returns to the same value—or completes a cycle—twice as fast as y_1. In general,

$$y = \sin(fx) \tag{2.56}$$

completes a cycle f times faster than $\sin(x)$. We call f the frequency of the function. Alternatively we can write (2.56) as

$$y = \sin\left(\frac{x}{p}\right) \tag{2.57}$$

Because $p = 1/f$, it is clear that every p periods the function will have the same value. In other words, the function completes a cycle in p periods or the cycle length is p. These concepts are better understood if we take the argument of the function to be time, measured in discrete values for a given time interval, that is, $t = 1, \ldots, T$. Let

$$y = \sin\left(\frac{2\pi t}{p}\right) \tag{2.58}$$

If $p = T$, then it takes T time periods to complete the cycle and the frequency is $1/T$. On the other hand, if frequency is $4/T$, then the length of the cycle is $T/4$. As an example, let time be measured in months and the period under consideration be a year, that is, $T = 12$. If $p = 3$, then we have four cycles per year and the frequency is 1/3 of a cycle per month. On the other hand, if $p = 1/2$, there are 24 cycles in a year and the frequency per month is two.

[11] A function $y = f(x)$ is called periodic if $f(x) = f(x + c), \quad c \neq 0$.

2.6.2 Exercises

E.2.16 Find the numerical value of the following:

$$i.\ \sin\frac{3}{2}\pi, \qquad ii.\ \frac{\cos^2\frac{7}{2}\pi - \sin\frac{2}{3}\pi}{2\cos\frac{5}{4}\pi}, \qquad iii.\ \frac{\sin^3\frac{5}{3}\pi - \tan\frac{3}{4}\pi}{\sin\frac{3}{4}\pi}$$

E.2.17 Graph the following functions in the interval $0 \leq x \leq 2\pi$.

$$i.\ \ y = \cos\left(x + \frac{\pi}{2}\right), \qquad ii.\ \ y = \sin\left(x + \frac{\pi}{2}\right)$$

$$iii.\ \ y = \cos\left(x + \frac{\pi}{2}\right) + \sin\left(x + \frac{\pi}{2}\right), \qquad iv.\ \ y = \cos\left(x + \frac{\pi}{2}\right) - \sin\left(x + \frac{\pi}{2}\right)$$

$$v.\ \ y = \tan\left(x + \frac{\pi}{2}\right), \qquad vi.\ \ y = \sin x + \sin 2x + \sin 3x$$

$$vii.\ \ y = \sin x + 0.5\sin 2x + 0.25\sin 3x$$

E.2.18 Show that

$$\tan 2\theta = \frac{2\tan\theta}{1 - \tan^2\theta}$$

E.2.19 Write $\sin 3x$ in terms of $\sin x$ and its powers.
E.2.20 Write $\sin 4x$ in terms of $\sin x$ and its powers.
E.2.21 Show that

$$i.\ \ \frac{1}{\tan\theta + \dfrac{1}{\tan\theta}} = \sin\theta\cos\theta \qquad ii.\ \ \frac{1 - \sin x}{\cos x} = \frac{\cos x}{1 + \sin x}$$

$$iii.\ \ \frac{1 + \sin\theta}{1 - \sin\theta} - \frac{1 - \sin\theta}{1 + \sin\theta} = 4\frac{1}{\cos\theta}\tan\theta \qquad iv.\ \ \frac{1 - \sin\theta}{1 + \sin\theta} = \left(\tan\theta - \frac{1}{\cos\theta}\right)^2$$

2.7 Complex Numbers

Complex numbers are two-dimensional numbers where one dimension is on the real axis and the other on the imaginary axis.[12] We are already familiar with real numbers and the real line. The imaginary number is

$$i = \sqrt{-1} \tag{2.59}$$

[12] It is customary to introduce complex numbers in the context of the solution to quadratic equations involving the square root of a negative number. This practice has the unfortunate consequence that students may get the impression that somewhere among the real numbers or along the real line there are caves where complex numbers are hiding and once in a while show their faces.

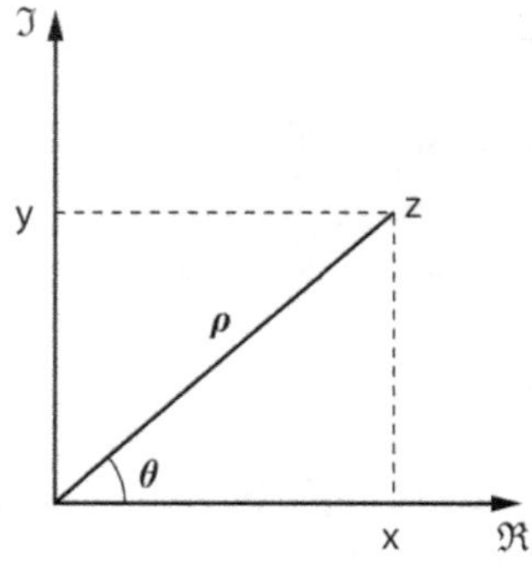

Fig. 2.5 Point z in the complex plane

i is an imaginary number because $i^2 = -1$, and there is no real number whose square is a negative number. Geometrically, a complex number is a point in the two-dimensional complex space. Any function of complex variables maps these variables into the two-dimensional complex plane.

Figure 2.5 depicts point z in the complex plane where the horizontal axis is the real line and the vertical axis the imaginary line. Thus, we can write z as

$$z = x + iy \tag{2.60}$$

Two complex numbers are equal if they are equal in both real and imaginary dimensions. That is, $z_1 = x_1 + iy_1$ is equal to $z_2 = x_2 + iy_2$ if $x_1 = x_2$ and $y_1 = y_2$. Real numbers are a special case of complex numbers when the imaginary dimension is set equal to zero. Similarly, an imaginary number is a complex number with its real dimension set equal to zero.

Example 2.22 The following are examples of complex numbers:

$$z_1 = 3 + i, \qquad z_2 = 5 - 3i, \qquad z_3 = 6 + 0.5i$$

Complex numbers come in pairs. Every complex number has its twin, called a *conjugate*. If $z = x + iy$, then its conjugate complex number is $\bar{z} = x - iy$. It follows that $\bar{\bar{z}} = z$. In other words, z is the conjugate of $\bar{z}$.

Example 2.23 The conjugates of the complex numbers in Example 2.22 are

$$\bar{z}_1 = 3 - i, \qquad \bar{z}_2 = 5 + 3i, \qquad \bar{z}_3 = 6 - 0.5i$$

Operations of addition, subtraction, and multiplication of complex variables are defined as

$$
\begin{aligned}
z_1 + z_2 &= (x_1 + x_2) + (y_1 + y_2)\, i \\
z_1 - z_2 &= (x_1 - x_2) + (y_1 - y_2)\, i \\
z_1 z_2 &= (x_1 + i y_1)\,(x_2 + i y_2) \\
&= (x_1 x_2 - y_1 y_2) + (x_1 y_2 + x_2 y_1)\, i
\end{aligned} \tag{2.61}
$$

Example 2.24

$$
\begin{aligned}
(3 - i) + (5 + 3i) &= 8 + 2i \\
(5 + 3i) - (6 + 0.5\, i) &= -1 + 2.5\, i \\
(6 - 0.5i)\,(3 + i) &= 18.5 + 4.5\, i
\end{aligned}
$$

Addition, subtraction, and multiplication of a complex number by its conjugate result in

$$
\begin{aligned}
z + \bar{z} &= 2x \\
z - \bar{z} &= 2iy \\
z\bar{z} &= x^2 + y^2 = \rho^2
\end{aligned} \tag{2.62}
$$

where the last equality refers to Fig. 2.5 and is based on the Pythagoras theorem. Division of complex numbers is a bit more involved:

$$
\frac{z_1}{z_2} = \frac{z_1 \bar{z}_2}{z_2 \bar{z}_2} = \frac{x_1 x_2 + y_1 y_2}{x_2^2 + y_2^2} + i \frac{x_2 y_1 - x_1 y_2}{x_2^2 + y_2^2} \tag{2.63}
$$

Example 2.25

$$
\frac{6 - 0.5i}{5 + 3i} = \frac{28.5}{34} - i\frac{20.5}{34} \approx 0.838 - 0.603i
$$

Referring again to Fig. 2.5, we observe that

$$
x = \rho \cos\theta \qquad \text{and} \qquad y = \rho \sin\theta
$$

which implies that

$$
\begin{aligned}
\rho^2 &= x^2 + y^2 \\
\tan\theta &= \tfrac{y}{x}
\end{aligned}
$$

These relationships enable us to write a complex variable either in terms of its Euclidean coordinates or in terms of ρ and θ, that is, its *polar coordinates*:

$$
z = x + iy = \rho(\cos\theta + i \sin\theta) \tag{2.64}
$$

where

$$
\begin{aligned}
\rho &= \sqrt{x^2 + y^2} \\
\theta &= \tan^{-1} \tfrac{y}{x}
\end{aligned} \tag{2.65}
$$

Example 2.26 Let us rewrite complex numbers in Example 2.22 using polar coordinates.[13] For

$$z_1 = 3 + i$$

we have

$$\begin{aligned}\rho &= \sqrt{3^2 + 1^2} = \sqrt{10}\\ \theta &= \tan^{-1}\frac{1}{3} = 0.32175 = 0.102416\pi\end{aligned}$$

therefore,

$$z_1 = \sqrt{10}(\cos 0.102416\pi + i \sin 0.102416\pi)$$

Similarly,

$$\begin{aligned}z_2 &= \sqrt{34}(\cos 0.54042 - i \sin 0.54042)\\ &= \sqrt{34}(\cos 0.17202\pi - i \sin 0.17202\pi)\end{aligned}$$

$$\begin{aligned}z_3 &= \sqrt{36.25}(\cos 0.08314 + i \sin 0.08314)\\ &= \sqrt{36.25}(\cos 0.02646\pi + i \sin 0.02646\pi)\end{aligned}$$

Note that because $\sin(-\theta) = -\sin\theta$, (2.64) implies

$$\bar{z} = \rho(\cos\theta - i \sin\theta)$$

We have a third way to write complex numbers. For this, we state without proof the following relationships[14]:

$$\begin{aligned}\exp(i\theta) &= \cos\theta + i \sin\theta\\ \exp(-i\theta) &= \cos\theta - i \sin\theta\end{aligned} \tag{2.66}$$

Therefore,

$$\begin{aligned}x + iy &= \rho \exp(i\theta)\\ x - iy &= \rho \exp(-i\theta)\end{aligned} \tag{2.67}$$

Where ρ and θ are as defined in (2.65).

---

[13] Angles are measured in radians. If you use a calculator, you need to set it in the radian mode to get the same numbers as in the text. If your calculator is in the degree mode, then in order to get the same numbers as in the text, $\theta = \tan^{-1}(x/y)$ needs to be converted into radians by multiplying it by $\pi/180$.

[14] We shall provide a proof of these relationships in Chap. 10.

Example 2.27 Again using complex numbers in Example 2.22, we have

$$z_1 = \sqrt{10}\exp(0.32175i)$$
$$z_2 = \sqrt{34}\exp(-0.54042i)$$
$$z_3 = \sqrt{36.25}\exp(0.08314i)$$

Using the following program, the reader could check the validity of the formulas in (2.66) for different values of the angle t.

Matlab code

```
% Set the value of the angle
t = pi./3;
% Trigonometric version
cos(t)+i*sin(t)
% Exponential version
exp(i*t)
% Trigonometric version
cos(t)-i*sin(t)
% Exponential version
exp(-i*t)
```

The idea of the equivalence of circular sine and cosine functions with the exponential function may bother the intuitive sense of some readers. But $\exp(i\theta)$ is indeed a circular function in the complex plane that traces a circle as θ changes from 0 to 2π. On the other hand, ρ determines the distance of the point from the origin. Indeed, we can define trigonometric functions in terms of the exponentials of complex numbers.

$$\begin{aligned}\cos\theta &= \frac{\exp(i\theta)+\exp(-i\theta)}{2}\\ \sin\theta &= \frac{\exp(i\theta)-\exp(-i\theta)}{2i}\end{aligned} \tag{2.68}$$

An important consequence of (2.66) is De Moivre's theorem.[15]

Theorem 2.1

$$\begin{aligned}z^k &= [\rho(\cos\theta + i\sin\theta)]^k\\ &= \rho^k(\exp(i\theta))^k\\ &= \rho^k\exp(ik\theta)\\ &= \rho^k(\cos k\theta + i\sin k\theta)\end{aligned} \tag{2.69}$$

[15]Abraham De Moivre (1667–1754), a French mathematician who spent most of his life in England, was a pioneer in the development of probability theory and analytic geometry. He was appointed to the commission set up to examine Newton's and Leibnitz's claims for the discovery of calculus.

Example 2.28

$$
\begin{aligned}
(\cos\theta + i\sin\theta)^3 &= \cos 3\theta + i\sin 3\theta \\
(\cos\theta - i\sin\theta)^3 &= \cos 3\theta - i\sin 3\theta
\end{aligned}
$$

Example 2.29 *Let* $\theta = \pi/4$, *then*

$$
\begin{aligned}
\left(\cos\frac{\pi}{4} + i\sin\frac{\pi}{4}\right)^5 &= (0.7071 + 0.7071i)^5 \\
&= -0.7071 - 0.7071i
\end{aligned}
$$

$$
\begin{aligned}
\cos\frac{5\pi}{4} + i\sin\frac{5\pi}{4} &= \cos\left(\pi + \frac{\pi}{4}\right) + i\sin\left(\pi + \frac{\pi}{4}\right) \\
&= -\cos\frac{\pi}{4} - i\sin\frac{\pi}{4} \\
&= -0.7071 - 0.7071i
\end{aligned}
$$

2.7.1 Exercises

E.2.22 Write the following complex numbers in alternative forms of (2.64) and (2.67)

i. $z_1 = 1 + i$ *ii.* $z_2 = 1 - i$
iii. $z_3 = 5i$ *iv.* $z_4 = 3.5 - 2.6i$
v. $z_5 = 7 + 4i$

E.2.23 Referring to E.2.21, compute

i. $z_1 z_2$ *ii.* $\dfrac{z_1}{z_2}$ *iii.* $z_3 z_4$
iv. $\dfrac{z_4}{z_5}$ *v.* $\bar{z}_3 z_5$ *vi.* $\dfrac{\bar{z}_5}{\bar{z}_3}$

E.2.24 We already know that $e^a e^b = e^{a+b}$ where a and b are real numbers. Show that for real numbers a and b,

$$
e^{ai+bi} = e^{ai} e^{bi}
$$

Chapter 3
Basic Concepts of Computation

3.1 Iterative Methods

The importance of computation in engineering, communication, economics, econometrics, finance, and everyday life needs no elaboration. Numerical analysis is the branch of mathematics dealing with computation. But one might question the need for such a specialized field. Computation is easy and we all know how to perform it. For instance, if we are given the equation

$$2x = 10$$

isn't it clear that $x = 5$? You are right. We are even able to obtain the solutions of a quadratic equation such as

$$x^2 - x + 6 = 0$$

We even have formulas, although more complicated ones, for equations involving polynomials of third and fourth degrees. You just plug in the numbers in the formula and get the exact answer. You wish life was that easy. It ain't!

To begin with, polynomials of degree $n \geq 5$ cannot be solved using a closed form general solution. By a closed form we mean a simple formula that involves only addition (and subtraction), multiplication (and division), and roots of expressions (radicals) and does not require the summation of an infinite number of terms. Because polynomials are the simplest types of equations that we may be interested in solving, it follows that no general closed form solutions exist for arbitrary nonlinear equations. But we still need to solve such equations.

Because an exact solution is not available, we approximate the solution in a number of steps. Starting from a set of inputs that include the knowledge of the equation and an initial solution (perhaps an informed guess) and applying a procedure, we get closer and closer to the solution. The steps that take us from the initial input to the solution or the output form an algorithm.[1] A numerical algorithm

---

[1]The word comes from the name of the Iranian mathematician, Mohammad Al-Khwarizmi (780–850), when it was translated into Latin as Algoritmi. He clearly spelled out a way to solve quadratic

K. Dadkhah, *Foundations of Mathematical and Computational Economics,* 2nd ed.,
DOI 10.1007/978-3-642-13748-8_3, © Springer-Verlag Berlin Heidelberg 2011

is judged by its stability, accuracy, efficiency, and reliability. Numerical algorithms are not confined to solving equations. We need algorithms to evaluate functions, to generate random numbers (called pseudo-random generating algorithms), to find the maximum and minimum of functions and to perform many other mathematical operations. Many algorithms use recursion. A simple recursive algorithm is of the form

$$x_{j+1} = g\left(x_j\right) \tag{3.1}$$

Here we need a starting point, say, x_0, and the function g, which is chosen for the problem at hand. Then it is straightforward to compute $x_1, x_2, \cdots$. But we need a stopping rule for the recursion. Usually we stop the calculation when the change in x from one step to the next is not appreciable. Thus, we stop when

$$\left|x_{j+1} - x_j\right| < \Delta \tag{3.2}$$

Where Δ is a preset number that determines the computation accuracy.

With present-day computers, we can reach very high levels of accuracy in most of our computational problems. Let us illustrate these concepts not with the solution of an equation (a subject we shall discuss in Chap. 10), but with a simpler example. Suppose we are interested in finding the square root of a number—say 76459. We need an initial guess; here we take $x_0 = 100$. We could have done better, but we choose this arbitrary number to show the power of iterative methods. In more complicated computations, mathematicians base the initial guess on some information contained in the problem or by using a simpler algorithm. If this guess of $x_0 = 100$ was a good one, then 76459/100 would be very close to our initial guess. But it is not. Nevertheless, we know that the correct answer lies between 100 and 76459/100. It seems reasonable to compute their average as a better guess in the sense of coming closer to a correct answer.

$$x_1 = \left(x_0 + \frac{76459}{x_0}\right) \Big/ 2$$

The next number is $x_1 = 432.295$. If we repeat the step above, we get

$$x_2 = \left(x_1 + \frac{76459}{x_1}\right) \Big/ 2 = 304.5813241$$

A few more iterations and we have our number $\sqrt{76459} = 276.5122059$. This procedure can be programmed in Matlab or Excel. In an Excel worksheet, make the entries shown in the Table 3.1. Highlight the squares B2 and C2 and drag them down a few rows.

equations, although it works only for the cases when the solutions are real. The word "algebra" is also taken from the title of his influential book *Hisab al-Jabr wal-Muqabala*, which was the first book on algebra.

Table 3.1 Using the iterative method in Excel to compute the square root

A	B	C
76459	100	=A$1/B1
	=(B1+C1)/2	=A$1/B2
	...	...
	...	...

You will get the answer.

Table 3.2 Computation results for the square root of a number using the iterative method in Excel

A	B	C
76459	100	764.59
	432.295	176.8676
	304.5813	251.0298
	277.8056	275.2249
	276.5152	276.5092
	276.5122	276.5122
	276.5122	276.5122
	276.5122	276.5122

You can check the accuracy of your results by typing in a cell in an Excel sheet: =sqrt(76459).

A short Matlab program accomplishes the same task.

Matlab code

```
% Specify the number
A = 76459
% Initialize x
x = 100
% Find the square root in 6 iterations
for j=1:6
     x = (x + A./x)./2
end
```

In finding the square root of a number, we arbitrarily chose six iterations, thus avoiding a crucial question: when to stop the iterative process. Alternatively, we could have chosen to stop the algorithm when $|x_j - x_{j-1}| < 0.0001$. In general, depending on the problem at hand, we should set a limit that reflects our desired precision of results and terminate the process when the results obtained in two consecutive iterations differ by less than our preset limit Δ. Of course, this limit should not be less than the precision of Matlab or any other software we may be using. To summarize, an iterative algorithm consists of the following steps:

1. Choose a Δ
2. Choose a starting point x_0

3. Calculate x_j based on x_{j-1}
4. Repeat step 3 until $\left|x_j - x_{j-1}\right| < \Delta$

In many areas of computation, solutions are obtained through approximation. In particular, whenever we have to deal with the sum of an infinite number of terms, we have to resort to approximation. Evaluation of definite integrals, about which we shall learn in Chap. 11, is a case in point.

Integrals play an important role in calculus, yet (for reasons that we will discuss in Chap. 11) direct computation of many integrals is impossible and we have to evaluate them using approximation. The same is true about many other areas of mathematics. Indeed, hardly an area of computation exists in which we do not use approximation. The fact that our results are approximately correct makes it necessary to be aware of computation error.

3.1.1 Naming Cells in Excel

In this section and in Chap. 2 we have used Excel to carry out computation. In specifying commands we referred to different cells as A39 or C$7. Such references are prone to error. It would be much better if we could refer to variable names in the formulas used. For instance in computing the sum of the terms in a geometric progression, we had

$$S = a\frac{1 - q^n}{1 - q}$$

It would be much better if we could refer to a or q in our computation rather than the cell number. This can be accomplished by naming cells in Excel.

Chose a cell (or even an array of cells) and then in Formulas click on Define Name. In the window that appears name the cell as you wish. You can even add a comment or description.

Let us assume that we have named the cells as follows:

Table 3.3 Name of variables and their designated cells in Excel

Cell	A2	B1	C1	D1
Name	S	a	q	n

Then we can enter any value we desire in those cells and carry out the computation as shown in Table 3.4:

Table 3.4 Computation with named cells in Excel

A	B	C	D	E
	12	0.85	20	
=a∗(1−q^n)/(1−q)				

3.2 Absolute and Relative Computation Errors

Computation error is a fact of life. In addition to the necessity of approximation, any computation involving irrational numbers will, of necessity, involve truncation error. Recall that an irrational number cannot be expressed in terms of the ratio of two integers. If we write it in decimals, we will have an infinite number of digits to the right of the decimal point. Therefore, no matter how many decimal places we include in our computation, a truncation occurs and the irrational number we use in our calculations will be an approximation. This point is of importance because many more irrational numbers exist than do rational numbers.

As the above sketch of an algorithm shows, in order not to fall into a never-ending loop of recalculation, we need to set a standard and accept some level of approximation error. All errors are not created equal. Suppose I have \$4.95 in my pocket and I am asked how much money I have. If I answer "five dollars," the answer is, of course, incorrect. On the other hand, it is a much better answer than "four dollars," and far better than "95 dollars."

It seems that the error in the first answer is forgivable, less so in the second answer, and the third is an outright lie. The first answer is off by $0.05/4.95 = 0.0101$ or slightly more than 1%, whereas the second answer is in error by 19%, and the last one by more than 1800%.

Thus, we need to be concerned with error of computation in relation to the actual value. The relative error of computation is defined as

$$\frac{|\tilde{x} - x|}{|x|} \tag{3.3}$$

where x is the actual value and $\tilde{x}$ its approximate value. We determine the acceptable level of the relative error by the nature of the problem at hand. For some problems, for example, computing the solvency of Social Security 40 years from now, a billion or even a 10 billion dollar discrepancy is nothing to worry about. But in some physics and engineering problems, an error involving a millionth of a millimeter may be catastrophic.

In setting the tolerance level for the error of a numerical algorithm, we should be aware of the capacity of the computer to distinguish between two numbers. The smallest difference between two numbers that software can detect should be taken into account in setting Δ. In Matlab, such a number is represented by `eps`, which in the current version is equal to 2.2204×10^{-16}. In writing a recursive algorithm, one has to make sure that $\Delta >$ `eps`. Needless to say, this level of error is more than acceptable for any economics or econometrics calculation.

We must also avoid the pitfall of infinite loops. This brings up the very important issue of the convergence of an algorithm. To implement an algorithm, we have to have assurances that it will converge to the true value of the variable we are seeking. In other words, we should be able to show that in successive iterations, the quantity $\left|x_j - x_{j-1}\right|$ gets smaller and smaller and this occurs relatively quickly. The proof of algorithm convergence starts with the initial value and shows that in a finite number of steps, the method reaches the desired result.

3.3 Efficiency of Computation

Today we have enormous computing powers—witness the fact that the simple calculator you carry around with you has more computing power than that which was available to all physicists working at Los Alamos to build the atomic bomb during World War II. Similarly, the computing power possessed by an average family today surpasses the computing power available to NASA when it landed a man on the moon. Yet efficiency of computation is still an issue, especially in complicated models in physics, economics, econometrics, and other disciplines. The smaller the number of additions and sign changes in an algorithm, the faster we can get the result and the more accurate our computation will be. One goal of numerical analysis is to devise efficient algorithms. Let us illustrate this with an example. Suppose we are interested in evaluating the following function:

$$f(x) = 5x^3 + 4x^2 - 7x + 9$$

at the point $x = 3$. Such a calculation can be performed in the following way:

$$\begin{array}{r} 5 \times 3 \times 3 \times 3 \\ + \\ 4 \times 3 \times 3 \\ + \\ -7 \times 3 \\ + \\ 9 \\ 159 \end{array}$$

As we can see, this calculation involves six multiplications, three additions, and one change of sign. We should remember that the operation of raising a number to a power is repeated multiplications and multiplication is repeated addition. Here, for simplicity, we count in terms of multiplications and additions. Now let us try an algorithm known as Horner's method.[2]

$$\begin{aligned} b_0 &= 5 \\ b_1 &= 4 + b_0 \times 3 = 19 \\ b_2 &= -7 + b_1 \times 3 = 50 \\ b_3 &= 9 + b_2 \times 3 = 159 \end{aligned}$$

In this simple calculation Horner's method reduces the number of multiplications by three. Because each multiplication consists of a number of additions, we save significant computation time. Of course, this is a very simple example and numerical algorithms are generally much more complicated, so the gains in efficiency become that much greater. Furthermore, to evaluate nonpolynomial functions, we

[2]Named after William Horner (1786–1837) although apparently the method was discovered centuries before by the Chinese mathematician, Zhu Shijie (~1260–~1320).

approximate them by a polynomial, hence the importance of an efficient method for evaluating polynomials.

To understand the rationale behind this method, note that we could write the function as

$$\begin{aligned} f(x) &= 5x^3 + 4x^2 - 7x + 9 \\ &= 9 + x(-7 + 4x + 5x^2) \\ &= 9 + x(-7 + x(4 + 5x)) \\ &= 9 + x(-7 + x(4 + x(5))) \end{aligned}$$

Thus, we start from the rightmost parenthesis and proceed to evaluate the terms inside each parenthesis and add it to the next term on the left. More generally, suppose we want to evaluate an n-th order polynomial function at the point $x = x_*$. We can write the function as

$$\begin{aligned} f(x_*) &= a_n x_*^n + a_{n-1} x_*^{n-1} + \ldots + a_1 x_* + a_0 \\ &= (a_0 + x_*(a_1 + x_*(a_2 + \ldots x_*(a_{n-1} + x_* a_n) \ldots) \end{aligned} \tag{3.4}$$

and the method takes the form

$$\begin{aligned} b_0 &= a_n \\ b_1 &= a_{n-1} + b_0 x_* \\ &\ldots \\ b_n &= a_0 + b_{n-1} x_* \end{aligned}$$

3.4 *o* and *O*

Two very useful symbols frequently used in mathematics and computation are o and O. They signify the order of magnitude of a term or function.[3] Consider the two series a_n and b_n:

	1	2	...	n
a_n	$\frac{1}{2}$	$\frac{1}{4}$	...	$\frac{1}{2^n}$
b_n	$\frac{1}{4}$	$\frac{1}{16}$	...	$\frac{1}{2^{2n}}$

The ratio of the two series takes the form

$$\frac{b_n}{a_n} = \frac{1}{2^n}$$

[3] These notations were introduced by the German mathematician Edmund Georg Hermann Landau (1877–1938).

It follows that

$$\frac{b_n}{a_n} \to 0, \qquad \text{as} \qquad n \to \infty \tag{3.5}$$

In such a case we write

$$b_n = o\,(a_n) \tag{3.6}$$

More formally, if for every $\varepsilon > 0$, which can be made arbitrarily small, we can find a number N such that when $n > N$, we have

$$\frac{b_n}{a_n} < \varepsilon \tag{3.7}$$

then we can write

$$b_n = o\,(a_n) \tag{3.8}$$

Note that N depends on ε.

$$x = o(x^2) \qquad x \to \infty$$

Indeed, we can write

$$x^\alpha = o(x^\beta) \qquad \alpha < \beta \qquad x \to \infty$$

because

$$\frac{x^\alpha}{x^\beta} = \frac{1}{x^{\beta-\alpha}}$$

and as long as $\beta - \alpha > 0$, we have

$$\lim_{x\to\infty} \frac{1}{x^{\beta-\alpha}} = 0$$

As another example consider

$$\ln x = o(x) \qquad x \to \infty$$

Again, we can write

$$\ln x = o(x^\alpha) \qquad \alpha > 0 \qquad x \to \infty$$

The concept is not always applied to the case of $x \to \infty$, as the following examples suggest.

Example 3.1

$$\cos x = 1 + o(x) \qquad x \to 0$$

and

$$e^{-1/x} = o(x^{\alpha}) \qquad \alpha > 0 \qquad x \to 0$$

Example 3.2 (Stirling's Formula). One of the most famous and important approximation formulas in mathematics and statistics is Stirling's formula:

$$\frac{n!}{\sqrt{2\pi}\, n^{n+1/2} e^{-n}} \to 1 \qquad \text{as} \qquad n \to \infty \tag{3.9}$$

In other words, for large n we can approximate $n!$ by

$$\sqrt{2\pi}\, n^{n+1/2} e^{-n} \tag{3.10}$$

Stirling's formula can be written as

$$\frac{n!}{\sqrt{2\pi}\, n^{n+1/2} e^{-n}} - 1 = o(1) \qquad \text{as} \qquad n \to \infty \tag{3.11}$$

The notation O is used to denote that the ratio of two series or two functions is bounded, provided that the denominator is not zero. Thus, when we write

$$a_n = O(b_n) \tag{3.12}$$

it means

$$|a_n| < A b_n, \qquad \forall n, \tag{3.13}$$

where A is some constant. Similarly

$$f(x) = O(\phi(x)) \tag{3.14}$$

means

$$|f(x)| < A\phi(x), \qquad \forall x, \tag{3.15}$$

again for some constant A.

Example 3.3

$$\begin{aligned} &x^2 = O(x) \qquad x \to 0 \\ &e^{-x} = O(1) \qquad x \to \infty \\ &e^x - 1 = O(x) \qquad -1 < x < 1 \\ &\tfrac{1}{\ln x} = O(1) \qquad x \to \infty \end{aligned}$$

3.5 Solving an Equation

On many occasions we have an equation of the form $f(x) = 0$ and we need to find the roots of this equation, that is, to find the value(s) of x that satisfy this equation.

Example 3.4 If we have the following equation

$$(x - a)(x + b) = 0$$

Clearly $x = a$ and $x = -b$ are its roots, because substitution of these numbers for x make the LHS of the equation equal to its RHS. Some equations, like the one above, can be solved easily. Others can be solved, but with more difficulty. Examples of the first kind are linear equations of the form

$$a + bx = 0$$

Then, provided $b \neq 0$,

$$x = -\frac{a}{b}$$

Note that this solution is general and applies to any equation with real coefficients a and b. Even the restriction $b \neq 0$ is not much of a restriction because if $b = 0$ so would $a = 0$ and we would have no equation to solve.

It is not surprising, therefore, that mathematicians have tried to find general solutions to polynomial equations. Of course as the degree of the polynomial increases, the problem gets harder. For example, for a quadratic equation

$$ax^2 + bx + c = 0 \tag{3.16}$$

the situation is slightly more involved. First, note that we should have $a \neq 0$ otherwise we will be dealing with a linear equation. Furthermore, divide (3.16) through by a

$$x^2 + \frac{b}{a}x + \frac{c}{a} = 0 \tag{3.17}$$

It is clear that the solution of (3.16) and (3.17) are the same. Finally, note that if we could turn the equation into a format like $(x + h)^2 - g = 0$, then the solution would be $x = -h \pm g^{\frac{1}{2}}$. This is what we will do:

$$x^2 + \frac{b}{a}x + \frac{b^2}{4a^2} - \frac{b^2}{4a^2} + \frac{c}{a} = 0$$

Rearranging the terms

$$\left(x + \frac{b}{2a}\right)^2 - \left(\frac{b^2}{4a^2} - \frac{c}{a}\right) = 0$$

Therefore

$$x = -\frac{b}{2a} \pm \sqrt{\frac{b^2 - 4ac}{4a^2}} = \frac{-b \pm \sqrt{b^2 - 4ac}}{2a} \tag{3.18}$$

We can try to solve cubic and quartic equations (polynomial equations of degrees three and four) in which we will encounter some difficulty. Still a general solution in terms of polynomials and radicals (the roots of polynomials, for example, square root, third root, etc.) could be found. But, that's it. In 1824, Abel[4] showed that quintic and higher order equations do not have a general solution in terms of radicals.

If polynomial equations don't have a general solution, other more involved equations could not possibly have such general solutions. But in applied work we need to solve polynomial equations of degree five and higher as well as equations involving exponential, logarithmic, and trigonometric functions. Such equations could be solved numerically, and we shall present a general framework for solving such equations in Chap. 10. Here we confine ourselves to a Matlab routine for solving polynomial equations.

Example 3.5 Suppose we would like to solve the equation

$$x^3 - 2.15x^2 - 0.67x + 0.7475 = 0$$

You may want to try the familiar tricks of checking for $x = 0$, $x = 1$, and so on before using Matlab. Let us save time; they won't work. But here is the function that will get us the solution.

Matlab code

```
% define the equation by specifying its coefficients.
% Note that order is important
g =[1 -2.15 -0.67 0.7475];
r = roots(g)
```

---

[4]Neils Henrik Abel (1802–1829) was a handsome man and a brilliant mathematician from Norway. Despite his short life he made great contributions to mathematics. He lived in poverty and died of tuberculosis. His teacher Bernt Holmboe helped Abel gain financial support for his studies, as did the German mathematician August Leopold Crelle (1780–1855), who founded the first journal devoted exclusively to mathematics and who published many of Abel's papers. Indeed Crelle secured a nice job for him, but it was too late. There is a sentence in the middle of one of Abel's notebooks that I have found it haunting: "Our Father who art in Heaven, give me bread and beer. Listen for once." The French mathematician Èveriste Galois (1811–1832) generalized Abel's work on solvability of equations. Galois also had a tragic life. He was imprisoned for his political views and lost his life in a duel; it is not clear if it was over love or politics. The modern or abstract algebra is based on the work of Abel and Galois. A good source on both men (but with some amount of math included) is Peter Pesic's *Abel's Proof* (2003); see also *Men of Mathematics* (1937) by E. T. Bell.

Matlab will return the roots $x_1 = 2.3$, $x_2 = -0.65$, and $x_3 = 0.5$. Roots of polynomials of second and higher orders are not all necessarily real. Many of them have complex roots. Matlab can handle both types of roots.

3.6 Exercises

E.3.1 Evaluate the following polynomials at $x = 5$ with and without Horner's method and determine how many multiplications and additions you save.

$$\begin{array}{ll} i. & f(x) = x^4 + 12x^3 - x^2 + 7x - 8 \\ ii. & f(x) = x^5 - 5x^4 + 10x^3 + 3x^2 - x + 15 \\ iii. & f(x) = x^6 - 6x^5 + 2x^4 + 9x^3 + 9x^2 - 11x - 56 \end{array}$$

E.3.2 Use Excel to calculate $n!$ directly and via Stirling's approximation. Calculate the ratio of the two and see if indeed the ratio approaches one. Also determine the highest n for which the calculation of $n!$ is feasible by either method.

E.3.3 Repeat Exercise E.3.1 using a Matlab program.

E.3.4 Solve the following equations using (3.18):

$$\begin{array}{ll} i. & x^2 - 2.25x + 2.25 = 0 \\ ii. & x^2 + 6x + 9 = 0 \\ iii. & x^2 - 4x + 5 = 0 \end{array}$$

E.3.5 Check your results for E.3.4 using the Matlab function `roots`.

E.3.6 Program the formula in (3.18) in Matlab using only the operators $+, -, *, \backslash$, and the exponent ^. Make sure that the program checks whether $b^2 - 4ac$ is positive, zero, or negative and writes out the solutions accordingly. Apply your program to problems in E.3.4.

E.3.7 Use the Matlab function `roots` to solve the following equations:

$$\begin{array}{ll} i. & x^3 - 4x^2 + 2x - 3 = 0 \\ ii. & x^4 + 2x^3 - 7x^2 + 5x + 17 = 0 \\ iii. & x^5 + x^4 + 8x^3 - 6x^2 + 6x - 33 = 0 \end{array}$$

Chapter 4
Basic Concepts and Methods of Probability Theory and Statistics

4.1 Probability

Probability theory is the branch of mathematics that deals with random events, that is, events whose occurrence we cannot predict with certainty. Random phenomena are a feature of every sphere of natural and social existence and of life. From the genetic makeup of plants, animals, and human beings to planetary configurations, from games such as poker and backgammon to movements in financial markets, and from weather patterns to social and political events (such as elections results), we witness stochastic or random phenomena.

We can think of three reasons for randomness. First, by nature, some or perhaps all features of our world are stochastic. Quantum mechanics and the evolution of such an immense variety of life forms from a single source are witnesses to unpredictability in the universe and environment. A second source of randomness is that many events are the result of a very large number of actions or decisions. For example, thousands and sometimes millions of individuals who decide to buy or sell a particular asset at a particular price and in a certain quantity—these decisions determine that asset's price. The same is true in an election result that is too close to call. Even if we could enumerate and measure each and every influence upon these processes, we could not predict the outcome with certainty. Also, some events, like a coin flip, are affected by a small number of forces wherein the initial force, shape, and weight of the coin, its initial position, and the force of gravity determine whether the coin lands on heads or tails. Yet the outcome of any one flip of a fair coin is random because we simply cannot determine the outcome a priori. Third, some variables may appear random because they are measured with error. You can measure your height or weight a number of times and, while you may get approximately the same number, each measurement will show a small discrepancy from previous measurements. The result is that your measured height or weight may resemble a random variable. The same happens with counting the population of a city or estimating the U.S. per capita consumption.

Each random event results in a set of outcomes, which we shall call the space of *elementary events* or the *sample space*.

K. Dadkhah, *Foundations of Mathematical and Computational Economics,* 2nd ed.,
DOI 10.1007/978-3-642-13748-8_4, © Springer-Verlag Berlin Heidelberg 2011

Example 4.1 Excluding the possibility of the coin standing on its edge or leaning on a wall, the set of possible outcomes of flipping a fair coin is

$$\Omega = \{H, T\}$$

The corresponding set for flipping two coins

$$\Omega = \{HT, TH, TT, HH\}$$

Next we form the set $\mathcal{F}$ that contains all elements of the set Ω as well as their unions and complements. In other words, if A and B are in $\mathcal{F}$, so are $A \cup B, A^c$, and B^c. Furthermore, this set contains both the certain events Ω and the impossible event ø We call $\mathcal{F}$, which is closed under the operations of union and complements, an algebra. An algebra that is closed under enumerable union is called a σ-algebra.

Example 4.2 The algebra for the outcome of the flip of one coin is

$$\mathcal{F} = \{\emptyset, \{H, T\}, \{H\}, \{T\}\}$$

and for the flip of two coins

$$\begin{aligned}\mathcal{F} = \{&\emptyset, \Omega, \{TT\}, \{HH\}, \{HT, TH\}, \{HH, TT\},\\ &\{HH, HT, TH\}, \{TT, HT, TH\}\}\end{aligned}$$

Whereas we are not sure about the outcomes of a random event, we can attach to each outcome a number called a probability. More formally, we can define a *probability measure* by assigning to each element of the *sample space*, Ω, a probability P.

Example 4.3 To each event in the flip of a coin we can attach a probability, for example:

$$P(H) = P(T) = 0.5$$

And in the case of flip of two coins:

$$P(HH) = P(HT) = P(TH) = P(TT) = 0.25$$

Definition 4.1 The set function P is called a probability measure if

$$\begin{aligned}&i. \quad P(\emptyset) = 0\\ &ii. \quad P(\Omega) = 1\\ &iii. \quad P(A \cup B) = P(A) + P(B),\\ &\qquad \forall A,\ B \in \Omega, \quad A \cap B = \emptyset\end{aligned} \tag{4.1}$$

The conditions stated in Definition 4.1 are the *axioms of probability theory*.[1] The triple of the set of outcomes, the algebra, and the probability measure, $(\Omega, \mathcal{F}, P)$ is referred to as a *probability model*.

So far so good, but what do we mean by these mathematical terms and what have they got to do with the real world? The fact is that probability theory is silent as to how to assign probabilities to different events. But we have three sources for attaching probabilities to the outcomes of random events: They can be based on equally likely events, based on long-run frequencies (classical or frequentist approach), and based on the degree of confidence (subjective or Bayesian approach). Note that however we assign probabilities to different events, the mathematical theory for dealing with random events and their probabilities remain the same.

1. *Equally likely events.* This is the oldest and simplest definition of probability. If an event has n equally likely outcomes, then each outcome has probability $1/n$. For instance, in flipping a fair coin, the probability of heads is the same as the probability of tails. Thus, both have the probability of 0.5. Similarly, in throwing a fair die, the probability of each side showing up is 1/6. The problem with this definition is that the notion of "equally likely" makes the definition circular. We define probability by referring to equally likely outcomes, and equally likely outcomes are those that have the same probability of occurring.

2. *Frequency of occurrence.* In the examples above, suppose we do not know if the coin or die is fair. In such cases, we can determine the probability of heads by flipping the coin a million times. If heads shows up 499,963 times, we can calculate the probability of heads as $499963/1000000 = 0.499963$. This definition or way of assigning probabilities is empirical and quite appropriate for events that can be repeated a large number of times under identical circumstances. In particular, it is appropriate for many phenomena in the physical sciences and engineering. This notion of probability should be applied with care to economics, social sciences, political life, or, indeed, to any discipline that studies human behavior. Some events and issues are amenable to being classified as one in a large, reasonably, homogeneous set of events. Household income and consumption in an economy, and prices of different goods and services fit this characterization. Even events like recessions and recoveries, if considered only as periods of output decline or growth, may be treated as appropriate to the application of probability theory. But each recession and recovery is a unique event and cannot be treated as one observation in a large set of similar events. In the same way we must be careful not to think that all wars can be lumped into one category. Wars certainly exhibit similarities, but even greater differences.

We can think of some aspects of elections—such as voter participation rates, the division of votes between parties in total, and divisions based on gender and race—that can be studied within the frequentist framework. But again, each election as a historical event is unique and not subject to repetition. Once we get to issues like

[1]The axioms of probability theory were formulated by the great Russian mathematician Andrey Nikolaevich Kolmogorov (1903–1987) and published in 1933.

the development path of a country or transition to democracy, we should note that such events are unique and one-time affairs; thinking about them in frequentist terms would be misleading. This observation leads us to the third definition of probability.

3. *Degree of confidence or subjective probability.* When political commentators predict an 80% chance that Governor Sarah Palin will run for president in 2012, they are expressing their degree of confidence or subjective probability in the event occurring. The same is true when researchers say that, at a 95% level of confidence, worsening economic conditions do not increase the crime rate, or forecasters say that next year's inflation rate will be between 2.5 and 3.5%. An important case happens in criminal trials. The jury is asked to find the defendant guilty only if they are sure "beyond a reasonable doubt." Now you readers can ask yourselves how confident you would be in the guilt of a defendant to call it beyond a reasonable doubt: 80, 90, or 95%? The important point to remember here: The mathematical theory of probability is equally applicable to all three definitions of probability.

4.2 Random Variables and Probability Distributions

Suppose we attach a number to each outcome of a random event. Because the event is random, so are the numbers. The rule that specifies what number should attach to which outcome is called a *random variable*. More formally, a random variable is a set function that maps the set of outcomes of a random event to the set of real numbers. Such a function is not unique, and depending on the purpose at hand, we may define one or many random variables corresponding to the same random event.

Example 4.4 Let us define the random variable X as the number of heads in the flip of two coins. We have

$$X(HH) = 2, \quad X(HT) = 1, X(TH) = 1, \quad X(TT) = 0$$

We could have defined the random variable as the number of tails, in which case

$$X(HH) = 0, \quad X(HT) = 1, X(TH) = 1, \quad X(TT) = 2$$

Example 4.5 In collecting labor statistics we are interested in the characteristics of respondents. For example, we may ask if a person is in the labor force or not, employed, or unemployed. In addition, we would like to know about the demographic characteristics of respondents such as gender, race, and age. For each answer we can define one or more binary variables. For example, we can define $X = 1$ if a respondent who is in the labor force is unemployed and $X = 0$ if employed. Alternatively, we can define $Y = 1$ if the person is employed and $Y = 0$ otherwise. We can define $X = 1$ if the respondent is a woman and employed and $X = 0$ otherwise. We could define a variable called "years of schooling," which would take values 1, 2,..., 11, 12, ... depending on how many years the respondent went to school. On the other hand, we may define a variable "high school," which would take the value of one if the respondent finished high school and zero, otherwise.

A random variable together with its probabilities is called a *probability distribution*.

Example 4.6 Let us define the random variable X as the number of heads in the flip of three coins. The probability distribution is

X	$P(X)$
0	0.125
1	0.375
2	0.375
3	0.125

This is a discrete probability distribution. Probability distributions become unwieldy if the number of outcomes is large or infinite, as in the case of continuous random variables (see Sect. 4.6).

One way to summarize the information about a probability distribution is through its *moments* such as the *mean*, which measures the central tendency, and *variance*, which measures the dispersion or variability of the distribution. Another moment reflects the *skewness* of the distribution to the right or left and yet another, its *kurtosis*, is an indicator of how peaked is the distribution of outcomes; the more values are concentrated near the mean, the taller is the peak of the distribution.

Definition 4.2 The first moment of the distribution around zero, which is the expected value or the mean of the distribution, is defined as[2]

$$E(X) = \mu = \sum_{i=1}^{n} x_i P(x_i) \tag{4.2}$$

Example 4.7 For the distribution of number of heads in Example 4.6, we have

$$\mu = 0(0.125) + 1(0.375) + 2(0.375) + 3(0.125) = 1.5$$

Definition 4.3 In the same fashion, we may define the r-th moment of a distribution around zero as

$$E(X^r) = m_r = \sum_{i=1}^{n} x_i^r P(x_i) \tag{4.3}$$

---

[2]We shall denote the random variable, that is, the rule that assigns a real number to a particular outcome, by the capital letter X. Every realization of this variable, that is, the particular value assigned to an outcome, shall be denoted by x_i, and by x when referring to such realizations generically.

Example 4.8 The second moment of the distribution in Example 4.6 is

$$E(X^2) = 0(0.125) + 1(0.375) + 4(0.375) + 9(0.125) = 3$$

Definition 4.4 A second measure, which is of great importance, is the variance or the second moment around the mean

$$E(X-\mu)^2 = \sigma^2 = \sum_{i=1}^{n}(x_i-\mu)^2 P(x_i) \tag{4.4}$$

Using the binomial expansion, we can write the formula for the variance in an alternative form, which would make the computation easier.

$$\begin{aligned} E(X-\mu)^2 &= \sum_{i=1}^{n}(x_i-\mu)^2 P(x_i) \\ &= \sum_{i=1}^{n} x_i^2 P(x_i) - 2\mu \sum_{i=1}^{n} x_i P(x_i) + \mu^2 \\ &= \sum_{i=1}^{n} x_i^2 P(x_i) - \mu^2 \end{aligned} \tag{4.5}$$

Example 4.9 Continuing with the distribution in Example 4.6 we have

$$\sigma^2 = E(X^2) - \mu^2 = 3 - 1.5^2 = 0.75$$

Mean is a measure of central tendency of a distribution showing its center of gravity, whereas variance and its square root, called *standard deviation*, measure the dispersion or volatility of the distribution. The advantage of standard deviation is that it measures dispersion in the same measurement unit as the original variable. But in probability theory and statistical analysis, variance is the main player because it is algebraically and statistically easier to handle. In finance, the variance of the return of an asset is used as a measure of risk.

4.3 Marginal and Conditional Distributions

A random event may give rise to a number of random variables, each defined by a different set function whose domains are the same set. Table 4.1 displays such a case where random variables X and Y and their probabilities are reported. We may think of Y as the annual income in \$1000 of a profession and X as gender, with

Table 4.1 Joint distribution of X and Y

X	Y	$P(X,Y)$
0	55	0.02
0	65	0.04
0	75	0.07
0	85	0.09
0	95	0.10
0	105	0.06
0	115	0.03
0	125	0.02
0	135	0.01
0	145	0.01
1	75	0.01
1	85	0.02
1	95	0.04
1	105	0.08
1	115	0.11
1	125	0.11
1	135	0.09
1	145	0.05
1	155	0.03
1	165	0.01

$X = 0$ denoting men and $X = 1$ denoting women. The information in Table 4.1 contains the probability of joint events, that is, the probability of X and Y each taking a particular value. For instance, the probability of $X = 1$ and $Y = 115$ is 0.11, which is denoted as

$$P(X = 1, Y = 115) = 0.11$$

Such a probability is referred to as *joint probability* because, in our example, it shows the probability of a woman making $115,000 a year. If we are interested only in X, then we can sum up over all relevant values of Y and get the *marginal probability* of X. For example,

$$\begin{aligned} P(X = 1) &= P(X = 1, Y = 75) + \cdots + P(X = 1, Y = 165) \\ &= 0.01 + 0.02 + 0.04 + 0.08 + 0.11 + 0.11 + 0.09 + 0.05 + 0.03 + 0.01 \\ &= 0.55 \end{aligned}$$

In general, we can write

$$P(X = x_k) = \sum_{j=1}^{n} P(X = x_k, Y = y_j) \tag{4.6}$$

In the same vein we can calculate the probability of $X = 0$, which would be 0.45. Thus, the *marginal distribution*of X is

X	P(X)
0	0.45
1	0.55

A similar procedure yields the marginal probability of Y. For example,

$$\begin{aligned} \mathrm{P}(\mathrm{Y}=85) &= \mathrm{P}(\mathrm{Y}=85,\ \mathrm{X}=0)+\mathrm{P}(\mathrm{Y}=85,\ \mathrm{X}=1) \\ &= 0.09+0.02 \\ &= 0.11 \end{aligned}$$

The marginal distribution of X shows the distribution of men and women in that profession (45% men, 55% women), whereas the marginal distribution of Y would show the distribution of income for both sexes, that is, the profession as a whole.

Definition 4.5 (Conditional probability) Consider two events A and B whose joint probability is $P(A \cap B)$. The conditional probability of A given B denoted by $P(A|B)$ is the probability of A given that the event B has already happened:

$$P(A|B) = \frac{P(A \cap B)}{P(B)} \tag{4.7}$$

If $P(A|B) = P(A)$, or equivalently $P(A \cap B) = P(A)P(B)$, then we say that the events A and B are independent.

Example 4.10 Continuing with our example of distribution of income of men and women, we may be interested to know the probability of a woman earning $105,000, that is, the probability of Y=105 when we already know that X=1. That is, we are interested in the *conditional probability* of Y=105, given that X=1.

$$\begin{aligned} P(Y=105|X=1) &= \frac{P(Y=105,\, X=1)}{P(X=1)} \\ &= \frac{0.08}{0.55} = 0.145 \end{aligned}$$

In general

$$P(Y=y_j|X=x_k) = \frac{P(Y=y_j,\, X=x_k)}{P(X=x_k)} \tag{4.8}$$

Indeed we can compute the *conditional distribution* of $Y|X = 0$ and $Y|X = 1$ (Table 4.2).

Example 4.11 Continuing with the data in Table 4.1, the conditional distribution of Y, given $X = 0$, is the distribution of income for the male population while $P(Y|X = 1)$ is the conditional distribution of income for the female population. A conditional distribution has a mean, variance, and other moments. We compute its mean as

Table 4.2 Conditional distribution of Y given X

Y	$P(Y \mid X=0)$	Y	$P(Y \mid X=1)$
55	0.044	75	0.018
65	0.089	85	0.036
75	0.156	95	0.073
85	0.200	105	0.145
95	0.222	115	0.200
105	0.133	125	0.200
115	0.067	135	0.164
125	0.044	145	0.091
135	0.022	155	0.055
145	0.022	165	0.018

$$E(Y|X = x_k) = \sum_{j=1}^{n} y_j P(y_j|X = x_k) \tag{4.9}$$

Variance and other moments of the conditional distribution are computed similarly.

Example 4.12 For the conditional distributions in Table 4.2, we have

$$E\,(Y|X = 0) = 91.4, \quad E\,(Y|X = 1) = 121.4$$

Example 4.13 (Expected Utility). In Chap. 2, we talked about utility function. In a world of certainty, and when choices are simple, one may be able to easily rank different alternatives. But we live in a world of uncertainty and we have to choose between prospects, not sure alternatives. In buying a car, you may be able to find out about all its characteristics such as gas mileage, speed, and reliability, but you would not know the future prices of gasoline or the resale value of the car. Similarly, in making an investment, you may have some idea about probable returns, but no one is promised tomorrow and any investment involves risk. To handle such uncertain or risky decisions, we introduce the concept of *expected utility*. To make matters simple, we shall assume that we are interested in only one variable: income.

Let Table 4.3 describe choices open to us. Two prospects or actions a_1 and a_2 are open to the decision maker or investor. Under the first, the investor receives \$50, regardless of the state of nature. Under the second, he would receive \$35 if s = 1

Table 4.3 Income prospects of two investment options

	States	
Actions	s = 1	s = 2
$a = 1$	$x_{11} = 50$	$x_{12} = 50$
$a = 2$	$x_{21} = 35$	$x_{22} = 85$

happens and \$85 if s = 2 happens. One can think of the states as boom and recession in the economy, or the Federal Reserves decision to raise or leave unchanged the interest rate, or the price of oil increasing and decreasing. As it stands, making a choice would be difficult. We need more information. Suppose s = 1 occurs with probability $p_1 = 0.7$ and s = 2 with probability $p_2 = 0.3$.

Before going further, make a note of two important points. First, probabilities are defined over states (of nature, market, etc.). States happen and you have no control over them; you only have information on the probability of their occurrence. Second, you decide on actions, therefore, we can rank the desirability or utility of each action. One way to decide is to look at the expected return of each action. But first we need to define a function representing our valuation or utility of each x_{as}. Let $U(x_{as})$ denote the utility of each outcome. Then the expected utility conditional on the action can be written as

$$E[U(x)|a] = p_1 U(x_{a1}) + p_2 U(x_{a2}), \qquad a = 1,\ 2 \tag{4.10}$$

To illustrate, let us assume in this case that

$$U(x_{as}) = \sqrt{x_{as}}, \qquad \forall a,\ s$$

Thus, for the two possible actions, we have

$$\begin{aligned} E[U(x)|a = 1] &= 0.7\,(\sqrt{50}) + 0.3\,(\sqrt{50}) = 7.071 \\ E[U(x)|a = 2] &= 0.7\,(\sqrt{35}) + 0.3\,(\sqrt{85}) = 6.907 \end{aligned}$$

Based on our utility function, clearly the first option is preferred to the second. Recall that in Chap. 2 we said that the utility function assigns ordinal values to each bundle of goods and services. In other words, the actual number is not important, but values should be assigned in such a way that the preferred bundle has a larger number. This is not true for the expected utility function; it has a cardinal value. The reason is that each U enters the computation of $E[U(x) \mid a]$ and the latter is the utility of a whole prospect and not just one particular amount of income. Some changes in U that will preserve the ordering of x_{as}'s will change the ordering of the prospects. To see this point, change the utility function to $U(x_{as}) = x_{as}$.

Now we have

$$U(a = 1) = 0.7(50) + 0.3(50) = 50$$

$$U(a = 2) = 0.7(35) + 0.3(85) = 50$$

We are indifferent between the two options. Indeed, the two utility functions represent two different behaviors. If $U(x_{xs}) = \sqrt{x_{as}}$ represents your utility function, then you are risk averse because you prefer a sure outcome—receiving \$50 regardless of the state of the economy—to a risky situation with the same expected value of return. On the other hand, an investor with a U function that makes the two investments equivalent is called risk neutral. Finally, if your U function is such that you prefer the second option to the first, then you are a risk-loving person.

4.4 The Bayes Theorem

The Bayes theorem is an extremely important proposition because not only it is the foundation of Bayesian statistics and econometrics but it describes the process by which we do science and in general understand the world. Its mathematical derivation is straightforward. Recall that

$$P(A \cap B) = P(A|B)P(B)$$

But also

$$P(A \cap B) = P(B|A)P(A)$$

It follows that

$$P(A|B) = \frac{P(B|A)P(A)}{P(B)} \tag{4.11}$$

Let $\bar{A}$ denote the complement of event A, then we have

$$P(B) = P(B|A)P(A) + P(B|\bar{A})P(\bar{A})$$

Thus,

$$P(A|B) = \frac{P(B|A)P(A)}{P(B|A)P(A) + P(B|\bar{A})P(\bar{A})} \tag{4.12}$$

More generally, if we have

$$P(A_i \cap B) = P(A_i|B)P(B), \qquad i = 1, \ldots, n$$

Then

$$P(A_j|B) = \frac{P(B|A_j)P(A_j)}{\sum_{i=1}^{n} P(B|A_i)P(A_i)} \tag{4.13}$$

Let us illustrate the theorem with an example and then we shall discuss its meaning and importance.

Example 4.14 Suppose we have two boxes, K and N. One is chosen by a blindfolded person and then she randomly picks a ball from that box, which turns out to be black (an event denoted by $B1$). What is the probability that she had picked box K?

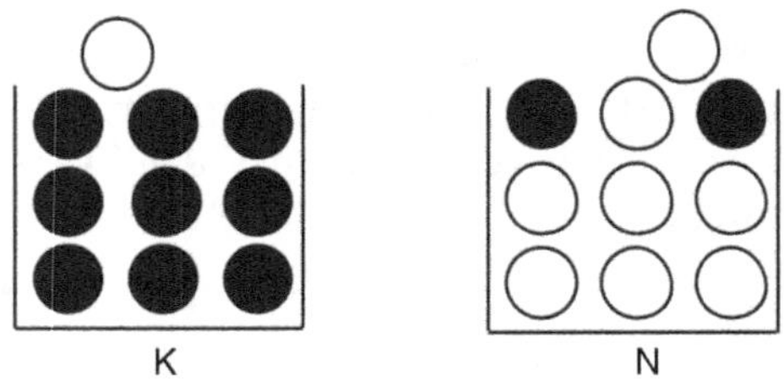

$$P(K|B1) = \frac{P(B1|K)P(K)}{P(B1|K)P(K) + P(B1|N)P(N)}$$

We know that $P(B1|K) = 0.9$, and $P(B1|N) = 0.2$. But what about $P(K)$ and $P(N)$? Of course, if we already *knew* these probabilities, there was nothing left to compute. The only thing we have is the data that the first ball was black. It is possible that we have some hunch or guess as to the value of these probabilities. Such information could serve as the *prior probability*. In our example we have no particular information. Therefore, we can assume that the events of K and N are equally likely. Such priors are called *uninformative* or *diffuse*. Now we have:

$$P(K|B1) = \frac{0.9 \times 0.5}{0.9 \times 0.5 + 0.2 \times 0.5} = 0.818$$

Thus, we have revised our probability of K being chosen from 0.5 to 0.818. Our new estimate is called *posterior probability*. Similarly

$$P(N|B1) = \frac{0.2 \times 0.5}{0.9 \times 0.5 + 0.2 \times 0.5} = 0.182$$

Now suppose the person still being blindfolded, replaces the ball, and again randomly chooses another ball from the same box. What is the probability that the chosen box is K if the second ball is also black (an event denoted by $B2$)? Note that we now have some information about the chosen box being K, that is, we have estimated the probability to be 0.818. This probability can now serve as our prior and we can revise our estimate as:

$$P(K|B2) = \frac{0.9 \times 0.818}{0.9 \times 0.818 + 0.2 \times 0.182} = 0.947$$

Thus, as a result of two consecutive drawing of black balls we are pretty sure (more than 90%) that the chosen box is K. Note that the Bayes theorem allows us to combine information from data $P(B1|K)$, which we shall call likelihood function, with the prior knowledge. Also note that since

$$P(K|B1) + P(N|B1) = 1$$

We only need the knowledge of numerator in (4.11) and (4.13). We can normalize the probabilities such that the sum of $P(K|B1)$ and $P(N|B1)$ is equal to one. Thus,

$$P(K|B1) \propto P(B1|K)P(K)$$

In words: Posterior probability (distribution) is proportional to likelihood function times prior probability (distribution).

The scheme depicted above, indeed describes the way we learn about different natural and social phenomena and revise our understanding of them. For instance, we can think of K as the Keynesian and N as the New Classical model, where black and white balls are statistical and historical evidence in favor of one or the other theory. The more evidence in favor of a theory the more confident we are that it indeed describes the real world.

As a final point, let us assume that after two consecutive black balls, our blindfolded individual randomly draws a white ball (after replacing the previous one). Then, denoting this event by W3, we have:

$$P(K|W3) = \frac{0.1 \times 0.947}{0.1 \times 0.947 + 0.9 \times 0.053} = 0.691$$

Thus, a contrary piece of evidence will shake our conviction

4.5 The Law of Iterated Expectations

The law of iterated expectations finds many applications in economics—for example, in rational expectations models—and in econometrics. The law of iterated expectations pertains to the relationship between conditional and unconditional expectations. In general,

$$E(Y) = E_X E(Y|X) = \sum_{j=1}^{n} E(Y|X = x_j)P(X = x_j) \tag{4.14}$$

Example 4.15 For the X and Y in Table 4.1, we have

$$\begin{aligned} E(Y) &= E(Y|X = 0)P(X = 0) + E(Y|X = 1)P(X = 1) \\ &= 91.4 \times 0.45 + 121.4 \times 0.55 \\ &= 107.9 \end{aligned}$$

The law of iterated expectations can be applied to cases when the expectation is conditional on more than one variable. As a matter of notation, the same way that we write $P\left(X = x_j\right)$ as $P(x_j)$or sometimes even $P(x)$, we write $E\left(Y|X = x_k\right)$ as $E(Y \mid x)$ which means the expectation of Y given all values of X. Suppose we know or assume that $E(Y|x) = 0$. This means that for all values of x, the conditional expectation of Y equals zero. Then it follows that

$$E\left(Y\right) \; = E_X E\left(Y|x\right) = 0 \tag{4.15}$$

Needless to say, the reverse is not true: $E(Y) = 0$ does not imply that $E(Y|x) = 0$ for all values of x.

4.6 Continuous Random Variables

So far we have been discussing only discrete random variables, that is, variables that take a countable number of values. Such variables need not be whole numbers as long as we are dealing with a finite number of them (see E.4.1). Many variables in nature, industry, science, economics, and society are continuous—their values cannot be enumerated. Heights and weights of different species, the amount of fuel a country uses, the amount of steel a region produces, and U.S. household incomes are examples of such variables. In assigning probabilities to continuous variables, we run into the problem that an interval, no matter how small, contains an infinite number of points. If we assign to each of them a positive probability, no matter how small, then the sum of an infinite series of positive numbers would not be bounded and the axiom of probability theory—that the sum of probabilities should be one—will be violated.

To overcome this problem, we assign probabilities to segments of the interval within which the random variable is defined. For example, we can write

$$P(X \le 12.6) = 0.59, \quad \text{or} \quad P(-1 < X \le 1.2) = 0.13$$

Example 4.16 A simple example of a continuous variable is the uniform distribution (see Fig. 4.1). Variable X can take any value between a and b, and the probability of X falling within the segment $[a, c]$ is proportional to the length of the interval compared to the interval $[a, b]$.

$$P(a < X \le c) = \frac{c-a}{b-a} \tag{4.16}$$

More generally,

$$P(X \le x) = \frac{x-a}{b-a} \tag{4.17}$$

Thus, the probability of an interval is equal to the area over the interval restricted from below by the x-axis and from above by the line at $1/(b-a)$. The straight line at $1/(b-a)$ is the *density function*. Whereas in this case the connection between the

Fig. 4.1 Uniform distribution

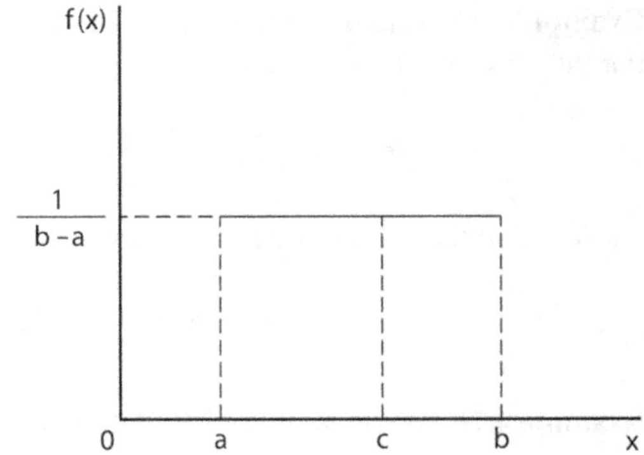

probability distribution and its density function is simple, the derivation of one from the other involves differentiation and integration, topics which we shall discuss in Part III (Chaps. 8, 9, 10, and 11) of this book.

Because an infinite number of points lie between a and b, it follows that each point will have probability zero.[3] Alternatively, we can say that the probability of point c is

$$P(x = c) = \frac{c - c}{b - a} = 0$$

That is, the width of a point is zero. The distribution function of X is denoted by

$$F(x) = P(X \leq x) \tag{4.18}$$

and has to conform to the following conditions:

1. $F(x)$ is continuous. We shall discuss the concepts of continuity and continuous functions in Chap. 8. Suffice it here to say that a continuous function has no break. In other words, you could draw the function without lifting the pen from the paper.
2. $F(x)$ is nondecreasing; that is,

$$F(x_1) \leq F(x_2), \quad \text{if} \quad x_1 < x_2 \tag{4.19}$$

3.

$$F(-\infty) = \lim_{x \to -\infty} F(x) = 0, \quad \text{and} \quad F(\infty) = \lim_{x \to \infty} F(x) = 1 \tag{4.20}$$

These conditions are the counterparts of the discrete case, which stipulates that probability is always positive and the sum of probabilities add up to one.

[3] An impossible event has probability zero, but an event with probability zero may still happen. Similarly, a sure event has probability one, but an event with probability one may still fail to happen.

Example 4.17 Note that conditions 1 and 2 are fulfilled for the uniform function. For the third condition, we have

$$F(a) = \frac{a-a}{b-a} = 0, \quad \text{and} \quad F(b) = \frac{b-a}{b-a} = 1$$

Note also that we can write

$$P(x_1 < X \leq x_2) = F(x_2) - F(x_1)$$

Example 4.18 For the uniform distribution, we have

$$\begin{aligned} P(x_1 < X \leq x_2) &= \frac{x_2 - x_1}{b-a} \\ &= \frac{x_2 - a}{b-a} - \frac{x_1 - a}{b-a} \\ &= F(x_2) - F(x_1) \end{aligned}$$

Example 4.19 Another example of a continuous distribution is the famous Gaussian or normal distribution, the bell-shaped curve that is depicted in Fig. 4.2. Its density function is

$$f(x) = \frac{1}{\sigma\sqrt{2\pi}} e^{-\frac{1}{2}\left(\frac{x-\mu}{\sigma}\right)^2} \tag{4.21}$$

Normal distributions have two parameters: mean μ and variance σ^2 Once we know these two parameters, the entire distribution is specified. In particular, when $\mu = 0$ and $\sigma^2 = 1$, we have the standardized normal distribution.

The definitions of mean and of variance of continuous distributions are similar to those of discrete distributions, but they require knowledge of integral calculus (the subject of Chap. 11) and, therefore, we do not discuss them here.

Fig. 4.2 Gaussian or normal distribution

In addition to uniform and normal distributions, several continuous distributions play important roles in probability theory, statistics, and econometrics. These include t, χ^2, F, beta, and exponential distributions. For all of them, we can write density functions some of which may be involved functions. Nevertheless, we should not assume that all continuous distributions have density functions.

Now that we have gained an intuitive understanding of continuous distributions, we may want to have a more formal definition of probability models that parallels the definition we had for discrete distributions. We start with the extended real line $\mathfrak{R} = (-\infty, \infty)$, which shall play the same role for continuous variables that Ω the set of outcomes, played for discrete random variables. Consider half-closed intervals on $\mathfrak{R}$:

$$(a, b] = \{x \in \mathfrak{R} : a < x \leq b\} \tag{4.22}$$

and form finite sums of such intervals provided the intervals are disjoint—that is, they do not overlap.

$$\mathcal{A} = \sum_{j=1}^{n} (a_j, b_j], \qquad n < \infty \tag{4.23}$$

A set consisting of all such sums plus the empty set $\emptyset$ is an algebra, because it is closed under the operations of union and complement. If the set is closed under enumerable union then it is a σ-algebra. The smallest σ-algebra that contains the set of half-closed intervals on $\mathfrak{R}$ is called the *Borel set* and denoted by $\mathcal{B}(\mathfrak{R})$. Finally, we define the probability measure

$$F(x) = P(-\infty, x] \tag{4.24}$$

The triple $(\mathfrak{R}, \mathcal{B}(\mathfrak{R}), P)$ is our probability model for continuous variables.

4.7 Correlation and Regression

Economics is an observational rather than an experimental science. Economists discover relationships between different sets of variables by examining data and using econometric techniques. The main workhorse of econometrics is regression analysis that allows economists to find correlation between different sets of variables. Examples are the relationship between consumption, income, and wealth; the relationship between investment, income, and the cost of capital; and the relationship between demand for different commodities and services and their prices and aggregate income.

The three reasons for observing randomness (Sect. 4.1)—randomness in nature and society, influence of many factors and decision makers, and measurement with error—apply to all economic variables. Thus, in studying economics we can treat all the variables as random and try to discern their connections using methods of probability theory and statistics.

If two or more random variables depend on the same random event, we may be interested in measuring the degree of their connection. This is done with two measures of association: covariance and correlation coefficient.

Definition 4.6 The covariance for two variables X and Y, whose probability distribution is $P(X, Y)$, is defined as

$$\begin{aligned}\mathrm{Cov}(X, Y) &= E[(X - E(X))(Y - E(Y))] \\ &= \sum_{j=1}^{n} (x_j - \mu_X)(y_j - \mu_Y)P(x_j, y_j)\end{aligned} \tag{4.25}$$

where $\mu_X = E(X)$ and $\mu_Y = E(Y)$. We can simplify this formula:

$$\begin{aligned}\mathrm{Cov}(X, Y) &= \sum_{j=1}^{n} (x_j - \mu_X)(y_j - \mu_Y)P(x_j, y_j) \\ &= \sum_{j=1}^{n} x_j y_j P(x_j, y_j) - \mu_X \sum_{j=1}^{n} y_j P(x_j, y_j) \\ &\quad - \mu_Y \sum_{j=1}^{n} x_j P(x_j, y_j) + \mu_X \mu_Y \sum_{j=1}^{n} P(x_j, y_j) \\ &= \sum_{j=1}^{n} x_j y_j P(x_j, y_j) - \mu_X \mu_Y\end{aligned} \tag{4.26}$$

Example 4.20 Table 4.4 shows the number of heads in five flips of a coin and the winning associated with each case. The cost of buying a ticket to play this game is fixed at \$8. The covariance of the number of heads and the amount of winnings is

$$\mathrm{Cov}(X, Y) = 25 - 2.5(7.5) = 6.25$$

A covariance greater than zero shows a positive connection between the two variables, whereas a covariance less than zero signifies a negative relationship. A

Table 4.4 Probability distribution of the number of heads in five flips of a coin

No. of heads X	Winnings Y	Probability $P(X, Y)$
0	0	0.03125
1	1	0.15625
2	4	0.31250
3	9	0.31250
4	16	0.15625
5	25	0.03125

zero covariance shows that the two variables are statistically unrelated. For example, the covariance between the ticket cost and the winnings equals zero, because the cost remains the same regardless of the number of heads or the amount of winnings. Whereas the covariance between two variables shows a statistical connection between them, it says nothing about the strength of that relationship. The *correlation coefficient* measures the strength of co-movement between two variables.

Definition 4.7 The correlation coefficient is obtained by normalizing the covariance, that is, dividing it by the product of the standard deviations of the two variables involved.

$$\mathrm{Corr}(X, Y) = \frac{\mathrm{Cov}(X, Y)}{\sigma_X \sigma_Y} \tag{4.27}$$

Correlation coefficients are always within the $[-1, 1]$ interval. The larger the absolute value of the coefficient, the stronger the connection between the two variables involved.

Example 4.21 In Example 4.20, the correlation coefficient between X and Y is

$$\mathrm{Corr}(X, Y) = \frac{6.25}{\sqrt{1.25}\sqrt{33.75}} = 0.96225$$

Definition 4.8 A simple regression model represents a relationship between two random variables[4] y and x. But this relationship is not exact; the magnitude of inexactness appears as a third random variable, u. As examples, we may think of y as the height of a son, which depends on the height of the father x, or consumption of a household y depending on its income x. We can observe variables x and y, but u is unobservable. Indeed, if we could observe u, then the relationship between y and x would be exact. Suppose we have n observations. If the relationship is linear, the simple regression model can be written as

$$y_i = \alpha + \beta x_i + u_i, \qquad i = 1, \cdots, n \tag{4.28}$$

[4]Some econometrics textbooks speak of x being fixed in repeated sampling. This cryptic assertion is a reference to the use of regression models in experimental sciences. Suppose you have 10 flowerpots filled with the same type of soil. In each you plant the same type of seed and water them the same amount. However, each plant is given a different amount of fertilizer x and the yield for each plant, y, is measured. You can repeat the same experiment keeping all variables affecting the growth of the plant in each pot constant. In particular, the amount of fertilizer in each pot remains constant. Such experiments have hardly anything to do with economic issues and the use of regression analysis in economics, so we speak more generally of two random variables.

$$E(u_i|x_i) = 0$$
$$E(u_i^2) = \sigma^2 \qquad \forall i,j \quad i \neq j$$
$$E(u_i u_j) = 0$$

We offer three reasons for the inexactitude of the relationship in (4.28). First, we may have left the effects of other variables out of the equation. A son's height may also be influenced by the height of the mother, nutrition, environment, and other variables. Consumption of a household may be affected by wealth, the number of individuals in the household, and so on. If these influences are numerous and none have a definite impact, then we may observe random variations around the exact part of the equation. Second, errors in measuring the variables x and y may arise. In economics, measured variables are often not the exact counterparts of theoretical constructs. For example, we do not measure consumption, but rather consumers' expenditures; we do not measure demand, but rather the amount purchased. Finally, it may be that the inexactitude or the random variation is intrinsic to the relationship.

A number of methods have been developed to estimate α and β based on observations of y and x. These include the least squares, maximum likelihood, and generalized method of moments.

The restrictions imposed on u are needed to make inference about α and β based on their estimated counterparts $\hat{\alpha}$ and $\hat{\beta}$. In addition, when we make statistical inference, we usually assume that u_i, $i = 1, \cdots, n$ are normally distributed. If the data violate any of the restrictions on u, such violations affect our inferences about α and β. Hence, a great part of classical econometrics seeks to remedy such deviations from the assumptions of the model in order to make inferences about α and β possible. A regression model need not have only one explanatory variable x. A multiple regression model will be of the form

$$y_i = \beta_0 + \beta_1 x_{1i} + \beta_2 x_{2i} + \cdots + \beta_k x_{ki} + u_i, \qquad i = 1, \cdots, n \tag{4.29}$$

The same restrictions as in (4.28) are imposed on u in (4.29). In addition, no exact or nearly exact linear relationship may exist between the explanatory variables, that is, between $x_1, x_2, \cdots, x_k$. We would use the same methods mentioned above to estimate the model parameters. Note that β_j measures the effect of x_j on y when all other variables are held constant.

4.8 Markov Chains

In recent years, Markov chains[5] have been applied to many economic problems including the study of business cycles. The subject is vast and very interesting. Here we introduce the reader to the rudiment of Markov chains theory and will come back to it in Chaps. 6 and 7.

[5]For the renowned Russian mathematician, Andrei Andreyevich Markov (1856–1922), who studied and made contributions to the subject.

Consider an individual's present employment situation and, for simplicity, let us assume that the person could be in only one of three states: employed, unemployed, and out of the labor force. If a person is employed, what is the probability that in the next period, say, next month, she still will be employed? What is the probability that she will be unemployed? Out of the labor force? We can list such probabilities for a hypothetical individual (see Table 4.5).

Table 4.5 Transition probabilities for different employment states

	Present state		
Next period	Employed	Unemployed	Out of labor force
Employed	0.960	0.350	0.060
Unemployed	0.039	0.600	0.120
Out of labor force	0.001	0.050	0.820

For example, if a person is employed in this period, the probability that she stays employed is 0.96, whereas the probability of her leaving the labor force altogether is 0.001. Thus, each entry (i, j) shows the probability, P_{ij} of the individual transiting from state i in the current period to state j in the next period. In general, we can write the matrix of transition probabilities as in Table 4.6.

Table 4.6 Matrix of transition probabilities

	Time t			
Time $t + 1$	1	2	...	n
1	P_{11}	P_{21}	...	P_{n1}
2	P_{12}	P_{22}	...	P_{n2}
⋮	⋮	⋮	⋱	⋮
n	P_{1n}	P_{2n}	...	P_{nn}

Note that regardless of its state at time t, the system will be in one of the n states at time t+1. Therefore, the sum of transition probabilities from any state to all other states should add up to one.

$$\sum_{j=1}^{n} P_{ij} = 1, \qquad \forall i$$

Some probabilities may equal one or zero. A probability zero, for example, $P_{ik} = 0$ signifies that no transition from state i to state k can take place, whereas $P_{is} = 1$ means that if we are in state i at time t, we will certainly be in state s at time t+1. In particular, $P_{ii} = 1$ means that once we are in state i, we shall remain in that state forever. To make these ideas more precise, let the variable $X_t = i$ denote that at time t a system is in state i. Then the transition probabilities are defined as

$$P(X_{t+1} = j | X_t = i) = P_{ij}, \qquad \forall i, j$$

Thus, $\{X_t\}$ is a stochastic (or random) process. It would be a Markov process if two conditions are met:

1. The number of states is finite, hence we have $n < \infty$ possible states.
2. Transition probabilities remain constant; hence $P(X_{t+1} = j | X_t = i) = P_{ij}$, and would not change over time.

Two parts attend the second property. First, the probability of transition to a state, j, at time t+1 depends only on X_t. Thus, previous values of X, that is, $X_{t-1}, X_{t-2}, \cdots$ have no bearing on the probability of X_{t+1} being in a particular state. This short memory is the essence of a Markov process. Second, the probability remains constant and does not change over time, which makes the process stationary.

Example 4.22 As another example, consider the American economy's cyclical behavior, which gives rise to periods of boom and recession. Between 1854 and 2010, the economy experienced 33 cycles. By a cycle, we mean a period of recession followed by a boom period. On average, recessions have lasted 16 months and booms 42 months. If we confine ourselves to the post-World War II era, the United States has experienced 11 cycles, with recessions lasting, on average, 11 months and booms lasting, on average, 59 months. Of course, as a boom continues, the probability of a peak increases. The same is true for a trough. Moreover, other indications in the economy warn of the impending onset of a recession or herald the beginning of a recovery. But for a moment suspend your disbelief and assume that the probability of transition from a recession to recovery and from boom to recession remains constant. Table 4.7 reflects these transition probabilities.

Table 4.7 Transition probabilities between booms and recessions in the U.S. economy

	t	
$t+1$	Boom	Recession
Boom	0.98	0.10
Recession	0.02	0.90

4.9 Exercises

E.4.1 Find first, second, third, and fourth moments around zero of the following distributions.

X	$P(X)$
−100	0.075
−50	0.244
10	0.328
60	0.236
110	0.095
175	0.020
245	0.002

Y	$P(Y)$
1.5	0.09
1.6	0.13
1.7	0.20
1.8	0.21
1.9	0.18
2.0	0.11
2.1	0.08

E.4.2 Calculate variances and standard deviations of the distributions in E.4.1.

E.4.3 Show that if X is a random variable and $Y = a + bX$, then

$$E(Y) = a + bE(X), \qquad \text{and} \qquad \sigma_Y^2 = b^2\sigma_X^2$$

E.4.4 In the simple regression model, show that

$$E(y_i|x_i) = \alpha + \beta x_i$$

and

$$\text{Var}(y_i|x_i) = \text{Var}(u_i)$$

E.4.5 You are given the joint distribution of X and Z in the following table.

X	Z	$P(X, Z)$
0	−5	0.08
0	−3	0.16
0	−1	0.17
0	0	0.07
0	1	0.04
1	−3	0.03
1	−1	0.08
1	0	0.19
1	1	0.14
1	3	0.04

i. Compute the marginal distribution of Z.
ii. Compute the conditional distributions $P(Z \mid X = 0)$ and $P(Z \mid X = 1)$.
iii. Compute the conditional expectations $E(Z \mid X = 0)$ and $E(Z \mid X = 1)$.
iv. Use this example to verify the law of iterated expectations.

E.4.6 Consider a test for a particular disease such as HIV, which detects the infection in 99% of those who have HIV. But it also has a 10% rate of false positive. Furthermore, assume that 2% of total population has HIV. What is the probability of having HIV if a person tests positive? Solve the problem once using the Bayes Theorem and once by making a frequency table assuming a total population of 100,000 people.

E.4.7 In a game show the host and contestant face three closed doors. Behind one door is a very valuable prize, say, a sports car. Behind the other two are duds like a baseball bat or a goat. The contestant chooses a door. Before the door is opened, the host opens one of the two doors not chosen by the contestant and shows that the main prize is not there. Then the contestant is offered the opportunity to change her choice. Should the contestant change the door chosen?

[*Hint*: Find the probability of winning with and without changing the door. Make the computation once using the Bayes Theorem and once assuming that the whole process is repeated 1,000 times.]

Height	Shoe size	# of siblings +1	Height	Shoe size	# of siblings +1
69	10	3	67	12	2
70	10	2	67	10	4
68	10	2	66	8	2
74	12	3	68	9.5	5
75	13	2	67	8.5	3
66	9	3	72	11	3
63	6.5	2	70	9.5	1
71	13	2	70	10	2
72	12	1	69	9.5	5
67	9	2	71	12	1
71	10.5	3	69	10	1
69	10	3	71	13	2
68	9	3	68	13	5
64	8	9	72	13	2
71	11	3	70	10.5	3

E.4.8 The above table gives the data on the heights and shoe sizes of 30 students. In addition, the number of students' siblings plus one is given in the last column.

i. Using Excel

```
Insert → Chart → XY(Scatter)
```

to make a scatter diagram of height against shoe size.

ii. Use the tool bar menu

```
Chart → Add Trendline
```

to draw the regression line.

iii Repeat i. and ii. for height and number of children in the students' families.

Part II
Linear Algebra

Chapter 5
Vectors

We have already encountered two-dimensional numbers and variables in the case of complex variables. Two, three,..., and n dimensional numbers are simply extensions of the concept of a number. For example, we can speak of the length of a string, which would be a one-dimensional number. Or we could talk of the length and width of a page, which would be two-dimensional. A three-dimensional number could represent length, width, and height of a room. Multidimensional numbers are of great importance in all sciences including economics. Such numbers provide economy in exposition and facilitate the manipulation and analysis of complex questions.

Multidimensional numbers and variables are indispensable in economics; many economic variables and quantities are essentially multidimensional. Consider any goods or service that you buy. Although we may use a simple word or expression such as computer, telephone service, or health care, each of these goods and services can only be described by an array of characteristics. In case of a computer, we have its price, its speed, technical support, warranty, and on-site service, to name only a few. A telephone service is even more complicated. In addition to its cost, there are features such as call waiting, call forwarding, caller ID, and long-distance rates.

The fact that we do not buy a simple goods, but a bundle of characteristics, is best seen in financial assets. Each financial asset is characterized by two variables: risk and return. If we include all assets including those with a thin market or even without daily functioning markets, then each asset will have at least three characteristics: risk, return, and liquidity.

But the story does not end here. In every facet of the economy opposing forces interact to determine, simultaneously or in rapid succession, several variables. Think of supply and demand forces that determine both price and quantity of a product. Ignoring this fact and treating some variables in isolation could potentially take us down a fallacious path and result in wrong arguments and conclusions. On many occasions you have heard on radio or television someone opines that "there was a selloff in the market and the index fell by so many points." But if some sold their stocks, didn't some others buy them? So why not speak of buy-off? The fact is that we are dealing with a two-dimensional variable that describes the situation in the market: the volume of trade and the change in the index.

K. Dadkhah, *Foundations of Mathematical and Computational Economics,* 2nd ed.,
DOI 10.1007/978-3-642-13748-8_5,

Another example is the question of oil reserves around the world. If we are speaking from a purely geological point of view, a number such as 900 or 1000 billion barrels may make sense. But from an economic point of view, the amount of crude oil reserves is not independent of the price of oil. If the price of oil is \$80 per barrel, it makes sense to extract oil from areas with extraction costs of \$25 per barrel. But if the price of crude falls to \$25, it does not. Under such circumstances, speaking of reserves in areas of high extraction costs is meaningless. An analogy would be if we include metal deposits on Mars or the Moon in the reserves of such metals because they are there and one day we may be able to get them.

Vectors are the mathematical equivalents of many-dimensional numbers and variables. They prove very handy in simplifying complicated notations and analysis. But if we are dealing with multidimensional numbers and variables, then we need devices for manipulating them. When dealing with one-dimensional variables, a function simply maps a point in one-dimensional space to another point in a one-dimensional space. Once we allow for more than one dimension, old transformations become more complicated and new possibilities present themselves. If we are mapping from a two-dimensional space to another two-dimensional space, then the question is how any of the two coordinates in the domain affect each coordinate in the range. A linear transformation requires four parameters. On the other hand, a linear mapping from a three-dimensional space to a two-dimensional space would require six parameters. Thus, we require new tools to handle these problems. To represent linear relationships we will use matrices, which are rectangular arrays of numbers and will greatly facilitate the presentation and manipulation of mathematical, economic, and statistical models. Matrix algebra, the subject that deals with these issues, will be covered in the next two chapters.

5.1 Vectors and Vector Space

A vector of order $n > 0$ is a set of ordered numbers.

Example 5.1 The following are examples of vectors:

$$\mathbf{a} = \begin{bmatrix} 4 \\ 3 \end{bmatrix}, \quad \mathbf{e}_1 = \begin{bmatrix} 1 \\ 0 \\ 0 \end{bmatrix}, \quad \mathbf{x} = \begin{bmatrix} x_1 \\ x_2 \\ x_3 \\ x_4 \end{bmatrix}$$

These are *column vectors.* In this book we follow the convention of writing all vectors as column vectors and write *row vectors* as the transpose of column vectors.

Example 5.2 If we transpose the column vectors in Example 5.1, we have the following row vectors:

$$\begin{aligned} \mathbf{a}' &= \begin{bmatrix} 4 & 3 \end{bmatrix} \\ \mathbf{e}'_1 &= \begin{bmatrix} 1 & 0 & 0 \end{bmatrix} \\ \mathbf{x}' &= \begin{bmatrix} x_1 & x_2 & x_3 & x_4 \end{bmatrix} \end{aligned}$$

Vectors have an intuitively appealing geometrical representation (see Fig. 5.1).

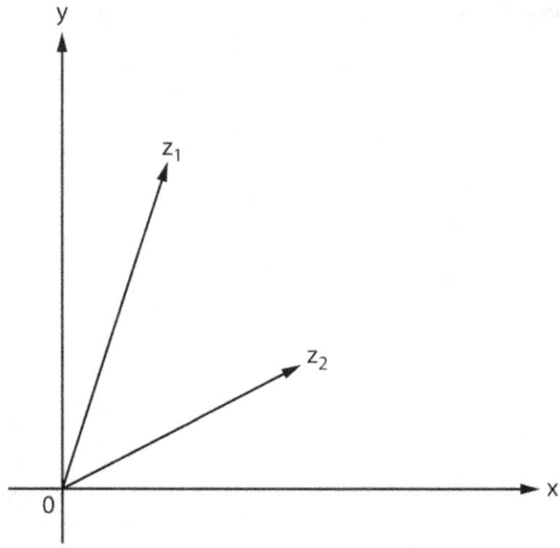

Fig. 5.1 Geometric representation of two-dimensional vectors

As can be seen, a vector has a direction. But so does a number. If on the real line we mark the point of zero and a number such as 5, we get a one-dimensional vector that starts at point zero and ends at point five and has a positive direction. On the other hand the number -3 has a negative direction. Vectors in-two dimensional space have a much larger variety of directions than just negative and positive. On the other hand a three-dimensional vector can have many more directions than a two-dimensional one.

Two vectors are said to be equal if they are of the same dimension and all their elements are equal. In this regard, we should hasten to add that a row vector does not have the same dimensions as a column vector. The former is a $1 \times n$ vector and the latter $n \times 1$. A number of operations on vectors, including addition of two or more vectors, multiplication of a vector by a scalar, and the difference of two vectors, are the generalization of operations on numbers.

Definition 5.1 Let

$$\mathbf{x} = \begin{bmatrix} x_1 \\ \vdots \\ x_n \end{bmatrix}, \qquad \mathbf{y} = \begin{bmatrix} y_1 \\ \vdots \\ y_n \end{bmatrix}, \qquad \mathbf{z} = \begin{bmatrix} z_1 \\ \vdots \\ z_n \end{bmatrix}$$

and λ be a scalar. Then

$$\mathbf{x} + \mathbf{y} = \begin{bmatrix} x_1 + y_1 \\ \vdots \\ x_n + y_n \end{bmatrix}, \qquad \lambda \mathbf{z} = \begin{bmatrix} \lambda z_1 \\ \vdots \\ \lambda z_n \end{bmatrix} \tag{5.1}$$

The difference of two vectors is multiplying one by -1 and adding it to the other (see Figs. 5.2 and 5.3).

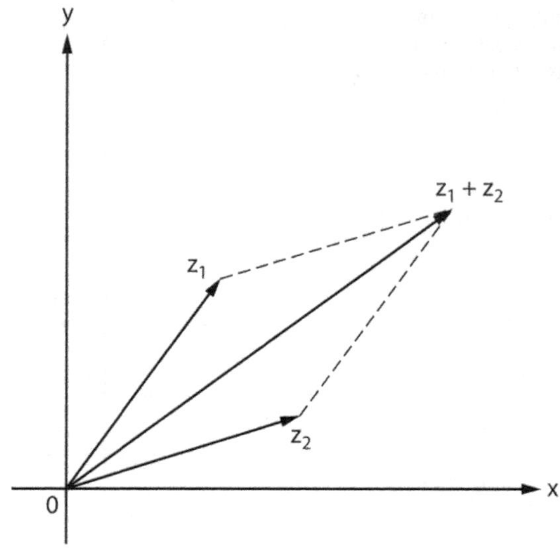

Fig. 5.2 Sum of two vectors

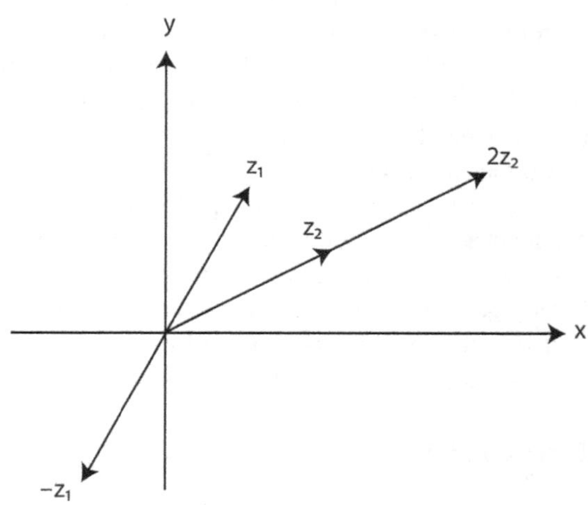

Fig. 5.3 Multiplication of vectors by scalars

Example 5.3 Let

$$\mathbf{a} = \begin{bmatrix} -14 \\ 8 \\ 3 \end{bmatrix}, \qquad \mathbf{b} = \begin{bmatrix} 6 \\ 0 \\ 32 \end{bmatrix}, \qquad \mathbf{c} = \begin{bmatrix} 7 \\ 9 \\ 55 \end{bmatrix}$$

then

$$\mathbf{a} + \mathbf{b} = \begin{bmatrix} -8 \\ 8 \\ 35 \end{bmatrix}, \qquad 2.5\mathbf{c} = \begin{bmatrix} 17.5 \\ 22.5 \\ 137.5 \end{bmatrix}, \qquad \mathbf{b} - 2\mathbf{c} = \begin{bmatrix} -8 \\ -18 \\ 78 \end{bmatrix}$$

The operations of addition and subtraction necessitate that we define a zero vector $\mathbf{0}$ whose elements are all zero with the following property:

$$\mathbf{x} + \mathbf{0} = \mathbf{x} \qquad \Rightarrow \qquad \mathbf{x} + (-\mathbf{x}) = \mathbf{0} \tag{5.2}$$

Two vectors can be multiplied together and, in order to do justice to the subject, we need to wait until the next chapter when we have become familiar with matrices and matrix operations. Here we first define one particular multiplication operation called the inner product of two vectors.

Definition 5.2 Let

$$\mathbf{x} = \begin{bmatrix} x_1 \\ \vdots \\ x_n \end{bmatrix}, \qquad \mathbf{y} = \begin{bmatrix} y_1 \\ \vdots \\ y_n \end{bmatrix}$$

Then the *inner product* of $\mathbf{x}$ and $\mathbf{y}$ denoted by $\langle \mathbf{x}, \mathbf{y} \rangle$ is defined as[1]

$$\langle \mathbf{x}, \mathbf{y} \rangle = \sum_{j=1}^{n} x_j y_j \tag{5.3}$$

Example 5.4 Going back to Example 5.3 we have

$$\langle \mathbf{a}, \mathbf{b} \rangle = -14 \times 6 + 8 \times 0 + 3 \times 32 = 12$$

and

$$\langle \mathbf{a}, \mathbf{c} \rangle = 139, \qquad \langle \mathbf{bc} \rangle = 1802$$

The inner product has the following properties:

$$\begin{aligned} &i.\ \langle \mathbf{x}, \alpha \mathbf{y} \rangle = \alpha \langle \mathbf{x}, \mathbf{y} \rangle \\ &ii.\ \langle \mathbf{x}, \mathbf{x} \rangle \begin{cases} > 0 & \text{if} \quad \mathbf{x} \neq \mathbf{0} \\ = 0 & \text{if and only if } \mathbf{x} = \mathbf{0} \end{cases} \end{aligned} \tag{5.4}$$

In the next chapter we shall define vector products $x'y$ and xy', which are two different entities. Here we only define the first and leave the second for the next chapter. Let

$$\mathbf{x} = [x_1, \ldots, x_n]', \qquad \mathbf{y} = [y_1, \ldots, y_n]'$$

[1] The inner product can be defined for both real and complex vectors. Here we confine ourselves to real vectors.

Then

$$\mathbf{x}'\mathbf{x} = \sum_{i=1}^{n} x_i^2, \qquad \mathbf{y}'\mathbf{y} = \sum_{i=1}^{n} y_i^2, \qquad \mathbf{x}'\mathbf{y} = \mathbf{y}'\mathbf{x} = \sum_{i=1}^{n} x_i y_i$$

5.1.1 Vector Space

Definition 5.3 Let $\mathbf{S}^n$ be a set of n-vectors. It is called a *vector space* if it is closed under the operations of addition and scalar multiplication. That is,

$$\mathbf{x},\ \mathbf{y} \in \mathbf{S}^n \quad \Rightarrow \quad \mathbf{x}+\mathbf{y} \in \mathbf{S}^n \tag{5.5}$$

and given a scalar λ

$$\mathbf{x} \in \mathbf{S}^n \quad \Rightarrow \quad \lambda\mathbf{x} \in \mathbf{S}^n \tag{5.6}$$

The two-dimensional space of Figs. 5.2 and 5.3 is an example of a vector space. As can be seen, both the sum of two vectors and the scalar product are in the same space. Similarly, we can have three-dimensional and, in general, n-dimensional vector spaces. To characterize $\mathbf{S}^2$ we need two vectors, for instance, the two vectors

$$\mathbf{e_1} = \begin{bmatrix} 1 \\ 0 \end{bmatrix} \quad \text{and} \quad \mathbf{e_2} = \begin{bmatrix} 0 \\ 1 \end{bmatrix}$$

because we can write every two-vector in terms of $\mathbf{e_1}$ and $\mathbf{e_2}$. We say that these two vectors *span* the vector space. To span an n-space we need n independent vectors.

Example 5.5

$$\begin{bmatrix} 5 \\ -2 \end{bmatrix} = 5\begin{bmatrix} 1 \\ 0 \end{bmatrix} - 2\begin{bmatrix} 0 \\ 1 \end{bmatrix} = 5\mathbf{e_1} - 2\mathbf{e_2}$$

Similarly, we can write

$$\begin{bmatrix} 6 \\ -5 \\ 2 \end{bmatrix} = 6\begin{bmatrix} 1 \\ 0 \\ 0 \end{bmatrix} - 5\begin{bmatrix} 0 \\ 1 \\ 0 \end{bmatrix} + 2\begin{bmatrix} 0 \\ 0 \\ 1 \end{bmatrix} = 6\mathbf{e_1} - 5\mathbf{e_2} + 2\mathbf{e_3}$$

We need not always use the unit vectors, $\mathbf{e_j}$'*s*, to span a vector space.

Example 5.6 The following vectors span $\mathbf{S^3}$:

$$\begin{bmatrix} 2 \\ 1 \\ 2 \end{bmatrix}, \quad \begin{bmatrix} 4 \\ -3 \\ 5 \end{bmatrix}, \quad \begin{bmatrix} 3 \\ 0 \\ 4 \end{bmatrix}$$

because they are independent and we can write any three-vector in terms of them. For example,

$$\begin{bmatrix} 6 \\ -5 \\ 2 \end{bmatrix} = 7 \begin{bmatrix} 2 \\ 1 \\ 2 \end{bmatrix} + 4 \begin{bmatrix} 4 \\ -3 \\ 5 \end{bmatrix} - 8 \begin{bmatrix} 3 \\ 0 \\ 4 \end{bmatrix}$$

Several times we made reference to the independence of vectors; what do we mean by independence?

Definition 5.4 A set of k-vectors $\mathbf{x_1}, \ldots, \mathbf{x_k}$ are *linearly dependent* if there are k constants $c_1, \ldots, c_k$ not all of them zero such that

$$c_1\mathbf{x_1} + \ldots + c_k\mathbf{x_k} = \mathbf{0} \tag{5.7}$$

Otherwise they are *independent*.

Example 5.7 The following vectors are dependent:

$$\mathbf{x_1} = \begin{bmatrix} 4 \\ -5 \\ 1 \end{bmatrix}, \quad \mathbf{x_2} = \begin{bmatrix} -3 \\ 7 \\ 2 \end{bmatrix}, \quad \mathbf{x_3} = \begin{bmatrix} -1 \\ 11 \\ 8 \end{bmatrix}$$

because

$$2\mathbf{x_1} + 3\mathbf{x_2} = \mathbf{x_3}$$

Vectors that spanned $\mathbf{S}^2$ and $\mathbf{S}^3$ in Examples 5.5 and 5.6 are independent. A way to test whether the set of k-vectors $\mathbf{x_1}, \ldots, \mathbf{x_k}$ are independent is to write the following system of equations:

$$c_1\mathbf{x_1} + \ldots + c_k\mathbf{x_k} = \mathbf{0}$$

and solve it for c_j's. If we get a unique solution with all c_j's equal to zero, then the vectors are independent; otherwise, if the solution is not unique or not all c_j's are zero, the vectors are dependent.

Example 5.8 The vectors

$$\begin{bmatrix} 5 \\ 2 \end{bmatrix}, \quad \begin{bmatrix} 10 \\ 4 \end{bmatrix}$$

are dependent because the system

$$\begin{aligned} 5c_1 + 10c_2 &= 0 \\ 2c_1 + \ \ 4c_2 &= 0 \end{aligned}$$

has infinitely many solutions. Any pair of c_1, c_2 that satisfies the condition $c_1 = -2c_2$ is a solution to the above system. On the other hand, the vectors

$$\begin{bmatrix} 4 \\ 1 \\ 6 \end{bmatrix}, \quad \begin{bmatrix} 3 \\ 2 \\ 5 \end{bmatrix}, \quad \begin{bmatrix} -4 \\ 7 \\ 0 \end{bmatrix}$$

are independent because the system

$$\begin{aligned} 4c_1 + 3c_2 - 4c_3 = 0 \\ c_1 + 2c_2 + 7c_3 = 0 \\ 6c_1 + 5c_2 = 0 \end{aligned}$$

has a unique solution, $c_1 = c_2 = c_3 = 0$.

In the next chapter we will find an easy way to determine if a set of vectors are independent and, if all are not independent, to find out how many among them are independent.

5.1.2 Norm of a Vector

It would make life much easier if we think of the norm of a vector as its length; indeed, the length called the *Euclidean norm* of a vector is a particular instance of a norm. If we consider Fig. 5.4, by the Pythagoras theorem, the length of vector **a** is equal to

$$\|\mathbf{a}\| = \sqrt{a_1^2 + a_2^2} \tag{5.8}$$

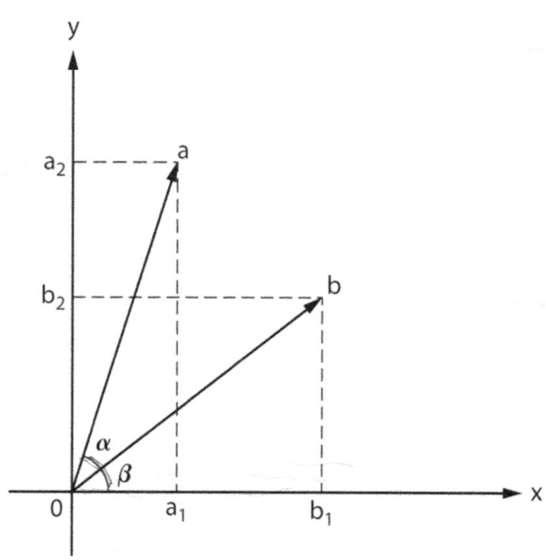

Fig. 5.4 Norm of a vector and angle between two vectors

where $\|\mathbf{a}\|$ denotes the norm of vector $\mathbf{a}$. In general, we can write

$$\|\mathbf{a}\| = \left(\sum_{j=1}^{n} a_j^2\right)^{\frac{1}{2}} = \langle \mathbf{a}, \mathbf{a} \rangle^{\frac{1}{2}} \tag{5.9}$$

Example 5.9 Let

$$\mathbf{x} = \begin{bmatrix} 3 \\ 4 \end{bmatrix}, \qquad \mathbf{y} = \begin{bmatrix} 6 \\ -6 \\ 7 \end{bmatrix}, \qquad \mathbf{z} = \begin{bmatrix} 4 \\ -10 \\ 0 \\ 6 \end{bmatrix}$$

$$\begin{aligned} \|\mathbf{x}\| &= \sqrt{9+16} = 5 \\ \|\mathbf{y}\| &= \sqrt{36+36+49} = 11 \\ \|\mathbf{z}\| &= \sqrt{16+100+0+36} = \sqrt{152} \approx 12.33 \end{aligned}$$

The length of a vector is only one possibility; there are other measures that can serve as a norm. In general, a norm should possess the following attributes:

$$\begin{aligned} &i. \ \ \|\mathbf{x}\| \geq 0 \\ &ii. \ \ \|\mathbf{x}\| = 0, \quad \text{iff} \quad \mathbf{x} = 0 \\ &iii. \ \ \|\alpha\mathbf{x}\| = |\alpha| \, \|\mathbf{x}\| \quad \forall\alpha \\ &iv. \ \ \|\mathbf{x}+\mathbf{y}\| \leq \|\mathbf{x}\| + \|\mathbf{y}\| \end{aligned} \tag{5.10}$$

Thus, the following measures can also serve as a norm:

$$\|\mathbf{x}\|_l = \sum_{j=1}^{n} |x_j| \tag{5.11}$$

which is referred to as the ℓ^1-*norm*, and

$$\|\mathbf{x}\|_\infty = \max_j\{|x_j|\} \tag{5.12}$$

which is called the *sup-norm*. In this book, however, we shall use the Euclidean norm as defined in (5.9). On occasions we need to normalize the length of a vector, that is, make sure that $\|\mathbf{x}\| = 1$. This can be accomplished by dividing each element of the vector by the norm.

Example 5.10 Consider the following vectors:

$$\mathbf{x} = \begin{bmatrix} 1 \\ 1 \end{bmatrix}, \qquad \mathbf{y} = \begin{bmatrix} 3 \\ -\sqrt{6} \\ 7 \end{bmatrix}$$

Then

$$\|\mathbf{x}\| = \sqrt{2}, \qquad \text{and} \qquad \|\mathbf{y}\| = 8$$

The normalized vectors are

$$\tilde{\mathbf{x}} = \begin{bmatrix} 1/\sqrt{2} \\ 1/\sqrt{2} \end{bmatrix}, \qquad \tilde{\mathbf{y}} = \begin{bmatrix} 3/8 \\ -\sqrt{6}/8 \\ 7/8 \end{bmatrix}$$

5.1.3 Metric

The concept of the norm immediately leads to the definition of a metric or distance between two vectors or points. The norm measures the distance of a point from the origin. Because the difference of two vectors is itself a vector, then its Euclidean norm measures the distance from the end of one vector to the end of the other, that is, the distance between two points.

$$d(\mathbf{x},\mathbf{y}) = \|\mathbf{x}-\mathbf{y}\| = \sqrt{\sum_{j=1}^{n}(x_j - y_j)^2} \tag{5.13}$$

Example 5.11 Let

$$\mathbf{x} = \begin{bmatrix} 3 \\ 1 \end{bmatrix}, \qquad \mathbf{y} = \begin{bmatrix} -4 \\ 7 \end{bmatrix}$$

Then

$$d(\mathbf{x},\mathbf{y}) = \sqrt{(3-(-4))^2 + (1-7)^2} \approx 9.22$$

Example 5.12 Let

$$\mathbf{x} = \begin{bmatrix} 2 \\ 8 \\ 1 \end{bmatrix}, \qquad \mathbf{y} = \begin{bmatrix} 9 \\ -5 \\ 5 \end{bmatrix}$$

Then

$$d(\mathbf{x},\mathbf{y}) = \sqrt{(2-9)^2 + (8-(-5))^2 + (1-5)^2} \approx 15.3$$

In one-dimensional analysis the distance between two points is measured as the straight line between them. Indeed, it is difficult to imagine any other measure that would not look strange. But once we are in two-, three-, ... dimensional spaces, there are other measures of distance that may be appealing. Of course, the most intuitively appealing measure is the length of a straight line between two points—the Euclidean metric discussed above. Any metric, however, has to have certain properties.

Definition 5.5 A metric is a function that maps $\mathbf{S^n} \times \mathbf{S^n}$ into $\Re$ and has the following characteristics:

$$\begin{aligned} &d(\mathbf{x}, \mathbf{y}) \geq 0 \\ &d(\mathbf{x}, \mathbf{y}) = 0 \quad \text{if} \quad \mathbf{x} = \mathbf{y} \\ &d(\mathbf{x}, \mathbf{y}) = d(\mathbf{y}, \mathbf{x}) \\ &d(\mathbf{x}, \mathbf{y}) \leq d(\mathbf{x}, \mathbf{z}) + d(\mathbf{z}, \mathbf{y}) \end{aligned} \tag{5.14}$$

Definition 5.6 Based on this definition, there are other measures that qualify as metrics. For example,

$$d_1(\mathbf{x}, \mathbf{y}) = \sum_{j=1}^{n} |x_j - y_j| \tag{5.15}$$

In probability theory we encounter another metric. Let $x, y \in \Re$; then

$$d_0(x, y) = \frac{|x - y|}{1 + |x - y|} \tag{5.16}$$

is a metric with two interesting properties. First it is confined to the interval [0, 1] and second it is defined for $(x - y) \rightarrow \pm\infty$, a property not shared by other metrics discussed here. While all of the above measures have their place in mathematical analysis, in this book, we will use the Euclidean metric. We have already encountered a vector space. A *metric space* is a vector space $\mathbf{S^n}$ with a metric.

5.1.4 Angle Between Two Vectors and the Cauchy-Schwarz Theorem

Two vectors are perpendicular if the angle between them is $\pi/2$ or $3\pi/2$, which means that the cosine of their angle is zero.

Example 5.13 The vectors

$$\mathbf{e}_1 = \begin{bmatrix} 1 \\ 0 \\ 0 \end{bmatrix}, \quad \text{and} \quad \mathbf{e}_3 = \begin{bmatrix} 0 \\ 0 \\ 1 \end{bmatrix}$$

are perpendicular and so are the vectors

$$\mathbf{a} = \begin{bmatrix} 1 \\ 1 \\ 0 \end{bmatrix}, \quad \text{and} \quad -\mathbf{e}_3 = \begin{bmatrix} 0 \\ 0 \\ -1 \end{bmatrix}$$

Note that in both cases, their inner product is zero. (Draw each pair of vectors and ascertain that they are perpendicular. Moreover, calculate their inner products and verify that they are zero.) This is interesting because $\cos(\pi/2) = \cos(3\pi/2) = 0$. Thus, we may suspect a connection between the inner product of two vectors and the angle between them. But what about the general case where the angle between two vectors can take any value and is not necessarily equal to 0 or $\pi/2$? Indeed, we can determine the angle between any two vectors. We derive the formula for the case of two-dimensional vectors, but the argument can be generalized to $n > 2$.

Consider vectors **a** and **b** in Fig. 5.4. We are interested in finding $\cos\alpha$. Let $\theta = \alpha + \beta$. Then

$$\begin{aligned} \cos\theta &= \frac{a_1}{\|\mathbf{a}\|}, \qquad \sin\theta = \frac{a_2}{\|\mathbf{a}\|} \\ \cos\beta &= \frac{b_1}{\|\mathbf{b}\|} \qquad \sin\beta = \frac{b_2}{\|\mathbf{b}\|} \end{aligned} \tag{5.17}$$

Now

$$\begin{aligned} \cos\alpha &= \cos(\theta - \beta) \\ &= \cos\theta\cos\beta + \sin\theta\sin\beta \\ &= \frac{a_1}{\|\mathbf{a}\|}\frac{b_1}{\|\mathbf{b}\|} + \frac{a_2}{\|\mathbf{a}\|}\frac{b_2}{\|\mathbf{b}\|} \\ &= \frac{\langle \mathbf{a}, \mathbf{b} \rangle}{\|\mathbf{a}\|\,\|\mathbf{b}\|} \end{aligned} \tag{5.18}$$

Example 5.14 The vector $\mathbf{x} = [1\ 1]'$ makes a 45° angle with the x-axis that is represented by $\mathbf{e}_1 = [1\ 0]'$.

$$\cos\alpha = \frac{1+0}{\sqrt{2} \times 1} = \frac{1}{\sqrt{2}} = \cos\frac{\pi}{4}$$

Example 5.15 Consider the vectors

$$\begin{bmatrix} 4 \\ 3 \end{bmatrix}, \qquad \begin{bmatrix} -3 \\ 1 \end{bmatrix}$$

The angle between them is

$$\cos\alpha = \frac{-9}{5\sqrt{10}} \approx -0.596 \qquad \Rightarrow \qquad \alpha \approx 0.69\pi$$

Example 5.16 For the vectors

$$\mathbf{x} = \begin{bmatrix} 3 \\ 2 \end{bmatrix}, \qquad \text{and} \qquad \mathbf{y} = \begin{bmatrix} 9 \\ 6 \end{bmatrix}$$

we have

$$\cos\alpha = \frac{27+12}{\sqrt{13}\sqrt{117}} = 1 \qquad \Rightarrow \qquad \alpha = 0$$

In other words, the two vectors are parallel (verify this by drawing the vectors).

The conclusion of Example 5.16 is quite general. The angle between two parallel vectors is either 0 (if they have the same direction) and π (if they have opposite directions). In the first case, $\cos\alpha = 1$ and in the latter $\cos\alpha = -1$. Consider two parallel vectors **x** and **y**. Because they are parallel, we have

$$\mathbf{y} = \alpha\mathbf{x} \tag{5.19}$$

and

$$\cos\alpha = \frac{\langle \mathbf{x}, \mathbf{y}\rangle}{\|\mathbf{x}\|\,\|\mathbf{y}\|} = \frac{\alpha\,\langle \mathbf{x}, \mathbf{x}\rangle}{|\alpha|\,\|\mathbf{x}\|\,\|\mathbf{x}\|} = \text{sgn}(\alpha) \times 1 \tag{5.20}$$

But

$$\text{sgn}(\alpha) \times 1 = \begin{cases} 1 & \text{if} \quad \alpha > 0 \\ -1 & \text{if} \quad \alpha < 0 \end{cases} \tag{5.21}$$

Example 5.17 Consider the vectors

$$\mathbf{a} = \begin{bmatrix} 3 \\ 4 \end{bmatrix}, \qquad \text{and} \qquad \mathbf{b} = \begin{bmatrix} -15 \\ -20 \end{bmatrix}$$

We have

$$\cos\alpha = \frac{-45 - 80}{\sqrt{25}\sqrt{625}} = -1 \qquad \Rightarrow \qquad \alpha = \pi$$

Calculation of the angle between two vectors, incidentally, leads us to a proof of the famous Cauchy-Schwarz inequality that finds many uses in mathematics and statistics.

Theorem 5.1 (Cauchy-Schwarz) *Let* **x** *and* **y** *be two vectors of order n. Then*

$$|\langle \mathbf{x}, \mathbf{y}\rangle| \leq \|\mathbf{x}\|\,\|\mathbf{y}\| \tag{5.22}$$

or, equivalently

$$\left|\sum x_j y_j\right| \leq \left(\sum x_j^2\right)^{\frac{1}{2}} \left(\sum y_j^2\right)^{\frac{1}{2}} \tag{5.23}$$

Proof Let α denote the angle between the two vectors. We have

$$\frac{\langle \mathbf{x}, \mathbf{y}\rangle}{\|\mathbf{x}\|\,\|\mathbf{y}\|} = \cos\alpha \leq 1 \tag{5.24}$$

Multiplying both sides of inequality by $\|\mathbf{x}\|\,\|\mathbf{y}\|$ proves the proposition.

5.1.5 Exercises

E.5.1 Given the following vectors,

$$\mathbf{a} = \begin{bmatrix} 6 \\ 11 \\ 2 \end{bmatrix}, \quad \mathbf{b} = \begin{bmatrix} 5 \\ 14 \\ 3 \end{bmatrix}, \quad \mathbf{c} = \begin{bmatrix} -1 \\ 0 \\ -1 \end{bmatrix}$$

$$\mathbf{d} = \begin{bmatrix} 7 \\ -4 \\ 9 \end{bmatrix}, \quad \mathbf{e} = \begin{bmatrix} 18 \\ -15 \\ 10 \end{bmatrix}$$

Find

$$\mathbf{a} + \mathbf{b} + \mathbf{c}, \qquad \mathbf{a} - 2\mathbf{d} + 3\mathbf{e}, \qquad 4\mathbf{e} - 4\mathbf{b} + \mathbf{c}$$

E.5.2 Compute the Euclidean norm of the vectors in E.5.1.

E.5.3 Find the distance between the following vectors in E.5.1.

a and **b**, **a** and **d**, **b** and **c**, **c** and **e**, **d** and **e**

E.5.4 Find the angle between the following vectors in E.5.1.

a and **b**, **a** and **d**, **b** and **c**, **c** and **e**, **d** and **e**

E.5.5 Normalize the vectors in E.5.1 to have unit length.

E.5.6 Prove the properties of the inner product in (5.4).

E.5.7 Show that the following vectors are independent and form the basis for a vector space.

$$\begin{bmatrix} 1 \\ 0 \\ 0 \\ 0 \end{bmatrix}, \quad \begin{bmatrix} 0 \\ 1 \\ 0 \\ 0 \end{bmatrix}, \quad \begin{bmatrix} 0 \\ 0 \\ 1 \\ 0 \end{bmatrix}, \quad \begin{bmatrix} 0 \\ 0 \\ 0 \\ 1 \end{bmatrix}$$

E.5.8 Check if the following vectors are independent.

$$\begin{bmatrix} 12 \\ 2 \\ 7 \end{bmatrix}, \quad \begin{bmatrix} 6 \\ 14 \\ -3 \end{bmatrix}, \quad \begin{bmatrix} 27 \\ -28 \\ 32 \end{bmatrix}$$

E.5.9 Show that

$$d_0(x, y) = \frac{|x - y|}{1 + |x - y|}$$

i. Satisfies the requirements of a metric,
ii. it is confined to the interval [0, 1], and
iii.

$$\lim_{(x-y)\to\pm\infty} d_0(x, y) = 1$$

5.2 Orthogonal Vectors

Definition 5.7 Two vectors that are perpendicular to each other, that is, their product is zero are called *orthogonal* and if, in addition their length is equal to one, they are called *orthonormal*.

Example 5.18 The vectors

$$\mathbf{x}_1 = \begin{bmatrix} 2 \\ -5 \end{bmatrix} \quad \text{and} \quad \mathbf{x}_2 = \begin{bmatrix} 5 \\ 2 \end{bmatrix}$$

are orthogonal and the vectors

$$\mathbf{x}_1 = \begin{bmatrix} 2/\sqrt{6} \\ 1/\sqrt{6} \\ 1/\sqrt{6} \end{bmatrix} \quad \text{and} \quad \mathbf{x}_2 = \begin{bmatrix} -1/\sqrt{5} \\ 0 \\ 2/\sqrt{5} \end{bmatrix}$$

are orthonormal.

Orthogonal vectors and matrices (see Chap. 7) play important roles in mathematics, econometrics, and computation. Here we shall discuss an algorithm to create a set of orthogonal and orthonormal vectors from a set of independent vectors.

5.2.1 Gramm-Schmidt Algorithm

Suppose we have a set of independent vectors $\mathbf{x}_1, \ldots, \mathbf{x}_k$. Can we construct a set of k orthogonal or orthonormal vectors from them? Let such vectors be $\mathbf{z}_1, \ldots, \mathbf{z}_k$. They are constructed as follows:

$$\begin{aligned}
\mathbf{z}_1 &= \mathbf{x}_1 \\
\mathbf{z}_2 &= \mathbf{x}_2 - \frac{\mathbf{z}_1'\mathbf{x}_2}{\|\mathbf{z}_1\|^2}\mathbf{z}_1 \\
\mathbf{z}_3 &= \mathbf{x}_3 - \frac{\mathbf{z}_1'\mathbf{x}_3}{\|\mathbf{z}_1\|^2}\mathbf{z}_1 - \frac{\mathbf{z}_2'\mathbf{x}_3}{\|\mathbf{z}_2\|^2}\mathbf{z}_2 \\
&\vdots \\
\mathbf{z}_k &= \mathbf{x}_k - \sum_{j=1}^{k-1}\frac{\mathbf{z}_j'\mathbf{x}_k}{\|\mathbf{z}_j\|^2}\mathbf{z}_j
\end{aligned} \tag{5.25}$$

The formulas may look a bit complicated, but as the examples below illustrate, their application is straightforward.

Example 5.19 Consider the following vectors

$$\mathbf{x}_1 = \begin{bmatrix} 1 \\ 2 \\ -1 \end{bmatrix}, \quad \mathbf{x}_2 = \begin{bmatrix} 4 \\ -7 \\ 1 \end{bmatrix}, \quad \mathbf{x}_3 = \begin{bmatrix} 5 \\ 0 \\ 6 \end{bmatrix}$$

We have

$$\mathbf{z}_1 = \mathbf{x}_1 = \begin{bmatrix} 1 \\ 2 \\ -1 \end{bmatrix}$$

To get $\mathbf{z}_2$, we first calculate

$$\mathbf{z}_1'\mathbf{x}_2 = -11$$

and

$$\|\mathbf{z}_1\|^2 = 6$$

and then

$$\mathbf{z}_2 = \begin{bmatrix} 4 \\ -7 \\ 1 \end{bmatrix} + \frac{11}{6}\begin{bmatrix} 1 \\ 2 \\ -1 \end{bmatrix} = \frac{1}{1.2}\begin{bmatrix} 7 \\ -4 \\ -1 \end{bmatrix}$$

To find $\mathbf{z}_3$, we calculate

$$\mathbf{z}_1'\mathbf{x}_3 = -1$$

$$\mathbf{z}_2'\mathbf{x}_3 = 24.1667$$

and

$$\|\mathbf{z}_2\|^2 = 6.77$$

Thus,

$$\mathbf{z}_3 = \begin{bmatrix} 5 \\ 0 \\ 6 \end{bmatrix} + \frac{1}{6}\begin{bmatrix} 1 \\ 2 \\ -1 \end{bmatrix} - \frac{24.1667}{6.77}\frac{1}{1.2}\begin{bmatrix} 7 \\ -4 \\ -1 \end{bmatrix} = \frac{1}{0.4783}\begin{bmatrix} 1 \\ 1 \\ 3 \end{bmatrix}$$

Our three vectors are

$$\mathbf{z}_1 = \begin{bmatrix} 1 \\ 2 \\ -1 \end{bmatrix}, \quad \mathbf{z}_2 = \frac{1}{1.2}\begin{bmatrix} 7 \\ -4 \\ -1 \end{bmatrix}, \quad \mathbf{z}_3 = \frac{1}{0.4783}\begin{bmatrix} 1 \\ 1 \\ 3 \end{bmatrix}$$

and normalizing them, we have a set of orthonormal vectors.

$$\tilde{\mathbf{z}}_1 = \frac{1}{\sqrt{6}}\begin{bmatrix} 1 \\ 2 \\ -1 \end{bmatrix}, \quad \tilde{\mathbf{z}}_2 = \frac{1}{\sqrt{66}}\begin{bmatrix} 7 \\ -4 \\ -1 \end{bmatrix}, \quad \tilde{\mathbf{z}}_3 = \frac{1}{\sqrt{11}}\begin{bmatrix} 1 \\ 1 \\ 3 \end{bmatrix}$$

5.2.2 Exercises

E.5.10 Check if the following sets of vectors are orthogonal. If they are not orthogonal, use the Gramm-Schmidt algorithm to find a set of orthogonal vectors based on them.

i. $$\mathbf{x}_1 = \begin{bmatrix} 3 \\ 2 \\ 5 \end{bmatrix}, \quad \mathbf{x}_2 = \begin{bmatrix} 1 \\ 0 \\ -1 \end{bmatrix}, \quad \mathbf{x}_3 = \begin{bmatrix} 6 \\ -1 \\ 4 \end{bmatrix}$$

ii. $$\mathbf{y}_1 = \begin{bmatrix} 1 \\ 0 \\ 0 \end{bmatrix}, \quad \mathbf{y}_2 = \begin{bmatrix} 0 \\ 1 \\ 1 \end{bmatrix}, \quad \mathbf{y}_3 = \begin{bmatrix} 3 \\ 3 \\ 3 \end{bmatrix}$$

iii. $$\mathbf{z}_1 = \begin{bmatrix} -7 \\ 12 \\ 15 \end{bmatrix}, \quad \mathbf{z}_2 = \begin{bmatrix} 11 \\ -2 \\ 8 \end{bmatrix}, \quad \mathbf{z}_3 = \begin{bmatrix} 14 \\ 1 \\ -9 \end{bmatrix}$$

Chapter 6
Matrices and Matrix Algebra

6.1 Basic Definitions and Operations

A matrix $\mathbf{A}$ consists of $m \times n$ numbers a_{ij}'s that are arranged in m rows and n columns and looks like the following:

$$\mathbf{A} = \begin{bmatrix} a_{11} & a_{12} & \dots & a_{1n} \\ a_{21} & a_{22} & \dots & a_{2n} \\ \vdots & \vdots & & \vdots \\ a_{m1} & a_{m2} & \dots & a_{mn} \end{bmatrix} \tag{6.1}$$

Example 6.1 The following are examples of matrices:

$$\mathbf{A} = \begin{bmatrix} 1 & 3 & -9 \\ 7 & 2 & 0 \end{bmatrix} \qquad \mathbf{B} = \begin{bmatrix} 14 & 6 & 5 \\ 13 & 18 & 3 \\ 7 & 21 & 44 \end{bmatrix} \qquad \mathbf{C} = \begin{bmatrix} c_{11} & c_{12} \\ c_{21} & c_{22} \\ c_{31} & c_{32} \end{bmatrix}$$

Definition 6.1 Two matrices $\mathbf{A} = [a_{ij}]$ and $\mathbf{B} = [b_{ij}]$ are equal, if they have the same dimensions and

$$\begin{aligned} &a_{ij} = b_{ij}, \\ &i = 1, \dots, m, \quad j = 1, \dots, n \end{aligned} \tag{6.2}$$

Definition 6.2 The *transpose* of the $m \times n$ matrix $\mathbf{A} = [a_{ij}]$ is the $n \times m$ matrix $\mathbf{A}' = [a_{ji}]$ whose rows are the columns of $\mathbf{A}$ and its columns are the rows of $\mathbf{A}$.

Example 6.2 The transposes of matrices in Example 6.1 are

$$\mathbf{A}' = \begin{bmatrix} 1 & 7 \\ 3 & 2 \\ -9 & 0 \end{bmatrix} \qquad \mathbf{B}' = \begin{bmatrix} 14 & 13 & 7 \\ 6 & 18 & 21 \\ 5 & 3 & 44 \end{bmatrix} \qquad \mathbf{C}' = \begin{bmatrix} c_{11} & c_{21} & c_{31} \\ c_{12} & c_{22} & c_{32} \end{bmatrix}$$

It should be clear that

$$(\mathbf{A}')' = \mathbf{A} \tag{6.3}$$

K. Dadkhah, *Foundations of Mathematical and Computational Economics*, 2nd ed.,
DOI 10.1007/978-3-642-13748-8_6,

Definition 6.3 A *square matrix*, that is, a matrix that has the same number of rows and columns, is called a *symmetric matrix*, if it is equal to its transpose:

$$\mathbf{A} = \mathbf{A}' \tag{6.4}$$

Thus, the elements below the diagonal are the mirror image of the elements above it.

Matrix operations of addition, subtraction, and multiplication by a scalar are quite straightforward.

Definition 6.4 The sum of two matrices $\mathbf{A} = [a_{ij}]$ and $\mathbf{B} = [b_{ij}]$ both of them of order $m \times n$ is defined as

$$\mathbf{A} + \mathbf{B} = \mathbf{C} \tag{6.5}$$

such that

$$\mathbf{C} = [c_{ij}], \qquad c_{ij} = a_{ij} + b_{ij} \tag{6.6}$$

$$i = 1, \ldots, m, \quad j = 1, \ldots, n$$

Example 6.3

$$\begin{bmatrix} 1 & 3 & -9 \\ 7 & 2 & 0 \end{bmatrix} + \begin{bmatrix} 14 & 6 & 5 \\ 13 & 18 & 3 \end{bmatrix} = \begin{bmatrix} 15 & 9 & -4 \\ 20 & 20 & 3 \end{bmatrix}$$

Definition 6.5 Let $\mathbf{A} = [a_{ij}]$ be a matrix and λ a scalar. Then

$$\lambda\mathbf{A} = [\lambda a_{ij}] \tag{6.7}$$

In other words, each and every element of $\mathbf{A}$ is multiplied by λ.

Example 6.4

$$3 \times \begin{bmatrix} -1 & 4 \\ 0 & 7 \\ 8 & -5 \end{bmatrix} = \begin{bmatrix} -3 & 12 \\ 0 & 21 \\ 24 & -15 \end{bmatrix}$$

It follows that subtraction of two matrices denoted by $\mathbf{A} - \mathbf{B}$ consists of multiplying $\mathbf{B}$ by $\lambda = -1$ and then adding it to $\mathbf{A}$.

Example 6.5

$$\begin{bmatrix} 10 & 26 \\ 14 & -31 \\ 48 & 19 \end{bmatrix} - \begin{bmatrix} 4 & 23 \\ 36 & 17 \\ -52 & 89 \end{bmatrix} = \begin{bmatrix} 6 & 3 \\ -22 & -48 \\ -100 & -70 \end{bmatrix}$$

The $m \times n$ matrix $\mathbf{0}$ all of whose elements are equal to zero plays the role of zero among matrices. It has the following property:

$$\mathbf{A} + \mathbf{0} = \mathbf{0} + \mathbf{A} = \mathbf{A} \tag{6.8}$$

Matrix addition obeys commutative and associative laws.

$$\begin{array}{rr} \text{Commutative law} & \mathbf{A}+\mathbf{B}=\mathbf{B}+\mathbf{A} \\ \text{Associative law} & (\mathbf{A}+\mathbf{B})+\mathbf{C}=\mathbf{A}+(\mathbf{B}+\mathbf{C}) \end{array} \tag{6.9}$$

Scalar multiplication obeys commutative, associative, and distributive laws.

$$\begin{array}{lr} \text{Commutative law} & \lambda\mathbf{A}=\mathbf{A}\lambda \\ \text{Associative law} & (\lambda\gamma)\mathbf{A}=\lambda(\gamma\mathbf{A}) \\ & \lambda(\mathbf{AB})=(\lambda\mathbf{A})\mathbf{B} \\ \text{Distributive law} & \lambda(\mathbf{A}+\mathbf{B})=\lambda\mathbf{A}+\lambda\mathbf{B} \\ & (\lambda+\gamma)\mathbf{A}=\lambda\mathbf{A}+\gamma\mathbf{A} \end{array} \tag{6.10}$$

Multiplication of two matrices is a bit tricky. First, not all matrices can be multiplied together. **AB** is defined only if the number of columns of **A** is the same as the number of rows of **B**. For instance if **A** is $m \times n$, then **B** must be $n \times r$ and the resulting matrix will be $m \times r$. Furthermore while in multiplication of numbers we had

$$ab = ba$$

that is not true in the case of matrices. Except for special cases,

$$\mathbf{AB} \neq \mathbf{BA} \tag{6.11}$$

Indeed, unless both matrices are square and of the same dimension, that is, both are of order $n \times n$, one of the two products will be undefined.

Definition 6.6 The product of two matrices

$$\mathbf{A} = [a_{ij}], \qquad i = 1, \ldots, m, j = 1, \ldots, n$$

and

$$\mathbf{B} = [b_{jk}], \qquad j = 1, \ldots, n, k = 1, \ldots, r$$

is the matrix

$$\mathbf{C} = [c_{ik}] = \left[\sum_{j=1}^{n} a_{ij} b_{jk}\right], \qquad i = 1, \ldots, m, \quad k = 1, \ldots, r \tag{6.12}$$

More completely

$$\mathbf{AB} = \begin{bmatrix} \sum_{j=1}^{n} a_{1j}b_{j1} & \sum_{j=1}^{n} a_{1j}b_{j2} & \dots & \sum_{j=1}^{n} a_{1j}b_{jn} \\ \sum_{j=1}^{n} a_{2j}b_{j1} & \sum_{j=1}^{n} a_{2j}b_{j2} & \dots & \sum_{j=1}^{n} a_{2j}b_{jn} \\ \vdots & \vdots & & \vdots \\ \sum_{j=1}^{n} a_{mj}b_{j1} & \sum_{j=1}^{n} a_{mj}b_{j2} & \dots & \sum_{j=1}^{n} a_{mj}b_{jn} \end{bmatrix} \tag{6.13}$$

Example 6.6 Given the matrices

$$\mathbf{A} = \begin{bmatrix} 12 & 3 & 6 \\ 9 & -1 & -4 \end{bmatrix}, \quad \mathbf{B} \begin{bmatrix} 7 & 8 \\ -2 & 0 \\ 1 & 11 \end{bmatrix}, \quad \mathbf{C} = \begin{bmatrix} 1 & -1 & 0 \\ -1 & 1 & 1 \\ 0 & 1 & 1 \end{bmatrix}$$

we have

$$\begin{aligned} \mathbf{AB} &= \begin{bmatrix} 12 \times 7 + 3 \times (-2) + 6 \times 1 & 12 \times 8 + 3 \times 0 + 6 \times 11 \\ 9 \times 7 + (-1) \times (-2) + (-4) \times 1 & 9 \times 8 + (-1) \times 0 + (-4) \times 11 \end{bmatrix} \\ &= \begin{bmatrix} 84 & 162 \\ 61 & 28 \end{bmatrix} \end{aligned}$$

$$\begin{aligned} \mathbf{CB} &= \begin{bmatrix} 1 \times 7 + (-1) \times (-2) + 0 \times 1 & 1 \times 8 + (-1) \times 0 + 0 \times 11 \\ (-1) \times 7 + 1 \times (-2) + 1 \times 1 & (-1) \times 8 + 1 \times 0 + 1 \times 11 \\ 0 \times 7 + 1 \times (-2) + 1 \times 1 & 0 \times 8 + 1 \times 0 + 1 \times 11 \end{bmatrix} \\ &= \begin{bmatrix} 9 & 8 \\ -8 & 3 \\ -1 & 11 \end{bmatrix} \end{aligned}$$

$$\mathbf{AC} = \begin{bmatrix} 9 & -3 & 9 \\ 10 & -14 & -5 \end{bmatrix}$$

$$\mathbf{C}'\mathbf{C} = \begin{bmatrix} 2 & -2 & -1 \\ -2 & 3 & 2 \\ -1 & 2 & 2 \end{bmatrix} \qquad \mathbf{CC}' = \begin{bmatrix} 2 & -2 & -1 \\ -2 & 3 & 2 \\ -1 & 2 & 2 \end{bmatrix}$$

The *identity matrix*, **I**, is a square matrix whose diagonal elements are all one and all off-diagonal elements zero. It comes in different sizes.

$$\begin{bmatrix} 1 & 0 \\ 0 & 1 \end{bmatrix}, \quad \begin{bmatrix} 1 & 0 & 0 \\ 0 & 1 & 0 \\ 0 & 0 & 1 \end{bmatrix}, \quad \begin{bmatrix} 1 & 0 & 0 & 0 \\ 0 & 1 & 0 & 0 \\ 0 & 0 & 1 & 0 \\ 0 & 0 & 0 & 1 \end{bmatrix}, \quad \cdots$$

The identity matrix plays the role of 1 in matrix algebra and has the property

$$\mathbf{IA} = \mathbf{A}, \qquad \mathbf{BI} = \mathbf{B} \tag{6.14}$$

assuming that the products exist.

Example 6.7

$$\begin{bmatrix} 1 & 0 \\ 0 & 1 \end{bmatrix} \begin{bmatrix} x_{11} & x_{12} & x_{13} \\ x_{21} & x_{22} & x_{23} \end{bmatrix} = \begin{bmatrix} x_{11} & x_{12} & x_{13} \\ x_{21} & x_{22} & x_{23} \end{bmatrix}$$

Example 6.8

$$\begin{bmatrix} 3 & 5 & 9 \\ 4 & 0 & 4 \\ 3 & 2 & 8 \end{bmatrix} \begin{bmatrix} 1 & 0 & 0 \\ 0 & 1 & 0 \\ 0 & 0 & 1 \end{bmatrix} = \begin{bmatrix} 3 & 5 & 9 \\ 4 & 0 & 4 \\ 3 & 2 & 8 \end{bmatrix}$$

Matrix multiplication, with some exceptions, does not obey the commutative law, but associative and distributive laws apply.

$$\begin{aligned} &\text{Associate law} \qquad (\mathbf{AB})\mathbf{C} = \mathbf{A}(\mathbf{BC}) \\ &\text{Distributive law} \quad \mathbf{A}(\mathbf{B} + \mathbf{C}) = \mathbf{AB} + \mathbf{AC} \end{aligned} \tag{6.15}$$

Another important characteristic of matrix multiplication relates to the transpose of two matrices

$$(\mathbf{AB})' = \mathbf{B}'\mathbf{A}' \tag{6.16}$$

This can be extended to several matrices

$$(\mathbf{ABC})' = \mathbf{C}'(\mathbf{AB})' = \mathbf{C}'\mathbf{B}'\mathbf{A}' \tag{6.17}$$

Vectors can be considered as matrices, one of whose dimensions is equal to one. Thus, we can define vector multiplication more generally. Consider the $n \times 1$ vectors **x** and **y**, then

$$\mathbf{x}'\mathbf{y} = \mathbf{y}'\mathbf{x} = \sum_{j=1}^{n} x_j y_j \tag{6.18}$$

Thus,

$$\mathbf{x}'\mathbf{y} = \langle \mathbf{x}, \mathbf{y} \rangle, \qquad \text{and} \qquad \mathbf{x}'\mathbf{x} = \langle \mathbf{x}, \mathbf{x} \rangle \tag{6.19}$$

But

$$\mathbf{xy}' = \begin{bmatrix} x_1y_1 & x_1y_2 & \dots & x_1y_n \\ x_2y_1 & x_2y_2 & \dots & x_2y_x \\ \vdots & \vdots & & \vdots \\ x_ny_1 & x_ny_2 & \dots & x_ny_n \end{bmatrix} \tag{6.20}$$

It is of interest to note that

$$\mathbf{xy}' \neq \mathbf{yx}' \tag{6.21}$$

but

$$\mathbf{xy}' = (\mathbf{yx}')' \tag{6.22}$$

Example 6.9 Let

$$\mathbf{x} = \begin{bmatrix} 3 \\ 1 \\ 5 \end{bmatrix}, \qquad \mathbf{y} = \begin{bmatrix} 2 \\ 4 \\ 7 \end{bmatrix}$$

Then

$$\mathbf{x}'\mathbf{y} = 45$$

and

$$\mathbf{xy}' = \begin{bmatrix} 6 & 12 & 21 \\ 2 & 4 & 7 \\ 10 & 20 & 35 \end{bmatrix}$$

6.1.1 Systems of Linear Equations

One of the great advantages of matrix algebra is allowing us to write many linear equations and relationships in a compact way and manipulate them. In other words, matrix notations can serve as shorthand. Let us illustrate this with the help of examples from mathematics, economics, and econometrics.

Example 6.10 Consider the following system of linear equations:

$$5x_1 + 3x_2 + 2x_3 = 4$$

$$x_1 - 2x_2 - x_3 = 5$$

$$4x_1 - 7x_2 + 2x_3 = -37$$

Writing it in matrix form we have

$$\begin{bmatrix} 5 & 3 & 2 \\ 1 & -2 & -1 \\ 4 & -7 & 2 \end{bmatrix} \begin{bmatrix} x_1 \\ x_2 \\ x_3 \end{bmatrix} = \begin{bmatrix} 4 \\ 5 \\ -37 \end{bmatrix}$$

or compactly

$$\mathbf{Ax} = \mathbf{d}$$

Example 6.11 Consider the Keynesian model:

$$\begin{aligned} Y &= C + I + G \\ C &= \alpha_0 + \alpha_1(Y - T) \qquad 0 < \alpha_0,\ \ 0 < \alpha_1 < 1, \\ I &= \beta_0 + \beta_1 Y + \beta_2 r \qquad 0 < \beta_0,\ \ 0 < \beta_1 < 1,\ \ \beta_2 < 0 \end{aligned}$$

where Y, C, I, G, T and r are, respectively, aggregate income, consumption, investment, government expenditures, taxes, and real interest rate. Government expenditures, taxes, and interest rates are given, that is, they are determined outside this model and are called *exogenous variables*. We are concerned with determining the *endogenous variables* of the system: income, consumption, and investment in terms of the exogenous variables. Indeed, the purpose of the model is to explain the endogenous variables in terms of the exogenous variables.

Of particular interest to us is finding the relationship between income and the interest rate, given government expenditures and taxes, that is, the IS curve (see Example 6.27). Moving the endogenous variables to the LHS and putting the system in matrix form, we have

$$\begin{bmatrix} 1 & -1 & -1 \\ -\alpha_1 & 1 & 0 \\ -\beta_1 & 0 & 1 \end{bmatrix} \begin{bmatrix} Y \\ C \\ I \end{bmatrix} = \begin{bmatrix} G \\ \alpha_0 - \alpha_1 T \\ \beta_0 + \beta_2 r \end{bmatrix}$$

Example 6.12 In econometrics we learn about the regression model

$$y_i = \beta_1 + \beta_2 x_{2i} + \ldots + \beta_k x_{ki} + u_i$$

where y is the dependent variable, $x_1, \ldots, x_k$, the explanatory variables, where $x_1 \equiv 1$ represents the constant term, β's, the parameters of the model, and u is the stochastic term. Let us assume that we have n observations on y and x's. We can organize our information and the model in the following way:

$$\mathbf{y} = \begin{bmatrix} y_1 \\ \vdots \\ y_n \end{bmatrix}, \quad \mathbf{X} = \begin{bmatrix} 1 & x_{21} & \ldots & x_{k1} \\ \vdots & \vdots & \ldots & \vdots \\ 1 & x_{2n} & \ldots & x_{kn} \end{bmatrix}, \quad \boldsymbol{\beta} = \begin{bmatrix} \beta_1 \\ \vdots \\ \beta_k \end{bmatrix}, \quad \mathbf{u} = \begin{bmatrix} u_1 \\ \vdots \\ u_n \end{bmatrix}$$

Now we can write all n equations in the following compact form which, as will be seen in later chapters, facilitates the derivation of estimators of the parameter vector.

$$\mathbf{y} = \mathbf{X}\boldsymbol{\beta} + \mathbf{u} \tag{6.23}$$

Example 6.13 In Chap. 4 we discussed the Markov chains. Matrix algebra provides a convenient way for presenting and manipulating the transition matrix.[1]

$$\mathbf{P} = \begin{bmatrix} p_{11} & p_{12} & \cdots & p_{n1} \\ p_{12} & p_{22} & \cdots & p_{n2} \\ \vdots & \vdots & & \vdots \\ p_{1n} & p_{n2} & \cdots & p_{nn} \end{bmatrix}$$

where p_{ij} denotes the probability of transition from state i to state j, and

$$\sum_{j=1}^{n} p_{ij} = 1, \qquad i = 1, \ldots, n.$$

Note also that the transition probabilities are assumed to remain constant over time. Now if we denote the vector of probabilities of being in each state i at time t by

$$\boldsymbol{\pi}_t = \begin{bmatrix} \pi_{1t} \\ \pi_{2t} \\ \vdots \\ \pi_{nt} \end{bmatrix}$$

then the vector of probabilities at time $t + 1$ would be

$$\boldsymbol{\pi}_{t+1} = \mathbf{P}\boldsymbol{\pi}_t \tag{6.24}$$

Again from Chap. 4, the transition matrix for the recession and boom in the U.S. economy can be written as

$$\mathbf{P} = \begin{bmatrix} 0.98 & 0.10 \\ 0.02 & 0.90 \end{bmatrix}$$

where state 1 is boom and state 2, recession. Thus, the first column shows the probabilities of the continuation of the boom or transition from boom to recession, whereas the second column shows the probabilities of transition from recession to boom or staying in recession. Thus, $p_{11} = 0.98$ shows the probability of the

[1] As can be seen, the notation for this matrix is different from our usual notation in that the first subscript of an element denotes its column and the second, its row. Some books reverse the notation and, as a result, the elements of each row add up to one. In such cases $\mathbf{P}'$ is the transition matrix. The reason for our choice of this notation is that it is preferred by econometricians.

continuation of the boom if the economy is in boom, whereas $p_{12} = 0.02$ is the probability of transition from boom to recession. Similarly, $p_{22} = 0.90$ is the probability of staying in recession if the economy is already in recession, whereas $p_{21} = 0.10$ is the probability of ending the recession. We should again repeat that the transition matrix here is only for illustration; by all accounts the transition probabilities from boom to recession and vice versa are duration dependent and do not stay constant. Also note that we rounded the probability of staying in a boom from 0.9831 to 0.98. Now suppose we are sure that we are in a recession. What are the probabilities of staying in recession or transiting to a boom period in the next month? We have

$$\pi_{t+1} = \mathbf{P}\pi_t = \begin{bmatrix} 0.98 & 0.10 \\ 0.02 & 0.90 \end{bmatrix} \begin{bmatrix} 0 \\ 1 \end{bmatrix} = \begin{bmatrix} 0.10 \\ 0.90 \end{bmatrix}$$

and after a year

$$\begin{bmatrix} 0.98 & 0.10 \\ 0.02 & 0.90 \end{bmatrix}^{12} \begin{bmatrix} 0 \\ 1 \end{bmatrix} = \begin{bmatrix} 0.6536 \\ 0.3464 \end{bmatrix}$$

In other words, if we are sure that we are in a recession today, there is a two-to-one-chance that we will be out of it in a year.

6.1.2 Computation with Matrices

Matlab is user-friendly for matrix operations. You can either specify the matrices in the program, which is appropriate if your matrices are small or they will be used only in one program. Alternatively, you can produce a file that contains the data and read your matrices from it. This is preferred for large matrices or when you will be using the same data in different programs. We take up each option in turn.

Matlab code

```
% You can either specify the matrices in the program
A = [12 3 6; 9 -1 -4]
B = [7 8; -2 0; 1 11]
C = [1 -1 0; -1 1 1; 0 1 1]
% Transpose of A
A'
% Add A' and B
D = A' + B
% Note that if you try to add A and B, Matlab
% will print an error message because the matrices
% are not of the same dimensions
% Premultiply B by C
E = C*B
% you can combine several operations
F = C*(A'+B)
```

Table 6.1 File containing data

A	B	C	D	E
1	5.0	12	15.32	1
2	6.5	18	18.45	1
8	8.2	27	19.77	0
12	10.3	32	26.60	0
⋮	⋮	⋮	⋮	⋮

Alternatively, you can read the data for matrices from a file and form your matrices inside the program. As an example suppose you have prepared an Excel file called `Ad.xls` (see Table 6.1).

Matlab code

```
% Read the data
D = xlsread('Ad.xls');
% Form the desired matrices. For instance, suppose we
% are interested in forming a matrix containing
% columns 1, 3, and 4 of the data
A = [D(:,1) D(:,3) D(:,4)]
% For the second matrix, we choose the first three
% rows of the data
B = [D(1,:); D(2,:); D(3,:)];
% Indeed we can form a matrix from any combinations
% of our data. For example, the following matrix will
% contains the first three elements in the first row
% and the last three elements in the third and fourth
% rows
C = [D(1,1) D(1,2) D(1,3); D(3,3) D(3,4) D(3,5);
     D(4,3) D(4,4) D(4,5)]
```

6.1.3 Exercises

E.6.1 Given the following matrices,

$$\mathbf{A} = \begin{bmatrix} 5 & -3 \\ 0 & 3 \end{bmatrix}, \quad \mathbf{B} = \begin{bmatrix} 3 & 0 \\ 0 & -2 \end{bmatrix}, \quad \mathbf{C} = \begin{bmatrix} 5 & 4 \\ -3 & 11 \end{bmatrix}$$

$$\mathbf{D} = \begin{bmatrix} 12 & 14 & 13 \\ 7 & -8 & -10 \end{bmatrix}, \quad \mathbf{E} = \begin{bmatrix} 1 & 1 & 1 \\ 4 & 2 & -7 \end{bmatrix}$$

$$\mathbf{F} = \begin{bmatrix} 2 & 7 \\ -6 & 3 \\ 4 & 1 \end{bmatrix}, \quad \mathbf{G} = \begin{bmatrix} 17 & 19 \\ 21 & 12 \\ 6 & 8 \end{bmatrix}$$

compute

i. $\mathbf{A}(\mathbf{B}+\mathbf{C})$ *ii.* $\mathbf{B}(\mathbf{D}+\mathbf{G}')$ *iii.* $\mathbf{G}\mathbf{F}'$
iv. $\mathbf{G}\mathbf{D}$ *v.* $\mathbf{E}\mathbf{F}$ *vi.* $\mathbf{B}'\mathbf{C}'$

E.6.2 Given the following matrices

$$\mathbf{A} = \begin{bmatrix} 1 & 5 & 4 \\ 0 & 3 & 2 \\ 0 & 0 & 1 \end{bmatrix}, \quad \mathbf{B} = \begin{bmatrix} 7 & 0 & 0 \\ 9 & 1 & 1 \\ 2 & 3 & 8 \end{bmatrix}, \quad \mathbf{C} = \begin{bmatrix} 1 & -5 & 14 \\ -7 & 1 & -6 \\ 14 & -3 & 1 \end{bmatrix}$$

compute

i. $\mathbf{AB}$ *ii.* $\mathbf{BCA}$ *iii.* $\mathbf{BA}$ *iv.* $\mathbf{CC}'$ *v.* $(\mathbf{CC}')'$

E.6.3 Check the results of your computation for E.6.1 and E.6.2 using Matlab.

6.2 The Inverse of a Matrix

We learned about addition, subtraction, and multiplication of matrices, but what about division? Recall that for ordinary numbers we have

$$\frac{a}{b} = ab^{-1}, \qquad b \neq 0$$

Is there a counterpart to b^{-1} for matrices? Indeed, there is.

Definition 6.7 Let $\mathbf{A}$ be a square matrix. Then its inverse $\mathbf{A}^{-1}$ is a square matrix[2] such that

$$\mathbf{A}\mathbf{A}^{-1} = \mathbf{A}^{-1}\mathbf{A} = \mathbf{I} \tag{6.25}$$

Not all square matrices have an inverse. Much the same way that in the case of real numbers we excluded $b = 0$, we exclude all *singular* matrices. Consider the following system of two equations:

$$\begin{aligned} 2x_1 + x_2 &= 5 \\ x_1 - 3x_2 &= -8 \end{aligned}$$

We can write it in matrix form as

$$\mathbf{A}\mathbf{x} = \mathbf{b} \tag{6.26}$$

where

$$\mathbf{A} = \begin{bmatrix} 2 & 1 \\ 1 & -3 \end{bmatrix}, \quad \mathbf{x} = \begin{bmatrix} x_1 \\ x_2 \end{bmatrix}, \quad \mathbf{b} = \begin{bmatrix} 5 \\ -8 \end{bmatrix}$$

[2]Non-square matrices do not have an inverse in the sense defined here. But in the next chapter we will define a *generalized inverse* for them.

If the inverse of $\mathbf{A}$ exists, and I can assure you that it exists, then we can premultiply (6.26) by $\mathbf{A}^{-1}$.

$$\mathbf{A}^{-1}\mathbf{A}\mathbf{x} = \mathbf{x} = \mathbf{A}^{-1}\mathbf{b} \tag{6.27}$$

In other words, finding the inverse of $\mathbf{A}$ and solving this system of equations are the same. Because once we have the inverse, we premultiply the RHS by $\mathbf{A}^{-1}$ and have the solution. Thus, let us solve the equations step by step the old-fashioned way, the way we know from elementary algebra.

1. The first step is to divide the first equation by 2 and make the coefficient of x_1 equal to 1.

$$\begin{aligned} x_1 + \frac{1}{2}x_2 &= \frac{5}{2} \\ x_1 - 3x_2 &= -8 \end{aligned}$$

2. Subtract the first equation from the second to eliminate x_1 from the second equation.

$$\begin{aligned} x_1 + \frac{1}{2}x_2 &= \frac{5}{2} \\ -\frac{7}{2}x_2 &= -\frac{21}{2} \end{aligned}$$

3. Multiply the second equation by $-2/7$.

$$\begin{aligned} x_1 + \frac{1}{2}x_2 &= \frac{5}{2} \\ x_2 &= 3 \end{aligned}$$

4. Finally, subtract one-half of the second equation from the first to obtain

$$\mathrm{x}_1 = 1, \qquad \mathrm{x}_2 = 3$$

The four steps above could be accomplished using matrices. But since solving the equations is equivalent to finding the inverse of $\mathbf{A}$, we also learn the method by which we could find the inverse of a matrix.

$$\begin{bmatrix} 2 & 1 \\ 1 & -3 \end{bmatrix}\begin{bmatrix} x_1 \\ x_2 \end{bmatrix} = \begin{bmatrix} 1 & 0 \\ 0 & 1 \end{bmatrix}\begin{bmatrix} 5 \\ -8 \end{bmatrix}$$

1 Divide the first row of $\mathbf{A}$ by 2 and do the same to the identity matrix on the right.

$$\begin{bmatrix} \frac{1}{2} & 0 \\ 0 & 1 \end{bmatrix}\begin{bmatrix} 2 & 1 \\ 1 & -3 \end{bmatrix}\begin{bmatrix} x_1 \\ x_2 \end{bmatrix} = \begin{bmatrix} \frac{1}{2} & 0 \\ 0 & 1 \end{bmatrix}\begin{bmatrix} 1 & 0 \\ 0 & 1 \end{bmatrix}\begin{bmatrix} 5 \\ -8 \end{bmatrix}$$

2. Subtract the first row from the second.

$$\begin{bmatrix} 1 & 0 \\ -1 & 1 \end{bmatrix}\begin{bmatrix} 1 & \frac{1}{2} \\ 1 & -3 \end{bmatrix}\begin{bmatrix} x_1 \\ x_2 \end{bmatrix} = \begin{bmatrix} 1 & 0 \\ -1 & 1 \end{bmatrix}\begin{bmatrix} \frac{1}{2} & 0 \\ 0 & 1 \end{bmatrix}\begin{bmatrix} 5 \\ -8 \end{bmatrix}$$

3. Multiply the second row by −2/7.

$$\begin{bmatrix} 1 & 0 \\ 0 & -\frac{2}{7} \end{bmatrix}\begin{bmatrix} 1 & \frac{1}{2} \\ 0 & -\frac{7}{2} \end{bmatrix}\begin{bmatrix} x_1 \\ x_2 \end{bmatrix} = \begin{bmatrix} 1 & 0 \\ 0 & -\frac{2}{7} \end{bmatrix}\begin{bmatrix} \frac{1}{2} & 0 \\ -\frac{1}{2} & 1 \end{bmatrix}\begin{bmatrix} 5 \\ -8 \end{bmatrix}$$

4. Subtract 1/2 of the second row from the first row.

$$\begin{bmatrix} 1 & -\frac{1}{2} \\ 0 & 1 \end{bmatrix}\begin{bmatrix} 1 & \frac{1}{2} \\ 0 & 1 \end{bmatrix}\begin{bmatrix} x_1 \\ x_2 \end{bmatrix} = \begin{bmatrix} 1 & -\frac{1}{2} \\ 0 & 1 \end{bmatrix}\begin{bmatrix} \frac{1}{2} & 0 \\ \frac{1}{7} & -\frac{2}{7} \end{bmatrix}\begin{bmatrix} 5 \\ -8 \end{bmatrix}$$

Voilà! We have both the solution to the equations and the inverse of **A**.

$$\begin{bmatrix} 1 & 0 \\ 0 & 1 \end{bmatrix}\begin{bmatrix} x_1 \\ x_2 \end{bmatrix} = \begin{bmatrix} x_1 \\ x_2 \end{bmatrix} = \begin{bmatrix} \frac{3}{7} & \frac{1}{7} \\ \frac{1}{7} & -\frac{2}{7} \end{bmatrix}\begin{bmatrix} 5 \\ -8 \end{bmatrix} = \begin{bmatrix} 1 \\ 3 \end{bmatrix}$$

In other words, there are a series of row operations that transform matrix **A** into an identity matrix. If we do the same exact transformations, and in the same order, on an identity matrix, we transform it into $\mathbf{A}^{-1}$. Of course, instead of row operations, we could have done column operations to turn **A** into an identity matrix and the resulting inverse would have been identical to the one obtained above. The row operations, in view of emulating the solution of a system of equations, appear intuitively appealing.

Example 6.14 Let us repeat the process for a matrix of order three. Below, R_i refers to the i-th row of the matrices.

We start with the matrix to be inverted and an identity matrix $\Rightarrow \begin{bmatrix} 2 & 1 & 2 \\ 4 & 2 & 0 \\ -1 & 5 & 3 \end{bmatrix} \Rightarrow \begin{bmatrix} 1 & 0 & 0 \\ 0 & 1 & 0 \\ 0 & 0 & 1 \end{bmatrix}$

Subtract2 × R_1 from R_2

$$\Rightarrow \begin{bmatrix} 2 & 1 & 2 \\ 0 & 0 & -4 \\ -1 & 5 & 3 \end{bmatrix} \Rightarrow \begin{bmatrix} 1 & 0 & 0 \\ -2 & 1 & 0 \\ 0 & 0 & 1 \end{bmatrix}$$

Add 2 × R_3 to R_1

$$\Rightarrow \begin{bmatrix} 0 & 11 & 8 \\ 0 & 0 & -4 \\ -1 & 5 & 3 \end{bmatrix} \Rightarrow \begin{bmatrix} 1 & 0 & 2 \\ -2 & 1 & 0 \\ 0 & 0 & 1 \end{bmatrix}$$

Exchange R_2 withR_3 then the new R_2 with R_1

$$\Rightarrow \begin{bmatrix} -1 & 5 & 3 \\ 0 & 11 & 8 \\ 0 & 0 & -4 \end{bmatrix} \Rightarrow \begin{bmatrix} 0 & 0 & 1 \\ 1 & 0 & 2 \\ -2 & 1 & 0 \end{bmatrix}$$

Multiply R_1 by -1, add 2 × R_3 to R_2, and divide R_3 by -4

$$\Rightarrow \begin{bmatrix} 1 & -5 & -3 \\ 0 & 11 & 0 \\ 0 & 0 & 1 \end{bmatrix} \Rightarrow \begin{bmatrix} 0 & 0 & -1 \\ -3 & 2 & 2 \\ \frac{2}{4} & -\frac{1}{4} & 0 \end{bmatrix}$$

Add 3 × R_3 to R_1, and divide R_2 by 11

$$\Rightarrow \begin{bmatrix} 1 & -5 & 0 \\ 0 & 1 & 0 \\ 0 & 0 & 1 \end{bmatrix} \Rightarrow \begin{bmatrix} \frac{6}{4} & -\frac{3}{4} & -1 \\ -\frac{3}{12} & \frac{2}{11} & \frac{2}{11} \\ \frac{2}{4} & -\frac{1}{4} & 0 \end{bmatrix}$$

Add 5 × R_2 to R_1

$$\Rightarrow \begin{bmatrix} 1 & 0 & 0 \\ 0 & 1 & 0 \\ 0 & 0 & 1 \end{bmatrix} \Rightarrow \begin{bmatrix} \frac{6}{44} & \frac{7}{44} & -\frac{4}{44} \\ -\frac{3}{11} & \frac{2}{11} & \frac{2}{11} \\ \frac{2}{4} & -\frac{1}{4} & 0 \end{bmatrix}$$

Cleaning up the inverse is

$$\frac{1}{44}\begin{bmatrix} 6 & 7 & -4 \\ -12 & 8 & 8 \\ 22 & -11 & 0 \end{bmatrix}$$

While the above operations may look to you "so twentieth century," they are actually good for the soul and you will want to make sure that you can carry out such operations with ease. They will be handy in manipulating linear economic models. But as far as numerical computations are concerned we definitely need something more efficient. First, in most applications the order of the matrices involved is much larger than 2 or 3. Once we go beyond a matrix of order 5, computation with a calculator becomes time consuming and prone to errors. Second, rarely in the real world do we encounter matrices with one-digit whole numbers. We usually have large numbers or numbers with decimals. This adds to the likelihood of errors in computation. But matrix inversion with Matlab is quite easy.

Matlab code

```
% Specify the matrix
A = [2 1 2; 4 2 0; -1 5 3]
% Calculate its inverse
A1 = inv(A)
```

The inverse of a matrix has the following properties:

$$\begin{aligned} (\mathbf{A}^{-1})^{-1} &= \mathbf{A} \\ (\mathbf{AB})^{-1} &= \mathbf{B}^{-1}\mathbf{A}^{-1} \\ (\mathbf{A}')^{-1} &= (\mathbf{A}^{-1})' \end{aligned} \tag{6.28}$$

Proof and numerical verification of the first and third properties are left to the reader (see E.6.8). Here we show the second property. Let

$$\mathbf{C} = (\mathbf{AB})^{-1} \tag{6.29}$$

Because $\mathbf{C}$ is the inverse of $\mathbf{AB}$, we have

$$\mathbf{C}(\mathbf{AB}) = \mathbf{I} \tag{6.30}$$

Postmultiplying both sides by $\mathbf{B}^{-1}\mathbf{A}^{-1}$ we get

$$\mathbf{C}(\mathbf{AB})\mathbf{B}^{-1}\mathbf{A}^{-1} = \mathbf{IB}^{-1}\mathbf{A}^{-1} \qquad \Rightarrow \qquad \mathbf{C} = \mathbf{B}^{-1}\mathbf{A}^{-1} \tag{6.31}$$

6.2.1 A Number Called the Determinant

Attached to every square matrix is a number called the *determinant*. It is denoted by

$$\Delta = |\mathbf{A}| \tag{6.32}$$

and its calculation for a determinant of order two is:

$$\Delta = \begin{vmatrix} a & b \\ c & d \end{vmatrix} = ad - bc$$

Example 6.15

$$\begin{vmatrix} 1 & 2 \\ 6 & 5 \end{vmatrix} = 1 \times 5 - 6 \times 2 = -7$$

For a determinant of order three the trick is to write it in terms of smaller determinants called *minors*. Note the alternate plus and minus signs. Thus,

$$\Delta = \begin{vmatrix} a_{11} & a_{12} & a_{13} \\ a_{21} & a_{22} & a_{23} \\ a_{31} & a_{32} & a_{33} \end{vmatrix} = a_{11}\begin{vmatrix} a_{22} & a_{23} \\ a_{32} & a_{33} \end{vmatrix} - a_{21}\begin{vmatrix} a_{12} & a_{13} \\ a_{32} & a_{33} \end{vmatrix} + a_{31}\begin{vmatrix} a_{12} & a_{13} \\ a_{22} & a_{23} \end{vmatrix}$$
$$= a_{11}a_{22}a_{33} + a_{21}a_{32}a_{13} + a_{31}a_{23}a_{12} - a_{13}a_{22}a_{31} - a_{23}a_{32}a_{11} - a_{33}a_{21}a_{12}$$

Example 6.16

$$\begin{vmatrix} 9 & 11 & 2 \\ 0 & 17 & 5 \\ -1 & 2 & 10 \end{vmatrix} = 9\begin{vmatrix} 17 & 5 \\ 2 & 10 \end{vmatrix} - 11\begin{vmatrix} 0 & 5 \\ -1 & 10 \end{vmatrix} + 2\begin{vmatrix} 0 & 17 \\ -1 & 2 \end{vmatrix}$$
$$= 9 \times 160 - 11 \times 5 + 2 \times 17 = 1419$$

When dealing with determinants of size three, it is easier to skip the expansion in terms of minors and directly compute the determinant. The schematic representation of Fig. 6.1 may help memorizing the rule for such a direct computation.

Computation of a determinant through expansion using minors can be extended to determinants of higher order. For a determinant of order four we have

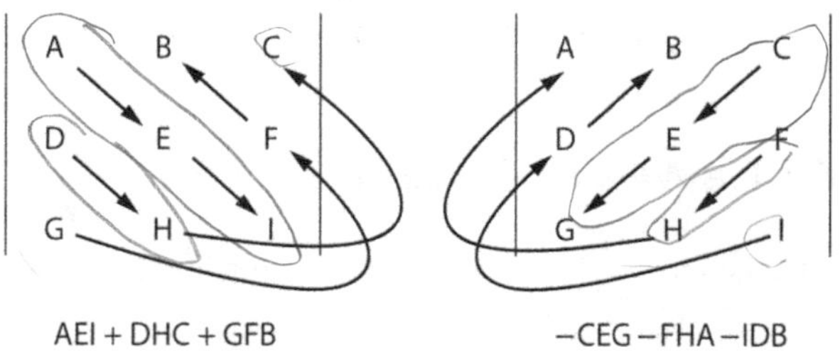

Fig. 6.1 Schematic calculation of determinants of order three

$$|\mathbf{A}| = \begin{vmatrix} a_{11} & a_{12} & a_{13} & a_{14} \\ a_{21} & a_{22} & a_{23} & a_{24} \\ a_{31} & a_{32} & a_{33} & a_{34} \\ a_{41} & a_{42} & a_{43} & a_{44} \end{vmatrix}$$

Expanding it in terms of minors,

$$\begin{aligned} |\mathbf{A}| &= a_{11} \begin{vmatrix} a_{22} & a_{23} & a_{24} \\ a_{32} & a_{33} & a_{34} \\ a_{42} & a_{43} & a_{44} \end{vmatrix} - a_{21} \begin{vmatrix} a_{12} & a_{13} & a_{14} \\ a_{32} & a_{33} & a_{34} \\ a_{42} & a_{43} & a_{44} \end{vmatrix} \\ &+ a_{31} \begin{vmatrix} a_{12} & a_{13} & a_{14} \\ a_{22} & a_{23} & a_{24} \\ a_{42} & a_{43} & a_{44} \end{vmatrix} - a_{41} \begin{vmatrix} a_{12} & a_{13} & a_{14} \\ a_{22} & a_{23} & a_{24} \\ a_{32} & a_{33} & a_{34} \end{vmatrix} \\ &= a_{11}M_{11} - a_{21}M_{21} + a_{31}M_{31} - a_{41}M_{41} \end{aligned} \quad (6.33)$$

Thus, the minor M_{ij}, which is a mini-determinant, is formed by deleting the i-th row and the j-th column. The same can be repeated for any of the minors. For instance, we can write

$$M_{11} = a_{22} \begin{vmatrix} a_{33} & a_{34} \\ a_{43} & a_{44} \end{vmatrix} - a_{23} \begin{vmatrix} a_{32} & a_{34} \\ a_{42} & a_{44} \end{vmatrix} + a_{24} \begin{vmatrix} a_{32} & a_{33} \\ a_{42} & a_{43} \end{vmatrix}$$

The rest of calculations should be clear to the reader. In expanding the determinant $|\mathbf{A}|$ we first used column one of the determinant and in expanding the minor M_{11} we used the first row. There is no magic about the first row or the first column. We could have used any row or column in each case. The important point, however, is to keep in mind the change in the sign of each term. The rule is that you start at the upper-left corner and move right, left, up or down, but not diagonally. There will be no change for the first element in the upper-left corner. But the next element will have its sign changed, that is, if it is positive, it will become negative and vice versa. The next one will undergo no sign change, and so on. The scheme of change-no change sign for each determinant or minor would be like the following:

$$\begin{vmatrix} \text{no change} & \text{change} & \text{no change} \\ \text{change} & \text{no change} & \text{change} \\ \text{no change} & \text{change} & \text{no change} \end{vmatrix}$$

Alternatively you can write the expansion of the determinant in (6.33) as

$$\begin{aligned}|\mathbf{A}| &= a_{11}(-1)^{1+1}M_{11} + a_{21}(-1)^{2+1}M_{21} + a_{31}(-1)^{3+1}M_{31} + a_{41}(-1)^{4+1}M_{41} \\ &= a_{11}C_{11} + a_{21}C_{21} + a_{31}C_{31} + a_{41}C_{41}\end{aligned} \tag{6.34}$$

Where

$$C_{ij} = (-1)^{i+j} M_{ij} \tag{6.35}$$

is called *cofactor*. We can generalize (6.33), and write

$$|\mathbf{A}| = \sum_{i=1}^{n} a_{ij}(-1)^{i+j}M_{ij} = \sum_{i=1}^{n} a_{ij}C_{ij} \tag{6.36}$$

Furthermore,

$$\sum_{i=1}^{n} a_{ij}C_{kj} = 0, \qquad \forall k \neq i \tag{6.37}$$

which is called expansion in terms of *alien cofactors*.

Example 6.17 Evaluate the following determinant:

$$|\mathbf{A}| = \begin{vmatrix} 3 & 6 & 7 & 12 \\ -9 & 0 & 15 & 2 \\ 0 & -9 & 1 & -4 \\ 8 & 8 & 5 & 1 \end{vmatrix}$$

It can be expanded into

$$\begin{aligned}|\mathbf{A}| &= -6\begin{vmatrix} -9 & 15 & 2 \\ 0 & 1 & -4 \\ 8 & 5 & 1 \end{vmatrix} + \begin{vmatrix} 3 & 7 & 12 \\ 0 & 1 & -4 \\ 8 & 5 & 1 \end{vmatrix} + 9\begin{vmatrix} 3 & 7 & 12 \\ -9 & 15 & 2 \\ 8 & 5 & 1 \end{vmatrix} + 8\begin{vmatrix} 3 & 7 & 12 \\ -9 & 15 & 2 \\ 0 & 1 & -4 \end{vmatrix} \\ &= -6(-685) + 0(-257) + 9(-1790) + 8(-546) = -16368\end{aligned}$$

Matlab can be used to evaluate a determinant:

Matlab code

```
% Specify the matrix
A = [9 11 2; 0 17 5; -1 2 10]
% Calculate its determinant
det(A)
```

Determinants have a number of properties that are quite interesting and prove useful in economics and econometrics (the reader is asked to prove and numerically verify these properties, see E.6.11 and E.6.12).

1. The determinant of a matrix and its transpose are the same:
$$|\mathbf{A}| = |\mathbf{A}'| \tag{6.38}$$
For example,
$$\begin{vmatrix} a & b \\ c & d \end{vmatrix} = ad - bc = \begin{vmatrix} a & c \\ b & d \end{vmatrix}$$
2. If we multiply a row (column) of a matrix by λ, its determinant is multiplied by λ. If we multiply all rows (columns) by λ or, equivalently, multiply the matrix by λ, then
$$|\lambda\mathbf{A}| = \lambda^n |\mathbf{A}| \tag{6.39}$$
where n is the order of the matrix. Consider the example
$$\begin{vmatrix} \lambda a_{11} & a_{12} & a_{13} \\ \lambda a_{21} & a_{22} & a_{23} \\ \lambda a_{31} & a_{32} & a_{33} \end{vmatrix} = \lambda a_{11} \begin{vmatrix} a_{22} & a_{23} \\ a_{32} & a_{33} \end{vmatrix} + \lambda a_{21} \begin{vmatrix} a_{12} & a_{13} \\ a_{32} & a_{33} \end{vmatrix} + \lambda a_{31} \begin{vmatrix} a_{12} & a_{13} \\ a_{22} & a_{23} \end{vmatrix}$$
$$= \lambda \begin{vmatrix} a_{11} & a_{12} & a_{13} \\ a_{21} & a_{22} & a_{23} \\ a_{31} & a_{32} & a_{33} \end{vmatrix}$$
This example also shows that if one of the rows (columns) consists of zeros, then the determinant is equal to zero because we can set $\lambda = 0$, which would make the elements of the first column zero. Also multiplying any of the rows (columns) by -1 will change the sign of the determinant.
3. If two rows (columns) are identical, then the determinant is equal to zero. For example,
$$\begin{vmatrix} a & a \\ b & b \end{vmatrix} = ab - ab = 0, \qquad \begin{vmatrix} 1 & 1 & 4 \\ 1 & 1 & -9 \\ 1 & 1 & 13 \end{vmatrix} = 0$$
Note that this property together with the previous one implies that if a column (row) is a multiple of another column (row), then the determinant is equal to zero.
4. If a column (row) is equal to a linear combination of some of the other columns (rows), then the determinant is equal to zero. I suggest that you make a note of

this property as it plays an important part in least squares estimation which we shall learn in the coming chapters. Verify that the following is correct:

$$\begin{vmatrix} a_1 + 2b_1 & a_1 & b_1 \\ a_2 + 2b_2 & a_2 & b_2 \\ a_3 + 2b_3 & a_3 & b_3 \end{vmatrix} = 0$$

5. If places of two columns (rows) are interchanged, the sign of the determinant changes. For example,

$$\begin{vmatrix} a & b \\ c & d \end{vmatrix} = ad - bc, \qquad \begin{vmatrix} b & a \\ d & c \end{vmatrix} = bc - ad$$

An application of the determinant is in determining if a matrix is invertible. If $|\mathbf{A}| = 0$, then the matrix $\mathbf{A}$ is not invertible and is called a *singular matrix*. A matrix whose determinant is not equal to zero is a *nonsingular matrix* and is invertible. Another application involves determining the independence of a set of vectors. Suppose we have a set of k-vectors. Needless to say, that only k of them could be independent. We can arrange them in a $k \times k$ matrix and calculate its determinant. If the determinant is nonzero, then they are independent; otherwise they are dependent.

6.2.2 Rank and Trace of a Matrix

Consider an $m \times n$ matrix, that is, a matrix that has m rows and n columns. How many of the rows are independent of each other? How many columns are independent? The number of independent rows and columns of a matrix is called the *rank* of that matrix. Denoting the rank of matrix $\mathbf{A}$ by $\rho(\mathbf{A})$, it is obvious that

$$\rho(\mathbf{A}) \leq \min(m, n) \tag{6.40}$$

The rank of a matrix is equal to the order of the largest nonzero determinant inside that matrix.

Example 6.18 Consider the following matrices:

$$\mathbf{A} = \begin{bmatrix} 0 & 0 & 0 \\ 0 & 0 & 0 \\ 0 & 0 & 0 \end{bmatrix}, \qquad \mathbf{B} = \begin{bmatrix} 2 & 3 & -9 \\ 7 & 14 & 1 \end{bmatrix}, \qquad \mathbf{C} = \begin{bmatrix} 1 & 0 & 3 \\ 0 & 5 & 0 \\ 2 & 7 & 8 \end{bmatrix}$$

Matrix $\mathbf{A}$ has rank zero. Indeed, if only all of the elements of a matrix are zero, then its rank is zero. $\mathbf{B}$ has rank 2, because the largest nonzero minor we can form, for example,

$$\begin{vmatrix} 2 & -9 \\ 7 & 1 \end{vmatrix}$$

is of order two. On the other hand, **C** has rank three because its determinant is not equal to zero.

If $\rho(\mathbf{A}) = m$, we say that the matrix has full row rank and if $\rho(\mathbf{A}) = n$, we say it has full column rank. In the case of a square matrix we simply speak of rank. A square matrix that has full rank has a nonzero determinant and, therefore, is invertible. The rank of a matrix has the following properties some of which will be quite handy in econometric analysis:

$$\begin{aligned} &i. \quad \rho(\mathbf{A}) = \rho(\mathbf{A}') = \rho(\mathbf{AA})' = \rho(\mathbf{A}'\mathbf{A}) \\ &ii. \quad \rho(\mathbf{AB}) \leq \min\{\rho(\mathbf{A}), \rho(\mathbf{B})\} \\ &iii. \quad \rho(\mathbf{A}+\mathbf{B}) \leq \rho(\mathbf{A}) + \rho(\mathbf{B}) \end{aligned} \tag{6.41}$$

The *trace* of a square matrix is defined as the sum of its diagonal elements. Let $\mathbf{A} = [a_{ij}]$ be of order n. Then its trace denoted by $tr(\mathbf{A})$ is defined as

$$tr(\mathbf{A}) = \sum_{i=1}^{n} a_{ii} \tag{6.42}$$

Rank and trace of a matrix can be obtained using Matlab

Matlab code

```
% Specify the matrix
A = [6 7 -4; -12 8 8; 22 -11 0]
% Calculate its rank
k = rank(A)
% Calculate its trace
s = trace(A)
```

6.2.3 Another Way to Find the Inverse of a Matrix

Whereas our previous method of inverting a matrix, which relied on manipulating rows and columns works fine, there is another way to find the inverse that may be appealing for the case of low-dimension matrices. First, recall that if we eliminate the i-th row and the j-th column of the determinant of a matrix, the resulting smaller determinant M_{ij} is called a minor. We defined the cofactor of the ij-th element in a matrix as

$$C_{ij} = (-1)^{i+j} M_{ij}$$

The following algorithm leads us to the inverse of a matrix.

1. Calculate the determinant of the matrix.
2. Transpose the matrix.

3. Replace each element of the transposed matrix by its cofactor.
4. Divide the resulting matrix by the determinant.

Example 6.19 Consider the matrix

$$\mathbf{A} = \begin{bmatrix} 2 & 7 \\ 1 & 5 \end{bmatrix}$$

Applying the algorithm we have

1. $|\mathbf{A}| = 3$
2. $\mathbf{A}' = \begin{bmatrix} 2 & 1 \\ 7 & 5 \end{bmatrix}$
3. $C_{11} = 5, \quad C_{12} = -7, \quad C_{21} = -1, \quad C_{22} = 2 \;\Rightarrow\; \begin{bmatrix} 5 & -7 \\ -1 & 2 \end{bmatrix}$
4. $\mathbf{A}^{-1} = \frac{1}{3}\begin{bmatrix} 5 & -7 \\ -1 & 2 \end{bmatrix}$

Example 6.20 Consider the matrix in Example 6.14:

$$\mathbf{B} = \begin{bmatrix} 2 & 1 & 2 \\ 4 & 2 & 0 \\ -1 & 5 & 3 \end{bmatrix}$$

Applying the algorithm

$$|\mathbf{B}| = 4$$

$$\mathbf{B}' = \begin{bmatrix} 2 & 4 & -1 \\ 1 & 2 & 5 \\ 2 & 0 & 3 \end{bmatrix}$$

and the inverse is

$$\mathbf{B}^{-1} = \frac{1}{44}\begin{bmatrix} \begin{vmatrix} 2 & 5 \\ 0 & 3 \end{vmatrix} & -\begin{vmatrix} 1 & 5 \\ 2 & 3 \end{vmatrix} & \begin{vmatrix} 1 & 2 \\ 2 & 0 \end{vmatrix} \\ -\begin{vmatrix} 4 & -1 \\ 0 & 3 \end{vmatrix} & \begin{vmatrix} 2 & -1 \\ 2 & 3 \end{vmatrix} & -\begin{vmatrix} 2 & 4 \\ 2 & 0 \end{vmatrix} \\ \begin{vmatrix} 4 & -1 \\ 2 & 5 \end{vmatrix} & -\begin{vmatrix} 2 & -1 \\ 1 & 5 \end{vmatrix} & \begin{vmatrix} 2 & 4 \\ 1 & 2 \end{vmatrix} \end{bmatrix} = \frac{1}{44}\begin{bmatrix} 6 & 7 & -4 \\ -12 & 8 & 8 \\ 22 & -11 & 0 \end{bmatrix}$$

Now it has become clear why

$|\mathbf{B}| = 0$ implies that $|\mathbf{B}|$ is not invertible. Note also that if $|\mathbf{B}|$ is too close to zero, a situation called near singularity, the elements of $\mathbf{B}^{-1}$ will be very large and their computation will be less precise. This problem is of importance in econometrics and we will come back to it in the next chapter.

6.2.4 Exercises

E.6.4 Given the following matrices:

$$\begin{bmatrix} 3 & 0 \\ 0 & 7 \end{bmatrix}, \quad \begin{bmatrix} 5 & 1 \\ 0 & -6 \end{bmatrix}, \quad \begin{bmatrix} 12 & 4 \\ 9 & 8 \end{bmatrix}, \quad \begin{bmatrix} 6 & -5 \\ -12 & 10 \end{bmatrix}$$

$$\begin{bmatrix} -11 & 2 \\ 4 & 5 \end{bmatrix}, \quad \begin{bmatrix} 14 & 3 \\ 21 & -12 \end{bmatrix}, \quad \begin{bmatrix} a & 2a \\ 2a & 4a \end{bmatrix}$$

i. Compute their determinants and traces
ii. Find their ranks
iii. Find their inverses

E.6.5 Given the following matrices:

$$\begin{bmatrix} 4 & 0 & 0 \\ 0 & 5 & 0 \\ 0 & 0 & 17 \end{bmatrix}, \quad \begin{bmatrix} -13 & 0 & 0 \\ 8 & 2 & 0 \\ 9 & 10 & -19 \end{bmatrix}, \quad \begin{bmatrix} 1 & 3 & 7 \\ -2 & 9 & 4 \\ 1 & -2 & 5 \end{bmatrix}$$

$$\begin{bmatrix} 1 & -2 & 1 \\ 3 & 9 & -2 \\ 7 & 4 & 5 \end{bmatrix}, \quad \begin{bmatrix} 17 & 15 & 11 \\ 22 & -30 & 19 \\ -10 & 29 & -18 \end{bmatrix}, \quad \begin{bmatrix} 11 & -3 & 21 \\ 2 & -4 & 12 \\ 7 & 5 & -3 \end{bmatrix}$$

i. Compute their determinants and traces
ii. Find their ranks
iii. Find their inverses

E.6.6 Given the following matrices:

$$\begin{bmatrix} -5 & 7 & 1 & -2 \\ 0 & 10 & 9 & 8 \\ -1 & 4 & 4 & 3 \\ 0 & 1 & 6 & -6 \end{bmatrix}, \quad \begin{bmatrix} 8 & 1 & 5 & 13 \\ -2 & 0 & 14 & 7 \\ -3 & -4 & 0 & 1 \\ 16 & 9 & 6 & 2 \end{bmatrix}$$

i. Compute their determinants and traces
ii. Find their ranks
iii. Find their inverses

E.6.7 Check your results for Exercises 6.4 – 6.6 using Matlab.

E.6.8 Show that the following equalities hold. Use Matlab and matrices of different orders to numerically verify them.

$$(\mathbf{A}^{-1})^{-1} = \mathbf{A}$$
$$(\mathbf{A}')^{-1} = (\mathbf{A}^{-1})'$$

E.6.9 Solve the following systems of equations using matrix inversion.

i. $3x + 4y = 11,\quad -2x + y = 0,$

ii. $2x - 9y = 2.8,\quad 5x + 4y = 1.7,$

iii. $x - y = -11.5\quad x + y = -1.5$

iv.
$$7x_1 + 10x_2 - 6x_3 = -167$$
$$5x_1 - 9x_2 + x_3 = 2677$$
$$2x_1 + 5x_2 - 4x_3 = -118$$

v.
$$x_1 + 6x_2 + 3x_3 = 0$$
$$x_1 - 4x_2 + 5x_3 = 0$$
$$12x_1 + 9x_2 - 13x_3 = 0$$

vi.
$$3x_1 - 8x_2 + 10x_3 = -36$$
$$7x_2 + 12x_3 = 11$$
$$x_3 = -2$$

vii.
$$x_1 + 10x_2 - x_3 = 21.00$$
$$4x_1 - 3x_2 + 2x_3 = 5.25$$
$$6x_1 + 9x_2 + 11x_3 = 64.25$$

E.6.10 Check your results for E.6.9 using Matlab.

E.6.11 Make up square matrices of order four and use Matlab to verify the following properties of the determinant:

i. $|\mathbf{A}| = |\mathbf{A}'|$
ii. If a column or a row of a matrix consists of zeros, then the determinant is equal to zero.
iii. If two rows (columns) are identical, then the determinant is equal to zero.
iv. If a column (row) is equal to a linear combination of some of the other columns (rows), then the determinant is equal to zero.
v. If a row (column) is multiplied by λ, the determinant is multiplied by λ. Repeat your computation with $\lambda = -1$.
vi. $|\lambda\mathbf{A}| = \lambda^n |\mathbf{A}|$, where n is the order of the matrix.
vii. If the place of two columns (rows) are interchanged, the sign of determinant changes.

E.6.12 Prove *iv* and *vi* in E.6.11. Show that *ii* and *iii* are special cases of *iv*. [*Hint*: Use (6.36) and (6.37).]

6.3 Solving Systems of Linear Equations Using Matrix Algebra

By a linear system we mean a group of linear equations that hold simultaneously. The meaning of a system is different from each equation in isolation or even if we consider all of the equations without considering their connections.

Example 6.21 Consider the equation

$$y - 5x = 2$$

It means that, whatever the value of x, then y is equal to 5 times x plus 2. For example, if $x = 3$, then $y = 17$. Similarly, if we know the value of y to be equal to 32, then $x = 6$. In other words, there are infinitely many combinations of x and y that satisfy this equation. But now suppose we have the following system of equations:

$$y - 5x = 2$$
$$3y + x = 38$$

This system means $x = 2$ and $y = 12$. There is only one pair of x and y that satisfies these equations.

Example 6.22 Systems of simultaneous equations play an important role in economics. The operation of the market for a product can be modeled as

$$Q_d = 100 - 2P + 0.1Y$$
$$Q_s = -10 + 4P$$
$$Q_d = Q_s$$

where Q_d, Q_s, P, and Y are, respectively, quantity demanded, quantity supplied, price, and income. The three equations depict the point of equilibrium in the market, where quantity demanded equals quantity supplied. Note that for each level of Y, we have a different point of equilibrium or combination of Q_d, Q_s, and P that satisfy the three equations.

In economics and in other sciences nonlinear systems play significant roles. In this section, however, we are concerned only with linear systems. The designation linear doesn't mean that the equations will not involve nonlinear functions of variables, but it requires that a variable appear in all equations in only one form.

Example 6.23 The Cobb-Douglas production function has the form

$$Q = AK^{\alpha}L^{\beta}$$

where Q is output, K capital, and L labor. Taking the logarithm of both sides

$$\ln Q = \ln A + \alpha \ln K + \beta \ln L$$

The equation is linear in the arguments ln Q, ln K, and ln L. But in all other equations they have to appear in this form and, for example, we cannot have K, L, or Q in any other equation.

In econometrics a system of equations that describes the economy or a part of it is referred to as a *structural model* because it describes the structure of the economy. Coefficients of the structural models are not known, and usually economic theory imposes very few specific restrictions on these coefficients. Thus, they are free parameters and their values have to be estimated from the data or somehow gleaned from available information. The variables of the model are divided into endogenous and exogenous variables.[3] The former are those variables that are determined within the system given the values of the latter variables and coefficients. That is, the structural system is designed to determine and explain endogenous variables. Exogenous variables are determined outside the system under study.

For example, if we are studying the market for oranges, price and quantity are endogenous but income is not because it is preposterous to think that the personal disposable income of the whole country is determined by supply and demand of oranges. On the other hand, in a macroeconomic model, aggregate income is endogenous whereas government expenditures, money supply, and income of other countries are exogenous.

Example 6.24 The Keynesian model in Example 6.11 is a structural model:

$$\begin{aligned} Y &= C + I + G \\ C &= \alpha_0 + \alpha_1(Y - T) \qquad 0 < \alpha_0,\ \ 0 < \alpha_1 < 1, \\ I &= \beta_0 + \beta_1 Y + \beta_2 r \qquad 0 < \beta_0,\ \ 0 < \beta_1 < 1,\ \ \beta_2 < 0 \end{aligned}$$

If we solve this model for the endogenous variables, we get the *reduced form* of the model

$$\begin{aligned} \begin{bmatrix} Y \\ C \\ I \end{bmatrix} &= \begin{bmatrix} 1 & -1 & -1 \\ -\alpha_1 & 1 & 0 \\ -\beta_1 & 0 & 1 \end{bmatrix}^{-1} \begin{bmatrix} G \\ \alpha_0 - \alpha_1 T \\ \beta_0 + \beta_2 r \end{bmatrix} \\ &= \frac{1}{1 - \alpha_1 - \beta_1} \begin{bmatrix} 1 & 1 & 1 \\ \alpha_1 & 1 - \beta_1 & \alpha_1 \\ \beta_1 & \beta_1 & 1 - \alpha_1 \end{bmatrix} \begin{bmatrix} G \\ \alpha_0 - \alpha_1 T \\ \beta_0 + \beta_2 r \end{bmatrix} \\ &= \frac{1}{1 - \alpha_1 - \beta_1} \begin{bmatrix} \alpha_0 + \beta_0 + \beta_2 r + G - \alpha_1 T \\ \alpha_0(1 - \beta_1) + \alpha_1(\beta_0 + \beta_2 r) + \alpha_1 G - \alpha_1(1 - \beta_1)T \\ \alpha_0\beta_1 + \beta_0(1 - \alpha_1) + \beta_2(1 - \alpha_1)r + \beta_1 G - \beta_1\alpha_1 T \end{bmatrix} \end{aligned}$$

[3] A better categorization is endogenous and *predetermined* variables. The latter includes exogenous and lagged endogenous variables. In dynamic models, past values of an endogenous variable, say, consumption, have an effect on its value at the present time. But these values are known at the time of analysis or forecasting and could be treated, for certain purposes, as exogenous variables.

As can be seen, we can go from the structural model to the reduced form. In addition, a knowledge of the magnitude of the coefficients of the structural model enables us to uniquely determine the coefficients of the reduced form.

As is discussed in econometrics, the estimation of structural models poses some problems. In particular, it requires imposing restriction on some of the coefficients. But if we know the reduced form, could we uniquely determine the coefficients of the structural model?[4] The answer is "not necessarily." We again require restrictions on structural coefficients. In all sciences, including economics, a reduced form could be consistent with a number of structural models. Because the structural model represents our theory, but the reduced form is what we usually observe or we are able to estimate its parameters, it may not be always possible to definitely choose among the competing theories.

6.3.1 Cramer's Rule

In the previous section we learned how to solve a system of equations using the inverse of a matrix. An alternative way is by *Cramer's rule*, which is easier to understand through examples.

Example 6.25 Let us write the system of equations

$$6x - 7y = 5$$
$$5x + 3y = 13$$

in matrix form

$$\begin{bmatrix} 6 & -7 \\ 5 & 3 \end{bmatrix} \begin{bmatrix} x \\ y \end{bmatrix} \begin{bmatrix} 5 \\ 13 \end{bmatrix}$$

and consider the following determinants:

$$\Delta = \begin{vmatrix} 6 & -7 \\ 5 & 3 \end{vmatrix} = 53 \quad \Delta_1 = \begin{vmatrix} 5 & -7 \\ 13 & 3 \end{vmatrix} = 106 \quad \Delta_2 = \begin{vmatrix} 6 & 5 \\ 5 & 13 \end{vmatrix} = 53$$

Δ is the determinant of the matrix of the coefficients on the LHS of the system of equations.Δ_1 is formed by replacing the first column of Δ with the vector of constants on the RHS of the system. Finally, Δ_2 is formed by replacing the second column of Δ with the column of constant coefficients. Cramer's rule states that

$$x = \frac{\Delta_1}{\Delta} = 2, \qquad y = \frac{\Delta_2}{\Delta} = 1$$

Now we can state Cramer's rule in more general terms. Consider the system of equations

$$\mathbf{Ax} = \mathbf{d} \tag{6.43}$$

[4]This is the famous *identification problem* in econometrics.

Write the matrix **A** in terms of its columns

$$\mathbf{A} = [\mathbf{A}_1 \ \dots \ \mathbf{A}_k] \tag{6.44}$$

Cramer's rule states

$$x_j = \frac{|\mathbf{A}_1 \ \dots \ \mathbf{A}_{j-1} \ \mathbf{d} \ \mathbf{A}_{j+1} \ \dots \ \mathbf{A}_k|}{|\mathbf{A}|}, \qquad j = 1, \dots, k \tag{6.45}$$

where x_j is the j-th element of **x**.

Example 6.26 Let

$$\begin{bmatrix} 2 & 9 & 4 \\ 7 & -8 & 6 \\ 5 & 3 & -6 \end{bmatrix} \begin{bmatrix} x_1 \\ x_2 \\ x_3 \end{bmatrix} = \begin{bmatrix} -13 \\ 63 \\ 6 \end{bmatrix}$$

Calculating the relevant determinants, we have

$$\Delta = \begin{vmatrix} 2 & 9 & 4 \\ 7 & -8 & 6 \\ 5 & 3 & -6 \end{vmatrix} = 952 \qquad \Delta_1 = \begin{vmatrix} -13 & 9 & 4 \\ 63 & -8 & 6 \\ 6 & 3 & -6 \end{vmatrix} = 4284$$

$$\Delta_2 = \begin{vmatrix} 2 & -13 & 4 \\ 7 & 63 & 6 \\ 5 & 6 & -6 \end{vmatrix} = -2856 \qquad \Delta_3 = \begin{vmatrix} 2 & 9 & -13 \\ 7 & -8 & 63 \\ 5 & 3 & 6 \end{vmatrix} = 1190$$

Thus,

$$x_1 = \frac{4284}{952} = 4.5, \quad x_2 = \frac{-2856}{952} = -3, \quad x_3 = \frac{1190}{952} = 1.25$$

Example 6.27 Let us use Cramer's rule to solve for Y in Example 6.11

$$Y = \frac{\begin{vmatrix} G & -1 & -1 \\ \alpha_0 - \alpha_1 T & 1 & 0 \\ \beta_0 + \beta_2 r & 0 & 1 \end{vmatrix}}{1 - \alpha_1 - \beta_1} = \frac{\alpha_0 + \beta_0}{1 - \alpha_1 - \beta_1} + \frac{\beta_2 r}{1 - \alpha_1 - \beta_1} + \frac{G - \alpha_1 T}{1 - \alpha_1 - \beta_1}$$

In obvious change of notation, we can write

$$Y = \phi_0 + \phi_1 r + \phi_2 (G - \alpha_1 T), \qquad 0 < \phi_0, \ \phi_1 < 0, \ \phi_2 > 0$$

which is the IS curve. Note that since $1-\alpha_1-\beta_1>0$, the effect of each variable on Y is theoretically known. For instance $\phi_1<0$ and the IS curve is downward sloping. Whereas an increase in government expenditures results in an increase in income, the same effect will be obtained by reducing taxes. Let us now add the LM curve

$$\gamma_1 Y+\gamma_2 r=\frac{M}{P}, \qquad 0<\gamma_1,\ \gamma_2<0$$

where M is the money supply and P the price level. Our complete system, that is the IS plus LM, in matrix notation will be

$$\begin{bmatrix} 1 & -\phi_1 \\ \gamma_1 & \gamma_2 \end{bmatrix}\begin{bmatrix} Y \\ r \end{bmatrix}=\begin{bmatrix} \phi_0+\phi_2(G-\alpha_1 T) \\ M/P \end{bmatrix}$$

Again we can solve for Y using Cramer's rule:

$$Y=\frac{\begin{vmatrix} \phi_0+\phi_2(G-\alpha_1 T) & -\phi_1 \\ M/P & \gamma_2 \end{vmatrix}}{\gamma_2+\phi_1\gamma_1}=\frac{\phi_0\gamma_2+\gamma_2\phi_2(G-\alpha_1 T)+\phi_1(M/P)}{\gamma_2+\phi_1\gamma_1}$$

Again the effect of each exogenous variable on the income can be unequivocally determined.

Before closing the subject of linear systems, we should state a few general results. The linear system

$$\mathbf{Ax}=\mathbf{b}$$

has a unique solution if $\mathbf{A}$ is square, that is, there are as many equations as the number of variables we want to determine and $|\mathbf{A}|\neq 0$. It follows that under such conditions, if $\mathbf{b}=\mathbf{0}$, then $\mathbf{x}=\mathbf{0}$. When $\mathbf{b}=\mathbf{0}$, the system is called *homogeneous*.

On the other hand, if the determinant of $\mathbf{A}$ is equal to zero, then we have infinitely many solutions. The same is true if the number of equations is less than the number of variables. In contrast, if there are more equations than unknowns, the system has a unique solution if there are enough redundant equations—that is, equations that are linear combinations of other equations in the system—so that by discarding them we can get a square matrix $\mathbf{A}$. But if equations are contradictory, then the system has no solution.

Example 6.28 Below (i) has a redundant equation, but (ii) is contradictory.

$$\text{i.}\quad \begin{aligned} 2x+3y&=4 \\ 7x+4y&=1 \\ -3x+2y&=7 \end{aligned} \qquad\qquad \text{ii.}\quad \begin{aligned} 2x+3y&=4 \\ 7x+4y&=1 \\ x+y&=5 \end{aligned}$$

Cramer's rule can be programmed in Matlab

Matlab code

```
% Specify the coefficients matrix and the RHS vector
A = [2 9 4; 7 -8 6; 5 3 -6]
d = [-13; 63; 6]
% Calculate the relevant determinants
delta = det(A)
delta1 = det([d A(:,2) A(:,3)])
delta2 = det([A(:,1) d A(:,3)])
delta3 = det([A(:,1) A(:,2) d])
% Find the solutions
x1 = delta1./delta
x2 = delta2./delta
x3 = delta3./delta
```

6.3.2 Exercises

E.6.13 Solve the problems in E.6.9 using Cramer's rule.

E.6.14 Check your results in E.6.13 using Matlab.

E.6.15 Use Cramer's rule to solve for C and I in the Keynesian model of Example 6.24.

E.6.16 Use Cramer's rule to solve for r in the IS-LM model:

$$\begin{bmatrix} 1 & -\phi_1 \\ \gamma_1 & \gamma_2 \end{bmatrix} \begin{bmatrix} Y \\ r \end{bmatrix} = \begin{bmatrix} \phi_0 + \phi_2(G - \alpha_1 T) \\ M/P \end{bmatrix}$$

Chapter 7
Advanced Topics in Matrix Algebra

In this chapter we build on the basic matrix theory we learned in Chap. 6 and present a number of advanced topics. The techniques and tools we acquire will prove quite useful in many areas of mathematics, economics, computation, and particularly, econometrics. The knowledge of eigenvalues and eigenvectors in Sect. 7.4 are crucial for understanding and solving systems of differential equations in Chap. 17. Such equations, in turn, play an important role in macroeconomic analysis. We shall learn several ways of factoring a matrix into two matrices, techniques which are of immense importance for efficient computation.

Topics discussed in this chapter include the Moore-Penrose generalized inverse of a matrix, positive and negative definite matrices, projection matrices, decomposition of a positive definite matrix, orthogonal complements, Cholesky factorization, and others, all of which find many applications in and are indeed indispensable tools of econometric analysis. The chapter, in addition to being expository, is also intended as a reference for students and practitioners of econometrics.

7.1 Quadratic Forms and Positive and Negative Definite Matrices

When the exponents of every term in an expression sum to the same number, the expression is referred to as a *form*. The following are forms:

$$f(x,y) = ax + by$$

$$f(x,y) = x^2 - 3xy + 2y^2$$

$$f(x,y,z) = a_{11}x^2 + a_{22}y^2 + a_{33}z^2 + 2a_{12}xy + 2a_{13}xz + 2a_{23}yz$$

$$f(x,y,z) = \alpha x^3 y + \beta y^2 z^2 + \gamma x^2 yz + \delta y^4 + \lambda xyz^2$$

The first is a linear form and the fourth a quartic (that is, of degree four) form. We are interested in the second and third expressions that are quadratic (i.e., of degree two) forms. We can rewrite them as

$$f(x,y) = \begin{bmatrix} x\ y \end{bmatrix} \begin{bmatrix} 1 & -3 \\ 0 & -2 \end{bmatrix} \begin{bmatrix} x \\ y \end{bmatrix}$$

K. Dadkhah, *Foundations of Mathematical and Computational Economics,* 2nd ed.,
DOI 10.1007/978-3-642-13748-8_7, © Springer-Verlag Berlin Heidelberg 2011

and

$$f(x, y, z) = \begin{bmatrix} x & y & z \end{bmatrix} \begin{bmatrix} a_{11} & a_{12} & a_{13} \\ a_{12} & a_{22} & a_{23} \\ a_{13} & a_{23} & a_{33} \end{bmatrix} \begin{bmatrix} x \\ y \\ z \end{bmatrix}$$

More generally, the quadratic form can be written as

$$f(\mathbf{x}) = \mathbf{x}'\mathbf{A}\mathbf{x} \tag{7.1}$$

where $\mathbf{x}$ is a k-vector, and $\mathbf{A}$ is a $k \times k$ matrix. It is always possible to make $\mathbf{A}$ symmetric. In the example above the second matrix is symmetric, and the first can be made symmetric by writing it as

$$f(x, y) = \begin{bmatrix} x & y \end{bmatrix} \begin{bmatrix} -1 & -\frac{3}{2} \\ -\frac{3}{2} & -2 \end{bmatrix} \begin{bmatrix} x \\ y \end{bmatrix}$$

Now we can define positive and negative definite matrices.

Definition 7.1 A symmetric matrix $\mathbf{A}$ is called

$$\begin{array}{ll} \text{positive definite if} & \mathbf{x}'\mathbf{A}\mathbf{x} > 0, \quad \forall \mathbf{x} \neq \mathbf{0} \\ \text{positive semidefinite if} & \mathbf{x}'\mathbf{A}\mathbf{x} \geq 0, \quad \forall \mathbf{x} \neq \mathbf{0} \\ \text{negative definite if} & \mathbf{x}'\mathbf{A}\mathbf{x} < 0, \quad \forall \mathbf{x} \neq \mathbf{0} \\ \text{negative semidefinite if} & \mathbf{x}'\mathbf{A}\mathbf{x} \leq 0, \quad \forall \mathbf{x} \neq \mathbf{0} \end{array} \tag{7.2}$$

But, how do we know if a matrix is positive or negative definite? For this we rely on *principal minors*. For any matrix

$$\mathbf{A} = \begin{bmatrix} a_{11} & a_{12} & \dots & a_{1n} \\ a_{21} & a_{22} & \dots & a_{2n} \\ \vdots & \vdots & \ddots & \vdots \\ a_{n1} & a_{n2} & \dots & a_{nn} \end{bmatrix}$$

the increasingly larger determinants

$$a_{11}, \qquad \begin{vmatrix} a_{11} & a_{12} \\ a_{21} & a_{22} \end{vmatrix}, \qquad \begin{vmatrix} a_{11} & a_{12} & a_{13} \\ a_{21} & a_{22} & a_{23} \\ a_{31} & a_{32} & a_{33} \end{vmatrix}, \qquad \dots, \qquad |\mathbf{A}|$$

are called the principal minors. The symmetric matrix **A** is positive definite if all its principal minors are positive.

$$a_{11} > 0, \quad \begin{vmatrix} a_{11} & a_{12} \\ a_{21} & a_{22} \end{vmatrix} > 0, \quad \begin{vmatrix} a_{11} & a_{12} & a_{13} \\ a_{21} & a_{22} & a_{23} \\ a_{31} & a_{32} & a_{33} \end{vmatrix} > 0, \quad \ldots \quad |\mathbf{A}| > 0 \tag{7.3}$$

If one or more (but not all) of the above are equal to zero, then the matrix is positive semidefinite. The symmetric matrix **A** is negative definite if its principal minors alternate in sign starting with negative.

$$a_{11} < 0 \quad \begin{vmatrix} a_{11} & a_{12} \\ a_{21} & a_{22} \end{vmatrix} > 0 \quad \begin{vmatrix} a_{11} & a_{12} & a_{13} \\ a_{21} & a_{22} & a_{23} \\ a_{31} & a_{32} & a_{33} \end{vmatrix} < 0 \quad \ldots \tag{7.4}$$

If one or more (but not all) of the above are equal to zero, then the matrix is negative semidefinite.

Example 7.1 Matrices

$$\begin{bmatrix} 9 & 2 \\ 2 & 7 \end{bmatrix} \quad \text{and} \quad \begin{bmatrix} 14 & 05 & -1 \\ 05 & 14 & -1 \\ -1 & -1 & 0.2 \end{bmatrix}$$

are positive definite whereas

$$\begin{bmatrix} -7 & 0.5 \\ 0.5 & -6 \end{bmatrix} \quad \text{and} \quad \begin{bmatrix} -29 & -4 & -13 \\ -\,04 & -6 & -02 \\ -\,13 & -2 & -11 \end{bmatrix}$$

are negative definite.

Example 7.2 Consider the n-vector of random variables

$$\mathbf{y} = \begin{bmatrix} y_1 \\ y_2 \\ \vdots \\ y_n \end{bmatrix} \tag{7.5}$$

and let

$$\tilde{\mathbf{y}} = \begin{bmatrix} y_1 - E(y_1) \\ y_2 - y(y_2) \\ \vdots \\ y_n - E(y_n) \end{bmatrix} \tag{7.6}$$

The covariance matrix of $\mathbf{y}$ is defined as

$$\mathbf{V} = E\begin{bmatrix} \tilde{y}_1^2 & \tilde{y}_1\tilde{y}_2 & \dots & \tilde{y}_1\tilde{y}_n \\ \tilde{y}_2\tilde{y}_1 & \tilde{y}_2^2 & \dots & \tilde{y}_2\tilde{y}_n \\ \vdots & \vdots & \ddots & \vdots \\ \tilde{y}_n\tilde{y}_1 & \tilde{y}_n\tilde{y}_2 & \dots & \tilde{y}_n^2 \end{bmatrix} \tag{7.7}$$

The covariance matrix is symmetric and positive semidefinite. To show the latter, first note that $\mathbf{V}$ can be written as

$$\mathbf{V} = E(\tilde{\mathbf{y}}\tilde{\mathbf{y}}') \tag{7.8}$$

Thus, for any $\mathbf{x} \neq \mathbf{0}$ we have

$$\begin{aligned} \mathbf{x}'\mathbf{V}\mathbf{x} &= \mathbf{x}'E(\tilde{\mathbf{y}}\tilde{\mathbf{y}}')\mathbf{x} \\ &= E\mathbf{x}'(\tilde{\mathbf{y}}\tilde{\mathbf{y}}')\mathbf{x} \\ &= E\left(\sum_{j=1}^{n} x_j\tilde{y}_j\right)^2 \\ &= E\left(\sum_{j=1}^{n} x_j(y_j - E(y_j))\right)^2 \geq 0 \end{aligned} \tag{7.9}$$

Note that for all practical purposes, the covariance matrix is positive definite. The equality to zero in the above formula can be ignored as the trivial case when there is no variation in $\mathbf{y}$. Such a situation is of no interest in statistics and econometrics.

7.1.1 Exercises

E.7.1 Determine which of the following matrices is positive or negative definite and which are positive or negative semidefinite.

$$\mathbf{A} = \begin{bmatrix} 6 & -6 & 12 \\ 6 & -6 & -12 \\ 12 & -12 & 24 \end{bmatrix}, \quad \mathbf{B} = \begin{bmatrix} -6 & +6 & -12 \\ -6 & -6 & 12 \\ -12 & 12 & -24 \end{bmatrix}, \quad \mathbf{C} = \begin{bmatrix} 3 & -6 & -2 \\ -6 & 18 & 5 \\ -2 & 5 & 14 \end{bmatrix}$$

$$\mathbf{D} = \begin{bmatrix} -3 & 6 & 2 \\ 6 & -18 & -5 \\ 2 & -5 & -14 \end{bmatrix}, \quad \mathbf{E} = \begin{bmatrix} 14 & 3 & 7 \\ 3 & 0 & 11 \\ 7 & 11 & 9 \end{bmatrix}$$

$$\mathbf{F} = \begin{bmatrix} 10 & 4 & -5 & 5 \\ 4 & 20 & 2 & 11 \\ -5 & 2 & 18 & -1 \\ 5 & 11 & -1 & 7 \end{bmatrix}, \quad \mathbf{G} = \begin{bmatrix} -10 & -4 & 5 & -5 \\ -4 & -20 & -2 & -11 \\ 5 & -2 & -18 & 1 \\ -5 & -11 & 1 & -7 \end{bmatrix}$$

E.7.2 Show that the identity matrix $\mathbf{I}$ is positive definite.

E.7.3 Show that any matrix of the form $\mathbf{X}'\mathbf{X}$ is positive semidefinite.

E.7.4 Show that if $\mathbf{A}$ is positive definite, then $-\mathbf{A}$ is negative definite and vice versa.

7.2 Generalized Inverse of a Matrix

In Chap. 6, we defined the inverse of a square nonsingular matrix $\mathbf{A}$ to be $\mathbf{A}^{-1}$ such that

$$\mathbf{A}\mathbf{A}^{-1} = \mathbf{A}^{-1}\mathbf{A} = \mathbf{I} \tag{7.10}$$

For the nonsquare matrix $\mathbf{A}$ with dimensions $n \times k$ where $k < n$, the *generalized inverse* is defined as $\mathbf{A}^{-}$ such that

$$\mathbf{A}\mathbf{A}^{-}\mathbf{A} = \mathbf{A} \tag{7.11}$$

The generalized inverse is not unique. Let $\mathbf{B}$ be any $n \times k$ matrix such that the matrix product $\mathbf{B}'\mathbf{A}$ is not singular and its inverse exists. Then

$$(\mathbf{B}'\mathbf{A})^{-1}\mathbf{B}' \tag{7.12}$$

is also a generalized inverse of $\mathbf{A}$.A more useful concept is the *Moore-Penrose generalized inverse* denoted by $\mathbf{A}^{+}$. The Moore-Penrose generalized inverse is defined only for matrices with full column rank. Such an inverse has to satisfy the following conditions:

$$\begin{aligned} \mathbf{A}\mathbf{A}^{+}\mathbf{A} &= \mathbf{A} \\ \mathbf{A}^{+}\mathbf{A}\mathbf{A}^{+} &= \mathbf{A}^{+} \\ (\mathbf{A}\mathbf{A}^{+})' &= \mathbf{A}\mathbf{A}^{+} \\ (\mathbf{A}^{+}\mathbf{A})' &= \mathbf{A}^{+}\mathbf{A} \end{aligned} \tag{7.13}$$

Based on these conditions, the Moore-Penrose generalized inverse of the $n \times k$ matrix $\mathbf{A}$ is uniquely defined as

$$\mathbf{A}^{+} = (\mathbf{A}'\mathbf{A})^{-1}\mathbf{A}' \tag{7.14}$$

where $\mathbf{A}$ has full column rank. It is left as an exercise for the reader to check that (7.14) indeed has the properties listed in (7.13) and that $\mathbf{A}^{+}$ is a special case of $\mathbf{A}^{-}$. See Exercises E.7.6 and E.7.8.

Example 7.3 The Moore-Penrose generalized inverse of the matrix

$$\mathbf{X} = \begin{bmatrix} 1 & -5 & 0 \\ 1 & 7 & 1 \\ 1 & -3 & 1 \\ 1 & 4 & 0 \\ 1 & -3 & 0 \\ 1 & -1 & 1 \end{bmatrix}$$

is

$$\mathbf{X}^{+} = \begin{bmatrix} -0.2848 & 0.0795 & -0.0530 & -0.4040 & -0.3113 & -0.0265 \\ -0.0364 & 0.0596 & -0.0397 & -0.0530 & -0.0166 & -0.0199 \\ -0.2483 & 0.1943 & -0.4260 & -0.4570 & -0.2947 & -0.3797 \end{bmatrix}$$

Use Matlab to verify the above result.

Example 7.4 (Least Squares Method). Least squares method is one of the principal techniques of estimation in econometrics. You may be familiar with its derivation as the solution to an optimization problem (see Chap. 12 for such a derivation). Here, we would like to introduce it somewhat differently, that is, as an approximation to the solution of a system of equations. Let us start with the standard classical model of regression

$$\mathbf{y} = \mathbf{X}\boldsymbol{\beta} + \mathbf{u} \tag{7.15}$$

where $\mathbf{y}$ is an $n \times 1$ vector of dependent variable, $\mathbf{x}$ an $n \times k$ matrix of explanatory variables, $\boldsymbol{\beta}$ a $k \times 1$ vector of unknown parameters, and $\mathbf{u}$ an $n \times 1$ vector of unobservable random disturbances. For instance, $\mathbf{y}$ may represent data on the consumption of n households in the country or in a particular city. The elements of the first column of $\mathbf{X}$ are all one, representing the constant term in the relationship between consumption and its determinants. Other columns of $\mathbf{X}$ are factors such as income, wealth, and the number of individuals in the household that determine the amount of a household's consumption.

The objective is to estimate $\boldsymbol{\beta}$ and get an order of magnitude of the effect of different variables on household consumption. Because $\mathbf{X}$ is not square, it cannot be inverted and we have to estimate $\boldsymbol{\beta}$ by some method. The least squares method achieves this by minimizing the quadratic objective function

$$(\mathbf{y} - \mathbf{X}\boldsymbol{\beta})'(\mathbf{y} - \mathbf{X}\boldsymbol{\beta}) \tag{7.16}$$

Alternatively, we could premultiply (7.15) by $\mathbf{X}'$

$$\mathbf{X}'y = \mathbf{X}'\mathbf{X}\boldsymbol{\beta} + \mathbf{X}'\mathbf{u} \tag{7.17}$$

and then solve for $\boldsymbol{\beta}$.

$$\boldsymbol{\beta} = (\mathbf{X}'\mathbf{X})^{-1}\mathbf{X}'\mathbf{y} - (\mathbf{X}'\mathbf{X})^{-1}\mathbf{X}'\mathbf{u} \tag{7.18}$$

In other words, we have solved for $\boldsymbol{\beta}$ using the Moore-Penrose generalized inverse of matrix $\mathbf{X}$:

$$\boldsymbol{\beta} = \mathbf{X}^{+}\mathbf{y} - \mathbf{X}^{+}\mathbf{u} \tag{7.19}$$

Because $\mathbf{u}$ is unobservable, there is no way that we can find the exact value of $\boldsymbol{\beta}$. But we can estimate (approximate) it by taking the first term in (7.18) and dropping the second term, which involves $\mathbf{u}$, and write

$$\hat{\boldsymbol{\beta}} = \mathbf{X}^{+}\mathbf{y} = (\mathbf{X}'\mathbf{X})^{-1}\mathbf{X}'\mathbf{y} \tag{7.20}$$

Under the following assumptions,

1. The stochastic process $\{\mathbf{X}, \mathbf{y}\}$ follows the linear model in (7.15).
2. $$E(\mathbf{u}|\mathbf{X}) = \mathbf{0} \tag{7.21}$$
3. Except for the first column of $\mathbf{X}$, no other variable is constant and $\mathbf{X}$ has full column rank (no perfect collinearity).

$\hat{\boldsymbol{\beta}}$ is an unbiased estimator of $\boldsymbol{\beta}$. That is,

$$E(\hat{\boldsymbol{\beta}}) = \boldsymbol{\beta} \tag{7.22}$$

Example 7.5 (Gauss[1]-Markov Theorem). If we add the following to assumptions 1–3 above,

$$E(\mathbf{u}\mathbf{u}') = \sigma^2\mathbf{I} \tag{7.23}$$

we shall have the Gauss-Markov theorem, which states that $\hat{\boldsymbol{\beta}}$ is the best linear unbiased estimator of $\boldsymbol{\beta}$ conditional on $\mathbf{X}$. That is, in the class of linear unbiased estimators, it has the lowest variance. More precisely, the covariance matrix of any other estimator in this class exceeds that of $\hat{\boldsymbol{\beta}}$ by a positive semidefinite matrix.

---

[1] The great German mathematician Johann Carl Friedrich Gauss (1777–1855) made contributions to many areas of mathematics including algebra, number theory, geometry, and differential equations.

7.2.1 Exercises

E.7.5 Using Matlab, find the Moore-Penrose generalized inverse of the following matrices:

$$\mathbf{A} = \begin{bmatrix} 3 & 12 & -1 \\ 3 & 14 & 0 \\ 3 & 18 & -1 \\ 5 & 23 & 0 \\ 5 & 28 & 0 \\ 7 & 34 & -1 \\ 7 & 40 & 1 \end{bmatrix}, \quad \mathbf{B} = \begin{bmatrix} 17 & 4 & 1 \\ 15 & 7 & 1 \\ 13 & 6 & 2 \\ 10 & 6 & 2 \\ 7 & 8 & 2 \\ 5 & 4 & 3 \\ 3 & 3 & 3 \end{bmatrix}, \quad \mathbf{C} = \begin{bmatrix} 1 & -32 & 10 & 3 \\ 1 & -33 & 15 & 7 \\ 1 & -33 & 20 & 0 \\ 1 & -34 & 25 & 4 \\ 1 & -38 & 30 & 9 \\ 1 & -34 & 35 & 0 \\ 1 & -30 & 40 & 1 \end{bmatrix}$$

E.7.6 Show that the matrix in (7.14) satisfies all the conditions in (7.13).

E.7.7 Verify (7.12).

E.7.8 Show that $\mathbf{X}^{+}$ is a special case of $\mathbf{X}^{-}$ by showing that if$\mathbf{x}$ has full column rank, then $\mathbf{X}^{-} = \mathbf{X}^{+}$. Further show that $\mathbf{X}^{-1}$ is a special case of $\mathbf{X}^{+}$ by showing that when $\mathbf{X}$ is not singular, then $\mathbf{X}^{+} = \mathbf{X}^{-1}$.

E.7.9 Prove the Gauss-Markov theorem in the following steps:

i. Define an alternative linear estimator of $\hat{\boldsymbol{\beta}}$ as

$$\boldsymbol{\beta}^{*} = \mathbf{A}\mathbf{y}$$

ii Derive the conditions under which $\boldsymbol{\beta}^{*}$ is an unbiased estimator.
iii. Find the covariance matrix of this estimator (call it$\sigma^{2}\mathbf{H}$).
iv. Show that $\mathbf{H} - (\mathbf{X}'X)^{-1}$ is equal to a positive semidefinite matrix.

7.3 Orthogonal Matrices

7.3.1 Orthogonal Projection

Orthogonal projection is related both to the Moore-Penrose generalized inverse and the least squares method, and perhaps it is easier to introduce it in the context of the latter. Recall that in (7.15) we tried to model an economic phenomenon, say, the household consumption. After estimating $\boldsymbol{\beta}$, we may want to see how the model predicts the amount of consumption, given the characteristics of household such as size and income. Denoting the model prediction of $\mathbf{y}$ by $\hat{\mathbf{y}}$, we have

$$\hat{\mathbf{y}} = \mathbf{X}\hat{\boldsymbol{\beta}} = \mathbf{X}(\mathbf{X}'\mathbf{X})^{-1}\mathbf{X}'\mathbf{y} \tag{7.24}$$

The matrix

$$\mathbf{P_x} = \mathbf{X}(\mathbf{X}'\mathbf{X})^{-1}\mathbf{X}' \tag{7.25}$$

is called *the orthogonal projector*, that is, it projects $\mathbf{y}$ on the matrix $\mathbf{X}$. In statistical sense, all the information that is common between $\mathbf{y}$ and $\mathbf{X}$ is reflected in $\hat{\mathbf{y}} = \mathbf{P_x y} = \mathbf{X}\hat{\boldsymbol{\beta}}$. The difference between $\mathbf{y}$ and its predicted value is called the *residual*; thus,

$$\begin{aligned} \mathbf{e} &= \mathbf{y} - \hat{\mathbf{y}} \\ &= \mathbf{y} - \mathbf{X}\hat{\boldsymbol{\beta}} \\ &= [\mathbf{I} - \mathbf{X}(\mathbf{X}'\mathbf{X})^{-1}\mathbf{X}']\,\mathbf{y} \end{aligned} \tag{7.26}$$

Vectors $\hat{\mathbf{y}}$ and $\mathbf{e}$ have several interesting characteristics that are left to the reader to verify (see E.7.11 and E.7.12).

$$\mathbf{X}'\mathbf{e} = \mathbf{0} \qquad \Rightarrow \qquad \hat{\mathbf{y}}\mathbf{e} = \mathbf{0} \tag{7.27}$$

That is, the orthogonal projector $\mathbf{P_X}$ breaks the vector $\mathbf{y}$ into two orthogonal vectors. Because $\mathbf{y} = \hat{\mathbf{y}} + \mathbf{e}$, it follows that

$$\mathbf{y}'\mathbf{y} = \hat{\mathbf{y}}'\hat{\mathbf{y}} + \mathbf{e}'\mathbf{e} \tag{7.28}$$

and

$$\|\mathbf{y}\| = \|\hat{\mathbf{y}}\| + \|\mathbf{e}\| \tag{7.29}$$

Example 7.6 Let

$$\mathbf{X} = \begin{bmatrix} 1 & -5 & 0 \\ 1 & -7 & 1 \\ 1 & -3 & 1 \\ 1 & -4 & 0 \\ 1 & -3 & 0 \\ 1 & -1 & 1 \end{bmatrix}, \qquad \mathbf{y} = \begin{bmatrix} -0.9 \\ -6.2 \\ -1.1 \\ -5.8 \\ -1.1 \\ -2.5 \end{bmatrix}$$

Then

$$\mathbf{P_X} = \begin{bmatrix} -0.4669 & -0.2185 & -0.1457 & -0.1391 & -0.3940 & -0.0728 \\ -0.4669 & -0.2185 & -0.1457 & -0.1391 & -0.3940 & -0.0728 \\ -0.2185 & -0.6909 & -0.0949 & -0.3179 & -0.0993 & -0.2141 \\ -0.1457 & -0.0949 & -0.4923 & -0.2119 & -0.0662 & -0.4128 \\ -0.1391 & -0.3179 & -0.2119 & -0.6159 & -0.2450 & -0.1060 \\ -0.3940 & -0.0993 & -0.0662 & -0.2450 & -0.3609 & -0.0331 \\ -0.0728 & -0.2141 & -0.4128 & -0.1060 & -0.0331 & -0.3731 \end{bmatrix}$$

and

$$\hat{\mathbf{y}} = \begin{bmatrix} -0.1927 \\ 6.8547 \\ 0.8746 \\ 5.1894 \\ 1.0033 \\ 2.0706 \end{bmatrix}, \qquad \mathbf{e} = \begin{bmatrix} 0.7073 \\ -0.6547 \\ 0.2254 \\ 0.6106 \\ 0.0967 \\ 0.4294 \end{bmatrix}$$

It is left to the reader to verify that the above results satisfy conditions (7.21) and (7.22).

7.3.2 Orthogonal Complement of a Matrix

The concept of *orthogonal complement* proves especially handy in the study of cointegration in time series analysis. Consider the matrix $\mathbf{X}$ whose dimensions are $n \times k$ where $k < n$ and has rank k. Then $\mathbf{X}_\perp$ whose dimensions are $n \times (n-k)$ and has the rank $n-k$ is called the *orthogonal complement* of $\mathbf{X}$ if

$$\mathbf{X}'\mathbf{X}_\perp = \mathbf{0} \tag{7.30}$$

Where $\mathbf{0}$ has dimensions $k \times (n-k)$. It follows that $\mathbf{X}'_\perp\mathbf{X} = \mathbf{0}$ where now $\mathbf{0}$ has dimensions $(n-k) \times k$.

Example 7.7 Let

$$\mathbf{X} = \begin{bmatrix} 1 & 2 \\ 1 & 3 \\ 1 & 4 \end{bmatrix}$$

Then an orthogonal complement of $\mathbf{X}$ is

$$\mathbf{X}_\perp = \begin{bmatrix} 0.1667 \\ -0.3333 \\ 0.1667 \end{bmatrix}$$

The orthogonal complement of a matrix is not unique. For example,

$$\mathbf{X}_\perp = \begin{bmatrix} -0.3333 \\ 0.6667 \\ -0.3333 \end{bmatrix}$$

is also an orthogonal complement of **X**. The practical way to find $\mathbf{X}_\perp$ for an $n \times k$ matrix where $n > k$ is to form the matrix

$$\mathbf{T} = \mathbf{I} - \mathbf{X}(\mathbf{X}'\mathbf{X})^{-1}\mathbf{X}' \tag{7.31}$$

Then the $n \times (n-k)$ matrix $\mathbf{X}_\perp$ can be formed from any $n-k$ columns of **T**.

7.3.3 Exercises

E.7.10 Find the orthogonal projectors of matrices **A**, **B**, and **C** in E.7.5.

E.7.11 Given the vector

$$\mathbf{y} = \begin{bmatrix} 28 \\ 28 \\ 24 \\ 19 \\ 16 \\ 10 \\ 6 \end{bmatrix}$$

find the projection of **y** on **A**, **B**, and **C** in E.7.5.

E.7.12 For each of the projections above, calculate the residual vector **e** and verify that relationships (7.27) and (7.28) hold.

E.7.13 Verify that relationships (7.27) and (7.28) hold for predicted values and residuals in Example 7.6.

E.7.14 Show that

$$\mathbf{X}'\mathbf{e} = \mathbf{0}$$

and

$$\mathbf{y}'\mathbf{y} = \hat{\mathbf{y}}'\hat{\mathbf{y}} + \mathbf{e}'\mathbf{e}$$

E.7.15 Find orthogonal complements of **A**, **B**, and **C** in E.7.5.

7.4 Eigenvalues and Eigenvectors

Dear reader, some mathematical concepts have immediate physical and economic counterparts. The concept of derivatives that we shall learn in the next chapter can be illustrated by the speed of a moving object or by marginal cost and marginal revenue in microeconomics. Not all mathematical concepts are so fortunate. Eigenvalues and eigenvectors that we shall learn about in this chapter have physical counterparts; for example, frequencies of a vibrating string are eigenvalues. But there are no easy and intuitively appealing (at least for economists) examples for them. Similarly, as we shall see, eigenvectors have a counterpart in Markov chains, but that requires some explanation. Therefore, the following material, at least in the beginning, will look too abstract. But fasten your seat belt and bear with me because this is one of the most fruitful parts of any book you may read.

Eigenvalues and eigenvectors in themselves are of importance and immense help in matrix algebra, and what is more important, they allow a factoring of matrices into three, and in the case of certain matrices into two (something akin to taking the square root of a matrix). This device greatly facilitates the understanding of many estimation techniques in econometrics and the solution of systems of linear differential equations (Chap. 17).

Definition 7.2 Let $\mathbf{A}$ be a square matrix of order n. If we could find a scalar λ and a vector $\mathbf{x}$ such that

$$\mathbf{Ax} = \lambda\mathbf{x} \tag{7.32}$$

or

$$(\mathbf{A} - \lambda\mathbf{I})\mathbf{x} = \mathbf{0} \tag{7.33}$$

then λ and $\mathbf{x}$ are called, respectively, eigenvalue and eigenvector of $\mathbf{A}$. They are also referred to as characteristic value and characteristic vector of $\mathbf{A}$.

As (7.33) shows, we have a system of linear homogeneous equations. Therefore, if the matrix $(\mathbf{A} - \lambda\mathbf{I})$ is invertible, then we have the trivial solution $\mathbf{x} = \mathbf{0}$. We don't want this; therefore, we force $\mathbf{A} - \lambda\mathbf{I}$ to be singular, that is, set its determinant equal to zero:

$$|\mathbf{A} - \lambda\mathbf{I}| = 0 \tag{7.34}$$

$P(\lambda) = |\mathbf{A} - \lambda\mathbf{I}|$ is a polynomial of degree n, called a characteristic polynomial of $\mathbf{A}$, and (7.34) determines all eigenvalues of $\mathbf{A}$. Corresponding to each eigenvalue is an eigenvector that is determined through (7.33). Note, however, that eigenvectors are not unique and they are determined up to a scalar. Before we hear ringing in our ears, let us have a few numerical examples, first, the old-fashioned way and then with the help of Matlab.

Example 7.8 Find all eigenvalues and eigenvectors of the matrix

$$\mathbf{A} = \begin{bmatrix} 3 & 6 \\ 4 & 1 \end{bmatrix}$$

Using (7.34), we have

$$|\mathbf{A} - \lambda\mathbf{I}| = \left| \begin{bmatrix} 3\,6 \\ 4\,1 \end{bmatrix} - \lambda \begin{bmatrix} 1\,0 \\ 0\,1 \end{bmatrix} \right| = \begin{vmatrix} 3-\lambda & 6 \\ 4 & 1-\lambda \end{vmatrix}$$

Setting the characteristic polynomial equal to zero, we have

$$P(\lambda) = \lambda^2 - 4\lambda - 21 = 0$$

therefore,

$$\lambda_1 = 7, \qquad \lambda_2 = -3$$

As long as we are dealing with 2×2 or 3×3 matrices, computation of eigenvalues is not difficult. But once we are dealing with matrices of higher order, especially when matrix elements are not all integers, the process borders on the macabre. Indeed, numerical calculation of eigenvalues follows an iterative procedure. Fortunately, Matlab has a simple routine by which eigenvalues (and later we will see eigenvectors) can be obtained.

Matlab code

```
% Define a matrix
A = [3 6; 4 1]
% Find its eigenvalues
D = eig(A)
```

Example 7.9 Try the code on the following matrix:

$$\mathbf{B} = \begin{bmatrix} 39 & 3 & 26 & 36 \\ 3 & 6 & 2 & 7 \\ 26 & 2 & 51 & 9 \\ 36 & 7 & 9 & 50 \end{bmatrix}$$

Example 7.10 Find the eigenvalues of the following matrix:

$$\mathbf{C} = \begin{bmatrix} 2 & 0 \\ 0 & 5 \end{bmatrix}$$

Again using (7.34), we find

$$\lambda_1 = 2 \qquad \text{and} \qquad \lambda_2 = 5$$

Although this is just an example, it illustrates the general proposition that eigenvalues of a diagonal matrix are its diagonal elements. The same is true for all upper and lower triangular matrices. Once we have the eigenvalues of a matrix, (7.33) enables us to compute its eigenvectors. Let us illustrate this with a few examples.

Example 7.11 Find the eigenvalues of matrix **A** in Example 7.8. Using (7.33), we have

$$\begin{bmatrix} 3-7 & 6 \\ 4 & 1-7 \end{bmatrix} \begin{bmatrix} x_{11} \\ x_{12} \end{bmatrix} = \begin{bmatrix} 0 \\ 0 \end{bmatrix}$$

or

$$-4x_{11} + 6x_{12} = 0$$
$$4x_{11} - 6x_{12} = 0$$

As can be seen, we have only one equation and two unknowns. We can choose one of the two elements of the eigenvector arbitrarily and the other will be determined. For instance, if we choose $x_{12} = 2$, then the eigenvector corresponding to $\lambda_1 = 7$ will be

$$\mathbf{x}_1 = \begin{bmatrix} 3 \\ 2 \end{bmatrix}$$

Repeating the procedure with the second eigenvalue $\lambda_2 = -3$, we get

$$x_{21} = -x_{22}$$

Setting $x_{22} = 1$ we have

$$\mathbf{x}_2 = \begin{bmatrix} -1 \\ 1 \end{bmatrix}$$

Example 7.12 The reader should repeat the procedure in Example 7.11 and show that the eigenvectors of **C** in Example 7.10 are

$$\mathbf{x}_1 = \begin{bmatrix} 1 \\ 0 \end{bmatrix} \quad \text{and} \quad \mathbf{x}_2 = \begin{bmatrix} 0 \\ 1 \end{bmatrix}$$

Because eigenvectors are not unique, sometimes a normalization rule is adopted to uniquely determine them. The most widely used normalization rule is to make the length of the vector unity. That is, for any eigenvector we have

$$\sqrt{\mathbf{x}'\mathbf{x}} = \left(\sum_{i=1}^{n} x_i^2\right)^{\frac{1}{2}} = 1 \tag{7.35}$$

Example 7.13 In Example 7.11, the first eigenvector was

$$\mathbf{x}_1 = \begin{bmatrix} 3 \\ 2 \end{bmatrix}$$

the length of which is

$$\sqrt{\mathbf{x}'_1\mathbf{x}_1} = \sqrt{13}$$

We apply the normalization rule in (7.35) by dividing the eigenvector by its length. Thus, the normalized eigenvector is

$$\mathbf{x}_1 = \frac{1}{\sqrt{13}}\begin{bmatrix} 3 \\ 2 \end{bmatrix} = \begin{bmatrix} \frac{3\sqrt{13}}{13} \\ \frac{2\sqrt{13}}{13} \end{bmatrix}$$

Repeating the same procedure for the second eigenvalue, we get

$$\mathbf{x}_2 = \begin{bmatrix} -\frac{\sqrt{2}}{2} \\ \frac{\sqrt{2}}{2} \end{bmatrix}$$

and putting the two eigenvectors together, we have

$$\mathbf{P} = \begin{bmatrix} \frac{3\sqrt{13}}{13} & -\frac{\sqrt{2}}{2} \\ \frac{2\sqrt{13}}{13} & \frac{\sqrt{2}}{2} \end{bmatrix}$$

If calculating eigenvalues is difficult, computation of eigenvectors is even more so. Again, a number of algorithms exist for this purpose, and one can program them or use a ready-made routine. Here, we utilize Matlab, which already has such a routine. Note that in calculating eigenvectors, Matlab normalizes their lengths to unity.

Matlab code

```
% Define a matrix
A = [3 6; 4 1]
% Find its eigenvalues and eigenvectors
[P, D]= eig(A)
[0]
```

If we run this program on Matlab, we get matrix **D** whose diagonal elements are eigenvalues of **A**:

$$\mathbf{D} = \begin{bmatrix} 7 & 0 \\ 0 & -3 \end{bmatrix}$$

and the matrix **P** made of eigenvectors

$$\mathbf{P} = \begin{bmatrix} 0.8321 & -0.7071 \\ 0.5547 & 0.7071 \end{bmatrix}$$

Eigenvectors associated with distinct eigenvalues are independent of each other and, therefore, form a linear vector space. This point can be checked by calculating the determinants of the matrices composed of eigenvectors in this and the next two subsections. A nonzero determinant signifies that the vectors are independent.

Example 7.14 (Markov Chains). In Chaps. 4 and 6 we became familiar with Markov chains and the transition probability matrix

$$\mathbf{P} = \begin{bmatrix} p_{11} & p_{12} & \cdots & p_{n1} \\ p_{12} & p_{22} & \cdots & p_{n2} \\ \vdots & \vdots & & \vdots \\ p_{1n} & p_{n2} & \cdots & p_{nn} \end{bmatrix} \tag{7.36}$$

Because columns of $\mathbf{P}$ add up to one, we have

$$\mathbf{P}'\mathbf{1} = \mathbf{1} \tag{7.37}$$

where $\mathbf{1}$ is a column vector of 1s. In other words, one eigenvalue of $\mathbf{P}'$ is equal to one and its associated eigenvector is

$$\mathbf{1} = \begin{bmatrix} 1 \\ \vdots \\ 1 \end{bmatrix} \tag{7.38}$$

Because $\mathbf{P}$ and its transpose have the same eigenvalues (see E.7.18), it follows that one eigenvalue of $\mathbf{P}$ is also unity. Let $\boldsymbol{\pi}^*$ be the eigenvector associated with this unity eigenvalue, we can write

$$\mathbf{P}\boldsymbol{\pi}^* = \boldsymbol{\pi}^* \tag{7.39}$$

Indeed, $\boldsymbol{\pi}^*$ is the vector of long-run probabilities of different states because (see E.7.20)

$$\mathbf{P}^k\boldsymbol{\pi}^* = \boldsymbol{\pi}^* \tag{7.40}$$

In other words, if we are to predict the probabilities of each state far into the future, we would pick $\boldsymbol{\pi}^*$. Moreover, these probabilities remain unchanged. In computing $\boldsymbol{\pi}^*$ we have to make sure that the probabilities add up to one. To ensure this, we should use the normalizing rule

$$\mathbf{1}'\boldsymbol{\pi}^* = 1 \tag{7.41}$$

For the transition matrix of boom and recession of Chaps. 4 and 6, we have

$$\begin{bmatrix} 0.98 & 0.10 \\ 0.02 & 0.90 \end{bmatrix} \begin{bmatrix} 0.8333 \\ 0.1667 \end{bmatrix} = \begin{bmatrix} 0.8333 \\ 0.1667 \end{bmatrix}$$

7.4.1 Complex Eigenvalues

So far we have been dealing with a matrix whose eigenvalues are real and distinct. This is not always the case. A matrix could have an eigenvalue with multiplicity r or have complex eigenvalues. Indeed, a matrix could have a combination of real eigenvalues with different multiplicities and complex eigenvalues. In this and the next subsections we will consider these cases.

Example 7.15 Find the eigenvalues of the matrix

$$\mathbf{C} = \begin{bmatrix} 1 & 9 \\ -1 & 1 \end{bmatrix}$$

The characteristic polynomial is

$$P(\lambda) = \lambda^2 - 2\lambda + 10$$

and the eigenvalues are

$$\lambda_1 = 1 + 3i, \qquad \lambda_2 = 1 - 3i$$

Note that complex eigenvalues come in pairs and are conjugate complex of each other. For eigenvectors, we have

$$\begin{bmatrix} -3i & 9 \\ -1 & -3i \end{bmatrix} \begin{bmatrix} x_{11} \\ x_{12} \end{bmatrix} = \begin{bmatrix} 0 \\ 0 \end{bmatrix}$$

Again, we have only one equation:

$$-3ix_{11} + 9x_{12} = 0$$

or

$$ix_{11} = 3x_{12}$$

Setting $x_{12} = i$, we have $x_{11} = 3$. Thus,

$$\mathbf{x}_1 = \begin{bmatrix} 3 \\ i \end{bmatrix} = \begin{bmatrix} 3 \\ 0 \end{bmatrix} + \begin{bmatrix} 0 \\ i \end{bmatrix} = \mathbf{u}_1 + i\mathbf{v}_1$$

The second eigenvector could be found in the same manner, and this time we have

$$3ix_{11} + 9x_{12} = 0$$

or

$$-ix_{11} = 3x_{12}$$

Letting $x_{22} = -i$, we have $x_{21} = 3$. The second eigenvector is

$$\mathbf{x}_2 = \begin{bmatrix} 3 \\ -i \end{bmatrix} = \begin{bmatrix} 3 \\ 0 \end{bmatrix} - \begin{bmatrix} 0 \\ i \end{bmatrix} = \mathbf{u}_2 - i\mathbf{v}_2$$

Putting them together in a matrix,

$$\mathbf{P} = \begin{bmatrix} 3 & 3 \\ i & -i \end{bmatrix}$$

or after normalizing,

$$\mathbf{P} = \begin{bmatrix} 0.9487 & 0.9487 \\ 0.3162i & -0.3162i \end{bmatrix}$$

The Matlab commands that would produce normalized eigenvectors are the same as before.

7.4.2 Repeated Eigenvalues

There are instances when an eigenvalue occurs with multiplicity $k > 1$. In such cases calculation of eigenvalue remains unaffected, but a complication arises in finding eigenvectors. That is, in some cases it is possible to find k independent eigenvectors, but, on many occasions this is not possible and the number of independent eigenvectors is less than k. This could cause a problem, and we need to think of a solution. We illustrate these possibilities and the solution with a few examples.

Example 7.16 Use Matlab to find eigenvalues and eigenvectors of the matrix

$$\mathbf{B} = \begin{bmatrix} 1 & 3 & 6 & 0 \\ -3 & -5 & -6 & 0 \\ 3 & 3 & 4 & 0 \\ 0 & 0 & 0 & 4 \end{bmatrix}$$

You will get

$$\mathbf{D} = \begin{bmatrix} -2 & 0 & 0 & 0 \\ 0 & -2 & 0 & 0 \\ 0 & 0 & 4 & 0 \\ 0 & 0 & 0 & 4 \end{bmatrix}$$

and

$$\mathbf{P} = \begin{bmatrix} -0.5774 & -0.5774 & -0.9086 & 0 \\ -0.5774 & 0.5774 & 0.2681 & 0 \\ 0.5774 & -0.5774 & 0.3203 & 0 \\ 0 & 0 & 0 & 1 \end{bmatrix}$$

As can be seen, there are four independent eigenvectors. Another example would be the identity matrix **I**, which has $\lambda = 1$ as its eigenvalue with multiplicity n (its dimension) and has n independent eigenvectors, namely, $\mathbf{e}_1, \ldots, \mathbf{e}_n$.

Example 7.17 Find eigenvalues and eigenvectors of the matrix

$$\mathbf{A} = \begin{bmatrix} 2 & -1 \\ 1 & 4 \end{bmatrix}$$

The characteristic polynomial is

$$P(\lambda) = (\lambda - 3)^2 = 0$$

and therefore,

$$\lambda_1 = \lambda_2 = 3$$

that is, **A** has eigenvalue 3 with multiplicity 2. As to the eigenvector, using $\lambda = 3$ we have the equation

$$x_{11} = -x_{12}$$

and the eigenvector is

$$\mathbf{x}_1 = \begin{bmatrix} -1 \\ 1 \end{bmatrix}$$

or after normalizing

$$\mathbf{x}_1 = \begin{bmatrix} -0.7071 \\ 0.7071 \end{bmatrix}$$

In this case only one independent eigenvector is associated with eigenvalue 3. To get a second one, we define a new concept.

Definition 7.3 (Generalized Eigenvector). Let the $n \times n$ matrix $\mathbf{A}$ have an eigenvalue λ with multiplicity $J \leq n$. Then any vector $\mathbf{v}_j \neq \mathbf{0},\ j = 1, \ldots, J$ is the generalized eigenvector of $\mathbf{A}$ if

$$(\mathbf{A} - \lambda\mathbf{I})^j \mathbf{v}_j = \mathbf{0} \tag{7.42}$$

Note that if we already have computed $\mathbf{v}_\mathrm{j}$, then we can find $\mathbf{v}_{j+1}$ as

$$(\mathbf{A} - \lambda\mathbf{I})\mathbf{v}_{j+1} = \mathbf{v}_j \tag{7.43}$$

There is no difference between (7.42) and (7.43) as the latter is a necessary and sufficient condition for the former. On the other hand, it is easier to use (7.43). Moreover, when there is only one eigenvalue and it has multiplicity n, then

$$(\mathbf{A} - \lambda\mathbf{I})^n = \mathbf{0} \tag{7.44}$$

Example 7.18 Now we can find a second independent eigenvector for the matrix in Example 7.17:

$$\begin{bmatrix} -1 & -1 \\ 1 & 1 \end{bmatrix} \begin{bmatrix} x_{21} \\ x_{22} \end{bmatrix} = \begin{bmatrix} -1 \\ 1 \end{bmatrix}$$

One solution for $\mathbf{x}_2$ would be

$$\mathbf{x}_2 = \begin{bmatrix} 1 \\ 0 \end{bmatrix}$$

Then we have two independent eigenvectors for $\mathbf{A}$.

When dealing with cases of repeated eigenvalues, the Matlab routine used above is not the appropriate one. Instead, if Symbolic Math Toolbox is available, one should use

Matlab code

```
% Define a matrix
A = [2 -1; 1 4]
% Find its eigenvalues and (generalized) eigenvectors
[P, J]= jordan(A)
```

$\mathbf{P}$ is the matrix of eigenvectors, which is not normalized, and $\mathbf{J}$ is an upper triangular matrix whose diagonal elements are eigenvalues of $\mathbf{A}$. The off-diagonal elements are equal to one. The next example shows that the same considerations apply to the cases of repeated complex roots.

Example 7.19 Find the eigenvalues and eigenvectors of the matrix

$$\mathbf{C} = \begin{bmatrix} 0 & -1 & 0 & 0 \\ 1 & 0 & 0 & 0 \\ 0 & 0 & 0 & -1 \\ 2 & 0 & 1 & 0 \end{bmatrix}$$

The characteristic polynomial

$$\lambda^4 + 2\lambda^2 + 1 = 0$$

or

$$(\lambda^2 + 1)^2 = 0$$

whose solutions are

$$\lambda = \pm i$$

Note that the solutions have multiplicity of two. For the first eigenvector we have

$$\begin{bmatrix} -i & -1 & 0 & 0 \\ 1 & -i & 0 & 0 \\ 0 & 0 & -i & -1 \\ 2 & 0 & 1 & -i \end{bmatrix} \begin{bmatrix} w_{11} \\ w_{12} \\ w_{13} \\ w_{14} \end{bmatrix} = \begin{bmatrix} 0 \\ 0 \\ 0 \\ 0 \end{bmatrix}$$

One solution would be

$$\mathbf{w}_1 = \begin{bmatrix} 0 \\ 0 \\ i \\ 1 \end{bmatrix}$$

or

$$\mathbf{u}_1 + \mathbf{v}_1 = \begin{bmatrix} 0 \\ 0 \\ 0 \\ 1 \end{bmatrix} + i \begin{bmatrix} 0 \\ 0 \\ 1 \\ 0 \end{bmatrix}$$

For the second eigenvector we have to use (7.38):

$$\begin{bmatrix} -i & -1 & 0 & 0 \\ 1 & -i & 0 & 0 \\ 0 & 0 & -i & -1 \\ 2 & 0 & 1 & -i \end{bmatrix} \begin{bmatrix} w_{21} \\ w_{22} \\ w_{23} \\ w_{24} \end{bmatrix} = \begin{bmatrix} 0 \\ 0 \\ i \\ 1 \end{bmatrix}$$

One solution would be

$$\mathbf{w}_2 = \begin{bmatrix} 1 \\ -i \\ 0 \\ -i \end{bmatrix}$$

or

$$\mathbf{u}_2 + \mathbf{v}_2 = \begin{bmatrix} 1 \\ 0 \\ 0 \\ 0 \end{bmatrix} + i \begin{bmatrix} 0 \\ -1 \\ 0 \\ -1 \end{bmatrix}$$

Needless to say, a matrix may have any combination of real, complex, and repeated eigenvalues. A matrix may have all of its roots real or complex, but it may also have some complex and some real eigenvalues. Similarly, each real or complex eigenvalue may have a different multiplicity.

7.4.3 Eigenvalues and the Determinant and Trace of a Matrix

Eigenvalues have many interesting properties. Among them are the connections between eigenvalues and the determinant and trace of a matrix. The determinant of a matrix is equal to the product of its eigenvalues. For the matrix $\mathbf{A} = [a_{ij}]$,

$$\prod_{i}^{n} \lambda_i = |\mathbf{A}| \tag{7.45}$$

Example 7.20 For the matrix

$$\mathbf{A} = \begin{bmatrix} 3 & 6 \\ 4 & 1 \end{bmatrix}$$

we have

$$\lambda_1 = 7, \quad \lambda_2 = -3$$

and

$$|\mathbf{A}| = \lambda_1 \lambda_2 = 7(-3) = -21$$

Example 7.21 For the matrix

$$\mathbf{B} = \begin{bmatrix} 3 & 16 \\ -1 & 3 \end{bmatrix}$$

we have

$$\lambda_1 = 3 + 4i, \qquad \lambda_2 = 3 - 4i$$

and

$$|\mathbf{B}| = (3 + 4i)(3 - 4i) = 25$$

Example 7.22 For the matrix

$$\mathbf{C} = \begin{bmatrix} 3 & 1 \\ -1 & 1 \end{bmatrix}$$

we have

$$\lambda_1 = \lambda_2 = 2$$

and

$$|\mathbf{C}| = 2^2 = 4$$

An implication of the above is that if any eigenvalue of a matrix is equal to zero, then the determinant is equal to zero, and the matrix is singular. The reverse is also true. The trace of a matrix is equal to the sum of its eigenvalues. For the matrix $\mathbf{A} = [a_{ij}]$,

$$\sum_i^n \lambda_i = \sum_i^n a_{ii} \tag{7.46}$$

Example 7.23 For the matrices **A**, **B**, and **C** in Examples 7.20, 7.21, and 7.22, above, we have

$$\begin{aligned} \mathrm{tr}(\mathbf{A}) &= 7 - 3 = 4 \\ \mathrm{tr}(\mathbf{B}) &= 3 + 4i + 3 - 4i = 6 \\ \mathrm{tr}(\mathbf{C}) &= 2 + 2 = 4 \end{aligned}$$

The following Matlab routine performs the numerical verification of the properties of eigenvalues discussed above. Try it with the matrix of your choice.

Matlab code

```
% Define a matrix:
B = [39 3 26 36;3 6 2 7; 26 2 51 9;36 7 9 50]
% Specify its dimension
n = 4
% Find its eigenvalues
D = eig(B)
% Let s1 and s2 be the sum and product of eigenvalues
% Initialize them
```

```
s1 = 0;
s2 = 1;
% Find the sum and product of eigenvalues
For i=1:n
        s1 = s1 + D(i);
        s2 = s2*D(i);
end
% Find the trace of B
trace(B)
% Retrieve s1
s1
% Find the determinant of B
det(B)
% Retrieve s2
s2
```

7.4.4 Exercises

E.7.16. Find eigenvalues and eigenvectors of the following matrices:

$$\begin{bmatrix} 1 & 1 \\ 3 & -1 \end{bmatrix} \quad \begin{bmatrix} 3 & 12 \\ -9 & 7 \end{bmatrix} \quad \begin{bmatrix} 2 & 3 \\ 0 & 2 \end{bmatrix} \quad \begin{bmatrix} 8 & 7 \\ -1 & 12 \end{bmatrix}$$

$$\begin{bmatrix} -2 & 4 & -1 \\ 3 & 2 & 0 \\ 1 & 6 & -1 \end{bmatrix} \quad \begin{bmatrix} 3 & -2 & 0 \\ 0 & 9 & -2 \\ 0 & 0 & 1 \end{bmatrix} \quad \begin{bmatrix} 6 & 5 & 21 \\ 5 & 10 & 13 \\ 21 & 13 & 114 \end{bmatrix}$$

E.7.17. Given the transition probability matrix of employment states from Chaps. 4 and 6

$$\mathbf{P} = \begin{bmatrix} 0.960 & 0.350 & 0.060 \\ 0.039 & 0.600 & 0.120 \\ 0.001 & 0.050 & 0.820 \end{bmatrix}$$

where state 1 is employed, state 2 unemployed, and state 3, out of the labor force. Compute the long-run probability of each state.

E.7.18. Show that a matrix and its transpose have the same eigenvalues.

E.7.19. Let λ and $\mathbf{x}$ denote the eigenvalue and eigenvector of $\mathbf{A}$. Show that if $\mathbf{A}^{-1}$ exists, then $1/\lambda$ and $\mathbf{x}$ are its eigenvalue and eigenvector.

E.7.20. Let λ and $\mathbf{x}$ denote the eigenvalue and eigenvector of $\mathbf{A}$. Show that λ^k and $\mathbf{x}$ are the eigenvalue and eigenvector of $\mathbf{A}^k$.

E.7.21. (Cayley-Hamilton Theorem). Using numerical examples of matrices of order 2, 3, and 4, verify that $\mathrm{P}(\mathbf{A}) = \mathbf{0}$ where $\mathrm{P}(\lambda) = 0$ is the characteristic equation of matrix $\mathbf{A}$.

7.5 Factorization of Symmetric Matrices

Symmetric matrices play important roles in optimization and statistical estimation. The foundation of microeconomic theory is maximization of utility by consumers and maximization of profit by producers. In order to establish that we have a maximum, we have to rely on the symmetric matrix of second order derivatives (see Chap. 13). In estimation techniques, the covariance matrix, which is a positive definite matrix, plays a significant role. At the same time symmetric matrices have a number of interesting characteristics that make manipulating them easier. These properties relate to their eigenvalues and eigenvectors. In particular, all eigenvalues of symmetric matrices are real and a symmetric matrix can be factored into three matrices. Let $\mathbf{A}$ be a symmetric matrix and $\mathbf{P}$ the matrix whose columns are the eigenvectors of $\mathbf{A}$. Then

$$\mathbf{A} = \mathbf{PDP}' \tag{7.47}$$

or equivalently

$$\mathbf{P}'\mathbf{AP} = \mathbf{D} \tag{7.48}$$

where $\mathbf{D}$ is a diagonal matrix with eigenvalues of $\mathbf{A}$ as its diagonal elements. Furthermore, if $\mathbf{A}$ is positive definite, then it can be factored into two matrices

$$\mathbf{A} = \mathbf{QQ}' \tag{7.49}$$

These properties will make it easier to determine if a matrix is positive or negative definite and facilitate derivation of many results in econometrics.

7.5.1 Some Interesting Properties of Symmetric Matrices

Theorem 7.1 *All eigenvalues of a real symmetric matrix are real.*

Proof Let $\mathbf{A}$ be a real symmetric matrix with eigenvalue λ and the associated eigenvector $\mathbf{x} = \mathbf{u} + i\mathbf{v}$. Then

$$\mathbf{A}(\mathbf{u} + i\mathbf{v}) = \lambda(\mathbf{u} + i\mathbf{v}) \tag{7.50}$$

Multiplying both sides by the conjugate complex of $\mathbf{x}$, we get

$$(\mathbf{u} - i\mathbf{v})'\mathbf{A}(\mathbf{u} + i\mathbf{v}) = \lambda(\mathbf{u} - i\mathbf{v})'(\mathbf{u} + i\mathbf{v}) \tag{7.51}$$

The LHS is

$$\mathbf{u}'\mathbf{A}\mathbf{u} + i\mathbf{u}'\mathbf{A}\mathbf{v} - i\mathbf{v}'\mathbf{A}\mathbf{u} + \mathbf{v}\mathbf{A}\mathbf{v} = \mathbf{u}'\mathbf{A}\mathbf{u} + \mathbf{v}\mathbf{A}\mathbf{v} \tag{7.52}$$

Because

$$i\mathbf{v}'\mathbf{A}\mathbf{u} = i\mathbf{u}'\mathbf{A}'\mathbf{v} = i\mathbf{u}'\mathbf{A}\mathbf{v} \tag{7.53}$$

the RHS is

$$\lambda(\mathbf{u}'\mathbf{u} + \mathbf{v}'\mathbf{v}) \tag{7.54}$$

Because all elements of the LHS as well as $\mathbf{u}'\mathbf{u} + \mathbf{v}'\mathbf{v}$ on the RHS are real, it follows that λ must be real.

Example 7.24 Eigenvalues of the symmetric matrix

$$\mathbf{A} = \begin{bmatrix} 2 & -1 \\ -1 & 2 \end{bmatrix}$$

are 1 and 3.

Example 7.25 Eigenvalues of the symmetric matrix

$$\mathbf{B} = \begin{bmatrix} 2 & 7 & -6 \\ 7 & 8 & 0 \\ -6 & 0 & 4 \end{bmatrix}$$

are -5.4516, 5.5366, and 13.9150.

The rank of a symmetric matrix is equal to the number of nonzero eigenvalues of that matrix (this does not necessarily hold for nonsymmetric matrices). We can illustrate this with two examples. The matrix in Example 7.24 has rank 2, and the one in Example 7.25 has rank 3. This can be ascertained by computing their determinants that are not equal to zero. On the other hand, the matrix

$$\mathbf{C} = \begin{bmatrix} 2 & 7 & 9 \\ 7 & 8 & 15 \\ 9 & 15 & 24 \end{bmatrix}$$

has only two nonzero eigenvalues. Computation of its determinant shows that it does not have rank 3. But, inside it we could find the 2×2 matrix

$$\begin{bmatrix} 2 & 7 \\ 7 & 8 \end{bmatrix}$$

that is nonsingular. Hence, its rank is 2.

Theorem 7.2 *For a symmetric matrix, eigenvectors associated with distinct eigenvalues are orthogonal to each other.*

Proof Let $\mathbf{x}_i$ and $\mathbf{x}_j$ be eigenvalues of the symmetric matrix $\mathbf{A}$ associated with λ_i and λ_j, respectively. Then we have

$$\mathbf{A}\mathbf{x}_i = \lambda_i \mathbf{x}_i$$
$$\mathbf{A}\mathbf{x}_j = \lambda_j \mathbf{x}_j$$

Multiplying both sides of the first equation by $\mathbf{x}'_j$ and both sides of the second by $\mathbf{x}'_i$ and then subtracting both sides of the second equation from the first, we have

$$\mathbf{x}'_j\mathbf{A}\mathbf{x}_i - \mathbf{x}'_i\mathbf{A}\mathbf{x}_j = \mathbf{x}'_j\lambda_i\mathbf{x}_i - \mathbf{x}'_i\lambda_j\mathbf{x}_j \tag{7.55}$$

Because $\mathbf{A}$ is symmetric, the LHS is equal to zero. Therefore,

$$\lambda_i\mathbf{x}'_j\mathbf{x}_i - \lambda_j\mathbf{x}'_i\mathbf{x}_j = (\lambda_i - \lambda_j)\mathbf{x}'_i\mathbf{x}_j = 0 \tag{7.56}$$

Because eigenvalues are distinct, it follows that

$$\mathbf{x}'_i\mathbf{x}_j = 0 \tag{7.57}$$

Example 7.26 The normalized eigenvectors of matrix $\mathbf{A}$ in Example 7.24 are

$$\mathbf{x}_1 = \begin{bmatrix} -0.7071 \\ -0.7071 \end{bmatrix}, \qquad \mathbf{x}_2 = \begin{bmatrix} -0.7071 \\ 0.7071 \end{bmatrix}$$

Because $\mathbf{x}'_1\mathbf{x}_2 = 0$, the two vectors are orthogonal.

Example 7.27 The normalized eigenvectors of $\mathbf{B}$ in Example 7.25 are

$$\mathbf{x}_1 = \begin{bmatrix} 0.7729 \\ -0.4022 \\ 0.4907 \end{bmatrix}, \quad \mathbf{x}_2 = \begin{bmatrix} 0.2028 \\ -0.5762 \\ -0.7918 \end{bmatrix}, \quad \mathbf{x}_3 = \begin{bmatrix} -0.6012 \\ -0.7115 \\ 0.3638 \end{bmatrix}$$

Multiplying them together shows that they are orthogonal.

Theorem 7.3 *Let* **P** *be a matrix whose columns are normalized eigenvectors of a symmetric matrix* **A**. *Further assume that all eigenvalues of* **A** *are distinct. Then*

$$\mathbf{P}^{-1} = \mathbf{P}' \tag{7.58}$$

Let $\mathbf{x}_i,\ i = 1, \ldots, n$ *be the normalized eigenvectors of a symmetric matrix. We have*

$$\begin{aligned} \mathbf{P}'\mathbf{P} &= \begin{bmatrix} \mathbf{x}'_1 \\ \vdots \\ \mathbf{x}'_n \end{bmatrix} [\mathbf{x}_1, \ldots \mathbf{x}_n] \\ &= [\mathbf{x}'_i \mathbf{x}_j] \qquad\qquad i, j = 1, \ldots, n \\ &= \mathbf{I} \end{aligned} \tag{7.59}$$

The last equality is based on the fact that because $\mathbf{x}_i,\ i = 1, \ldots, n$ *are normalized vectors,* $\mathbf{x}'_i\mathbf{x}_i = 1$ and, *by Theorem 7.2,* $\mathbf{x}'_i\mathbf{x}_j = 0$ *when* $i \neq j$.

Example 7.28 For the matrix in Example 7.26, we have

$$\mathbf{P}^{-1} = \mathbf{P}' = \begin{bmatrix} -0.7071 & -0.7071 \\ -0.7071 & 0.7071 \end{bmatrix}$$

Example 7.29 For matrix **B** in Example 7.27,

$$\mathbf{P}^{-1} = \mathbf{P}' = \begin{bmatrix} 0.7729 & -0.4022 & 0.4907 \\ 0.2028 & -0.5762 & -0.7918 \\ -0.6012 & -0.7115 & 0.3638 \end{bmatrix}$$

The reader may want to use Matlab to check that indeed $\mathbf{P}' = \mathbf{P}^{-1}$.

7.5.2 Factorization of Matrix with Real Distinct Roots

Here we first show how a matrix with real distinct roots can be factored into three matrices, and then consider the special case of symmetric matrices. Let **A** be a matrix whose eigenvalues are all real and distinct. Recall that

$$\mathbf{A}\mathbf{x}_i = \lambda_i \mathbf{x}_i \tag{7.60}$$

Therefore,

$$\mathbf{A}[\mathbf{x}_1 \ldots \mathbf{x}_i \ldots \mathbf{x}_n] = [\lambda_1\mathbf{x}_1 \ldots \lambda_i\mathbf{x}_i \ldots \lambda_n\mathbf{x}_n] \tag{7.61}$$

or

$$\mathbf{AP} = \mathbf{PD} \tag{7.62}$$

where

$$\mathbf{D} = \begin{bmatrix} \lambda_1 & & & \\ & \ddots & & \\ & & \lambda_j & \\ & & & \ddots \end{bmatrix} \tag{7.63}$$

Postmultiplying (7.62) by $\mathbf{P}^{-1}$, we have

$$\mathbf{A} = \mathbf{PDP}^{-1} \tag{7.64}$$

or equivalently

$$\mathbf{D} = \mathbf{P}^{-1}\mathbf{AP} \tag{7.65}$$

Example 7.30 From Examples 7.8 and 7.13, we had

$$\mathbf{A} = \begin{bmatrix} 3 & 6 \\ 4 & 1 \end{bmatrix}, \quad \mathbf{P} = \begin{bmatrix} 0.8321 & -0.7071 \\ 0.5547 & 0.7071 \end{bmatrix}, \quad \mathbf{D} = \begin{bmatrix} 7 & 0 \\ 0 & -3 \end{bmatrix}$$

and

$$\mathbf{P}^{-1} = \begin{bmatrix} 0.7211 & 0.7211 \\ -0.5657 & 0.8485 \end{bmatrix}$$

Direct calculation shows that both (7.64) and (7.65) hold.

We showed that in the case of symmetric matrices $\mathbf{P}^{-1} = \mathbf{P}'$. Therefore, in case of $\mathbf{A}$ being a symmetric matrix, (7.64) and (7.65) become

$$\mathbf{A} = \mathbf{PDP}' \tag{7.66}$$

and

$$\mathbf{D} = \mathbf{P}'\mathbf{AP} \tag{7.67}$$

Example 7.31 Continuing with Examples 7.26 and 7.28, we have

$$\begin{bmatrix} 2 & -1 \\ -1 & 2 \end{bmatrix} = \begin{bmatrix} -0.7071 & -0.7071 \\ -0.7071 & 0.7071 \end{bmatrix} \begin{bmatrix} 1 & 0 \\ 0 & 3 \end{bmatrix} \begin{bmatrix} -0.7071 & -0.7071 \\ -0.7071 & 0.7071 \end{bmatrix}$$

and

$$\begin{bmatrix} 1 & 0 \\ 0 & 3 \end{bmatrix} = \begin{bmatrix} -0.7071 & -0.7071 \\ -0.7071 & 0.7071 \end{bmatrix} \begin{bmatrix} 2 & -1 \\ -1 & 2 \end{bmatrix} \begin{bmatrix} -0.7071 & -0.7071 \\ -0.7071 & 0.7071 \end{bmatrix}$$

The reader is encouraged to use Matlab and check the above relationships in the cases of the matrices of Examples 7.28 and 7.29. Now we can revisit positive and negative definite and semidefinite matrices and use (7.64) and (7.65) to identify them based on their eigenvalues. Let $\mathbf{A}$ be a symmetric matrix, then for any vector $\mathbf{x} \neq \mathbf{0}$, we can write

$$\mathbf{x}'\mathbf{A}\mathbf{x} = \mathbf{x}'\mathbf{P}\mathbf{D}\mathbf{P}'\mathbf{x} = \mathbf{z}'\mathbf{D}\mathbf{z} = \sum_{j=1}^{n} \lambda_j z_j^2 \tag{7.68}$$

where $\mathbf{z} = \mathbf{P}'\mathbf{x}$ and z_j's are elements of $\mathbf{z}$. Because $z_j^2,\ j = 1, \ldots, n$ are positive, it follows that the sign of $\mathbf{x}'\mathbf{A}\mathbf{x}$ is determined by the sign of λ_j's. Thus,

$$\mathbf{x}'\mathbf{A}\mathbf{x} > 0 \qquad \forall \mathbf{x} \neq \mathbf{0}. \tag{7.69}$$

and $\mathbf{A}$ is positive definite, if all its eigenvalues are positive. Similarly, $\mathbf{A}$ is positive semidefinite if all its eigenvalues are nonnegative with at least one, but not all, being equal to zero. On the other hand, a symmetric matrix is negative definite if all its eigenvalues are negative, and it is negative semidefinite if all its eigenvalues are nonpositive with at least one, but not all, being equal to zero.

7.5.3 Factorization of a Positive Definite Matrix

In the previous section, we showed that for a symmetric matrix $\mathbf{B}$, we have

$$\mathbf{B} = \mathbf{P}\mathbf{D}\mathbf{P}' \tag{7.70}$$

For a positive definite matrix, $\mathbf{D}$ is a diagonal matrix with positive elements. Therefore, we can write

$$\mathbf{B} = \mathbf{P}\mathbf{D}^{1/2}\mathbf{D}^{1/2}\mathbf{P}' \tag{7.71}$$

where $\mathbf{D}^{1/2}$ is a diagonal matrix with its diagonal elements equal to the square roots of the diagonal elements of $\mathbf{D}$. Let $\mathbf{Q} = \mathbf{P}\mathbf{D}^{1/2}$ then

$$\mathbf{Q}'\mathbf{Q} = \mathbf{D} \tag{7.72}$$

and

$$\mathbf{Q}\mathbf{Q}' = \mathbf{B} \tag{7.73}$$

Furthermore,

$$\mathbf{Q}'\mathbf{B}^{-1}\mathbf{Q} = \mathbf{I} \tag{7.74}$$

Example 7.32 Consider the following matrix:

$$\mathbf{B} = \begin{bmatrix} 9 & 2 \\ 2 & 6 \end{bmatrix}$$

and then

$$\mathbf{D} = \begin{bmatrix} 5 & 0 \\ 0 & 10 \end{bmatrix} \quad \text{and} \quad \mathbf{P} = \begin{bmatrix} 0.4472 & -0.8944 \\ -0.8944 & -0.4472 \end{bmatrix}$$

Simple computation results in

$$\mathbf{D}^{\frac{1}{2}} = \begin{bmatrix} 2.2361 & 0 \\ 0 & 3.1623 \end{bmatrix}$$

and

$$\mathbf{Q} = \begin{bmatrix} 1.0000 & -2.8284 \\ -2.0000 & -1.4142 \end{bmatrix}$$

To check our results, we have

$$\mathbf{QQ}' = \begin{bmatrix} 1.0000 & -2.8284 \\ -2.0000 & -1.4142 \end{bmatrix} \begin{bmatrix} 1.0000 & -2.0000 \\ -2.8284 & -1.4142 \end{bmatrix} = \begin{bmatrix} 9 & 2 \\ 2 & 6 \end{bmatrix}$$

and

$$\mathbf{Q}'\mathbf{Q} = \begin{bmatrix} 1.0000 & -2.0000 \\ -2.8284 & -1.4142 \end{bmatrix} \begin{bmatrix} 1.0000 & -2.8284 \\ -2.0000 & -1.4142 \end{bmatrix} = \begin{bmatrix} 5 & 0 \\ 0 & 10 \end{bmatrix}$$

Finally, because

$$\mathbf{B}^{-1} = \begin{bmatrix} 0.12 & -0.04 \\ -0.04 & 0.18 \end{bmatrix}$$

we have

$$\begin{aligned} \mathbf{Q}'\mathbf{B}^{-1}\mathbf{Q} &= \begin{bmatrix} 1.0000 & -2.0000 \\ -2.8284 & -1.4142 \end{bmatrix} \begin{bmatrix} 0.12 & -0.04 \\ -0.04 & 0.18 \end{bmatrix} \begin{bmatrix} 1.0000 & -2.8284 \\ -2.0000 & -1.4142 \end{bmatrix} \\ &= \begin{bmatrix} 1 & 0 \\ 0 & 1 \end{bmatrix} \end{aligned}$$

Example 7.33 Run the following Matlab code to verify (7.72)–(7.74) for a positive definite matrix. Try the program on other positive definite matrices.

Matlab code

```
% Define a matrix
B = [39 -3 14 36; -3 6 2 0; 14 2 51 -7; 36 0 -7 50]
[P, D]= eig(B)
% Check that P' is the inverse of P
P'*P
% Define Q
Q = P*sqrt(D)
% Check (7.72) and (7.73)
Q'*Q
Q*Q'
% Check (7.74)
Q'*inv(B)*Q
```

Example 7.34 (Generalized Least Squares). We noted that a necessary condition for the Gauss-Markov theorem is

$$E(\mathbf{uu}') = \sigma^2\mathbf{I} \tag{7.75}$$

Now, suppose that this assumption is violated and we have

$$E(\mathbf{uu}') = \sigma^2\mathbf{V} \tag{7.76}$$

where $\mathbf{V}$ is a positive definite matrix. The Gauss-Markov theorem does not hold anymore. But because $\mathbf{V}$ is positive definite, so is its inverse (see E.7.26). Based on what we learned in factoring a positive definite matrix, we can find a matrix $\mathbf{Q}$ such that

$$\mathbf{QQ}' = \mathbf{V}^{-1} \tag{7.77}$$

Premultiplying the regression model in (7.15) by $\mathbf{Q}'$, we have

$$\mathbf{Q}'\mathbf{y} = \mathbf{Q}'\mathbf{X}\boldsymbol{\beta} + \mathbf{Q}'\mathbf{u} \tag{7.78}$$

The new regression equation meets all the assumptions of the Gauss-Markov theorem. In particular, we have

$$\begin{aligned} E(\mathbf{Q}'\mathbf{uu}'\mathbf{Q}) &= \mathbf{Q}'E(\mathbf{uu}')\mathbf{Q} \\ &= \sigma^2\mathbf{Q}'\mathbf{VQ} \\ &= \sigma^2\mathbf{I} \end{aligned} \tag{7.79}$$

Following the same line of reasoning as in Example 7.4 we obtain the generalized least squares estimator as

$$\begin{aligned}\hat{\beta}_g &= (\mathbf{X}'\mathbf{Q}\mathbf{Q}'\mathbf{X})^{-1}\mathbf{X}'\mathbf{Q}\mathbf{Q}'\mathbf{y} \\ &= (\mathbf{X}'\mathbf{V}^{-1}\mathbf{X})^{-1}\mathbf{X}'\mathbf{V}^{-1}\mathbf{y}\end{aligned} \tag{7.80}$$

which will be both unbiased and best.

The generalized least squares estimator is applicable to many situations including models with heteroscedasticity or serial correlation in the error term, pooling cross-section and time series data, and seemingly unrelated regressions. These models differ only in the structure of **V**. In practice **V** is an $n \times n$ unknown matrix. But because it is symmetric, there are $n(n+1)/2$ parameters to be estimated. Given n observations, it is ludicrous to even attempt such estimation. Therefore, in each case the structure of the model should contain enough restrictions on the elements of **V** to enable its efficient estimation. For example, in case of first-order serial correlation in the error term, one needs to estimate only the autocorrelation parameter ρ in addition to σ^2.

Example 7.35 (Ridge Regression). The least squares method requires the inversion of the matrix $\mathbf{X}'\mathbf{X}$; therefore, a basic assumption of the regression model is that **X** has full column rank. If, however, a linear relationship exists between different columns of **X**, then we have the problem of multicollinearity and the least squares method breaks down. If the relationship between columns of **X** is not exact, but there is a high level of correlation, we face the problem of near collinearity. In this case, our estimated coefficients are imprecise and their variances are quite high. We know that in case of multicollinearity the matrix $\mathbf{X}'\mathbf{X}$ is singular and, in the case of near collinearity, it is near singular. Thus, at least one of its eigenvalues is either zero or near zero. In case of near collinearity a way to improve the efficiency of the estimates is to add a small number to the eigenvalues of the matrix. Because the matrix $\mathbf{X}'\mathbf{X}$ is positive definite, we can write

$$\mathbf{X}'\mathbf{X} = \mathbf{P}\mathbf{D}\mathbf{P}' \tag{7.81}$$

where again **D** is a diagonal matrix whose diagonal elements are eigenvalues and **P** the matrix whose columns are eigenvectors of $\mathbf{X}'\mathbf{X}$. Adding a quantity k to each eigenvalue results in

$$\begin{aligned}\mathbf{P}(\mathbf{D}+k\mathbf{I})\mathbf{P}' &= \mathbf{P}\mathbf{D}\mathbf{P}' + k\mathbf{P}\mathbf{I}\mathbf{P}' \\ &= \mathbf{X}'\mathbf{X} + k\mathbf{I}\end{aligned} \tag{7.82}$$

Therefore, the ridge regression would be

$$\hat{\boldsymbol{\beta}}_R = (\mathbf{X}'\mathbf{X} + k\mathbf{I})^{-1}\mathbf{X}'\mathbf{y} \tag{7.83}$$

Note that the new estimator is biased, but compared to the ordinary least squares, variances of the estimated coefficients are smaller. Thus, ridge regression trades off unbiasedness for efficiency.

7.5.4 Exercises

E.7.22 Show that the following matrices are positive definite, and then factor them into $\mathbf{QQ}'$ form.

$$\begin{bmatrix} 25 & -6 \\ -6 & 13 \end{bmatrix} \quad \begin{bmatrix} 100 & -90 \\ -90 & 85 \end{bmatrix} \quad \begin{bmatrix} 2 & -1 \\ -1 & 13 \end{bmatrix} \quad \begin{bmatrix} 8 & 0 \\ 0 & 2 \end{bmatrix}$$

$$\begin{bmatrix} 3 & 0 & 0 \\ 0 & 6 & 1 \\ 0 & 1 & 6 \end{bmatrix} \quad \begin{bmatrix} 98 & -8 & -10 \\ -8 & 3 & 1 \\ -10 & 1 & 10 \end{bmatrix} \quad \begin{bmatrix} 10 & -5 & 3 \\ -5 & 9 & 1 \\ 3 & 1 & 10 \end{bmatrix}$$

E.7.23 Show that $\mathbf{X}'\mathbf{X}$ where $\mathbf{X}$ is $n \times k$ is always positive semidefinite. Moreover, show that it is positive definite if $\mathbf{X}$ has full column rank.

E.7.24 Show that Eqs. (7.64) and (7.65) are equivalent.

E.7.25 Show that if

$$\mathbf{QQ}' = \mathbf{B}$$

then

$$\mathbf{Q}'\mathbf{B}^{-1}\mathbf{Q} = \mathbf{I}$$

and vice versa.

E.7.26 Show that if $\mathbf{V}$ is positive definite, so is $\mathbf{V}^{-1}$. [*Hint*: Use the result of E.7.19.]

7.6 LU Factorization of a Square Matrix

The motivation for this factorization is the interesting and useful characteristics of *triangular matrices*. An *upper triangular matrix* is a square matrix with all of its below diagonal elements equal to zero:

$$\mathbf{T} = \begin{bmatrix} t_{11} & t_{12} & \dots & t_{1n} \\ 0 & t_{22} & \dots & t_{2n} \\ \vdots & \vdots & \ddots & \vdots \\ 0 & 0 & \dots & t_{nn} \end{bmatrix} \tag{7.84}$$

Similarly, a *lower triangular matrix* is a square matrix with all of its above diagonal elements equal to zero:

$$\mathbf{S} = \begin{bmatrix} s_{11} & 0 & \dots & 0 \\ s_{21} & s_{22} & \dots & 0 \\ \vdots & \vdots & \ddots & \vdots \\ s_{n1} & s_{n2} & \dots & s_{nn} \end{bmatrix} \tag{7.85}$$

The determinant of a triangular matrix is equal to the product of its diagonal elements, and the sum and product of several upper (lower) triangular matrices are also an upper (a lower) triangular matrix. But most important, the inverse of an upper (a lower) triangular matrix is also an upper (a lower) triangular matrix. The last characteristic facilitates the computation of the inverse of a triangular matrix as shown in the following example.

Example 7.36 Consider the upper diagonal matrix (see Exercise E.7.27 for the case of a lower triangular matrix)

$$\mathbf{T} = \begin{bmatrix} 1 & 3 & 5 \\ 0 & 2 & -3 \\ 0 & 0 & 1 \end{bmatrix}$$

Denote its inverse by

$$\mathbf{T}^{-1} = \begin{bmatrix} t^{11} & t^{12} & t^{13} \\ 0 & t^{22} & t^{23} \\ 0 & 0 & t^{33} \end{bmatrix}$$

we then have

$$\begin{bmatrix} 1 & 3 & 5 \\ 0 & 2 & -3 \\ 0 & 0 & 1 \end{bmatrix} \begin{bmatrix} t^{11} & t^{12} & t^{13} \\ 0 & t^{22} & t^{23} \\ 0 & 0 & t^{33} \end{bmatrix} = \begin{bmatrix} 1 & 0 & 0 \\ 0 & 1 & 0 \\ 0 & 0 & 1 \end{bmatrix}$$

It follows that

$$\begin{aligned} &t^{11} = 1, \qquad t^{12} + 3\,t^{22} = 0, \qquad t^{13} + 3t^{23} + 5t^{33} = 0, \\ &2t^{22} = 1, \qquad 2t^{23} - t^{33} = 0, \qquad t^{33} = 1 \end{aligned}$$

Solving the equations recursively from last to first, we get

$$t^{33} = 1, \quad t^{23} = \frac{3}{2}, \quad t^{22} = \frac{1}{2}, \quad t^{13} = -\frac{19}{2}, \quad t^{12} = -\frac{3}{2}, \quad t^{11} = 1$$

Thus, the inverse is

$$\mathbf{T}^{-1} = \begin{bmatrix} 1 & -\frac{3}{2} & -\frac{19}{2} \\ 0 & \frac{1}{2} & \frac{3}{2} \\ 0 & 0 & 1 \end{bmatrix} = \frac{1}{2}\begin{bmatrix} 2 & -3 & -19 \\ 0 & 1 & 3 \\ 0 & 0 & 2 \end{bmatrix}$$

As can be seen, the inverse of a triangular matrix can be found with a few simple additions and multiplications. But now we arrive at the pièce de résistance of the topic. Any square matrix that satisfies certain conditions can be written as the product of a lower triangular matrix **L** and an upper triangular matrix **U**. Furthermore, if we fix the diagonal elements of either matrix, then the factorization is unique. As we shall see, this operation is rather simple and recursive, therefore, for a large class of matrices, we can first factor them into two triangular matrices, find the inverses of **U** and **L**, and finally obtain the inverse of the original matrix as the product of the two inverses. We next present these ideas as a theorem. We shall skip the proof but illustrate the process with an example.

Theorem 7.4 *Let* **A** *be a square matrix*

$$\mathbf{A} = \begin{bmatrix} a_{11} & a_{12} & \dots & a_{1n} \\ a_{21} & a_{22} & \dots & a_{2n} \\ \vdots & \vdots & \ddots & \vdots \\ a_{n1} & a_{n2} & \dots & a_{nn} \end{bmatrix} \tag{7.86}$$

such that its principal diagonal minors are not zero. That is,

$$a_{11} \neq 0, \qquad \begin{vmatrix} a_{11} & a_{12} \\ a_{21} & a_{22} \end{vmatrix} \neq 0, \qquad \dots, \qquad |\mathbf{A}| \neq 0 \tag{7.87}$$

Then **A** *can be written as*

$$\mathbf{A} = \mathbf{LU} \tag{7.88}$$

where **L** *is a lower triangular matrix and* **U** *is an upper triangular matrix. Thus,*

$$\mathbf{A} = \begin{bmatrix} \ell_{11} & 0 & \dots & 0 \\ \ell_{21} & \ell_{22} & \dots & 0 \\ \vdots & \vdots & \ddots & \vdots \\ \ell_{n1} & \ell_{n2} & \dots & \ell_{nn} \end{bmatrix} \begin{bmatrix} u_{11} & u_{12} & \dots & u_{1n} \\ 0 & u_{22} & \dots & u_{2n} \\ \vdots & \vdots & \ddots & \vdots \\ 0 & 0 & \dots & u_{nn} \end{bmatrix} \tag{7.89}$$

If we let $\ell_{11} = \ell_{22} = \ell_{nn} = 1$*, then this factorization is unique. The same would also be true if we set the diagonal elements of* **U** *all equal to 1.*

Example 7.37 Let

$$\mathbf{A} = \begin{bmatrix} 2 & 4 \\ -6 & 1 \end{bmatrix}$$

To find matrices **L** and **U**, we write

$$\begin{bmatrix} 2 & 4 \\ -6 & 1 \end{bmatrix} = \begin{bmatrix} 1 & 0 \\ \ell_{21} & 1 \end{bmatrix} \begin{bmatrix} u_{11} & u_{12} \\ 0 & u_{22} \end{bmatrix}$$

which gives us the following four equations to calculate the elements of matrices **L** and **U**:

$$\begin{aligned} &u_{11} = 2 && u_{12} = 4 \\ &\ell_{21}u_{11} = -6 && \Rightarrow \quad \ell_{21} = -3 \\ &\ell_{21}u_{12} + u_{22} = 1 && \Rightarrow \quad u_{22} = 13 \end{aligned}$$

Thus,

$$\mathbf{L} = \begin{bmatrix} 1 & 0 \\ -3 & 1 \end{bmatrix}, \qquad \mathbf{U} = \begin{bmatrix} 2 & 4 \\ 0 & 13 \end{bmatrix}$$

Example 7.38 Let

$$\mathbf{B} = \begin{bmatrix} 2 & -1 & 3 \\ -1 & 5 & 6 \\ 3 & 6 & 14 \end{bmatrix}$$

To find matrices **L** and **U**, we write

$$\begin{bmatrix} 2 & -1 & 3 \\ -1 & 5 & 6 \\ 3 & 6 & 14 \end{bmatrix} = \begin{bmatrix} 1 & 0 & 0 \\ \ell_{21} & 1 & 0 \\ \ell_{31} & \ell_{32} & 1 \end{bmatrix} \begin{bmatrix} u_{11} & u_{12} & u_{13} \\ 0 & u_{22} & u_{23} \\ 0 & 0 & u_{33} \end{bmatrix}$$

which gives us the following nine equations to compute the elements of matrices **L** and **U**:

$$\begin{aligned} &u_{11} = 2, && u_{12} = -1, && u_{13} = 3 \\ &\ell_{21}u_{11} = -1 && \Rightarrow && \ell_{21} = -\tfrac{1}{2} \\ &\ell_{21}u_{12} + u_{22} = 5 && \Rightarrow && u_{22} = \tfrac{9}{2} \\ &\ell_{21}u_{13} + u_{23} = 6 && \Rightarrow && u_{23} = \tfrac{15}{2} \\ &\ell_{31}u_{11} = 3 && \Rightarrow && \ell_{31} = \tfrac{3}{2} \\ &\ell_{31}u_{12} + \ell_{32}u_{22} = 6 && \Rightarrow && \ell_{32} = \tfrac{5}{3} \\ &\ell_{31}u_{13} + \ell_{32}u_{23} + u_{33} = 14 && \Rightarrow && u_{33} = -3 \end{aligned}$$

Thus,

$$\mathbf{L} = \begin{bmatrix} 1 & 0 & 0 \\ -\frac{1}{2} & 1 & 0 \\ \frac{3}{2} & \frac{5}{3} & 1 \end{bmatrix}, \qquad \mathbf{U} = \begin{bmatrix} 2 & -1 & 3 \\ 0 & \frac{9}{2} & \frac{15}{2} \\ 0 & 0 & -3 \end{bmatrix}$$

LU factorization with Matlab is a bit tricky. For computational reasons, Matlab factors a matrix into an upper triangular matrix and "psychologically lower triangular" matrix. The latter is a lower triangular matrix whose rows are interchanged. Thus, using the following routine on matrix **A** of Example 7.37,

Matlab code

```
% define the matrix
A =[2 4; -6 1]
[L,U] = lu(A)
```

we get the following matrices:

$$\mathbf{L} = \begin{bmatrix} -0.3333 & 1 \\ 1 & 0 \end{bmatrix}, \quad \mathbf{U} = \begin{bmatrix} -6 & 1 \\ 0 & 4.3333 \end{bmatrix}$$

An alternative would be to use the routine

Matlab code

```
% define the matrix
A =[2 4; -6 1]
[L,U,P] = lu(A)
```

in which case we get the matrices

$$\mathbf{L} = \begin{bmatrix} 1 & 0 \\ -0.3333 & 1 \end{bmatrix}, \quad \mathbf{U} = \begin{bmatrix} -6 & 1 \\ 0 & 4.3333 \end{bmatrix}$$

and

$$\mathbf{P} = \begin{bmatrix} 0 & 1 \\ 1 & 0 \end{bmatrix}$$

where **P** is the transformation matrix such that

$$\mathbf{LU} = \mathbf{PA} \tag{7.90}$$

Similarly, if we apply the Matlab function `lu()` to matrix **B** of Example 7.38, we get the following matrices

$$\mathbf{L} = \begin{bmatrix} 1 & 0 & 0 \\ -0.3333 & 1 & 0 \\ 0.6667 & -0.7143 & 1 \end{bmatrix}, \quad \mathbf{U} = \begin{bmatrix} 3 & 6 & 14 \\ 0 & 7 & 10.6667 \\ 0 & 0 & 1.2857 \end{bmatrix}$$

and the transformation matrix

$$\mathbf{P} = \begin{bmatrix} 0 & 0 & 1 \\ 0 & 1 & 0 \\ 1 & 0 & 0 \end{bmatrix}$$

7.6.1 Cholesky Factorization

Cholesky factorization can be considered a special case of LU factorization when applied to positive definite matrices. Cholesky factorization is quite useful in econometric analysis, particularly in VAR modeling of time series. This should not be surprising because as we showed in Example 7.2 all covariance matrices are positive definite.

Let **A** be a positive definite matrix. Recall that such matrices are symmetric. Then we can write **A** as the product of an upper triangular matrix and its transpose. Thus,

$$\mathbf{A} = \mathbf{C}'\mathbf{C} \tag{7.91}$$

that is,

$$\begin{bmatrix} a_{11} & a_{12} & \dots & a_{1n} \\ a_{21} & a_{22} & \dots & a_{2n} \\ \vdots & \vdots & \ddots & \vdots \\ a_{n1} & a_{n2} & \dots & a_{nn} \end{bmatrix} = \begin{bmatrix} c_{11} & 0 & \dots & 0 \\ c_{12} & c_{22} & \dots & 0 \\ \vdots & \vdots & \ddots & \vdots \\ c_{1n} & c_{2n} & \dots & c_{nn} \end{bmatrix} \begin{bmatrix} c_{11} & c_{12} & \dots & c_{1n} \\ 0 & a_{22} & \dots & a_{2n} \\ \vdots & \vdots & \ddots & \vdots \\ 0 & 0 & \dots & c_{nn} \end{bmatrix} \tag{7.92}$$

Example 7.39 Let

$$\mathbf{A} = \begin{bmatrix} 4 & 2 \\ 2 & 10 \end{bmatrix}$$

Check that this matrix is positive definite. It can be factored into

$$\mathbf{A} = \begin{bmatrix} 2 & 0 \\ 1 & 3 \end{bmatrix} \begin{bmatrix} 2 & 1 \\ 0 & 3 \end{bmatrix}$$

Example 7.40 Let

$$\mathbf{B} = \begin{bmatrix} 9 & -3 & 6 \\ -3 & 2 & 3 \\ 6 & 3 & 45 \end{bmatrix}$$

Again check that the matrix is positive definite. It can be factored into

$$\mathbf{B} = \begin{bmatrix} 3 & 0 & 0 \\ -1 & 1 & 0 \\ 2 & 5 & 4 \end{bmatrix} \begin{bmatrix} 3 & -1 & 2 \\ 0 & 1 & 5 \\ 0 & 0 & 4 \end{bmatrix}$$

Calculation of Cholesky factorization with Matlab is straightforward.

Matlab code

```
% define the matrix
B = [9 -3 6; -3 2 3; 6 3 45]
% Calculate the Cholesky factorization
C = chol(B)
% Program returns the upper triangular matrix C
```

7.6.2 Exercises

E.7.27 Find the inverses of the following triangular matrices using the method outlined in Example 7.36. Check your results using Matlab.

$$\begin{bmatrix} 2 & 0 & 0 \\ -3 & 10 & 0 \\ -4 & -9 & -1 \end{bmatrix}, \quad \begin{bmatrix} 7 & -0 & 0 \\ 2 & -6 & 0 \\ 5 & -8 & 2 \end{bmatrix}, \quad \begin{bmatrix} 0.7 & 0 & 0 \\ 0.5 & -0.8 & 0 \\ 0.2 & -0.9 & -0.3 \end{bmatrix}$$

$$\begin{bmatrix} 1 & 2 & -1 \\ 0 & 11 & -4 \\ 0 & 0 & -1 \end{bmatrix}, \quad \begin{bmatrix} 5 & -6 & 2 \\ 0 & -1 & 9 \\ 0 & 0 & 3 \end{bmatrix}, \quad \begin{bmatrix} 0.6 & -0.5 & -0.4 \\ 0 & -0.9 & -0.7 \\ 0 & 0 & -1 \end{bmatrix}$$

E.7.28 Compute the LU factorization of the following matrices first by direct calculation and then using Matlab.

$$\begin{bmatrix} 10 & 3 \\ -1 & 7 \end{bmatrix}, \quad \begin{bmatrix} 14 & 8 \\ 9 & 11 \end{bmatrix}, \quad \begin{bmatrix} 6 & 2 \\ 2 & 1 \end{bmatrix}, \quad \begin{bmatrix} -13 & -1 \\ -7 & -8 \end{bmatrix}$$

$$\begin{bmatrix} 8 & 8 & 12 \\ -11 & 9 & 4 \\ -3 & 0 & 7 \end{bmatrix}, \quad \begin{bmatrix} 3 & -1 & 9 \\ -1 & 11 & 5 \\ 9 & 5 & -7 \end{bmatrix}$$

E.7.29 Compute the Cholesky factorization of the following matrices.

$$\begin{bmatrix} 10 & 4 \\ 4 & 8 \end{bmatrix}, \quad \begin{bmatrix} 5 & 1 \\ 1 & 10 \end{bmatrix}$$

$$\begin{bmatrix} 3 & 0 & -1 \\ 0 & 2 & 2 \\ -1 & 2 & 5 \end{bmatrix}, \quad \begin{bmatrix} 62 & 32 & 24 \\ 32 & 59 & 40 \\ 24 & 40 & 32 \end{bmatrix}$$

7.7 Kronecker Product and Vec Operator

The Kroenker[2] product of two matrices is a useful concept in some econometrics estimation. Consider the two matrices $\mathbf{A} = [a_{ij}]$ with dimensions $m \times p$ and $\mathbf{B} = [b_{kl}]$ with dimensions $n \times q$. Their Kronecker products are defined as

$$\mathbf{A} \otimes \mathbf{B} = \begin{bmatrix} a_{11}\mathbf{B} & a_{12}\mathbf{B} & \dots & a_{1p}\mathbf{B} \\ a_{21}\mathbf{B} & a_{22}\mathbf{B} & \dots & a_{2p}\mathbf{B} \\ \vdots & \vdots & \ddots & \vdots \\ a_{m1}\mathbf{B} & a_{m2}\mathbf{B} & \dots & a_{mp}\mathbf{B} \end{bmatrix} \tag{7.93}$$

and

$$\mathbf{B} \otimes \mathbf{A} = \begin{bmatrix} b_{11}\mathbf{A} & b_{12}\mathbf{A} & \dots & b_{1q}\mathbf{A} \\ b_{21}\mathbf{A} & b_{22}\mathbf{A} & \dots & b_{2q}\mathbf{A} \\ \vdots & \vdots & \ddots & \vdots \\ b_{n1}\mathbf{A} & b_{n2}\mathbf{A} & \dots & b_{nq}\mathbf{A} \end{bmatrix} \tag{7.94}$$

Example 7.41 Let

$$\mathbf{A} = \begin{bmatrix} a_{11} & a_{12} & a_{13} \\ a_{21} & a_{22} & a_{23} \end{bmatrix} \qquad \mathbf{I} = \begin{bmatrix} 1 & 0 \\ 0 & 1 \end{bmatrix}$$

---

[2]Named after the German mathematician Leopold Kronecker (1823–1891), who studied and made contributions to number theory and elliptic functions. He also held views on mathematics contrary to those of his contemporaries such as Cantor and Dedekind.

Then their Kronecker products are

$$\mathbf{I} \otimes \mathbf{A} = \begin{bmatrix} a_{11} & a_{12} & a_{13} & 0 & 0 & 0 \\ a_{21} & a_{22} & a_{23} & 0 & 0 & 0 \\ 0 & 0 & 0 & a_{11} & a_{12} & a_{13} \\ 0 & 0 & 0 & a_{21} & a_{22} & a_{23} \end{bmatrix}$$

and

$$\mathbf{A} \otimes \mathbf{I} = \begin{bmatrix} a_{11} & 0 & a_{12} & 0 & a_{13} & 0 \\ 0 & a_{11} & 0 & a_{12} & 0 & a_{13} \\ a_{21} & 0 & a_{22} & 0 & a_{23} & 0 \\ 0 & a_{21} & 0 & a_{22} & 0 & a_{23} \end{bmatrix}$$

You can numerically calculate Kroneker product with Matlab.

Matlab code

```
% Specify the matrices
A = [9 11 2; 0 17 5; -1 2 10]
B = [3 2; -1 -4]
% To compute
kron(A, B)
% and to compute
kron(B, A)
```

Several properties of the Kroneker product will prove useful in econometric analysis. We state them here without proof. The reader is encouraged to verify these properties (see E.7.32).

$$\begin{array}{ll} i. & (\mathbf{A} \otimes \mathbf{B})' = \mathbf{A}' \otimes \mathbf{B}' \\ ii. & (\mathbf{A} \otimes \mathbf{B})(\mathbf{C} \otimes \mathbf{D}) = \mathbf{AC} \otimes \mathbf{BD} \\ iii. & \mathbf{A} \otimes (\mathbf{B} + \mathbf{C}) = \mathbf{A} \otimes \mathbf{B} + \mathbf{A} \otimes \mathbf{C} \\ iv. & (\mathbf{B} + \mathbf{C}) \otimes \mathbf{A} = \mathbf{B} \otimes \mathbf{A} + \mathbf{C} \otimes \mathbf{A} \\ v. & \mathbf{A} \otimes (\mathbf{B} \otimes \mathbf{B}) = (\mathbf{A} \otimes \mathbf{B}) \otimes \mathbf{B} \\ vi. & (\mathbf{A} \otimes \mathbf{B})^{-1} = \mathbf{A}^{-1} \otimes \mathbf{B}^{-1} \end{array} \tag{7.95}$$

In (7.95*i*) it is assumed that **AC** and **BD** exist. Similarly in (7.95*iii*) and (7.95*iv*) it is assumed that **B**+**C** exists, and in (7.95*vi*) both **A** and **B** are square and nonsingular matrices. Note that (7.95*vi*) and (7.95*ii*) imply

$$\begin{aligned} (\mathbf{A} \otimes \mathbf{B})(\mathbf{A}^{-1} \otimes \mathbf{B}^{-1}) &= \mathbf{AA}^{-1} \otimes \mathbf{BB}^{-1} \\ &= \mathbf{I}_{m\times m} \otimes \mathbf{I}_{n\times n} \\ &= \mathbf{I}_{mn\times mn} \end{aligned} \tag{7.96}$$

7.7.1 Vectorization of a Matrix

Vectorization of a matrix, which proves useful in both econometric analysis and differentiation of vectors and matrices, is the simplest of the operations discussed in this chapter. To vectorize an $m \times n$ matrix $\mathbf{A} = [a_{ij}]$, we stack its columns on top of each other. Column 1 will go on top and column n to the bottom. Thus,

$$\text{Vec}(\mathbf{A}) = \begin{bmatrix} a_{11} \\ \vdots \\ a_{m1} \\ a_{12} \\ \vdots \\ a_{m2} \\ \vdots \\ a_{1n} \\ \vdots \\ a_{mn} \end{bmatrix} \tag{7.97}$$

Example 7.42 Let

$$\mathbf{A} = \begin{bmatrix} a\ b\ c \\ d\ e\ f \end{bmatrix}, \quad \mathbf{B} = \begin{bmatrix} 12 & -6 \\ 3 & 15 \end{bmatrix}$$

Then

$$\text{Vec}(\mathbf{A}) = \begin{bmatrix} a \\ d \\ b \\ e \\ c \\ f \end{bmatrix}, \qquad \text{Vec}(\mathbf{B}) = \begin{bmatrix} 12 \\ 3 \\ -6 \\ 15 \end{bmatrix}$$

7.7.2 Exercises

E.7.30 Given the following matrices

$$\mathbf{A} = \begin{bmatrix} 4 & 17 & -9 \\ 12 & 0 & -1 \end{bmatrix}, \quad \mathbf{B} = \begin{bmatrix} -6 & 08 \\ -1 & 11 \\ -5 & 08 \end{bmatrix}, \quad \mathbf{C} = \begin{bmatrix} 3 & 3 \\ 8 & 7 \end{bmatrix}, \quad \mathbf{I} = \begin{bmatrix} 1 & 0 \\ 0 & 1 \end{bmatrix}$$

find

i. $\mathbf{A} \otimes \mathbf{B}$,	*ii*. $\mathbf{C} \otimes \mathbf{B}$,	*iii*. $\mathbf{A} \otimes \mathbf{B} \otimes \mathbf{C}$
iv. $\mathbf{I} \otimes \mathbf{B}$,	*v*. $\mathbf{C} \otimes \mathbf{I}$,	*vi*. $\mathbf{A} \otimes \mathbf{B} \otimes \mathbf{I}$
vii. $\text{Vec}(\mathbf{C})$,	*viii*. $\text{Vec}(\mathbf{B} \otimes \mathbf{I})$,	*ix*. $\text{Vec}(\mathbf{A} \otimes \mathbf{B} \otimes \mathbf{C})$

E.7.31 Use Matlab to verify your results for *i*- *vi* in E.7.30.

E.7.32 Given the following matrices, verify the relationships in (7.95*i*)–(7.95*vi*). Check your results using Matlab.

$$\mathbf{A} = \begin{bmatrix} -1 & 3 & 1 \\ -2 & 5 & 2 \\ -3 & 6 & 1 \end{bmatrix}, \quad \mathbf{B} = \begin{bmatrix} -3 & -2 & -1 \\ -1 & -1 & 0 \\ -5 & -6 & -1 \end{bmatrix},$$
$$\mathbf{C} = \begin{bmatrix} -7 & 4 & 8 \\ -9 & 2 & 2 \\ -3 & 7 & 4 \end{bmatrix}, \quad \mathbf{D} = \begin{bmatrix} 1 & 1 & 1 \\ 0 & 1 & 1 \\ 0 & 0 & 1 \end{bmatrix}$$

Part III
Calculus

Chapter 8
Differentiation: Functions of One Variable

8.1 Marginal Analysis in Economics

In the early 1870s three economists, William Stanley Jevons, Carl Menger, and Léon Walras, in three different countries, England, Austria, and Switzerland, simultaneously, but independently, made discoveries that profoundly changed economics. They broke with classical economics in terms of the basis for valuation of goods and services.[1]

The classical economists based the value of a commodity on its production cost, thus bestowing on it a kind of intrinsic value. Similarly, from the consumer point of view, the value depended on the amount of need for a commodity or pleasure derived from its consumption. But the marginalists or neoclassical economists, as they came to be known, based the value of a commodity or service on the cost of producing the last unit produced or the utility of the last unit consumed—hence the ideas of marginal cost and marginal utility. This way of thinking explained why a good for which there was no demand would be disposed of regardless of its production cost, and why air that is so vital for our survival is free, but diamonds that we can live without are so expensive.

The main idea of neoclassical economics was the maximization of utility by consumers and profit by producers. At the same time marginal cost and marginal utility had a ready-made counterpart in calculus in the form of the derivative. Maximizing utility or minimizing cost was a special case of optimization. Hence the use of mathematics in economics intensified. In 1892 Irving Fisher published his dissertation, which extensively used mathematics. Many economists used the tools of calculus and tackled different economic problems. It was Paul Samuelson who, in his *Foundations of Economic Analysis* (1947), presented the marginal analysis in a unified manner. Today, marginal analysis and the use of calculus permeate every corner of economics.

The importance of differentiation is not confined to microeconomics. Differentiation is the foundation of dynamic optimization and differential equations,

[1] These discoveries had their origins in the early nineteenth century and the ideas of Johan Heinrich von Thünen, Antoine Augustin Cournot, and Hermann Heinrich Gossen.

K. Dadkhah, *Foundations of Mathematical and Computational Economics*, 2nd ed.,
DOI 10.1007/978-3-642-13748-8_8, © Springer-Verlag Berlin Heidelberg 2011

which are the main tools of dynamic analysis in macroeconomics. Finally, as we will see in Chaps. 12 and 13, derivatives and optimization play a significant role in statistical and econometric estimation and inference.

8.1.1 Marginal Concepts and Derivatives

We shall illustrate the connection between the derivative of a function and marginal concepts in economics such as marginal cost, marginal revenue, and marginal utility, through a few examples.

Example 8.1 Suppose a company's cost function can be written as

$$C = C(Q) = a + bQ \tag{8.1}$$

where C represents total cost, Q, output, and a and b are known constants. If the company produces 350 units of output, its total cost will be

$$C_1 = a + 350\,b$$

Now suppose the company increases its output to 400. The cost will be

$$C_2 = a + 400b$$

The additional cost of the extra 50 units is

$$\Delta C = C_2 - C_1 = a + 400b - a - 350b = 50b$$

The total additional cost is $50b$ and because we have 50 additional units of output, b represents the additional cost of each unit.

$$\frac{\Delta C}{\Delta Q} = \frac{C_2 - C_1}{Q_2 - Q_1} = \frac{50b}{50} = b \tag{8.2}$$

Thus, the ratio of the additional cost to additional output is constant. This makes life easy because whether we add 50, 100, or 1 or even one-tenth of a unit of output, the ratio of additional cost to additional output remains constant and equal to b. Let us make the game a bit more interesting. Suppose the total cost function is

$$C = C(Q) = a + bQ + cQ^2$$

Now if we go from 350 units to 400 units, the ratio of additional cost to additional output is

$$\begin{aligned}\frac{\Delta C}{\Delta Q} &= \frac{a + 400b + 160000c - a - 350b - 122500c}{400 - 350} \\ &= \frac{50b + 37500c}{50} \\ &= b + 750c\end{aligned}$$

This ratio is not constant and depends on the number of additional units produced. For instance, if output is increased from 350 to 360, we have

$$\frac{\Delta C}{\Delta Q} = b + 710c$$

The idea of marginal cost is to measure the additional cost resulting from an infinitesimal increase in output. To see how this works, let us consider the general case of output increasing from Q to $Q + \Delta Q$. We have

$$\begin{aligned}\frac{\Delta C}{\Delta Q} &= \frac{a + b(Q + \Delta Q) + c(Q + \Delta Q)^2 - a - bQ - cQ^2}{\Delta Q} \\ &= \frac{b\Delta Q + 2cQ\Delta Q + c(\Delta Q)^2}{\Delta Q}\end{aligned}$$

The idea of marginal cost is to make ΔQ as small as possible, indeed make it go to zero and then calculate the ratio. At this point you might say, wait a minute, something divided by zero goes to infinity. That is true, but we have one more trick up our sleeve. Let us simplify the ratio. We get

$$\frac{\Delta C}{\Delta Q} = b + 2cQ + c\Delta Q$$

Now if we let ΔQ to approach zero, the limit (in the next section we shall clarify what we mean by this) of the ratio is the marginal cost:

$$MC = b + 2cQ$$

Note that the marginal cost, MC, depends on the level of output. Thus, as soon as we know the level of output, we can calculate the additional cost of increasing output by an infinitesimal amount. As we shall see in this chapter, the limit of the ratio$\Delta C/\Delta Q$ when ΔQ goes to zero is the derivative of the cost function with respect to output. What we showed in the case of cost function can be repeated for revenue function (see Exercise E.8.1).

Example 8.2 Consider the utility function of a worker or consumer, which solely depends on income, x. We can write

$$U = U(x) \tag{8.3}$$

When income increases from x_1 to x_2, the utility increases from $U(x_1)$ to $U(x_2)$, and the marginal utility is

$$\frac{U(x_2) - U(x_1)}{x_2 - x_1} \tag{8.4}$$

Because both numerator and denominator are positive, marginal utility is also positive. For example, let

$$U = \ln x$$

then, marginal utility will be

$$\frac{\ln x_2 - \ln x_1}{x_2 - x_1} = \frac{1}{\Delta x} \ln\left(\frac{x_2}{x_1}\right) = \frac{1}{\Delta x} \ln\left(1 + \frac{\Delta x}{x_1}\right)$$

Because $\Delta x > 0$ and the logarithm of any number greater than one is also positive, it follows that marginal utility is positive.

Now, you may wonder why we need these concepts in economics in the first place. Calculus deals with continuous variables and infinitesimal changes. But most economic variables such as the number of cars produced or houses sold, as well as the number of workers hired are discrete variables. Furthermore, no one hires an infinitesimal amount of extra labor and no one cares about an infinitesimal extra income.

It is true that a factory uses discrete units of input and produces discrete units or even discrete batches of output. But the range of possible amounts of inputs and output is quite wide. In a factory the installed capacity may allow for the production of between 1 to 1.5 million units of output. Even if the amount of actual production could be manipulated at batches of 100 units, the available range—500,000 divided by 100—involves 5000 possible points that can practically be treated as a continuous variable (make a graph with even 500 data points to see this).

Moreover, in economics we are usually interested in the behavior of an average firm in the industry or average consumer in the economy. Although the output or input of one particular firm may take only discrete values, the average for the whole industry or economy very much resembles a continuous variable. The same is true for the demand of an average consumer.

Finally, recall that science is an approximation of reality and marginal analysis has proved a potent approximation to economic reality. Calculus is an immensely powerful tool both in representing different marginal concepts and in optimization, which is the backbone of economic analysis (see Part IV). On the other hand, working with discrete changes could be quite cumbersome with little or no advantage in furthering economic analysis.

8.1.2 Comparative Static Analysis

We are familiar with the concept of equilibrium in economics, which describes the state where there is no tendency for change in the system. In other words, the forces in the system are in such a configuration that their resultant is zero. The analysis of equilibrium points in isolation is referred to as *static analysis* because it studies the system when it is at rest and has no tendency to change. Calculus and the concept of derivatives allow us to study the effects of a change in one or more forces acting on the system.

Suppose the supply of and demand for a product are in equilibrium. What happens if consumers' incomes are increased? Or what happens if the government imposes a 5% tax on this product? By using calculus, we can investigate the effects of a change in one variable on other variables. The method is called *comparative static* because it compares two or more equilibrium points.

Comparative static does not concern itself with how we got from one equilibrium to the other or what were the features of the system when in disequilibrium. Comparative static looks at the characteristics of the system after all changes are affected and there is no more tendency for change. Because most economic models consist of several equations and many variables, the method also allows for tracing the effects of a change in one variable through a multitude of equations to determine its effects on other variables. We discuss this latter method in the next chapter in the context of functions of several variables.

Dynamic analysis, which we shall discuss in Chaps. 14 through 17, introduces time in the model in an essential way. In dynamic analysis we shall study the movement of a system over time and from one equilibrium to another, as well as the behavior of the system in disequilibrium.

8.1.3 Exercises

E.8.1 Let the revenue function of a monopolistic firm be

$$R = (A - Q)Q, \qquad A > 0$$

where R denotes revenue and Q, output. Use the same method as in Example 8.1 to derive the marginal revenue of the firm.

E.8.2 Profit is defined as the difference between revenue and costs:

$$\Pi = R(Q) - C(Q)$$

Using the method of Example 8.1 write the "marginal profit" in terms of marginal revenue and marginal cost. Under what conditions does marginal profit equal zero?

E.8.3 Suppose the cost function of a firm can be written as

$$C = \alpha + \beta Q + \gamma Q^2 + \delta Q^3$$

Using the method described in the text, derive the marginal cost of the firm.

8.2 Limit and Continuity

In the above discussion we came across the concepts of *limit* and *continuity* of a function. Later in this chapter, we shall encounter the concept of *differentiability* of a function. These concepts are quite important in mathematics and in the advanced study of economics. We shall first present them intuitively and then more formally and rigorously.

8.2.1 Limit

In Chap. 2, we became familiar with the limit of a series. For example, if $|q| < 1$,

$$\lim_{n\to\infty} q^n = 0 \tag{8.5}$$

We briefly discussed the limiting behavior of a function of a continuous variable, that is, a variable that takes all real values and not just integers.

$$\lim_{r\to\infty}\left(1+\frac{1}{r}\right)^r = e \tag{8.6}$$

In this chapter we discuss in more detail the idea of limits of functions of continuous variables. Consider the function $f(x) = x^2 + 3x - 7$ and the point $x_0 = 2$. We can allow x to gradually approach x_0. Table 8.1 reflects the process of approaching x_0 from values both larger and smaller than 2. As x gets closer and closer to $x_0 = 2$, $f(x)$ gets closer and closer to $f(x_0) = 3$. This occurs whether we approach x_0 from above or from below.

Table 8.1 Approaching the limit of $f(x)$ at $x = x_0$

X	$f(x)$	x	$f(x)$
3	11	1	−3
2.5	6.75	1.5	−0.25
2.2	4.44	1.8	1.64
2.1	3.71	1.9	2.31
2.05	3.3525	1.95	2.6525
2.03	3.2109	1.97	2.7909
2.01	3.0701	1.99	2.9301
2.001	3.007001	1.999	2.993001
2.0001	3.00070001	1.9999	2.99930001

When we approach x_0 from above, we write

$$\lim_{x\to x_0^+} f(x) = f(x_0) = 3$$

and approaching it from below

$$\lim_{x \to x_0^-} f(x) = f(x_0) = 3$$

In this case, the two limits are identical, and $f(x)$ has a limit at x_0. We denote the limit as

$$\lim_{x \to x_0} f(x) = f(x_0) = 3$$

We can put the above process in more general terms. At each stage, we can make the difference between $f(x)$ and $f(x_0)$ smaller than any prespecified positive number ε (for example, in the last stage we had $\varepsilon = 0.00070002$). Note that the difference could be in both the positive and negative direction:

$$|f(x) - f(x_0)| < \varepsilon \tag{8.7}$$

But no matter how small ε is, we can find a positive number δ such that the difference between x and x_0 is less than it. Again, the difference could be in both the positive and negative direction (in the last stage we had $\delta = 0.0001001$):

$$|x - x_0| < \delta \tag{8.8}$$

Once we have established that, we can say that the $\lim_{x \to x_0} f(x)$ exists. Such a case is depicted in Fig. 8.1.

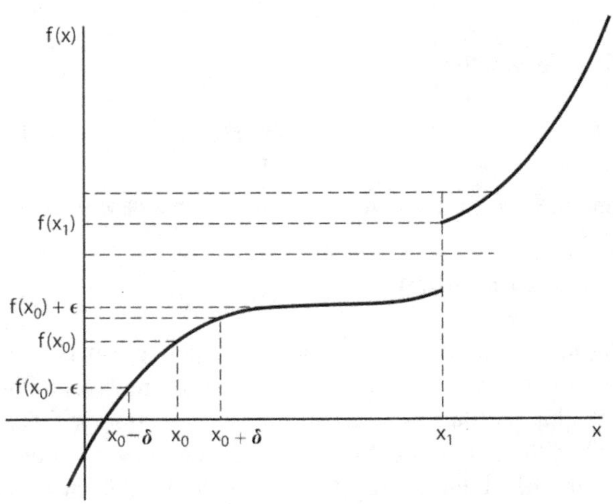

Fig. 8.1 Continuity and discontinuity of a function

We can state this concept in a slightly different language. If we consider a neighborhood of size δ on either side of x_0, we can find a corresponding neighborhood of size ε around $f(x_0)$ that encompasses the interval $[f(x_0 - \delta), f(x_0 + \delta)]$. We can make the first neighborhood as small as possible, but there is still a neighborhood $[f(x_0 - \delta), f(x_0 + \delta)]$ that contains $f(x_0)$. In mathematics, we state this condition the other way around. We say that $f(x)$ has a limit at the point $f(x_0)$, if for every

ε neighborhood around it (no matter how small), we can find a corresponding δ neighborhood around x_0.

Not all functions have a limit for all points. Consider the function

$$f(x) = \frac{x+3}{x-2}$$

As x gets closer and closer to the point $x_0 = 2$, the numerator gets closer and closer to 5. But the ratio gets larger and larger. For example, for $x = 2.1$, we get $f(x) = 51$ and when $x = 2.01, f(x) = 501$. Thus,

$$\lim_{x \to 2} \frac{x+3}{x-2} \to \infty$$

The function does not have a limit at $x_0 = 2$. This is true whether we approach the point from below or above. In some cases the limit from above and the limit from below do not coincide. For example, at the point $x = x_1$, the function

$$f(x) = \begin{cases} x^2 + 3x - 7 & \text{if } x \leq x_1 \\ x^2 + 3x - 15 & \text{if } x \geq x_1 \end{cases}$$

has the left limit

$$\lim_{x \to x_1^-} f(x) = x_1^2 + 3x_1 - 7$$

and the right limit

$$\lim_{x \to x_1^+} f(x) = x_1^2 + 3x_1 - 15$$

and the two limits do not coincide. Such a case is also depicted in Fig. 8.1.

8.2.2 Continuity

Intuitively, *continuity* is an easy concept to grasp. For example, the function depicted in Fig. 8.1 is clearly continuous everywhere except at the point x_1. One can simply say that as long as you can graph a function without lifting the pencil from the paper, the function is continuous. If you have to lift the pencil to get from one point to another, then the function is not continuous at that point. But we need a more precise definition.

Definition 8.1 (Continuity). The function $f(x)$ is continuous at the point x_0 if for any real number $\varepsilon > 0$ there exists a real number $\delta > 0$ such that $|x - x_0| < \delta$ implies that $|f(x) - f(x_0)| < \varepsilon$.

In other words, all the points within the ϵ radius of $f(x_0)$ have been mapped by f from the points within δ radius of x_0 no matter how small we choose ϵ to be. This is not possible if the function is not continuous at x_0. If a function is discontinuous at a point, say x_1, as we make ϵ smaller and smaller, we reach a point where the points around $f(x_1)$ on the y-axis are not mapped from the points within δ radius of x_1. (see Fig. 8.1). The function is continuous at point x_0, but it is not continuous at the point x_1.

A function that is continuous at all points in its domain is called a *continuous function*. A function that is continuous over different segments of its domain, but discontinuous at the connecting points of these segments, is called *piecewise continuous*.

Another way of looking at continuity is quite instructive. Consider the function $y = f(x)$ and assume that x_0 is in its domain and $f(x_0)$ in its range. Furthermore, let

$$\begin{aligned} \Delta x &= x - x_0 \\ \Delta y &= f(x_0 + \Delta x) - f(x_0) \end{aligned} \tag{8.9}$$

Then continuity implies

$$\lim_{\Delta x \to 0} \Delta y = \lim_{\Delta x \to 0} [f(x_0 + \Delta x) - f(x_0)] = 0 \tag{8.10}$$

Because $x = x_0 + \Delta x$, we have

$$\lim_{\Delta x \to 0} f(x_0 + \Delta x) = \lim_{\Delta x \to 0} f(x) = f(x_0) \tag{8.11}$$

or

$$\lim_{\Delta x \to 0} f(x) = f(\lim_{\Delta x \to 0} x) \tag{8.12}$$

because

$$x_0 = \lim_{\Delta x \to 0} x \tag{8.13}$$

Thus, the limit of a continuous function is the function evaluated at the limit. To summarize, the function f is continuous at the point x_0 if this point is in its domain, $f(x_0)$ is in the function's range, and $\lim_{x \to x_0} f(x)$ exists and is equal to $f(x_0)$.

Example 8.3 The function $y = x^2$ is continuous at the arbitrary point x_0:

$$\Delta y = (x_0 + \Delta x)^2 - x_0^2 = 2x_0 \Delta x + (\Delta x)^2$$

Therefore,

$$\lim_{\Delta x \to 0} \Delta y = 2x \lim_{\Delta x \to 0} \Delta x + \lim_{\Delta x \to 0} (\Delta x)^2 = 0$$

Example 8.4 $y = e^x$ is continuous at the arbitrary point x_0.

$$\Delta y = e^{x_0+\Delta x} - e^{x_0} = e^{x_0}(e^{\Delta x} - 1)$$

Therefore,

$$\lim_{\Delta x \to 0} \Delta y = \lim_{\Delta x \to 0} e^{x_0}(e^{\Delta x} - 1) = e^{x_0} \lim_{\Delta x \to 0} (e^{\Delta x} - 1) = 0$$

8.2.3 Exercises

E.8.4 Find the limits of the following functions at the point $x_0 = 5$.

i. $f(x) = x^3 - 2x^2 + 5x + 17$

ii. $f(x) = \dfrac{x+4}{x-5}$

iii. $f(x) = 3\ln x$

iv. $f(x) = xe^{-2x}$

v. $f(x) = \begin{cases} x - 8 & \text{if } x \leq 4 \\ x - 5 & \text{if } x > 4 \end{cases}$

E.8.5 Graph the functions in E.8.4 and determine if they are continuous.

8.3 Derivatives

The derivative is one of the most fundamental concepts in calculus and indeed in mathematics. It measures the ratio of change in one variable to the change in another when the change in the former has been caused by an infinitesimal change in the latter. In other words, a derivative is the instantaneous rate of change of one variable per change in the other.

As an illustration, consider an automobile that has traveled the 200 miles from Boston to New York City in 4 hours. Evidently it has traveled at an average speed of 50 mph. Had the car traveled at a constant speed, we could say that its speed at every moment had been 50 mph. But it is conceivable, indeed more than likely, that the car's speed varied during the trip. We can compute its average speed during any 1 hour, half hour, minute, or even second. Indeed, the odometer measures the instantaneous speed of the car at every moment during the trip. What the odometer shows corresponds to the idea of the derivative of distance traveled with respect to time. Of course it is expressed in terms of mph, that is, it shows the distance that the car would have traveled if it continued at that speed for an hour.

Mathematically, the derivative of a function is the change in the value of the function (in the dependent variable) as a result of an infinitesimal change in its argument (the independent variable). Consider the continuous function

$$y = f(x) \tag{8.14}$$

and the two points (x_1, y_1) and (x_2, y_2), where $y_1 = f(x_1)$ and $y_2 = f(x_2)$. The ratio of the change in $y = f(x)$ to the change in x is

$$\frac{y_2 - y_1}{x_2 - x_1} = \frac{\Delta y}{\Delta x}$$

Now we let Δx get smaller and smaller and indeed tend to zero. The limit of the ratio when $\Delta x \to 0$ (assuming the limit exists) is the derivative of the function at the point x_1. or the derivative of y with respect to x at the point $x = x_1$:

$$\frac{dy}{dx} = \lim_{\Delta x \to 0} \frac{\Delta y}{\Delta x} \tag{8.15}$$

We shall use dy/dx, $f'(x)$, and y' interchangeably to denote the derivative of a function.

Example 8.5 Consider the function

$$f(x) = a + bx$$

then

$$f(x + \Delta x) - f(x) = a + b(x + \Delta x) - a - bx$$

it follows that

$$\frac{f(x + \Delta x) - f(x)}{\Delta x} = \frac{b\Delta x}{\Delta x} = b$$

Because the limit of a constant term is itself, we have

$$f'(x) = b$$

Example 8.6 Let

$$y = a + bx + cx^2$$

we have

$$\begin{aligned}\frac{\Delta y}{\Delta x} &= \frac{a + b(x + \Delta x) + c(x + \Delta x)^2 - a - bx - cx^2}{\Delta x} \\ &= \frac{b\Delta x + 2cx\Delta x + c(\Delta x)^2}{\Delta x} \\ &= b + 2cx + c\Delta x\end{aligned}$$

taking the limit

$$\frac{dy}{dx} = \lim_{\Delta x \to 0} \frac{\Delta y}{\Delta x} = \lim_{\Delta x \to 0} (b + 2cx + c\Delta x) = b + 2cx$$

A comparison of the above result with Example 8.1 shows that marginal cost is the derivative of the total cost function with respect to output.

Example 8.7 Letting $y = x^3$, we have

$$\begin{aligned}\frac{\Delta y}{\Delta x} &= \frac{(x+\Delta x)^3 - x^3}{\Delta x} \\ &= \frac{3x^2\Delta x + 3x(\Delta x)^2 + (\Delta x)^3}{\Delta x} \\ &= 3x^2 + 3x\Delta x + (\Delta x)^2\end{aligned}$$

Taking the limit, we have

$$\frac{dy}{dx} = \lim_{\Delta x \to 0} \frac{\Delta y}{\Delta x} = 3x^2 + 3x \lim_{\Delta x \to 0} \Delta x + \lim_{\Delta x \to 0} (\Delta x)^2 = 3x^2$$

Example 8.8 Let the consumption function be of the form

$$C = C(Y)$$

where C denotes consumption and Y, income. We define *marginal propensity to consume* as the ratio of change in consumption divided by the change in income:

$$MPC = \frac{\Delta C}{\Delta Y}$$

In the limit when $\Delta Y \to 0$, we have:

$$MPC = \lim_{\Delta Y \to 0} \frac{\Delta C}{\Delta Y} = C'(Y)$$

8.3.1 Geometric Representation of Derivative

Let us make the concept of derivatives clearer with the help of geometric representation. In Fig. 8.2

$$\frac{y_2 - y_1}{x_2 - x_1} = \frac{\Delta y}{\Delta x}$$

Measures the ratio of change in y to change in x.

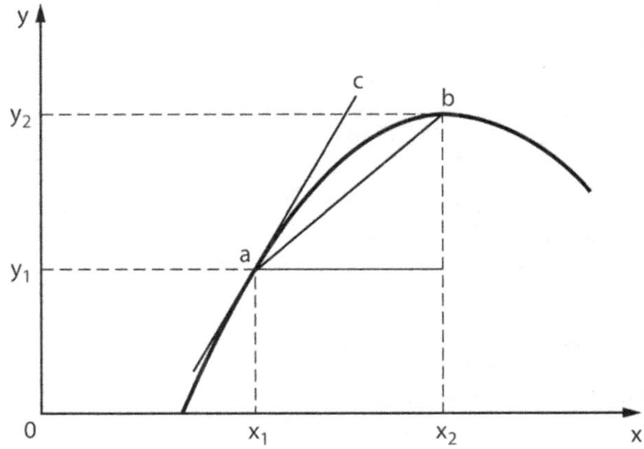

Fig. 8.2 Geometric representation of derivative

If we move the point x_2 closer and closer to point x_1, point b slides down along the curve and gets closer and closer to point a. In the limit, when the distance between x_2 and x_1. approaches zero, point b moves to point a and the line ab coincides with ac, which is the tangent to the curve at point a. This process is the same as letting Δx approach zero. If the limit exists, we will have the derivative of y with respect to x

$$\frac{dy}{dx} = \lim_{\Delta x \to 0} \frac{\Delta y}{\Delta x} \tag{8.16}$$

which is the equation of the tangent to the curve at the point a. Because the derivative is the equation of the tangent to the curve, it follows that when the derivative is positive, the function is increasing and when it is negative, the function is decreasing. The case of a zero derivative, which will prove quite important in later chapters, signifies a flat point (or segment) in a function.

8.3.2 *Differentiability*

Continuity is a necessary but not a sufficient condition for differentiability. That is, a differentiable function is continuous, but the reverse is not necessarily true. In Fig. 8.3a the function is continuous at x_0 but is not differentiable at that point. The reason is the sharp point of the function at x_0. On the other hand, the function in Fig. 8.3b is differentiable at the point x_0. Intuitively, therefore, a differentiable function is smooth.

Definition 8.2 The continuous function $y = f(x)$ is differentiable at the point x_0 if

$$\lim_{\Delta x \to 0} \frac{\Delta y}{\Delta x} = \lim_{\Delta x \to 0} \frac{f(x_0 + \Delta x) - f(x_0)}{\Delta x} \tag{8.17}$$

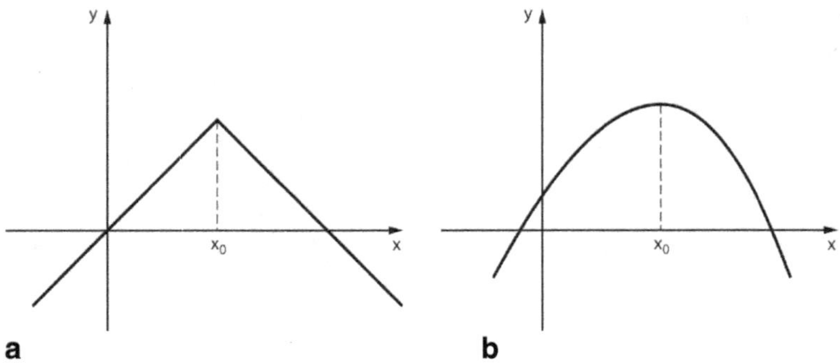

Fig. 8.3 (**a**) Continuous but not differentiable at x_0; (**b**) continuous and differentiable at x_0

exists. This means that for every $\varepsilon > 0$, there exists a number $\delta > 0$ such that

$$|\Delta x| < \delta$$

implies

$$\left|\frac{\Delta y}{\Delta x} - f'(x_0)\right| < \varepsilon$$

The limit denoted by $f'(x_0)$ is the derivative of f at the point x_0. A function that is differentiable at all points in its domain is called continuously differentiable. Within the domain where f is differentiable, the derivative itself is a function and is denoted in one of the following ways:

$$\frac{dy}{dx} = f'(x) = y'$$

Recall that as $\Delta x \to 0$ so does Δy. Therefore, the derivative exists if $\Delta y/\Delta x$ does not take the form 0/0 or the indeterminacy could be resolved. In particular, we require that the limit of $\Delta y/\Delta x$ be the same whether Δx approaches zero from above or below. We illustrate this point with two examples.

Example 8.9 In Fig. 8.3a we have

$$y = \begin{cases} x & \text{if } x \le x_0 \\ 2x_0 - x & \text{if } x \ge x_0 \end{cases}$$

Therefore, when approaching x_0 from below, we have

$$\lim_{\Delta x \to 0^+} \frac{\Delta y}{\Delta x} = \lim_{\Delta x \to 0^+} \frac{x_0 + \Delta x - x_0}{\Delta x} = 1$$

whereas when approaching it from above, we get

$$\lim_{\Delta x\to 0^-}\frac{\Delta y}{\Delta x}=\lim_{\Delta x\to 0^-}\frac{(2x_0-x_0-\Delta x)-(2x_0-x_0)}{\Delta x}=-1$$

Thus the two limits do not coincide, and the function is not differentiable at the point x_0.

Example 8.10 Letting $y = x^2$, we have

$$\frac{\Delta y}{\Delta x}=\frac{(x+\Delta x)^2-x^2}{\Delta x}=\frac{2x\Delta x+(\Delta x)^2}{\Delta x}=2x+\Delta x$$

Taking the limit, we have

$$\lim_{\Delta x\to 0}\frac{\Delta y}{\Delta x}=2x+\lim_{\Delta x\to 0}\Delta x=2x$$

Here the limit exists irrespective of how we approach x_0. As mentioned before, a differentiable function is continuous. Let us first get an intuitive understanding of this point. Recall that

$$\lim_{\Delta x\to 0}\frac{\Delta y}{\Delta x}=f'(x_0)$$

Multiplying through by Δx, we have

$$\lim_{\Delta x\to 0}\Delta y=\lim_{\Delta x\to 0}[f(x_0+\Delta x)-f(x_0)]\;=f'(x_0)\lim_{\Delta x\to 0}\Delta x=0$$

which is the same as (8.1). More rigorously, we have the following theorem.

Theorem 8.1 *If f is differentiable at x_0, then f is continuous at x_0.*

Proof From the definition of differentiability we have

$$\left|\Delta y-\Delta x f'(x_0)\right|<\varepsilon\left|\Delta x\right|$$

$$|\Delta y|\le\left|\Delta y-f'(x_0)\Delta x\right|+\left|f'(x_0)\Delta x\right|<(\varepsilon+\left|f'(x_0)\right|)\left|\Delta x\right|$$

Now define

$$\delta'=\min\left(\delta,\;\frac{\varepsilon'}{\varepsilon+|f'(x_0)|}\right)$$

Then whenever $|\Delta x|<\delta'$, we have $|\Delta y|<\varepsilon'$. Therefore, f is continuous at x_0.

8.3.3 Rules of Differentiation

There are a number of basic rules for differentiation, the mastery of which is essential for both macroeconomic and microeconomic analysis. We shall introduce these rules and prove some of them. Given the importance of differentiation in all fields of economics, I recommend that the reader memorize the rules and apply them to a large number of problems, starting with those at the end of this section. For easy reference, the most important rules are collected in Table 8.2.

Table 8.2 Rules of differentiation and properties of derivatives

$y = c$	$y' = 0$
$y = bx$	$y' = b$
$y = x^n$	$y' = nx^{n-1}$
$y = \ln x$	$y' = \frac{1}{x}$
$y = a^x$	$y' = a^x \ln a$
$y = e^x$	$y' = e^x$
$y = \sin x$	$y' = \cos x$
$y = \cos x$	$y' = -\sin x$
$y = u(x) + v(x) + w(x)$	$y' = u' + v' + w'$
$y = f(u), \quad u = \varphi(x)$	$y' = f'(u)\varphi'(x)$
$y = u(x)v(x)$	$y' = u'v + v'u$
$y = \frac{u(x)}{v(x)}$	$y' = \frac{u'v - v'u}{v^2}$

c, b, n, a are all real constants and $a > 0$

The first rule is that the derivative of a constant function is zero. The reason is that a constant function does not change, hence its derivative (or its rate of change) is zero. In Examples 8.5, 8.6, and 8.7, we have shown how derivatives of linear, quadratic, and cubic functions are calculated. Here we show the general rule for the function $y = x^n$. The formula shown in Table 8.2 is quite general and pertains to all real values of n for which x^n is also real. But we will prove the rule for the case where n is a positive integer. The case of a negative integer can be similarly proved.

Theorem 8.2 *Let $y = x^n$ where n is a positive integer. Then $y' = nx^{n-1}$.*

Proof Consider

$$\Delta y = (x + \Delta x)^n - x^n$$

Using binomial expansion, we have

$$\Delta y = x^n + \binom{n}{1} x^{n-1}\Delta x + \binom{n}{2} x^{n-2}(\Delta x)^2 + \cdots + (\Delta x)^n - x^n$$
$$= nx^{n-1}\Delta x + \binom{n}{2} x^{n-2}(\Delta x)^2 + \cdots + (\Delta x)^n$$

where

$$\binom{n}{r} = \frac{n!}{(n-r)!r!}$$

Dividing through by Δx

$$\frac{\Delta y}{\Delta x} = nx^{n-1} + \binom{n}{2} x^{n-2}(\Delta x) + \cdots + (\Delta x)^{n-1}$$

and letting $\Delta x \to 0$

$$\frac{dy}{dx} = \lim_{\Delta x \to 0} \frac{\Delta y}{\Delta x} = nx^{n-1} + \lim_{\Delta x \to 0} \sum_{i=2}^{n} \binom{n}{i} x^{n-i}(\Delta x)^{i-1} = nx^{n-1}$$

Example 8.11 Using rules of differentiation, the derivative of the function

$$y = x^3$$

is

$$\frac{dy}{dx} = 3x^2$$

Example 8.12 Derivative of the function

$$y = x^{1/2}$$

is

$$\frac{dy}{dx} = \frac{1}{2}x^{-1/2}$$

Example 8.13 (Marginal Cost). In microeconomics we learn that cost function relates the total cost of production to the amount of output. In the short run, cost function consists of two parts: fixed cost, which is constant irrespective of the amount of production, and variable cost, which depends on output. In the long run all costs are variable costs. Letting C, C_0, and Q denote total cost, fixed cost, and output, respectively, the short-run cost function will be

$$C = C_0 + C(Q)$$

The average cost will be

$$AC = \frac{C}{Q} = \frac{C_0 + C(Q)}{Q}$$

The marginal cost is the incremental change in total cost as a result of an incremental change in output. That is,

$$MC = \frac{dC}{dQ} = C'(Q)$$

Note that marginal cost is marginal variable cost. We return to the rules of differentiation and prove the following theorem.

Theorem 8.3 *Let* $f(x) = \ln(x)$, *then* $\dfrac{dy}{dx} = \dfrac{1}{x}$.

Proof

$$\begin{aligned}\frac{\Delta y}{\Delta x} &= \frac{\ln(x + \Delta x) - \ln(x)}{\Delta x} \\ &= \frac{1}{\Delta x}\ln\left(\frac{x + \Delta x}{x}\right) \\ &= \frac{1}{x}\frac{x}{\Delta x}\ln\left(1 + \frac{\Delta x}{x}\right) \\ &= \frac{1}{x}\ln\left(1 + \frac{\Delta x}{x}\right)^{\frac{x}{\Delta x}}\end{aligned}$$

Taking the limit

$$\frac{dy}{dx} = \frac{1}{x}\ln \lim_{\Delta x \to 0}\left(1 + \frac{\Delta x}{x}\right)^{\frac{x}{\Delta x}} = \frac{1}{x}\ln e = \frac{1}{x}$$

Example 8.14 If the utility function depends on income alone, we can find the marginal utility as

$$\frac{dU}{dx} = U'(x)$$

For example, if $U = \ln x$, we have

$$U'(x) = \frac{d}{dx} \ln x = \frac{1}{x}$$

Note that marginal utility is positive, as the logarithm is defined only for $x > 0$. But as income increases, marginal utility declines.

8.3.4 Properties of Derivatives

Differentiation will become much simpler once we learn some properties of derivatives.

Theorem 8.4 *Let* $y = u(x)+v(x)+w(x)$ *where* u, v, and w *are differentiable functions of* x. *Then*

$$\frac{dy}{dx} = y'(x) = u'(x) + v'(x) + w'(x) \tag{8.18}$$

Indeed, the theorem is true for any number of functions of x. Proof of this theorem is left to the reader.

Example 8.15 Let

$$y = 2 + 5x - x^{1/2}$$

Applying the rules of differentiation

$$y' = 5 - \frac{1}{2}x^{-1/2}$$

Example 8.16 Let

$$y = x^6 - e^x$$

Again applying the rules of differentiation

$$y' = 6x^5 - e^x$$

Theorem 8.5 (Chain Rule).[2] *Let* $y = f(u)$ *and* $u = \phi(x)$. *Then*

$$\frac{dy}{dx} = \frac{dy}{du}\frac{du}{dx} = f'(u)\phi'(x) \tag{8.19}$$

---

[2]Also known as the composite function rule.

Proof We know that

$$\frac{dy}{du} = \lim_{\Delta x \to 0} \frac{\Delta y}{\Delta u}$$

Therefore, we can write

$$\frac{\Delta y}{\Delta u} = \frac{dy}{du} + \eta$$

or

$$\Delta y = f'(u)\Delta u + \eta \Delta u$$

where η depends on Δu. Dividing through by Δx and taking the limit and noting that

$$\Delta x \to 0 \quad \Rightarrow \quad \Delta u \to 0 \quad \Rightarrow \quad \eta \to 0$$

we have

$$\frac{dy}{dx} = \lim_{\Delta x \to 0} \frac{\Delta y}{\Delta x} = f'(u) \lim_{\Delta x \to 0} \frac{\Delta u}{\Delta x} + \lim_{\Delta x \to 0} \eta \lim_{\Delta x \to 0} \frac{\Delta u}{\Delta x} = f'(u)\phi'(x)$$

Example 8.17 Find dy/dx given

$$y = u^4, \qquad u = x^2 - 1$$

Because

$$\frac{dy}{du} = 4u^3, \qquad \frac{du}{dx} = 2x$$

we have

$$\frac{dy}{dx} = 4u^3(2x) = 8x(x^2 - 1)^3$$

Example 8.18 Given

$$y = 5e^{4x^2}$$

Find *dy/dx*. To solve the problem, let $u = 4x^2$ and $v = e^u$. Then

$$y = 5v$$

and using the chain rule

$$\frac{dy}{dx} = \frac{dy}{dv}\frac{dv}{du}\frac{du}{dx} = 5e^u(8x) = 40xe^{4x^2}$$

The chain rule is a useful device for finding derivatives and one has to practice it often so that the intermediate steps can be skipped.

Theorem 8.6 (Product Rule). *Let $y = uv$ where u and v are both differentiable functions of x. Then*

$$y' = u'v + v'u \tag{8.20}$$

Proof

$$\begin{aligned}
\frac{dy}{dx} &= \lim_{\Delta x \to 0} \frac{\Delta y}{\Delta x} \\
&= \lim_{\Delta x \to 0} \frac{(u + \Delta u)(v + \Delta v) - uv}{\Delta x} \\
&= \lim_{\Delta x \to 0} \frac{u\Delta v + v\Delta u + \Delta u \Delta v}{\Delta x} \\
&= \lim_{\Delta x \to 0} v\frac{\Delta u}{\Delta x} + \lim_{\Delta x \to 0} u\frac{\Delta v}{\Delta x} + \lim_{\Delta x \to 0} \frac{\Delta u}{\Delta x} \lim_{\Delta x \to 0} \Delta v \\
&= v\frac{du}{dx} + u\frac{dv}{dx} \\
&= vu' + uv'
\end{aligned}$$

Example 8.19 Consider the function

$$y = xe^{2x}$$

Let

$$u = x, \quad v = e^{2x}$$

Then

$$y' = u'v + v'u = e^{2x} + 2xe^{2x}$$

Example 8.20

$$y = (x^3 - 2x)\ln x$$

Let

$$\begin{aligned}
u &= (x^3 - 2x), \qquad v = \ln x \\
y' &= u'v + v'u \\
&= (3x^2 - 2)\ln x + \frac{x^3 - 2x}{x} \\
&= (3x^2 - 2)\ln x + x^2 - 2
\end{aligned}$$

Example 8.21 The quantity theory of money provides a link between money, M, output, Y, and price level, P:

$$MV = PY$$

where V is the velocity of circulation. Let us assume that all variables in the equation are functions of time, t, except V, which is assumed to be constant. Then

$$V\frac{dM}{dt} = Y\frac{dP}{dt} + P\frac{dY}{dt}$$

Dividing both sides of the above equation by the $MV = PY$, we have

$$\frac{dM/dt}{M} = \frac{dP/dt}{P} + \frac{dY/dt}{Y}$$

Thus, the rate of growth of the money supply is equal to the real growth rate of the economy plus the inflation rate. We could derive the same relationship by first taking the logarithm of both sides of the quantity theory equation:

$$\ln M + \ln V = \ln P + \ln Y$$

and then differentiating both sides with respect to t

$$\frac{dM/dt}{M} = \frac{dP/dt}{P} + \frac{dY/dt}{Y}$$

In other words, the derivative of the logarithm of a variable with respect to time equals its rate of growth.

The product rule can be extended to the case of three and more functions. Consider

$$y = uvw \tag{8.21}$$

And let $z = uv$, then

$$y' = z'w + w'z$$

Because $z' = u'v + v'u$, we have

$$\begin{aligned} y' &= z'w + w'z \\ &= \left(u'v + v'u\right) w + w'z \\ &= u'vw + v'uw + w'uv \end{aligned} \tag{8.22}$$

Example 8.22 Let

$$y = uvw$$

and

$$u = t^2, \quad v = e^t, \quad w = \frac{1-t}{1+t}$$

Because

$$u' = 2t, \quad v' = e^t, \quad w' = \frac{-2}{(1+t)^2}$$

we have

$$\begin{aligned} y' &= u'vw + v'uw + w'uv \\ &= 2te^t \frac{1-t}{1+t} + e^t t^2 \frac{1-t}{1+t} + \frac{-2}{(1+t)^2} t^2 e^t \\ &= -\frac{te^t}{(1+t)^2}(t^3 + 2t^2 + t - 2) \end{aligned}$$

The product rule also allows us to prove another rule of differentiation.

Theorem 8.7 *Let*

$$y = u^v \tag{8.23}$$

where u and v are differentiable functions of x and $u > 0$. Then

$$\frac{dy}{dx} = u'vu^{v-1} + v'u^v \ln u \tag{8.24}$$

Proof Let

$$z = \ln y$$

which implies

$$\frac{dz}{dx} = \frac{1}{y}\frac{dy}{dx}$$

On the other hand, because

$$\ln y = v \ln u$$

we have

$$\frac{dz}{dx} = v' \ln u + \frac{u'}{u} v$$

Therefore,

$$\begin{aligned} \frac{dy}{dx} &= y \frac{dz}{dx} \\ &= y(v' \ln u + \frac{u'}{u} v) \\ &= u^v (v' \ln u + \frac{u'}{u} v) \\ &= v'u^v \ln u + u'vu^{v-1} \end{aligned}$$

Note that rules proved in Theorems 8.2 and 8.3 are special cases of this theorem.

Example 8.23 For the function

$$y = (2x^2 + 3)^{5x+1}$$

we have

$$\frac{dy}{dx} = 5(2x^2 + 3)^{5x+1} \ln(2x^2 + 3) + 4x(5x + 1)(2x^2 + 3)^{5x}$$

Theorem 8.8 (Quotient Rule). *Let u and v be differentiable functions of x and*

$$y = \frac{u}{v} \tag{8.25}$$

Then

$$\frac{dy}{dx} = \frac{u'v - v'u}{v^2} \tag{8.26}$$

Proof Write the function as

$$y = uv^{-1}$$

and let $w = v^{-1}$.

Then

$$y = uw$$

Applying the product rule and the chain rule, we have

$$\frac{dy}{dx} = u'w + w'u = u'v^{-1} + (-v'v^{-2})u = \frac{u'}{v} - \frac{uv'}{v^2} = \frac{u'v - v'u}{v^2}$$

Example 8.24 We use the quotient rule to find the derivative of the following function:

$$y = \frac{x^2 - 3}{1 + 2x}$$

Let

$$u = x^2 - 3, \qquad v = 1 + 2x$$

We have

$$y' = \frac{u'v - v'u}{v^2} = \frac{2x(1 + 2x) - 2(x^2 - 3)}{(1 + 2x)^2} = \frac{2x^2 + 2x + 6}{(1 + 2x)^2}$$

Example 8.25 The derivative of the function

$$y = \frac{\sin x}{\cos x}$$

is

$$y' = \frac{\cos^2 x + \sin^2 x}{\cos^2 x} = \frac{1}{\cos^2 x}$$

Recall that

$$\frac{\sin x}{\cos x} = \tan x$$

Therefore, $1/\cos^2 x$ is the derivative of the tangent of an angle.

Example 8.26 Let Y denote national income and P the population of a country. Per capita income is defined as

$$y = \frac{Y}{P} \tag{8.27}$$

Let us assume that both Y and P are functions of time t. Then

$$\frac{dy}{dt} = \frac{P\dfrac{dY}{dt} - Y\dfrac{dP}{dt}}{P^2} \tag{8.28}$$

Divide both sides of (8.28) by (8.27)

$$\frac{1}{y}\frac{dy}{dt} = \frac{1}{Y}\frac{dY}{dt} - \frac{1}{P}\frac{dP}{dt} \tag{8.29}$$

In words, the instantaneous rate of growth of per capita income is equal to the rate of growth of national income less the growth rate of population.

A very important theorem pertaining to derivatives is the *mean value theorem*.

Theorem 8.9 (Mean Value). *Suppose $y = f(x)$ is a continuously differentiable function on the interval $[a, b]$. Then there is a point $a < x^* < b$ such that*

$$f(b) - f(a) = f'(x^*)(b - a)$$

We will not prove this theorem and instead illustrate it with two examples.

Example 8.27 Let $f(x) = x^2$ and $a = 2$, $b = 5$. We have

$$5^2 - 2^2 = 2x^*(5 - 2)$$

and

$$x^* = \frac{21}{6} = 3.5$$

Example 8.28 Let $f(x) = e^{2x}$ and $a = 0$, $b = 1$.

We have

$$e^2 - 1 = 2e^{2x^*}$$

and

$$x^* = \frac{1}{2} \ln\left(\frac{e^2 - 1}{2}\right) = 0.58072$$

8.3.5 l'Hôpital's Rule

Consider the following limit:

$$\lim_{x \to 2} \frac{x - 2}{x^2 - 4}$$

This limit is not defined because we get 0/0. We can however find the limit by noting that

$$\lim_{x \to 2} \frac{x - 2}{x^2 - 4} = \lim_{x \to 2} \frac{1}{x + 2} = \frac{1}{4}$$

Such a simple solution doesn't always exit, that is why l'Hôpital's rule[3] is so useful. It says that faced with a limit that is undetermined, we can compute it as

$$\lim_{x \to a} \frac{u(x)}{v(x)} = \lim_{x \to a} \frac{u'(x)}{v'(x)} \tag{8.30}$$

In other words, in such cases, the limit can be obtained by taking the derivatives of numerator and denominator separately and then compute the imit. In the above example, the derivative of numerator is 1 and that of the denominator $2x$, thus

$$\lim_{x \to 2} \frac{x - 2}{x^2 - 4} = \lim_{x \to 2} \frac{1}{2x} = \frac{1}{4}$$

Example 8.29 Consider the following limit

$$\lim_{x \to 0} \frac{x}{e^x - 1}$$

[3]After the French mathematician Guillaume l'Hôpital (1661–1704) who was a student of john Bernoulli.

which is undetermined. Using l'Hôpital's rule we have

$$\lim_{x\to 0} \frac{x}{e^x - 1} = \lim_{x\to 0} \frac{1}{e^x} = 1$$

Example 8.30 l'Hôpital's rule finds an interesting application in econometrics. One issue in estimating a model is its functional form. For instance, should the variables enter in linear or log form in the equation. In other words, which of the following equations fits the data best?

$$y = \alpha_1 + \beta_1 x + u, \qquad \text{or} \qquad \log y = \alpha_2 + \beta_2 \log x + v$$

Needless to say there can be more variables in the equation. One way to test between the two models is to use Box-Cox transformation and estimate the model:

$$\frac{y^\lambda - 1}{\lambda} = \alpha + \beta \frac{x^\lambda - 1}{\lambda} + w \tag{8.31}$$

Using nonlinear methods, we can estimate λ, α, and β. If we cannot reject the null hypothesis of $\lambda = 1$, the appropriate model is linear. On the other hand, if we cannot reject the null hypothesis of $\lambda = 0$, then the logarithmic model is preferred. The reason is that

$$\frac{y^0 - 1}{0} = \frac{0}{0}$$

But using l'Hôpital's rule and taking the derivative of numerator and denominator with respect to λ, we have

$$\lim_{x\to 0} \frac{y^\lambda - 1}{\lambda} = \lim_{x\to 0} \frac{y^\lambda \ln y}{1} = \ln y \tag{8.32}$$

8.3.6 Exercises

E.8.6 Find the derivative of the following functions:

i. $y = \dfrac{1}{x^2}$ *ii.* $y = \sqrt{2x}$

iii. $y = \dfrac{2x - 1}{\sqrt{x}}$ *iv.* $y = 3x^{1/2} + x^{1/3} + x^{-1}$

v. $y = \dfrac{x}{1 + x^2}$ *vi.* $y = \dfrac{\sin x}{1 + \cos x}$

vii. $y = \ln(x^3 + 2x - 5) \quad x > 2$

viii. $y = xe^{-x}$ *ix.* $y = 5(x - 4)\sin x$

E.8.7 Using l'Hôpital's rule show that $\lim_{n\to\infty} n\lambda^n = 0$.

8.4 Monotonic Functions and the Inverse Rule

A monotonic function is always heading up or down and does not reverse course. Such functions play an important role in economic analysis, particularly in utility theory. A monotonically increasing function is always upward sloping, that is, as x increases so does $f(x)$. With a monotonically decreasing function as we move rightward along the x-axis, we observe lower and lower $f(x)$. A nondecreasing function is the same as an increasing function except that it contains flat segments where the function is neither increasing nor decreasing. Similarly, a nonincreasing function is the same as a decreasing function except that it has flat segments. More formally,

Definition 8.3 Let $y = f(x)$ be a function defined over the domain $D \subset \Re$ and let x_1 and x_2 be two arbitrary points such that $x_1 < x_2$. Let $y_1 = f(x_1)$ and $y_2 = f(x_2)$. Then f is called

Monotonically increasing	if	$y_2 > y_1$	$\forall x_1, x_2,$	$x \in D$
Monotonically decreasing	if	$y_2 < y_1$	$\forall x_1, x_2,$	$x \in D$
Monotonically nondecreasing	if	$y_2 \geq y_1$	$\forall x_1, x_2,$	$x \in D$
Monotonically nonincreasing	if	$y_2 \leq y_1$	$\forall x_1, x_2,$	$x \in D$

By graphing a function, we can see if it is monotonic or not as well as determining if it is increasing, nondecreasing, decreasing, or nonincreasing. Therefore, I encourage the reader to use Matlab, Maple, or Excel to graph the functions in the examples below and verify that they are correctly classified. But we cannot solely rely on this device. Recall that if a function is increasing, its derivative is positive, and vice versa. Similarly, a decreasing function has a negative derivative and a negative derivative signifies a decreasing function. Thus, we have a ready-made tool to determine the monotonicity of a function and its nature.

Monotonically increasing	if	$\Leftrightarrow f'(x) > 0,$	$x \in D$
Monotonically decreasing	if	$\Leftrightarrow f'(x) < 0,$	$x \in D$
Monotonically nondecreasing	if	$\Leftrightarrow f'(x) \geq 0,$	$x \in D$
Monotonically nonincreasing	if	$\Leftrightarrow f'(x) \leq 0,$	$x \in D$

Example 8.31 The function

$$y = a + bx \qquad b > 0$$

is monotonically increasing because

$$\frac{dy}{dx} = b > 0$$

On the other hand the function

$$y = a + kx \qquad k < 0$$

is monotonically decreasing because

$$\frac{dy}{dx} = k < 0$$

Example 8.32 The function

$$f(x) = e^{2x}$$

is monotonically increasing because

$$f'(x) = 2e^{2x} > 0, \qquad \forall x$$

On the other hand, the function

$$f(x) = e^{-2x}$$

is monotonically decreasing because

$$f'(x) = -2e^{-2x} < 0, \qquad \forall x$$

Example 8.33 Trigonometric functions are periodic and, therefore, are not monotonic. Let us check this fact for $f(x) = \sin x$.

$$\frac{d}{dx}\sin x = \cos x \begin{cases} > 0 & -\frac{\pi}{2} < x < \frac{\pi}{2} \\ < 0 & \frac{\pi}{2} < x < \frac{3\pi}{2} \end{cases}$$

Example 8.34 A utility function is a monotonically increasing function because it is assumed that more is preferred to less. Let us assume that utility depends solely on income x. If $x_2 > x_1$, then we have $U(x_2) > U(x_1)$. Any monotonically increasing transformation of the utility function could also serve as the utility function because it leaves this characteristic unchanged. Let f be a monotonic transformation and $V(x) = f(U(x))$. Then

$$V' = f'U'$$

Because both f' and U' are positive, so is V', which means that V is monotonically increasing and, if $U(x_2) > U(x_1)$, then $V(x_2) > V(x_1)$. Because utility is an ordinal variable, it would make no difference whether we measure it in units of U or V.

Thus, if U represents the utility function of a consumer so does $\ln U$, e^U and any other monotonically increasing function of U. We should note that the same is not true for an expected utility function.

Definition 8.4 (Inverse Function). *Let* $y = f(x)$. Then the function $x = f^{-1}(y) = \varphi(y)$ is called the inverse function of f, which when applied to f we have

$$\varphi(f(x)) = f^{-1}(f(x)) = f^{-1}(y) = x \tag{8.33}$$

We state, without proof, the following results regarding monotonic functions.

Theorem 8.10 *If f is monotonic (increasing or decreasing) and continuous in the domain D, then its inverse is continuous over the range of f.*

Theorem 8.11 (Inverse Function Rule). *Let $y = f(x)$ be a differentiable function within its domain D, and suppose that its inverse $x = \phi(y)$ exists. Further assume that $f'(x) \neq 0,\ x \in D$. Then*

$$\frac{dx}{dy} = \phi'(y) = \frac{1}{dy/dx} = \frac{1}{f'(x)} \tag{8.34}$$

Example 8.35 Let

$$y = 3x + 7$$

which implies

$$x = \frac{y-7}{3}$$

Thus,

$$\frac{dx}{dy} = \frac{1}{3} = \frac{1}{\dfrac{dy}{dx}}$$

Example 8.36 Consider the function

$$y = \frac{1}{x}, \qquad x > 1$$

Taking the derivative

$$\frac{dy}{dx} = -\frac{1}{x^2}$$

Thus,

$$\frac{dx}{dy} = -x^2 = -\frac{1}{y^2}$$

The result could be verified by solving for x in terms of y

$$x = \frac{1}{y}, \qquad 0 < y < 1$$

and taking the derivative of x with respect to y.

As these examples show, for a monotonically increasing or decreasing function

$$f'(x)\phi'(y) = 1 \tag{8.35}$$

A corollary of the theorem on inverse function is the following theorem on the derivative of exponential functions.

Theorem 8.12 *Let* $y = e^x$, then $dy/dx = e^x$.

Proof First note that

$$x = \ln y$$

which implies

$$\frac{dx}{dy} = \frac{1}{y}$$

Using the inverse function rule, we have

$$\frac{dy}{dx} = \frac{1}{dx/dy} = \frac{1}{1/y} = y = e^x$$

8.4.1 Exercises

E.8.8 Determine if the following functions are monotonic, and if they are, determine the nature of monotonicity.

i. $y = \ln x, \quad x > 0$
ii. $y = 3x^3 + 2x + 5$
iii. $y = -\ln x$
iv. $y = -3x^3 - 2x + 5$

E.8.9 Determine the domain over which the following functions are monotonically increasing or decreasing.

i. $y = 5x^2$ *ii.* $y = \cos x$ *iii.* $y = \tan x$

E.8.10 Graph the functions in E.8.8 and E.8.9 to verify your answers to these problems.

8.5 Second- and Higher-Order Derivatives

The derivative of a function is itself a function. As such it could be continuous and differentiable. If the derivative is differentiable, we could take its derivative, which would be the second derivative of the original function.

Example 8.37 Let

$$y = 5x^6 - 3e^{2x}$$

Then

$$y' = \frac{dy}{dx} = 30x^5 - 6e^{2x}$$

and

$$y'' = \frac{d}{dx}\frac{dy}{dx} = \frac{d^2y}{dx^2} = 150x^4 - 12e^{2x}$$

We need not stop at the second derivative. In this example the second derivative can be differentiated

$$y''' = \frac{d^3y}{dx^3} = 600x^3 - 24e^{2x}$$

A function may be once, twice, three times, four times, and indeed n times differentiable. For $n > 3$ we write the derivative as

$$y^{(n)} = f^{(n)}(x) = \frac{d^ny}{dx^n}$$

Recall that we could learn about the behavior of a function by examining its derivative. In particular, a positive derivative signifies an increasing function and a negative derivative a decreasing one. The second derivative tells us the same thing about the first derivative. For example, if the function determines the distance traveled as a function of time, then the first derivative is the speed and the second acceleration. A positive (negative) second derivative means that the speed is increasing (decreasing). Figure 8.4 shows a few configurations of the first and second derivatives and the resulting behavior of the function.

Example 8.38 (Marginal Cost). Marginal cost is assumed to be first decreasing and then increasing. The supply curve of a firm is the increasing portion of its marginal cost function. Thus, for the first portion of the marginal cost curve we have $C''(Q) < 0$ and for the second portion or the supply function we have $C''(Q) > 0$.

Definition 8.5 (Concave and Convex Functions). Consider a function $y = f(x),\ x \in D$ and let $D_1 \subset D$ be a subset of its domain. If $f''(x) > 0$, whenever $x \in D_1$, the

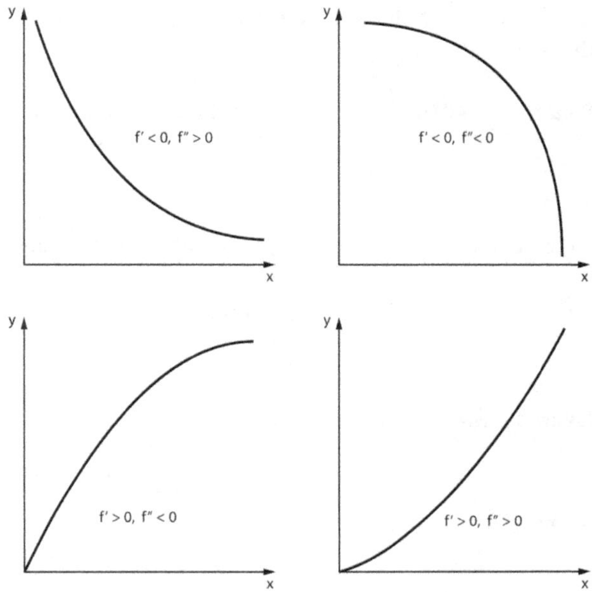

Fig. 8.4 First and second derivatives and shapes of functions

function is convex in D_1. If $f''(x) < 0$, whenever $x \in D_1$, the function is concave in D_1.

A function can be globally convex (concave) if $D_1 = D$ or it can be convex (concave) in some subsets of its domain and concave (convex) in others. We will have more to say on this subject in Chaps. 12 and 13.

Example 8.39 (Utility Function). Within the relevant range for economic analysis, marginal utility is positive, but decreasing. That is, $U'' < 0$.

It follows that the utility function is convex.

8.5.1 Exercises

E.8.11 Find second-order derivatives of the functions in E. 8.6.

8.6 Differential

So far we have treated dy/dx as one entity. In other words we have applied the operator d/dx to a function $y = f(x)$ to obtain its derivative. This need not be the case; we can separate the two. dy, called the *differential* of y, measures the change in y as a result of an infinitesimal change in x. It is equal to

$$dy = f'(x)\, dx \tag{8.36}$$

The rules of differentiation can be modified to apply to differentials. We illustrate this with a few examples.

Example 8.40 To find the differential of the function

$$y = 3(x - 9)^2$$

we note that $f'(x) = 6(x - 9)$. Therefore, the differential is

$$dy = f'(x)\ dx = 6(x - 9)\, dx$$

Example 8.41 For the function

$$f(x) = 7e^{3x}$$

we have

$$df(x) = f'(x)dx = 21e^{3x}dx$$

Rules of differentiation are all carried to differentials.

$$\begin{aligned} y &= u(x) + v(x), & dy &= u'dx + v'dx \\ y &= u(x)v(x), & dy &= vu'dx + v'udx \\ y &= \frac{u(x)}{v(x)}, & dy &= \frac{vu' - v'u}{v^2}dx \end{aligned} \tag{8.37}$$

Example 8.42 The differential of the function

$$y = x^3 - 3x$$

is

$$dy = 3x^2dx - 3dx$$

Example 8.43 To find the differential of the function $f(x) = 5x \ln x$, we let

$$u = 5x, \qquad v = \ln x$$

Then

$$\begin{aligned} dy &= vu'dx + v'udx \\ &= 5 \ln x dx + \frac{1}{x} 5x dx \\ &= 5(\ln x + 1)dx \end{aligned}$$

Example 8.44 Consider the function

$$y = \frac{1-3x}{(x+1)^2}$$

and let

$$u = 1 - 3x, \qquad v = (x+1)^2$$

Then

$$\begin{aligned} dy &= \frac{vu'dx - v'udx}{v^2} \\ &= \frac{-3(x+1)^2dx - 2(x+1)(1-3x)\,dx}{(x+1)^4} \\ &= \frac{3x-5}{(x+1)^3}dx \end{aligned}$$

Note that we can derive the inverse function rule from (8.36) because if $f'(x) \neq 0$ it follows that

$$\frac{dx}{dy} = \frac{1}{f'(x)} \tag{8.38}$$

provided that $f'(x) \neq 0$ in the interval for which we define dx/dy.

8.6.1 Second- and Higher-Order Differentials

Similar to derivatives, we can compute second- and higher-order differentials.

$$d^2y = f''(x)dx^2 \tag{8.39}$$

and, in general,

$$d^ny = f^{(n)}(x)dx^n \tag{8.40}$$

assuming that second- and higher-order derivatives exist.

Example 8.45 The second-order differential of the function

$$y = 3(x-9)^2$$

is

$$d^2y = d(6(x-9)dx) = 6dx^2$$

Example 8.46 For the functions

$$f(x) = 7e^{3x}$$

We have

$$d^2f(x) = df'(x)dx = d(21e^{3x}dx) = 63e^{3x}dx^2$$

Example 8.47 Consider the function

$$f(x) = 5x\ln x$$

We have

$$d^2f(x) = d[5(\ln x + 1)dx] = \frac{5}{x}\,dx^2$$

8.6.2 Exercises

E.8.12 Write the differentials of the functions in E.8.6.

E.8.13 Write the second-order differentials of the functions in E.8.6.

8.7 Computer and Numerical Differentiation

8.7.1 Computer Differentiation

According to a joke dating back to a time when universities had a single large computer, a university president had complained that students weren't learning anything anymore, because the computer solved all their problems. It's not such a joke anymore. If you need to find the derivative or integral (see Chap. 11) of a function, you *can* simply ask the computer to find it for you. Here are examples of the Maple code for finding the derivative of a function. But it should be noted that you still need to know how to find derivatives and integrals and their meanings and applications.

Maple code

```
Specify the function
f := 5*x^2 - 3*x + 14
# Find the first derivative
diff(f, x)
# But if you type
Diff(f, x)
# you get
```

$$\frac{d}{dx}(5x^2 - 3x + 14)$$

```
# Find the second derivative
diff(f, x, x)
# Higher order derivatives could be found in a
# similar way.
```

8.7.2 Numerical Differentiation

There are rare occasions when numerical calculation of a derivative at a particular point is needed for its own sake. But as will be seen in Chaps. 10, 12, and 13, both for solving nonlinear equations and for optimization, we need to evaluate derivatives of a function at different points. As it turns out this is a simple task. Suppose we are interested in evaluating the derivative of $y = f(x)$ at the point x_0. Let

$$y_{-1} = f(x_0 - h), \qquad y_1 = f(x_0 + h) \tag{8.41}$$

where h is a predetermined constant. Then

$$f'(x_0) \approx \frac{y_1 - y_{-1}}{2h} \tag{8.42}$$

The precision of this formula for different values of h is depicted in Tables 8.3 and 8.4. Table 8.3 shows derivatives of polynomials, exponential, and logarithmic functions at the point $x_0 = 3$, for different values of h. Table 8.4 displays the same information for derivatives of trigonometric functions, sin, cos, and tan, evaluated at $\pi/5$.

Table 8.3 Precision of computed derivative for selected functions

h	x	x^2	x^3	exp(x)	ln(x)
1.00	1	6	28.0000	23.6045470	0.3465736
0.50	1	6	27.2500	20.9329580	0.3364722
0.25	1	6	27.0625	20.2954161	0.3341082
0.10	1	6	27.0100	20.1190296	0.3334569
0.05	1	6	27.0025	20.0939069	0.3333642
0.02	1	6	27.0004	20.0868760	0.3333383
0.01	1	6	27.0001	20.0858717	0.3333346
Exact value	1	6	27.0000	20.0855369	0.3333333

Table 8.4 Precision of computed derivative for selected trigonometric functions

h	sin(x)	cos(x)	tan(x)
$\pi/5$	0.7568267	−0.5498668	2.4491427
$\pi/10$	0.7957747	−0.5781642	1.6734541
$\pi/20$	0.8056941	−0.5853711	1.5612290
$\pi/50$	0.8084848	−0.5873986	1.5330811
$\pi/100$	0.8088839	−0.5876886	1.5291641
$\pi/250$	0.8089957	−0.5877698	1.5280719
$\pi/500$	0.8090117	−0.5877814	1.5279160
Exact value	0.8090170	−0.5877853	1.5278640

As can be seen, one can decrease the error of computation to the desired degree by the choice of appropriate h value. The precision of this simple formula should not be surprising. Recall the mean value theorem

$$f'(x^*) = \frac{f(b) - f(a)}{b - a} \qquad a < x^* < b$$

and let

$$a = x_0 - h, \qquad b = x_0 + h$$

By making the interval $b - a$ smaller and smaller, we make the arbitrary point x^* closer and closer to the midpoint of $[a, b]$, that is x_0, and thus reduce the error of computation. Intuitively, for a small value of h the line connecting two points y_{-1} and y_1 is more or less parallel to the tangent at the midpoint (x_0, y_0).

For the second derivative, the formula is

$$f''(x_0) \approx \frac{f(x_2) - 2f(x_0) + f(x_{-2})}{4\,h^2} \tag{8.43}$$

where $x_2 = x_0 + 2\,h$, and $x_{-2} = x_0 - 2\,h$. Checking the accuracy of this formula for different functions is left to the reader (see E. 8.16).

8.7.3 Exercises

E.8.14 Use Maple to verify your solutions to E.8.6.

E.8.15 Use formulas in (8.42) and (8.43) on Matlab or Excel to numerically evaluate the first and second derivatives of the following functions at the indicated points. Use the exact formulas to check the accuracy of your results.

$$y = x^3 - 2x^2 + 3x - 8 \qquad \text{at} \qquad x = 6.75$$
$$y = \frac{x}{1 + x^2} \qquad \text{at} \qquad x = 2.2$$
$$y = \ln(x^3 + 2x - 5) \qquad \text{at} \qquad x = 5.8$$

E.8.16 Evaluate the accuracy of (8.43) for computing the second derivatives of functions in Tables 8.3 and 8.4. Use the same values of x_0 and h as in those tables.

Chapter 9
Differentiation: Functions of Several Variables

In this chapter we shall extend the concept and methods of differentiation to functions of several variables. Most economic relationships involve more than one variable and their analysis require the methods of this chapter. In addition, many economic models consist of several equations. Tracing the effect of a change in one variable in an equation throughout the model is a preoccupation of economics. In later sections we shall learn how to solve such problems. Differentiation plays an important role in statistics and econometrics as well. In particular, derivation of least squares and maximum likelihood estimators rely on differentiation. Differentiation of functions of several variables brings up the issues of continuity and differentiability. We discussed these subjects in the previous chapters and their extension to the case of functions of several variables, using notations developed in this chapter, is straightforward. Therefore, we will not revisit these topics in this chapter.

9.1 Partial Differentiation

The idea of differentiation of a function of one variable can be extended to functions of several variables. Consider the function

$$y = f(x_1,\ x_2) \tag{9.1}$$

The partial derivative of y with respect to x_1 is the derivative of y with respect to x_1 when x_2 remains constant and, therefore, can be treated as such. The same is true for the derivative of y with respect to x_2 holding x_1 constant. Partial derivatives are written as

$$\begin{aligned} \frac{\partial y}{\partial x_1} &= f_1(x_1, x_2) \\ \frac{\partial y}{\partial x_2} &= f_2(x_1, x_2) \end{aligned} \tag{9.2}$$

Example 9.1 Consider the function

$$y = 3x_1^2 - 6x_1x_2 + 4x_2$$

K. Dadkhah, *Foundations of Mathematical and Computational Economics*, 2nd ed.,
DOI 10.1007/978-3-642-13748-8_9, © Springer-Verlag Berlin Heidelberg 2011

We have

$$\frac{\partial y}{\partial x_1} = 6x_1 - 6x_2$$
$$\frac{\partial y}{\partial x_2} = -6x_1 + 4$$

Example 9.2 The partial derivatives of the function $z = x^2 \sin^2 y$ are

$$\begin{aligned}\frac{\partial z}{\partial x} &= 2x \sin^2 y \\ \frac{\partial z}{\partial y} &= 2x^2 \sin y \cos y \\ &= x^2 \sin 2y\end{aligned}$$

Partial differentiation is not confined to the functions of two variables. The function may have many variables.

Example 9.3 The function

$$u = xe^{yz}$$

has three partial derivatives:

$$\frac{\partial u}{\partial x} = e^{yz}, \qquad \frac{\partial u}{\partial y} = zxe^{yz}, \qquad \frac{\partial u}{\partial z} = yxe^{yz}$$

All the rules of differentiation for derivatives apply to partial derivatives and we need not repeat them here. As mentioned before, the derivative is an important tool of economic analysis. The following examples illustrate this point.

Example 9.4 (Marginal Utility). The utility function measures total satisfaction derived from the consumption of different bundles of goods and services. Letting U denote utility and x_i, $i = 1, \ldots, n$ the amount of commodities 1 through n consumed, we have

$$U = U(x_1, x_2, \ldots, x_n) \tag{9.3}$$

The marginal utility of good i is the additional satisfaction derived from the consumption of an incremental amount of that good. Thus,

$$U_i = \frac{\partial U}{\partial x_i} \tag{9.4}$$

Marginal utilities are assumed to be positive:

$$U_i > 0, \qquad i = 1, \ldots, n \tag{9.5}$$

Because marginal utilities of all goods and service are positive, the utility function will be a monotonically increasing function in all of its arguments.

Example 9.5 (Marginal Product). The production function relates the maximum attainable output Q from a combination of inputs $\mathbf{x} = [x_1, \ldots, x_k]$ to those inputs

$$Q = f(\mathbf{x}) \tag{9.6}$$

The function f is nondecreasing. The partial derivative of output with respect to each input is called the marginal product of that input, that is, the increment to output resulting from a small change in an input. All marginal products are assumed to be positive:

$$\frac{\partial Q}{\partial x_i} > 0, \qquad i = 1, \ldots, k \tag{9.7}$$

In a production function with two inputs, say, labor (L) and capital[1] (K),

$$Q = f(K, L)$$

the marginal product of labor is $f_L = \partial Q/\partial L$ and the marginal product of capital is $f_K = \partial Q/\partial K$.

In experimental sciences such as physics and chemistry, in order to measure the effect of a change in one variable on another, all other variables are held constant. Economics is mainly an observational science and is denied the luxury of "holding everything else constant." But suppose that somehow we are able to approximate the connection between the variable of interest and a group of variables affecting it by a function. Further assume that we have data on the variables of interest, are able to specify the form of the function, and estimate its parameters using econometric methods. Then the partial derivatives measure the effect of each RHS variable on the variable of interest while everything else is held constant. Needless to say, this measurement is an approximation because the function on which it is based is an approximation and its parameters are estimates.

Example 9.6 The process of Money supply can be formalized as[2]

$$M = \frac{cu + 1}{(cu + x)}(F + G + H(i - i^d))$$

where M is money supply, cu currency deposit ratio, x required reserves ratio, F net foreign assets of the central bank, G net government borrowing, i market interest rate, and i^d the discount rate. Thus, H would be the net borrowing by commercial banks from the central bank.

[1] We mean, of course, services rendered by labor and capital because neither workers nor capital goods themselves go into the output.

[2] See Kamran Dadkhah, *The Evolution of Macroeconomic Theory and Policy*, Springer, 2009, Chap. 9.

Suppose we are interested in finding the effects of the central bank's policy instruments on money supply. If the required reserves ratio is increased, we have

$$\frac{\partial M}{\partial x} = -\frac{cu+1}{(cu+x)^2}(F+G+H(i-i^d)) < 0$$

In other words, an increase in the required reserves ratio will decrease money supply. On the other hand, since

$$\frac{\partial M}{\partial G} = \frac{cu+1}{cu+x} > 0$$

An open market operation that increases the central bank's holding of government debt will increase M. But if the central bank sells government bonds and reduces its holding of government debt, money supply will decrease. If the discount rate is increased:

$$\frac{\partial M}{\partial i^d} = -\frac{cu+1}{cu+x}H' < 0$$

$H' > 0$ because as the difference between market rate of interest and discount rate increases, banks will lend more. Therefore, the effect of an increase in the discount rate is a reduction of money supply. Finally, let us consider the effect of a change in the public's attitude toward cash and deposits:

$$\frac{\partial M}{\partial cu} = \frac{x-1}{(cu+x)^2}(F+G+H(i-i^d)) < 0$$

In other words, as the public relies less on cash and more on deposits, money supply shows a gradual upward trend.

9.1.1 Second-Order Partial Derivatives

In the same manner that we had second and higher order derivatives of functions of one variable, we have second- and higher-partial derivatives for functions of several variables. Note, however, that whereas a function of one variable has at most one first order, one second order, ..., and one n-th order derivative, a function of two variables has at most two first-order, at most four second order, ..., and at most 2^n n-th order derivatives. In general, a function of k variables has at most k^i i-th order derivatives. In the case of a function of two variables $y = f(x_1, x_2)$, we have

$$\begin{aligned}
\frac{\partial^2 y}{\partial x_1 \partial x_1} &= \frac{\partial^2 y}{\partial x_1^2} = f_{11}(x_1,x_2), & \frac{\partial^2 y}{\partial x_2 \partial x_2} &= \frac{\partial^2 y}{\partial x_2^2} = f_{22}(x_1,x_2) \\
\frac{\partial^2 y}{\partial x_1 \partial x_2} &= \frac{\partial^2 y}{\partial x_1 \partial x_2} = f_{12}(x_1,x_2), & \frac{\partial^2 y}{\partial x_2 \partial x_1} &= \frac{\partial^2 y}{\partial x_2 \partial x_1} = f_{21}(x_1,x_2)
\end{aligned} \tag{9.8}$$

These derivatives are not necessarily all distinct. Indeed, under fairly general conditions, many cross derivatives of the same order are equal. Specifically, for the function $z = f(x, y)$, if first-order derivatives f_x and f_y and cross derivatives f_{xy} and f_{yx} all exist and are continuous, then

$$\frac{\partial^2 z}{\partial x \partial y} = f_{xy} = f_{yx} = \frac{\partial^2 z}{\partial y \partial x} \tag{9.9}$$

Thus, assuming the existence and continuity of first-order and second-order cross derivatives, a function of two variables has three and a function of three variables has six distinct second-order derivatives. In general, a function of k variables has $\binom{k}{2} + k$ distinct second-order derivatives.

Example 9.7 For the function in Example 9.2, we have

$$\frac{\partial^2 y}{\partial x_1^2} = 6, \qquad \frac{\partial^2 y}{\partial x_2^2} = 0, \qquad \frac{\partial^2 y}{\partial x_1 \partial x_2} = \frac{\partial^2 y}{\partial x_2 \partial x_1} = -6$$

Example 9.8 In Example 9.5 we have

$$\frac{\partial^2 u}{\partial x^2} = 0, \qquad \frac{\partial^2 u}{\partial y^2} = z^2 x e^{yz}, \qquad \frac{\partial^2 u}{\partial z^2} = y^2 x e^{yz}$$

$$\frac{\partial^2 u}{\partial x \partial y} = \frac{\partial^2 u}{\partial y \partial x} = z e^{yz} \qquad \frac{\partial^2 u}{\partial x \partial z} = \frac{\partial^2 u}{\partial z \partial x} = y x e^{yz}$$

$$\frac{\partial^2 u}{\partial y \partial z} = \frac{\partial^2 u}{\partial z \partial y} = x e^{yz} + yzxe^{yz}$$

Example 9.9 In Example 9.5 we discussed marginal utilities and noted that they are positive. We should add now that they are decreasing, that is,

$$\frac{\partial^2 U}{\partial x_i^2} = U_{ii} < 0 \qquad i = 1, \ldots, n \tag{9.10}$$

On the other hand if

$$\frac{\partial^2 U}{\partial x_i \partial x_j} = U_{ij} < 0 \tag{9.11}$$

then goods i and j are substitutes

Example 9.10 Returning to Example 9.6, recall that marginal products are positive. An increase in the utilization of an input is assumed to result in a decrease in its marginal output:

$$\frac{\partial^2 Q}{\partial x_i^2} < 0 \qquad i = 1, \ldots, k \tag{9.12}$$

On the other hand an increase in one input enhances the marginal product of others:

$$\frac{\partial^2 Q}{\partial x_i \partial x_j} > 0, \qquad i,j = 1, \ldots, k, \;\; j \neq i \tag{9.13}$$

9.1.2 *Differentiation of Functions of Several Variables Using Computer*

In Chap. 8 we became familiar with differentiation using Maple program. Maple can be used for partial differentiation too. Consider the function:

$$z = x\exp(2y)$$

Using the following code we can compute its first and second derivatives:

Maple code

```
# Specify a function
z := x*exp(2*y)
# Find the partial derivative of z with respect to y
  diff(z, y)
  # and x
  diff(z, x)
  # Second derivatives with respect to x and y can be
  # obtained by the following commands
  diff(z, x, x)
  diff(z, y, y)
  # We can also find the cross derivative ∂²z/∂x∂y using
diff(z, y, x)
```

9.1.3 *The Gradient and Hessian*

The gradient and Hessian are two important concepts in analyzing functions of several variables as well as systems of equations. They also provide compact notations when dealing with functions of several variables.

Definition 9.1 (Gradient). Let

$$y = f(\mathbf{x}) \tag{9.14}$$
$$\mathbf{x} = [x_1, x_2, \ldots, x_n]'$$

Then the vector of first-order partial derivatives of f with respect to all its arguments

$$\nabla f(\mathbf{x}) = \begin{bmatrix} \partial f/\partial x_1 \\ \vdots \\ \partial f/\partial x_n \end{bmatrix} \tag{9.15}$$

is called the gradient of f. The gradient is a vector and as such it has both a direction and length (see Chap. 5). We illustrate these points with an example.

Example 9.11 Let

$$z = x + y$$

$$\nabla z = \begin{bmatrix} \partial z/\partial x \\ \partial z/\partial y \end{bmatrix} = \begin{bmatrix} 1 \\ 1 \end{bmatrix}$$

In this case ∇z is a vector pointing at the (1,1) direction and it has the length of $\sqrt{2}$.

Example 9.12 For the equation

$$f(x,y,z) = xz - 2xy + y^2 + 6yz^2$$

we have

$$\nabla f = \begin{bmatrix} z - 2y \\ -2x + 2y + 6z^2 \\ x + 12yz \end{bmatrix}$$

Definition 9.2 (Hessian). Let f be the same as in Definition 9.1. The matrix of its second-order partial derivatives is called its Hessian[3] matrix.

$$\nabla^2 f(\mathbf{x}) = \left[\frac{\partial^2 f}{\partial x_i \partial x_j}\right] \qquad i,j = 1,\ldots n \tag{9.16}$$

Example 9.13 Let

$$y = x_1^2 x_2^2$$

Then

$$\nabla f(\mathbf{x}) = \begin{bmatrix} 2x_1 x_2^2 \\ 2x_1^2 x_2 \end{bmatrix}$$

[3] After German mathematician Ludwig Otto Hesse (1811–1874), who was a student of Jacobi.

and

$$\nabla^2 f(\mathbf{x}) = \begin{bmatrix} 2x_2^2 & 4x_1x_2 \\ 4x_1x_2 & 2x_1^2 \end{bmatrix}$$

Example 9.14 For the equation

$$f(x,y) = x\sin^2 y$$

we have

$$\nabla f = \begin{bmatrix} \sin^2 y \\ x\sin 2y \end{bmatrix}$$

and

$$\nabla^2 f = \begin{bmatrix} 0 & \sin 2y \\ \sin 2y & 2x\cos 2y \end{bmatrix}$$

9.1.4 Exercises

E.9.1 Find all partial derivatives of the following functions:

i. $u = x^2 + 2xy + y^3$ *ii.* $z = \dfrac{x}{x^2+y^2}$ *iii.* $z = x\sin^2 y$

iv. $y = x_1^3 - 4x_1^2x_2 + 5x_2x_3^2 + 2x_3^3$ *v.* $z = \ln(x+y)$

vi. $z = \dfrac{x^2y^2}{x+y}$ *vii.* $u = e^{x^2+y^2+z^2}$ *viii.* $y = \ln(x_1+x_2+x_3)$

ix. $u = zx^y$ *x.* $z = e^x \ln y$ *xi.* $u = \ln(x^2+y^2+z^2)$

xii. $f(x,y) = x^3 + 3x^2 + 4xy + y^2$

E.9.2 Find all second-order partial derivatives of the functions in E.9.1.

E.9.3 Check your results for E.9.1 and E.9.2 using Maple.

E.9.4 Write your results in E.9.1 and E.9.2 using the gradient and Hessian notations.

E.9.5 Find the marginal products of labor and capital using the following widely used production functions and show that they are all positive.

$Q = \min(\alpha K, \beta L)$	Leontief Production Function
$Q = AK^\alpha L^\beta$	Cobb-Douglas Production Function
$Q = (\alpha K^\rho + \beta L^\rho)^{1/\rho}$	Constant Elasticity of Substitution (CES) Production Function

$$\alpha,\ \beta,\ \rho, A > 0$$

E.9.6 Take the second partial derivatives of the production functions in E.9.5. Show that the marginal product of a factor decreases when the utilization of a factor increases but increases when the utilization of another factor is increased.

9.2 Differential and Total Derivative

9.2.1 Differential

In Chap. 8 we became familiar with the concept of differentials:

$$dy = f'(x)\,dx \tag{9.17}$$

Here we extend it to functions of several variables. The differential becomes more important when dealing with functions of several variables. In such instances it measures the change in the dependent variable y when all independent variables $\mathbf{x} = [x_1, \ldots, x_k]$ change infinitesimally and simultaneously. Let

$$y = f(\mathbf{x}) \tag{9.18}$$

Then

$$dy = \sum_{i=1}^{k} \frac{\partial f}{\partial x_i} dx_i \tag{9.19}$$

Thus, all independent variables $\mathbf{x} = [x_1, \ldots, x_k]'$ have changed and the amount of change is $d\mathbf{x} = [dx_1, \ldots, dx_k]'$ The response of the dependent variable y is equal to the weighted sum of these changes where weights are partial derivatives of f with respect to each x_i.

Example 9.15 The differential dy of the function

$$y = x_1^2 x_2^2$$

$$dy = 2x_1 x_2^2 dx_1 + 2x_1^2 x_2 dx_2$$

Example 9.16 For the function

$$y = \frac{x_1 - 3}{x_2 + 3}$$

we have

$$dy = \frac{(x_2 + 3)dx_1 - (x_1 - 3)dx_2}{(x_2 + 3)^2}$$

Example 9.17 The differential of the function

$$z = xe^y$$

is

$$dz = e^y dx + xe^y dy$$

Example 9.18 For the utility function

$$U = U(x_1, \ldots, x_n) \tag{9.20}$$

the differential is

$$dU = \sum_{j=1}^{n} \frac{\partial U}{\partial x_j} dx_j \tag{9.21}$$

Thus, the change in total utility is the sum of changes in the consumption of each good and service in the utility function weighted by its marginal utility.

Example 9.19 For the production function

$$Q = f(x_1, \ldots, x_k) \tag{9.22}$$

the differential is

$$dQ = \sum_{j=1}^{k} \frac{\partial Q}{\partial x_j} dx_j \tag{9.23}$$

The change in total output is the sum of changes in the utilization of each input weighted by its marginal product.

Similar to second- and higher-order derivatives, we can compute second- and higher-order differentials. For the function

$$y = f(x_1, \ldots, x_k) \tag{9.24}$$

the second-order differential is

$$\begin{aligned} d^2y &= \frac{\partial^2 f}{\partial x_1^2} dx_1^2 + \frac{\partial^2 f}{\partial x_1 \partial x_2} dx_1 dx_2 + \ldots + \frac{\partial^2 f}{\partial x_{k-1} \partial x_k} dx_{k-1} dx_k + \frac{\partial^2 f}{\partial x_k^2} dx_k^2 \\ &= \sum_{i=1}^{k} \sum_{j=1}^{k} \frac{\partial^2 f}{\partial x_i \partial x_j} dx_i dx_j \end{aligned} \tag{9.25}$$

Note that when $i = j$, $dx_i dx_j = dx_i^2$.

Example 9.20 The second-order differential of the function

$$y = x_1^2 x_2^2$$

is

$$\begin{aligned} d^2y &= 2x_2^2dx_1^2 + 4x_1x_2dx_1dx_2 + 4x_1x_2dx_2dx_1 + 2x_1^2dx_2^2 \\ &= 2x_2^2dx_1^2 + 8x_1x_2dx_1dx_2 + 2x_1^2dx_2^2 \end{aligned}$$

Example 9.21 For the function in Example 9.16, we have

$$y = \frac{x_1 - 3}{x_2 + 3}$$

we have

$$\begin{aligned} d^2y &= \frac{-1}{(x_2+3)^2}dx_2dx_1 + \frac{-1}{(x_2+3)^2}dx_1dx_2 + \frac{2(x_1-3)}{(x_2+3)^3}dx_2^2 \\ &= \frac{-2}{(x_2+3)^2}dx_1dx_2 + \frac{2(x_1-3)}{(x_2+3)^3}dx_2^2 \end{aligned}$$

and for the function in Example 9.17

$$z = xe^y$$

the second-order differential is

$$\begin{aligned} d^2z &= e^y dydx + e^y dxdy + xe^y dy^2 \\ &= 2e^y dxdy + xe^y dy^2 \end{aligned}$$

9.2.2 Total Derivative

We noted that the differential measures the change in the dependent variable when all independent variables change. Now suppose that the change in the independent variables is caused by a change in another variable. More specifically let

$$x_1 = x_1(t), \ldots, x_k = x_k(t) \tag{9.26}$$

and therefore

$$dx_i = x_i'(t)dt \qquad i = 1, \ldots, k \tag{9.27}$$

Substituting (9.27) in (9.26) and dividing through by *dt*, we will have the total derivative of y with respect to t:

$$\frac{dy}{dt} = \sum_{i=1}^{k} \frac{\partial f}{\partial x_i}\frac{dx_i}{dt} = \sum_{i=1}^{k} \frac{\partial f}{\partial x_i}x_i'(t) \tag{9.28}$$

A total derivative measures the ratio of change in the dependent variable to an infinitesimal change in the ultimate independent variable. It is called a total derivative as opposed to a partial derivative. The latter measures the ratio of the change in the dependent variable to an infinitesimal change in one of the independent variables. Variable t could be time in which case dy/dt signifies the instantaneous change in y as all independent variables change and affect it. Alternatively, t could be a policy variable such as money supply. An increase in money affects the GDP, but the influence may be transmitted through several different channels.

Example 9.22 Consider the function

$$y = x_1^2 x_2^2$$

and let

$$x_1 = 2t, \qquad x_2 = \frac{1}{t}$$

We have

$$\frac{dy}{dt} = 2x_1 x_2^2 \frac{dx_1}{dt} + 2x_1^2 x_2 \frac{dx_2}{dt}$$

Because

$$\frac{dx_1}{dt} = 2, \quad \text{and} \quad \frac{dx_2}{dt} = -\frac{1}{t^2}$$

the total derivative of y with respect to t is

$$\frac{dy}{dt} = 4x_1 x_2^2 - \frac{2x_1^2 x_2}{t^2} = 0$$

Example 9.23 For the function

$$z = xe^y, \qquad x = t, \qquad y = \ln t$$

we get

$$\frac{dz}{dt} = e^y \frac{dx}{dt} + xe^y \frac{dy}{dt} = e^{\ln t} + te^{\ln t}\frac{1}{t} = 2t$$

Example 9.24 Consider the following model depicting the connection between foreign assets of a country and domestic price index.

$$F = F_0 + \theta V$$
$$M = \mu[F + G + H(i - i^d)]$$
$$P = \frac{k}{y} M$$

The second equation is the money supply process of Example 9.6 where $\mu = (cu + 1)(cu + x)$ is the money multiplier and here we have added an equation that links the amount of net foreign assets F to a constant V and an exogenously determined variable θ. For example, V could be the capacity of oil exports and θ the internationally determined oil price. The model would be applicable to the case of Saudi Arabia. Or V could be the exchange rate fixed by the government and θ the exogenously determined net exports, whereas F_0 is the previously accumulated foreign assets. Such a model may depict the case of China. Finally, the third equation is Cambridge version of the quantity theory of money.

Suppose θ increases. The implication for the economy would be an increase in net foreign assets leading to an increase in money supply, which would result in an increase in the price level. Can the central bank do something about this situation?

Let us assume that μ is fixed and the bank does not want to change the discount rate. The only policy instrument would be open market operation and the goal is to make $dM/d\theta = 0$, that is, to neutralize the effect of an increase in oil price or net exports on the price level. Then we should have

$$\frac{dM}{d\theta} = \mu\left(\frac{dF}{d\theta} + \frac{dG}{d\theta}\right) = 0$$

which implies

$$\frac{dG}{d\theta} = -\frac{dF}{d\theta}$$

In other words, the central bank should decrease its net holding of government securities by the same amount that its net foreign assets are increased and avoid any adverse effects on the domestic economy. Such a policy is called *sterilization*.

It may not always be possible to reduce the central bank's holdings of government securities by the same amount as an increase in net foreign assets. In such cases the bank may resort to selling foreign exchange to the public and companies or encourage investment abroad. Of course, the central bank can use more than one instrument to neutralize the effect of an increase in θ. Let us consider a combination of open market operation and increase in the discount rate:

$$\frac{dM}{d\theta} = \mu\left(\frac{dF}{d\theta} + \frac{dG}{d\theta} - H'\frac{di^d}{d\theta}\right) = 0$$

which implies

$$\frac{dG}{d\theta} - H'\frac{di^d}{d\theta} = -\frac{dF}{d\theta}$$

Whereas decreasing the central bank holdings of government assets by the same amount as an increase in F involves a simple calculation, devising a policy to partly restrict credit to the private sector and partly sell government bonds requires an estimate of H' through econometric analysis.

9.2.3 Exercises

E.9.7 Find the differentials of the functions in E.9.1.

E.9.8 Find the second-order differentials of the functions in E.9.1.

E.9.9 Find the total derivatives of the following functions:

i. $y = \dfrac{x_1 - 3}{x_2 + 3}$ where $x_1 = \gamma_1 t, \quad x_2 = \gamma_2 t$

ii. $y = x_1^3 - 4x_1^2 x_2 + 5x_2 x_3^2 + 2x_3^3$
where $x_1 = h_1(t), \quad x_2 = h_2(t), \quad x_3 = h_3(t)$

iii. $u = zx^y$ where $x = t, \quad y = t^2, \quad z = \dfrac{1}{t}$

9.3 Homogeneous Functions and the Euler Theorem

Homogeneous functions play an important role in economics, especially in the theory of production and growth. Such functions serve as a counterpart to economic concepts of decreasing, constant, and increasing returns to scale. The Euler theorem brings out the economic implications of modeling production with homogeneous functions.

Definition 9.3 A function $f(\mathbf{x}) = f(x_1, \ldots, x_n)$ is said to be homogeneous of degree γ if for all $\lambda > 0$, we have

$$f(\lambda \mathbf{x}) = f(\lambda x_1, \ldots, \lambda x_n) = \lambda^{\gamma} f(\mathbf{x}) \tag{9.29}$$

Example 9.25 $y = axz$, where a is a constant, is homogeneous of degree two because

$$a(\lambda x)(\lambda z) = \lambda^2 axz = \lambda^2 y$$

Example 9.26 The Cobb-Douglas production function

$$Q = AK^{\alpha}L^{\beta}$$

is homogeneous of degree $\alpha + \beta$ because

$$A(\lambda K)^{\alpha}(\lambda L)^{\beta} = \lambda^{\alpha+\beta}AK^{\alpha}L^{\beta}$$

Now if

$\alpha+\beta<1$, the production function shows decreasing returns to scale.
$\alpha+\beta=1$, the production function shows constant returns to scale.
$\alpha+\beta>1$, the production function has increasing returns to scale.

An interesting property of homogeneous functions is the Euler theorem.

Theorem 9.1 (Euler Theorem). *Let the function $f(\mathbf{x}) = f(x_1, x_2, \ldots, x_n)$ be homogeneous of degree γ. Then*

$$\gamma f(\mathbf{x}) = \mathbf{x}'\nabla f(\mathbf{x}) = \sum_{j=1}^{n} x_j \frac{\partial f}{\partial x_j}, \qquad \gamma > 0 \tag{9.30}$$

The proof of this theorem is instructive and we prove it first for a special case and then for the general case. Let $f(x)$ be the function of one variable and homogeneous of degree one. Thus, for any given value of x, say x^* we can write

$$\frac{1}{\lambda} f(\lambda x^*) = c \tag{9.31}$$

where c is a constant. Taking the derivative of the LHS with respect to λ, we have

$$\frac{d}{d\lambda}\left[\frac{1}{\lambda} f(\lambda x^*)\right] = 0 \tag{9.32}$$

Therefore,

$$-\frac{1}{\lambda^2} f(\lambda x^*) + \frac{x^*}{\lambda} f'(\lambda x^*) = 0$$

letting $\lambda = 1$,

$$f(x^*) = x^* f'(x^*) \tag{9.33}$$

Note that whereas we showed this for a particular value of x, namely x^* the proposition can be proved for all values of x. Thus, without causing confusion we shall drop the superscript * in the proof of the general case. Let $f(x_1, \ldots, x_n)$ be a homogeneous function of degree γ. Then

$$\frac{d}{d\lambda}\left[\frac{1}{\lambda^\gamma} f(\lambda x_1, \ldots, \lambda x_n)\right] = 0$$

Carrying out the differentiation, we have

$$-\frac{\gamma}{\lambda^{\gamma+1}} f(\lambda x_1, \ldots, \lambda x_n) + \frac{1}{\lambda^{\gamma}} \sum_{j=1}^{n} x_j \frac{\partial f}{\partial x_j} = 0$$

Again, setting $\lambda = 1$ *and rearranging the equation, we get*

$$\gamma f(x_1, \ldots, x_n) = \sum_{j=1}^{n} x_j f_j \tag{9.34}$$

where $f_j = \partial f / \partial x_j$.

Example 9.27 In the case of Cobb-Douglas production function, we have

$$\frac{\partial Q}{\partial K} = \alpha A K^{\alpha-1} L^{\beta}, \qquad \frac{\partial Q}{\partial L} = \beta A K^{\alpha} L^{\beta-1}$$

and

$$\begin{aligned} K\frac{\partial Q}{\partial K} + L\frac{\partial Q}{\partial L} &= K\alpha A K^{\alpha-1} L^{\beta} + L\beta A K^{\alpha} L^{\beta-1} \\ &= (\alpha + \beta) A K^{\alpha} L^{\beta} \\ &= (\alpha + \beta) Q \end{aligned}$$

Economically, the Euler theorem has an important implication for production functions with constant returns to scale. As we shall learn in Chap. 13, the optimal decision for a firm is to employ each factor of production to the point where the marginal product of that factor equals its remuneration. Specifically, the work force of the firm is expanded to the point where the value of the marginal product of labor becomes equal to the wage rate. Similarly, capital is employed by the firm to the point where the value of its marginal product is equal to the rate of interest. The same would be true for all factors of production.

If the production process has constant returns to scale, that is, the production function is homogeneous of degree one, then by virtue of (9.30) and setting $\gamma = 1$, paying each factor the value of its marginal product will exhaust the revenue of the firm. We shall return to this point in Chap. 12. Another implication of the Euler theorem is that if a function is homogeneous of degree γ, then its partial derivatives are homogeneous functions of degree $\gamma - 1$. Again, we show this first for the case of a function of one variable that is homogeneous of degree one. In (9.33), we replace x with λx on both sides of the equation:

$$f(\lambda x) = \lambda x f'(\lambda x) \tag{9.35}$$

Because $f(\lambda x) = \lambda f(x)$, (9.35) can be rewritten as

$$f(x) = x f'(\lambda x) \tag{9.36}$$

Comparing (9.36) and (9.33), we conclude that

$$f'(\lambda x) = f'(x) = \lambda^0 f'(x) \tag{9.37}$$

which means that the derivative is homogeneous of degree zero. Extending the same argument to (9.34), we have

$$\gamma f(\lambda \mathbf{x}) = \gamma \lambda^\gamma f(\mathbf{x}) = \sum_{j=1}^{n} \lambda x_j f_j(\lambda \mathbf{x}) \tag{9.38}$$

Dividing through by λ^γ, we get

$$\gamma f(\mathbf{x}) = \sum_{j=1}^{n} \frac{1}{\lambda^{\gamma-1}} x_j f_j(\lambda \mathbf{x}) \tag{9.39}$$

Comparing (9.39) and (9.34) and noting that both equations have to hold for all values of $\mathbf{x}$, we conclude that

$$f_j(\lambda \mathbf{x}) = \lambda^{\gamma-1} f_j(\mathbf{x}), \qquad j = 1, \ldots, n \tag{9.40}$$

which shows that partial derivatives of a homogeneous function are also homogeneous but of a degree one less than the original function. Again, this implication of the Euler theorem has an important economic interpretation that we shall return to in Chap. 12.

9.3.1 Exercises

E.9.10 Determine if the following functions are homogeneous and, if so, of what degree.

i. $f(x_1, x_2, x_3) = x_1 x_2 x_3$

ii. $f(x_1, x_2, x_3) = (x_1 - a_1)(x_2 - a_2)(x_3 - a_3)$,

iii. $f(x, y) = \dfrac{x^a}{y^b}$, $\quad y \neq 0$

iv. $f(x, y) = xe^y$

v. $f(x, y, z) = \ln(x + y + z)$

E.9.11 Verify that the Euler theorem holds for the homogeneous functions in E.9.9.

E.9.12 Determine if the Leontief and the CES production functions of E.9.4 are homogeneous and of what degree. Verify that the Euler theorem holds for these functions.

9.4 Implicit Function Theorem

Let us first illustrate this theorem with an example and then give it a formal treatment.

Example 9.28 Consider the following function

$$xy = D, \qquad x, y \neq 0$$

This function has two branches (as an exercise, graph this function), with domains

$$x > 0 \qquad \text{and} \qquad x < 0$$

Suppose that we are interested in finding dy/dx. One way is to solve for y in terms of x

$$y = \frac{D}{x}$$

and take the derivative

$$\frac{dy}{dx} = -\frac{D}{x^2} = -\frac{y}{x}$$

But there are cases where it is not possible to explicitly solve for y in terms of x. Would it still be possible to find dy/dx? Consider taking the total differential of the function

$$xdy + ydx = 0$$

which implies

$$\frac{dy}{dx} = -\frac{y}{x}$$

This gives us the same answer as before without requiring an explicit solution of y in terms of x.

To make our answer more general, let us write the function as

$$F(x, y) = 0, \qquad F_y \neq 0 \tag{9.41}$$

Under what conditions can we write

$$\frac{dy}{dx} = -\frac{F_x}{F_y} \tag{9.42}$$

The answer: Whenever a function $y = \phi(x)$ exists, even if we cannot explicitly find it. But then under what conditions does $\phi(x)$ exist? The answer is given by the implicit function theorem.

Theorem 9.2 *Consider the real-valued continuous function F(x, y) on an open set $V \in \Re^2$. Further assume that it is continuously differentiable with respect to its arguments and*

$$\begin{aligned} &F(x_0, y_0) = 0, && (x_0, y_0) \in V \\ &\frac{\partial F}{\partial y}(x_0, y_0) = F_y(x_0, y_0) \neq 0 \end{aligned} \tag{9.43}$$

Then there exists an open set $W \in \Re^2$ such that $(x_0, y_0) \in W$ and a unique continuously differentiable function $\phi(x)$ such that

$$y_0 = \phi(x_0) \tag{9.44}$$

and

$$F(x, \phi(x)) = 0 \qquad x \in W \tag{9.45}$$

The implicit function theorem can be extended to functions of more than two variables. We simply need to replace x with the vector $\mathbf{x} = [x_1, \ldots, x_k]'$ and the Theorem holds for $F(\mathbf{x}, y)$.

There are two circumstances under which we may not be able to find the explicit form of $\phi(x)$. The first is when the form of the function won't allow us to explicitly solve for y in terms of x. An example would be the function

$$y + \ln y - x_1 x_2 = 0$$

The second, which is particularly true in economics, is when we are reluctant to specify the exact functional form of the relations between different variables. The reason may be uncertainty about such forms or a desire to make the analysis as general as possible.

Example 9.29 Find dy/dx for the function

$$y^3 - 3y + 5x = 0 \qquad y > 1$$

We have

$$3y^2 dy - 3dy + 5dx = 0$$

Therefore,

$$\frac{dy}{dx} = \frac{-5}{3y^2 - 3}$$

Example 9.30 Suppose we are interested in finding dy/dx for the function

$$e^y - e^{2x} + xy = 0 \qquad x > 0$$

Taking the differential of both sides of the function, we get

$$e^y dy - 2e^{2x}dx + ydx + xdy = 0$$

and the derivative is

$$\frac{dy}{dx} = \frac{2e^{2x} - y}{e^y + x}$$

Example 9.31 For the function

$$\ln(x^2 + y^2) - 1 = 0 \qquad y > 0$$

we have

$$\frac{2xdx + 2ydy}{x^2 + y^2} = 0$$

and

$$\frac{dy}{dx} = -\frac{x}{y}$$

Example 9.32 (The Slope of an Indifference Curve). For the utility function

$$U = U(x_1, x_2) \tag{9.46}$$

If we set the level of utility at a certain value U_0, we get the indifference curve

$$U_0 = U(x_1, x_2) \tag{9.47}$$

which can be rearranged into

$$U_0 - U(x_1, x_2) = 0 \tag{9.48}$$

Because marginal utility functions are assumed to be continuous, the conditions of the implicit function theorem are met for every point on the indifference curve. Thus,

$$\frac{dx_2}{dx_1} = -\frac{\partial U/\partial x_1}{\partial U/\partial x_2} \tag{9.49}$$

which shows that the indifference curves for normal goods are downward sloping, because the marginal utilities of such goods are positive. The slope of the indifference curve is called the marginal rate of substitution and is equal to the ratio of marginal utilities of the goods involved.

Example 9.33 (Marginal Product and Marginal Rate of Technical Substitution). Let us concentrate on a production function with two inputs, say, labor L and capital K. If we consider all combinations of the inputs that give rise to a given level of output $\bar{Q}$, then we have a curve depicting the trade-off between the inputs. These curves are called isoquants and are similar to indifference curves of the consumer choice theory. They are also downward sloping and convex. Differentiating the production function

$$\bar{Q} = f(K, L) \tag{9.50}$$

we get

$$f_K dK + f_L \, dL = 0 \tag{9.51}$$

Therefore,

$$\frac{dK}{dL} = -\frac{f_L}{f_K} < 0 \tag{9.52}$$

The slope of the isoquants, which is called the marginal rate of technical substitution (MRTS), measures the amount of one input needed to replace a unit of the other input so as to keep the amount of output constant. The higher the MRTS, the more flexible the production process in substituting one input for another. Another measure of the flexibility of production is the elasticity of substitution

$$\sigma = \frac{d\ln(K/L)}{d\ln(f_L/f_K)} = \frac{d(K/L)}{K/L}\frac{f_L/f_K}{d(f_L/f_K)} \tag{9.53}$$

$\sigma = 0$ means no substitution between inputs and $\sigma \to \infty$ signifies perfect substitution. In general the higher the σ, the higher is the substitutability between inputs. In case of the Cobb-Douglas production function with constant returns to scale,

$$Q = AK^{\alpha}L^{1-\alpha}$$

We have

$$f_K = \alpha AK^{\alpha-1}L^{1-\alpha} \, f_L = (1-\alpha)AK^{\alpha}L^{-\alpha}$$

Letting $k = K/L$, we have

$$\begin{aligned} f_K &= \alpha Ak^{-(1-\alpha)} \\ f_L &= (1-\alpha)Ak^{\alpha} \end{aligned}$$

Thus,

$$\text{MRTS} = \frac{\alpha - 1}{\alpha}k$$

and

$$\sigma = 1$$

9.4.1 Exercises

E.9.13 Find the slope of the indifference curves for the utility functions

i. $U = (x_1 - \alpha)(x_2 - \beta)$

ii. $U = x_1^{\alpha} x_2^{\beta}$

where α and β are positive constants.

E.9.14 Find the slope of isoquants (MRTS) and elasticity of substitution of the production functions in E.9.5.

9.5 Differentiating Systems of Equations

In many sciences and particularly in economics, a phenomenon or process is modeled not by just one equation, but by a system of equations. Thus, any change in an independent or exogenous variable is transmitted through several channels to dependent or endogenous variables. Therefore, in taking derivatives in the context of a system of equations we should deal with the interdependencies among the variables. It is not enough for the derivative to reflect only the direct impact of a variable on another. It should be the sum total of all direct and indirect effects of the change in one variable on another. When dealing with a system of m equations with n variables, some of the equations may be redundant or inconsistent with each other. Surely if $m > n$, unless $m - n$ equations are redundant, we will have inconsistency among the equations.

When dealing with linear equations, we checked for the independence of equations by inspecting the determinant of the matrix of their coefficients. Could we check for the independence among m functions of n variables when $m = n$? The answer is yes. But for this purpose we need to define the Jacobian matrix and determinant.

9.5.1 The Jacobian and Independence of Nonlinear Functions

Definition 9.4 (Jacobian). Let there be m functions

$$y_i = g_i(\mathbf{x}), \qquad i = 1, \ldots, m \tag{9.54}$$

where $\mathbf{x}$ is an n-vector. Then the Jacobian[4] matrix is defined as

[4]Carl Gustav Jacob Jacobi (1804–1851) was one of the nineteenth century's great mathematicians.

$$\mathbf{J} = \begin{bmatrix} \frac{\partial g_1}{\partial x_1} & \cdots & \frac{\partial g_1}{\partial x_n} \\ \vdots & \ddots & \vdots \\ \frac{\partial g_m}{\partial x_1} & \cdots & \frac{\partial g_m}{\partial x_n} \end{bmatrix} \tag{9.55}$$

Now we can answer the independence question. If the determinant of the Jacobian matrix $|\mathbf{J}| \neq 0$, then the functions are independent.

Example 9.34 Consider the functions

$$z_1 = xe^y$$
$$z_2 = 3y^3 - 2x^4$$

Then

$$|\mathbf{J}| = \begin{vmatrix} e^y & xe^y \\ -8x^3 & 9y^2 \end{vmatrix} = e^y(9y^2 + 8x^4) \neq 0$$

As can be seen the determinant is not zero regardless of the values taken by x and y. Of course this is not the case for all systems of equations. We may encounter situations where the determinant is not zero for some values of the independent variables and equal to zero for some other values. On the other hand, if two functions are dependent, then the determinant of the Jacobian will be identically zero.

Example 9.35 Let

$$z_1 = xe^y$$
$$z_2 = \ln x + y$$

$$|\mathbf{J}| = \begin{vmatrix} e^y & xe^y \\ 1/x & 1 \end{vmatrix} = e^y - e^y = 0$$

Therefore, the two functions are not independent. But that should not be surprising because $z_2 = \ln z_1$. To understand the rationale behind the use of the Jacobian to test for the independence of functions, note that two functions are equal if they intersect at least in one point and their derivatives are identical. Thus starting at the point of intersection, they move together in both directions. Even if the functions do not intersect at a point, the fact that their derivatives are the same means that over their domain they will be moving together in the same direction and, therefore, are not independent. We can generalize this notion to the case of functions of several variables. Note that the Jacobian of a system of equations can be written as

$$\mathbf{J} = [\nabla g_1(\mathbf{x}) \ldots \nabla g_n(\mathbf{x})]' \tag{9.56}$$

Each of the vectors specifies the direction in which the function will move. If $|\mathbf{J}| = 0$, it means that these directions do not span an n-space and at least one of them is dependent on (or is a combination of) the others. Hence, we do not have n independent functions.

9.5.2 Differentiating Several Functions

In some cases tracing the effects of a change through the system may be straightforward. Consider our model in Example 9.24. To find the effect of an increase in θ on P, we need only follow the chain rule

$$\frac{\partial P}{\partial \theta} = \frac{\partial P}{\partial M}\frac{\partial M}{\partial F}\frac{dF}{d\theta} = \frac{k}{y}\mu V$$

The reason is that we have a recursive model. When dealing with systems of simultaneous equations, the situation is a bit more involved. Consider the system of equations

$$\begin{aligned} f_1(y_1, \ldots, y_m, x_1, \ldots, x_k) &= 0 \\ \vdots \qquad\qquad & \vdots \\ f_m(y_1, \ldots, y_m, x_1, \ldots, x_k) &= 0 \end{aligned} \tag{9.57}$$

There are m endogenous variables and m equations. Therefore, in principle, we should be able to solve for the endogenous variables $y_1, \ldots, y_m$ in terms of the exogenous variables. But it may be that either because of the complexity of the equations or our reluctance to specify the forms of equations, we are unable to do so. Could we find the effects of a change in an exogenous variable, say x_j, on the endogenous variables? The answer is yes, provided the conditions of the general form of the implicit function theorem for systems of equations are satisfied.

These conditions require the determinant of the Jacobian composed of partial derivatives of the functions with respect to endogenous variables be different from zero. We will show the method for the case of three equations with two exogenous variables. But it will be clear that the method can be extended to any number of equations and any number of exogenous variables.

Note also that we need not have all our equations in implicit form. Indeed, we may have functions for which either some or all variables could be separated. The implicit form is used here to show the generality of the method. Consider the following system of equations:

$$\begin{aligned} f(y_1, y_2, y_3, x, z) &= 0 \\ g(y_1, y_2, y_3, x, z) &= 0 \\ h(y_1, y_2, y_3, x, z) &= 0 \end{aligned}$$

Partially differentiating with respect to z, we have

$$\begin{aligned} f_1\frac{\partial y_1}{\partial z} + f_2\frac{\partial y_2}{\partial z} + f_3\frac{\partial y_3}{\partial z} + f_z &= 0 \\ g_1\frac{\partial y_1}{\partial z} + g_2\frac{\partial y_2}{\partial z} + g_3\frac{\partial y_3}{\partial z} + g_z &= 0 \\ h_1\frac{\partial y_1}{\partial z} + h_2\frac{\partial y_2}{\partial z} + h_3\frac{\partial y_3}{\partial z} + h_z &= 0 \end{aligned}$$

Or in matrix form

$$\begin{bmatrix} f_1 & f_2 & f_3 \\ g_1 & g_2 & g_3 \\ h_1 & h_2 & h_3 \end{bmatrix} \begin{bmatrix} \partial y_1/\partial z \\ \partial y_2/\partial z \\ \partial y_3/\partial z \end{bmatrix} = \begin{bmatrix} -f_z \\ -g_z \\ -h_z \end{bmatrix} \tag{9.58}$$

Now we solve these equations for $\partial y_1/\partial z$, $\partial y_2/\partial z$, and $\partial y_3/\partial z$ using the methods of Chap. 6. The solutions will depend on the partial derivatives f_i's, g_i's, h_i's, and f_z, g_z, h_z. These in turn may be simple coefficients or themselves functions of variables in the equations. A numerical solution will require a knowledge of the exact form of these partial derivatives and their numerical values at the point where we wish to evaluate our solutions. On the other hand, if we have qualitative information regarding the signs of these partial derivatives, in some cases we may be able to determine the directions of the effects of exogenous variables on the dependent variables. This is especially true when we have a simple model with a few variables and equations.

When we allow for different channels of influence between variables, the determination of the direction of effects of a change in one variable on others will depend on the relative magnitude of partial derivatives. Thus, it may not be possible to determine such effects without a knowledge of the numerical values of coefficients and partial derivatives. As a result, the effects of a change in exogenous variables on endogenous variables will be ambiguous.

Example 9.36

$$\begin{aligned} y_1 + y_2 - 2x &= 0 \\ y_1 - y_2 + 5 &= 0 \end{aligned}$$

Differentiating the equations with respect to x,

$$\begin{aligned} \frac{\partial y_1}{\partial x} + \frac{\partial y_2}{\partial x} &= 2 \\ \frac{\partial y_1}{\partial x} - \frac{\partial y_2}{\partial x} &= 0 \end{aligned}$$

or in matrix notation,

$$\begin{bmatrix} 1 & 1 \\ 1 & -1 \end{bmatrix} \begin{bmatrix} \partial y_1/\partial x \\ \partial y_2/\partial x \end{bmatrix} = \begin{bmatrix} 2 \\ 0 \end{bmatrix}$$

Solving the equations,

$$\frac{\partial y_1}{\partial x} = 1, \qquad \frac{\partial y_2}{\partial x} = 1$$

Example 9.37

$$\begin{aligned} y_1 + y_2 + y_3 + x - z &= 0 \\ y_1 - 2y_2 + \ln x &= 0 \\ y_2^2 + 5y_3 + z^2 &= 0 \end{aligned}$$

In this model there are three endogenous variables y_1, y_2, and y_3 and two exogenous variables x and z. Suppose we want to find the effects of x on the endogenous variables. To isolate these effects, we set $dz = 0$, that is, we hold z constant. Differentiating the equations with respect to x and writing the results in matrix form, we get

$$\begin{bmatrix} 1 & 1 & 1 \\ 1 & -2 & 0 \\ 0 & 2y_2 & 5 \end{bmatrix} \begin{bmatrix} \partial y_1/\partial x \\ \partial y_2/\partial x \\ \partial y_3/\partial x \end{bmatrix} = \begin{bmatrix} -1 \\ -1/x \\ 0 \end{bmatrix}$$

Example 9.38 (A Model of Supply and Demand). Consider the simple demand and supply function of a particular commodity:

$$\begin{aligned} Q^D &= D(P, Y) \qquad D_P < 0, \;\; D_Y > 0 \\ Q^S &= S(P, R) \qquad S_P > 0, \;\; S_R < 0 \\ Q^D &= Q^S \end{aligned}$$

where Q, P, Y, and R are, respectively, quantity demanded, quantity supplied, price, income, and the price of a resource, say oil, used in the production of the good under study. Combining the equations and letting Q be the equilibrium quantity demanded and supplied,

$$\begin{aligned} Q - D(P, Y) &= 0 \\ Q - S(P, R) &= 0 \end{aligned}$$

We may be interested in analyzing the effects of a change in exogenous variables on quantity and price in this market. Let us set $dY = 0$ and analyze the effect of an increase in the price of oil on quantity and price (analyzing the effect of income is left as an exercise, see E.9.15).

$$\frac{\partial Q}{\partial R} - D_P \frac{\partial P}{\partial R} = 0$$
$$\frac{\partial Q}{\partial R} - S_P \frac{\partial P}{\partial R} = S_R$$

Solving the equations

$$\frac{\partial Q}{\partial R} = \frac{-S_R D_P}{S_P - D_P} < 0$$
$$\frac{\partial P}{\partial R} = \frac{-S_R}{S_P - D_P} > 0$$

In other words an increase in the price of oil or any factor of production will result in a higher price and lower quantity of the good. Note that assuming continuously differentiable demand and supply functions, and D_p and S_p having opposite signs, the conditions of the implicit function theorem are met for the model. If D_p and S_p did not have the opposite signs, we could not rule out the possibility of a zero Jacobian determinant at the point of equilibrium. In such an eventuality, we could not have solved for $\partial Q/\partial R$ and $\partial P/\partial R$ as the denominator would be zero.

Example 9.39 (An Open Economy Macroeconomic Model). Consider the following model of an open economy

$$\begin{aligned} &Y - C(Y) - I(r) - X(e) = G && \text{IS} \\ &L(Y, r) = M && \text{LM} \\ &X(e) + F(r) = 0 && \text{BOP} \\ 0 < C' < 1 \quad & X',\ L_Y,\ F' > 0, && I',\ L_r < 0 \end{aligned}$$

where it is assumed that the net exports (exports minus imports), X, depends only on the exchange rate, e, and the net inflow of capital, F, depends on the interest rate r. The exchange rate is the price of foreign currency in terms of domestic currency. Thus, an increase in e means the depreciation of domestic currency. The balance of payments equation states that the sum of the net current account and net capital account is zero. Let us analyze the effects of an expansionary monetary policy on the three endogenous variables Y, r, and e. Letting $dG = 0$, while allowing M to change, we have

$$\begin{aligned}\frac{\partial Y}{\partial M}-C'\frac{\partial Y}{\partial M}-I'\frac{\partial r}{\partial M}-X'\frac{\partial e}{\partial M}&=0\\ L_Y\frac{\partial Y}{\partial M}+L_r\frac{\partial r}{\partial M}&=1\\ X'\frac{\partial e}{\partial M}+F'\frac{\partial r}{\partial M}&=0\end{aligned}$$

Rearranging the equations in matrix format,

$$\begin{bmatrix}1-C' & -I' & -X'\\ L_Y & L_r & 0\\ 0 & F' & X'\end{bmatrix}\begin{bmatrix}\partial Y/\partial M\\ \partial r/\partial M\\ \partial e/\partial M\end{bmatrix}=\begin{bmatrix}0\\1\\0\end{bmatrix}$$

The determinant of the matrix on the LHS is

$$\Delta=(1-C')L_rX'-L_YF'X'+X'L_YI'<0$$

Because the determinant is not zero, we can solve for the vector of unknowns as

$$\begin{bmatrix}\partial Y/\partial M\\ \partial r/\partial M\\ \partial e/\partial M\end{bmatrix}=\frac{1}{\Delta}\begin{bmatrix}X'L_r & I'X'-F'X' & X'L_r\\ -X'L_Y & X'(1-C') & -X'L_Y\\ L_YF' & -(1-C')F' & L_r(1-C')+I'L_Y\end{bmatrix}\begin{bmatrix}0\\1\\0\end{bmatrix}$$

Thus, an expansionary monetary policy results in an increase in income, a lowering of the interest rate, and the depreciation of domestic currency because

$$\begin{aligned}\frac{\partial Y}{\partial M}&=\frac{I'X'-F'X'}{\Delta}>0,\\ \frac{\partial r}{\partial M}&=\frac{X'(1-C')}{\Delta}<0,\\ \frac{\partial e}{\partial M}&=\frac{-(1-C')F'}{\Delta}>0\end{aligned}$$

In Example 9.39 we arrived at unequivocal conclusions regarding the effects of monetary policy on variables of interest. This was possible because of our restrictive assumptions. For example, if we assume that net exports depend on both income and the exchange rate, we cannot determine the sign of $\partial e/\partial M$. It will depend on the value of partial derivatives involved.

9.5.3 Exercises

E.9.15 Referring to our model of supply and demand, assume that income and the price of oil are constant. Under what conditions are functions representing supply and demand independent? What is the economic meaning of these conditions?

E.9.16 Using the model of supply and demand, find the effects of an increase in income on quantity and price.

E.9.17 Show that in the $y - r$ plane the IS curve is downward sloping.

E.9.18 Show that in the $y - r$ plane the LM curve is upward sloping.

E.9.19 Show the effects of an increase in government expenditures on income, the interest rate, and the exchange rate.

E.9.20 Change the net export function to $X = X(Y, \ e)$ with $X_Y < 0$ and $X_e > 0$. Find the effects of money supply and government expenditures on income, the interest rate, and the exchange rate.

Chapter 10
The Taylor Series and Its Applications

Apparently it started with a discussion in Child's Coffeehouse where Brook Taylor (1685–1731) got the idea for the now famous series. He was talking with his friend John Machin about solving Kepler's problem. As it turned out, the Taylor series was of such importance that Lagrange called it "the basic principle of differential calculus." Indeed, it plays a very important part in calculus as well as in computation, statistics, and econometrics. As it is well known, a calculator or computer can only add and, in fact, can deal only with 0s and 1s. So how is it possible that you punch in a number and then press a button, and the calculator finds the logarithm or exponential of that number? Similarly, how can a machine capable of only adding give you the sine and cosine of an angle, find solutions to an equation, and find the maxima and minima of a function? All these and more can be done due to the Taylor series.

Furthermore, frequently in statistics and econometrics we estimate a set of parameters but need to make an inference on nonlinear functions of them. For example, we may estimate the reduced form of a system of simultaneous equations, but we need to make inference about the structural parameters. The inference has to be based on the distributions of the estimators of the latter parameters, the derivation of which may be difficult or impossible. One alternative is to approximate moments of these distributions via the Taylor series.

Still there are other issues in economics—such as measuring risk aversion or the connection of expected utility function and mean-variance analysis in finance—that can be fully understood only with the help of the Taylor expansion.

10.1 The Taylor Expansion

Consider a function $f(x)$ that is differentiable $n+1$ times, that is, its $(n+1)$-th derivative exists. Would it be possible to find a polynomial $P_n(x)$ of degree less than or equal to n with the following properties:

1. Its value at the point a is $f(a)$.
2. The value of its j-th derivative $j = 1, \ldots n$ at point a is the same as the derivatives of $f(x)$. In other words,

K. Dadkhah, *Foundations of Mathematical and Computational Economics,* 2nd ed.,
DOI 10.1007/978-3-642-13748-8_10, © Springer-Verlag Berlin Heidelberg 2011

$$\left.\frac{d^jP_n}{dx^j}\right|_{x=a} = f^{(j)}(a), \quad j = 1, \ldots, n \tag{10.1}$$

Because the polynomial has to pass through point $(a, f(a))$, we may write it in terms of $x - a$ as

$$P_n(x) = C_0 + C_1(x-a) + C_2(x-a)^2 + \ldots + C_n(x-a)^n \tag{10.2}$$

From condition (1) above, it follows that

$$C_0 = P_n(a) = f(a) \tag{10.3}$$

and from condition (2),

$$\begin{aligned} C_1 &= f'(a) \\ C_2 &= \frac{f''(a)}{2} \\ C_3 &= \frac{f'''(a)}{3!} \\ \cdots &= \cdots \\ C_n &= \frac{f^{(n)}(a)}{n!} \end{aligned} \tag{10.4}$$

The reason is that

$$P'_n(x) = C_1 + 2C_2(x-a) + 3C_3(x-a)^2 + \ldots + nC_n(x-a)^{n-1}$$

and evaluating $P'_n(x)$ at point a causes all the terms involving $x - a$ to vanish. Therefore,

$$P'_n(a) = C_1 \qquad \Rightarrow \qquad C_1 = f'(a)$$

The second equality is condition (2) above. Similarly

$$P''_n(a) = 2C_2 \qquad \Rightarrow \qquad C_2 = \frac{f''(a)}{2}$$

In general,

$$\begin{aligned} P_n^{(j)}(x) = {} & j \times (j-1) \times \ldots \times 2\, C_j + (j+1) \times j \times \ldots \times 2\, C_{j+1}(x-a) \\ & + (j+2) \times (j+1) \times \ldots \times 3\, C_{j+2}(x-a)^2 + \ldots \\ & + n \times (n-1) \times \ldots \times (n-j+1) C_n(x-a)^{n-j} \end{aligned}$$

At point $x = a$, we have

$$P_n^{(j)}(a) = j!C_j \qquad \Rightarrow \qquad C_j = \frac{f^{(j)}(a)}{j!}$$

Thus we can write

$$\begin{aligned} P_n(x) &= f(a) + f'(a)(x-a) + \frac{1}{2}f''(a)(x-a)^2 \\ &+ \frac{1}{3!}f'''(a)(x-a)^3 + \ldots + \frac{1}{n!}f^{(n)}(a)(x-a)^n \end{aligned} \tag{10.5}$$

Let us call the difference between the polynomial $P_n(x)$ and $f(x)$, the remainder, that is,

$$f(x) = P_n(x) + R_n(x) \tag{10.6}$$

If $R_n(x)$ is small, then $P_n(x)$ is a good approximation for $f(x)$. Indeed, we could derive the Taylor formula as an approximation of the difference between $f(x)$ and $f(a)$. Geometrically, we could try to approximate the segment of the curve between the two points $(x, f(x))$ and $(a, f(a))$ by a line (see Fig. 10.1). Of course it would not be a good approximation. Perhaps a parabola or a cubic function, or more generally, a polynomial of degree n would be better. Incidentally note that the segment $[(x, f(x)), (b, f(b))]$ can reasonably be approximated by a line because the two points are much closer. Because we are approximating the difference between the values of the function at points x and a, the polynomial has to be in terms of their difference $x - a$. Therefore,

$$f(x) - f(a) = \sum_{j=1}^{n} C_j(x-a)^j + R_n(x) \tag{10.7}$$

Using the same procedure as before we have

$$f(x) - f(a) = \sum_{j=1}^{n} f^{(j)}(a)\frac{(x-a)^j}{j!} + R_n(x) \tag{10.8}$$

or

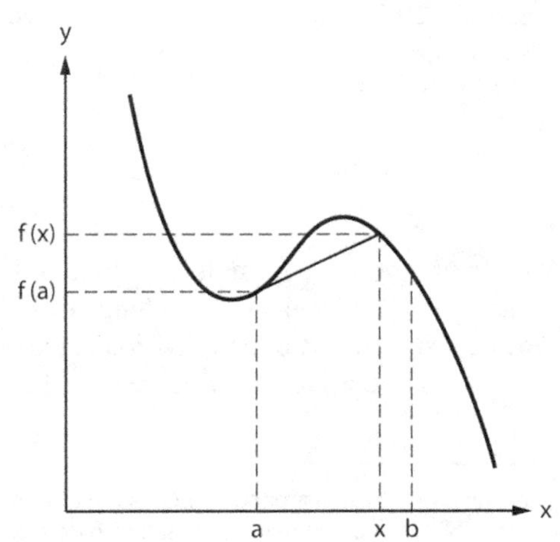

Fig. 10.1 Approximating $f(x)$ near point a with a polynomial

$$f(x) \approx \sum_{j=0}^{n} f^{(j)}(a) \frac{(x-a)^j}{j!} \tag{10.9}$$

Example 10.1 Finding the Taylor expansion of a polynomial function is pointless in that we already have the expansion. Nevertheless, such an exercise is quite useful in terms of illustrating the procedure and its objective. Here we write the Taylor expansion[1] of the function $y = x^3 - 2x$, first near point $x_0 = 0$ and then near point $x_1 = 1$. We start with the derivatives of the function

$$\begin{aligned} y' &= 3x^2 - 2 \\ y'' &= 6x \\ y''' &= 6 \end{aligned}$$

Therefore, the Taylor series near point x_0 is

$$y = 0 - 2x + 0 + \frac{6}{6}x^3 = x^3 - 2x$$

and near x_1

$$y = 1 - 2 + (x-1) + \frac{6}{2}(x-1)^2 + \frac{6}{6}(x-1)^3 = x^3 - 2x$$

Example 10.2 Write the first five terms of the Taylor series of $y = \ln x$ near point $x_0 = 1$.

$$y' = \frac{1}{x}, \quad y'' = -\frac{1}{x^2}, \quad y''' = \frac{2}{x^3}, \quad y^{(4)} = -\frac{6}{x^4}$$

Thus,

$$\ln x = 0 + (x-1) - \frac{1}{2}(x-1)^2 + \frac{1}{3}(x-1)^3 - \frac{1}{4}(x-1)^4$$

Example 10.3 Consider the Taylor expansion of $f(x) = e^x$ near point $a = 0$. We already know that $f^{(j)}(x) = e^x$, $j = 1, \ldots, n$, and $e^0 = 1$.

Therefore,

$$e^x \approx \sum_{j=0}^{n} \frac{x^j}{j!} \tag{10.10}$$

We can program this formula in Matlab or Excel and calculate the value of e by letting $x = 1$. The formula uses nothing but addition because multiplication is addition repeated so many times and division is the inverse of multiplication. The results are reported in Table 10.1. As n increases, the precision of the computation increases.

[1]The Taylor expansion around point 0 is referred to as Maclaurin expansion after Colin Maclaurin (1698–1746), a brilliant mathematician who derived it as a special case of Taylor series.

Table 10.1 Evaluating e using the Taylor expansion

n	e
0	1.00000000000000
1	2.00000000000000
2	2.50000000000000
3	2.66666666666667
4	2.70833333333333
5	2.71666666666667
6	2.71805555555556
7	2.71825396825397
8	2.71827876984127
9	2.71828152557319
10	2.71828180114638
11	2.71828182619849
12	2.71828182828617
13	2.71828182844676
14	2.71828182845823
15	2.71828182845899
16	2.71828182845904
17	2.71828182845905

Check the result obtained against the value of e that you can get from your calculator. Note that the result of the Matlab expression `factorial()` is accurate for the first 15 digits, that is, for $n \leq 21$.

Matlab code

```
% Initialize an array of length 20
e = zeros(20,1);
% Set the first entry equal to 1 which
% corresponds to e^0
  e(1) = 1;
% calculate e by adding in each step 1/i! to the
% result of the previous step. Note that we have
% to adjust the counter because Matlab counters
% cannot be zero.
for j=2:20
       e(j) = e(j-1) + 1/(factorial(j-1));
end
% We will use the long format to get as many decimals
% as possible
format long
% Print e
e
```

Table 10.2 Computing e using Excel

A	B
0	=1/FACT(A1)
1	=B1+1/FACT(A2)
2	=B2+1/FACT(A3)
3	=B3+1/FACT(A4)
4	=B4+1/FACT(A5)
⋮	⋮

For the Excel program, we have (Table 10.2).

Example 10.4 (Mean-Variance Analysis in Finance). Following the seminal work of Harry Markowitz, financial economists and financial analysts have used the mean-variance analysis. The idea is to represent the risk of an asset or portfolio with the variance of its rate of return. By balancing expected return against risk, the investor or analyst can choose an optimal portfolio. At first glance such an analysis seems ad hoc. But indeed it has its roots in the expected utility analysis. Let

$$U = U(x) \tag{10.11}$$

where U denotes utility and x is the return of an asset that we assume to be a random variable distributed with mean $\mu = E(x)$ and variance $\sigma^2 = E(x-\mu)^2$. The return of an asset is considered a random variable because it cannot be forecast with certainty. It depends on many factors, including the economic condition of the country and the world as well as the specific workings of the firm issuing the security. In the case of U.S. government bonds, the probability distribution is concentrated at one point, namely the yield that has probability one. Using the Taylor formula, we can write

$$U(x) = U(\mu) + (x-\mu)U'(\mu) + \frac{1}{2}(x-\mu)^2 U''(\mu) + R \tag{10.12}$$

Assuming that the remainder is negligible, we can write the expected utility as

$$E(U) = U(\mu) + \frac{1}{2}\sigma^2 U''(\mu) \tag{10.13}$$

Thus, the expected utility depends only on the mean and variance of the return. If the assumptions made are valid, then decisions based on mean-variance analysis are equivalent to decisions based on expected utility maximization. Finally, note that for a fixed value of an expected utility, say, $E(U) = U_0$, we have

$$U'd\mu + \frac{1}{2}U''d\sigma^2 = 0 \tag{10.14}$$

because all higher derivatives of U are assumed to vanish. Thus,

$$\frac{d\mu}{d\sigma^2} = -\frac{1}{2}\frac{U''}{U'} \tag{10.15}$$

The above equation shows the trade-off between risk and return. In order to remain indifferent to an infinitesimal change in risk, σ^2, the investor requires a $-U''/2U'$ increase in return. The higher the required compensation, the higher the degree of risk aversion on the part of the investor. Thus,

$$\nu = -\frac{U''}{U'} \tag{10.16}$$

is taken as the measure of risk aversion. Furthermore, if we normalize it by the amount of income, x, or the expected return μ, then we have the measure of relative risk aversion:

$$\varepsilon = -\mu\frac{U''}{U'} \tag{10.17}$$

Example 10.5 Consider a random variable X with mean $E(X) = \mu$ and variance $E(X - \mu)^2 = \sigma^2$. Suppose we are interested in the variable $Y = f(X)$ where f is a nonlinear function. What are the mean and variance of Y?

If the distribution of X is known and if we could explicitly derive the resulting distribution of Y, then we can come up with the mean and variance of Y. Usually, however, the derivation of the resulting distribution is rather involved and we may not end up with a known distribution. In such cases, if we know the distribution of the original variable still we can simulate the new distribution and compute its mean and variance. If we don't know the distribution of X then this option is also unavailable. In such cases we can resort to Taylor approximation.

Let X have mean and variance μ and σ^2, and let $Y = f(X)$. We can write:

$$Y \approx f(\mu) + (X - \mu)f'(\mu) \tag{10.18}$$

Taking the expected value of both sides, we have

$$E(Y) \approx f(\mu) \tag{10.19}$$

Further,

$$E[Y - E(Y)]^2 \approx \sigma^2 f'(\mu)^2 \tag{10.20}$$

This, procedure, as the next example will show, proves quite useful in the theory and practice of econometrics.

Example 10.6 (Nonlinear Functions in Econometrics). On many occasions we estimate a parameter, say, θ by $\hat{\theta}$, but we are interested in making inference about $f(\theta)$

where f is nonlinear. In such cases we may try to derive the distribution of $f(\hat{\theta})$ based on the distribution of $\hat{\theta}$. This is not always easy, nor is it guaranteed that the resulting distribution is tractable. An alternative would be to approximate moments of $f(\hat{\theta})$, in particular its mean and variance, using the Taylor expansion near the point $\hat{\theta}$.

$$f(\hat{\theta}) \approx f(E(\hat{\theta})) + (\hat{\theta} - E(\hat{\theta}))f'(E(\hat{\theta})) \tag{10.21}$$

Taking the expected value of both sides of (10.21) and dropping the approximation sign, we have

$$E[f(\hat{\theta})] = f(E(\hat{\theta})) \tag{10.22}$$

Moreover,

$$\begin{aligned}\operatorname{Var} f(\hat{\theta}) &= E[f(\hat{\theta}) - Ef(\hat{\theta})]^2 \\ &= E[\hat{\theta} - E(\hat{\theta})]^2[f'(\hat{\theta})]^2 \\ &= \operatorname{Var}(\hat{\theta})[f'(\hat{\theta})]^2\end{aligned} \tag{10.23}$$

Example 10.7 In elementary econometrics, the following model is used to illustrate the issues regarding identification and estimation of systems of simultaneous equations:

$$\begin{aligned}Y &= C + I \\ C &= \beta Y + \varepsilon\end{aligned}$$

where Y, C, and I are, respectively, income, consumption, and investment, or autonomous expenditures. β is the marginal propensity to consume, and ε is a stochastic term. Because Y is endogenous, if we estimate β directly using the second equation, the result will be a biased and inconsistent estimate. To get a consistent estimate of β we write the reduced form of the equation

$$C = \frac{\beta}{1-\beta} I + \frac{\varepsilon}{1-\beta} = kI + \eta$$

where $k = \beta/(1-\beta)$ and $\eta = \varepsilon/(1-\beta)$. Now we can obtain an unbiased and consistent estimate of k together with $\operatorname{Var}(\hat{k})$. But we still want to make inference about $\beta = k/(1+k)$. We can estimate β as

$$\hat{\beta} = \frac{\hat{k}}{1+\hat{k}}$$

and its variance as

$$\operatorname{Var}(\hat{\beta}) = \operatorname{Var}(\hat{k})\left(\frac{d\beta}{dk}\right)^2 = \operatorname{Var}(\hat{k})\frac{1}{(1+\hat{k})^4}$$

$\hat{\beta}$ is a biased but consistent estimator because

$$plim\ \hat{\beta} = \frac{plim\ \hat{k}}{1 + plim\ \hat{k}} = \frac{k}{1+k}$$

plim is short for *probability limit* and $plim\,\hat{k} = k$, means that $\hat{k}$ converges to k in probability. That is,

$$\lim_{N\to\infty} P\left[|k - \hat{k}| < \varepsilon\right] = 1 \tag{10.24}$$

Example 10.8 (Equivalence of Different Forms of Complex Variables). In Chap. 2 we noted that a complex variable can be written in three forms:

$$\begin{aligned} z &= x \pm iy \\ &= \rho(\cos\theta \pm i \sin\theta) \\ &= \rho e^{\pm i\theta} \end{aligned} \tag{10.25}$$

where

$$\rho = \sqrt{x^2 + y^2} \tag{10.26}$$

and

$$\tan\theta = \frac{y}{x} \qquad \text{or} \qquad \theta = \tan^{-1}\frac{y}{x} \tag{10.27}$$

Whereas the first equality in (10.28) follows from the Pythagoras theorem and definitions of trigonometric functions (see Fig. 10.2), the second equality is not self-evident. Here we will show the second equality with the help of the Taylor expansion. First, let us write the Maclaurin expansions of the $\sin\theta$ and $\cos\theta$.

$$\cos\theta = \cos(0) - \theta\,\sin(0) - \frac{\theta^2}{2}\cos(0) + \frac{\theta^3}{3!}\sin(0) + \frac{\theta^4}{4!}\cos(0) - \cdots$$

Noting that cos(0) =1 and sin(0) =0, we have

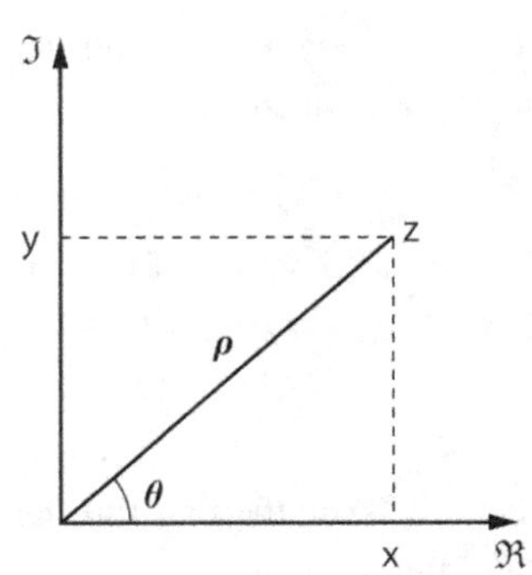

Fig. 10.2 Point z in the complex plane

$$\cos\theta = 1 - \frac{\theta^2}{2} + \frac{\theta^4}{4!} - \frac{\theta^6}{6!} - \cdots = \sum_{k=0}^{\infty} (-1)^k \frac{\theta^{2k}}{(2k)!} \tag{10.28}$$

Similarly,

$$\sin\theta = \sin(0) + \theta\cos(0) - \frac{\theta^2}{2}\sin(0) - \frac{\theta^3}{3!}\cos(0) + \frac{\theta^4}{4!}\sin(0) + \cdots$$

and

$$\sin\theta = \theta - \frac{\theta^3}{3!} + \frac{\theta^5}{5!} - \frac{\theta^7}{7!} + \cdots = \sum_{k=1}^{\infty} (-1)^{k-1} \frac{\theta^{2k-1}}{(2k-1)!} \tag{10.29}$$

Therefore,

$$\cos\theta + i\sin\theta = 1 + i\theta - \frac{\theta^2}{2} - i\frac{\theta^3}{3!} + \frac{\theta^4}{4!} + i\frac{\theta^5}{5!} - \cdots$$

Recalling that

$$i^0 = 1, \qquad i^2 = -1, \qquad i^4 = 1, \qquad \ldots$$

we can write

$$\cos\theta + i\sin\theta = \sum_{k=0}^{\infty} \frac{(i\theta)^k}{k!} \tag{10.30}$$

But we already know that

$$e^{i\theta} = \sum_{k=0}^{\infty} \frac{(i\theta)^k}{k!} \tag{10.31}$$

Hence the equality is verified. In a similar fashion we can show that $\cos\theta - i\sin\theta = e^{-i\theta}$. Incidentally, we have shown how the sine and cosine functions are calculated.

10.1.1 Exercises

E.10.1 Write the first four terms of the Taylor expansion of the following functions near the point x_0

i. $f(x) = 5x^4 - 3x^3 - x^2 + 7x + 14$
ii. $f(x) = \sqrt{2x}$
iii. $f(x) = e^{3x}$
iv. $f(x) = xe^x$

E.10.2 Write the first four terms of the Taylor expansion of the following functions near the point $x_0 = \pi/2$:

i. $f(x) = \cos 2x$, *ii.* $f(x) = \sin 2x$, *iii.* $f(x) = \tan 3x$

E.10.3 Use the Taylor series to evaluate the following expressions:

i. $\sqrt{54}$ *ii.* e^2 *iii.* $\ln 6$

E.10.4 Program the Taylor formula in Matlab to calculate a table of logarithms for numbers between 0.001 and 9.999.

E.10.5 Write a program to calculate the sine and cosine functions of different angles based on the Taylor expansion.

E.10.6 Given the mean μ, and variance σ^2 of the variable $X > 0$, find the mean and variance of variable $Y = \ln(X)$.

10.2 The Remainder and the Precision of Approximation

Although by increasing n the precision of the computation is increased, the process cannot continue indefinitely. Therefore we might ask, what is the nature of this approximation or, to put it differently, what is the order of magnitude of $R_n(x)$? We have a good approximation formula and we can replace $f(x)$ by $P_n(x)$ if $R_n(x) \to 0$ as $n \to \infty$. To find an expression for $R_n(x)$, we need the basic result of Rolle's theorem.

Theorem 10.1 (Rolle's[2] Theorem). *Let $f(x)$ be a continuous function on the interval $[a, b]$ with $f(a) = f(b) = 0$. Further, assume that $f(x)$ is differentiable on the open interval (a, b). Then there exists at least one point x^* such that $a < x^* < b$ and $f'(x^*) = 0$.*

Proof Because the function f is continuous in the interval $[a, b]$, then it has a maximum and a minimum. First note that if the maximum and minimum coincide, then we have a horizontal line and $f'(x) = 0$ for all values of x in the interval and the theorem holds. If the maximum and minimum are not equal, then at least one of them is not equal to zero. For the sake of simplicity, let us assume that the maximum $M > 0$. Note that this excludes the maximum occurring either at a or b. Thus, let

$$f(x^*) = M$$

It follows that at all points on both sides of the point $x = x^*$, we have:

$$f(x^* + \Delta x) - f(x^*) \leq 0 \tag{10.32}$$

(10.32) holds both when $\Delta x < 0$ and $\Delta x > 0$. Therefore,

$$\begin{aligned} &\frac{f(x^* + \Delta x) - f(x^*)}{\Delta x} \leq 0, && \Delta x > 0 \\ &\frac{f(x^* + \Delta x) - f(x^*)}{\Delta x} \geq 0, && \Delta x < 0 \end{aligned} \tag{10.33}$$

[2] The self-taught French mathematician Michel Rolle (1652–1719) is best known for this theorem.

Taking the limits of both relationships in (10.33), we have

$$\begin{aligned} \lim_{\Delta x \to 0+} \frac{f(x^* + \Delta x) - f(x^*)}{\Delta x} &= f'(x^*) \leq 0 \\ \lim_{\Delta x \to 0-} \frac{f(x^* + \Delta x) - f(x^*)}{\Delta x} &= f'(x^*) \geq 0 \end{aligned} \tag{10.34}$$

which implies $f'(x^*) = 0$.
Going back to our question, we note that because the last term of (10.9) is

$$f^{(n)} \frac{(x-a)^n}{n!}$$

we can surmise the remainder to be of the form

$$R_n(x) = \frac{(x-a)^{n+1}}{(n+1)!} Q(x) \tag{10.35}$$

$Q(x)$ depends on x and a. Therefore, for particular values of x, and a, $Q(x)$ has a fixed value. Let us denote it by Q^* and try to determine it. Consider the expansion of $f(x)$ about a point z that lies somewhere between a and x.

$$f(x) \approx \sum_{j=0}^{n} f^{(j)} \frac{(x-z)^j}{j!} \tag{10.36}$$

and define

$$F(z) = f(x) - \sum_{j=0}^{n} f^{(j)} \frac{(x-z)^j}{j!} - \frac{(x-z)^{n+1}}{(n+1)!} Q^* \tag{10.37}$$

Note that for fixed values of a and x, F is a function of z alone and is differentiable. Therefore,

$$\begin{aligned} F'(z) = &-f'(z) + f'(z) - (x-z)f''(z) + \frac{2(x-z)}{2!} f''(z) \\ &- \frac{(x-z)^2}{2!} f^{(3)}(z) + \ldots + \frac{(x-z)^{n-1}}{(n-1)!} f^{(n)}(z) \\ &+ \frac{n(x-z)^{n-1}}{n!} f^{(n)}(z) - \frac{(x-z)^n}{n!} f^{(n+1)}(z) \\ &+ \frac{(n+1)(x-z)^n}{(n+1)!} Q^* \end{aligned} \tag{10.38}$$

Simplifying,

$$F'(z) = -\frac{(x-z)^n}{n!} f^{(n+1)}(z) + \frac{(x-z)^n}{n!} Q^* \tag{10.39}$$

Furthermore, because $F(x) = F(a) = 0$, the conditions of Rolle's theorem hold and for some value of z, say, ξ, $F'(\xi) = 0$. Therefore,

$$Q^* = f^{(n+1)}(\xi) \tag{10.40}$$

and

$$R_n(x) = \frac{(x-a)^{n+1}}{(n+1)!} f^{(n+1)}(\xi) \tag{10.41}$$

Recall that ξ lies somewhere between x and a. Therefore we can write

$$\xi = a + \theta(x-a) \qquad 0 < \theta < 1 \tag{10.42}$$

Substituting (10.42) in (10.41) we have Lagrange's[3] form of the remainder:

$$R_n(x) = \frac{(x-a)^{n+1}}{(n+1)!} f^{(n+1)}(a + \theta(x-a)) \tag{10.43}$$

and when $a = 0$

$$R_n(x) = \frac{x^{n+1}}{(n+1)!} f^{(n+1)}(\theta x) \tag{10.44}$$

Now we can assess the accuracy of the Taylor approximation. For example, in the case of calculating e, the remainder will be

$$R_n = \frac{x^{n+1}}{(n+1)!} e^{\theta x}$$

The first question is whether $R_n \to 0$ as $n \to \infty$, that is, if the Taylor series of e is convergent. For a fixed x the value of $e^{\theta x}$ is constant. Therefore we should look at

$$\lim_{n\to\infty} \frac{x^{n+1}}{(n+1)!} = \lim_{n\to\infty} \frac{x}{1} \times \frac{x}{2} \times \ldots \times \frac{x}{n} \times \frac{x}{n+1}$$

For a finite fixed x, the last term approaches zero as $n \to \infty$. Therefore, the series is convergent. We can estimate the error of calculation by finding an upper limit for it. For $x = 1$ and $\theta < 1$ we have

$$e^{\theta x} < 3$$

Therefore, the upper limit of the error of calculation is

$$R_n < \frac{1}{(n+1)!} 3$$

[3] After Joseph-Louis Lagrange (1736–1813), who was considered a great mathematician at 23 and whom Napoleon Bonaparte referred to as "The Lofty Pyramid of the mathematical sciences."

Thus,

$$R_n < \begin{cases} 8.2672 \times 10^{-6} \\ 4.68576 \times 10^{-16} \\ 3.64838 \times 10^{-34} \end{cases} \quad \text{for} \quad n = \begin{cases} 8 \\ 17 \\ 30 \end{cases}$$

10.2.1 Exercises

E.10.7 Write the remainder term for the function in E.10.1 and E.10.2.

E.10.8 Compute an order of magnitude for the accuracy of your results in E.10.3.

10.3 Finding the Roots of an Equation

Finding the roots of an equation, that is, computing x^* such that $f(x^*) = 0$, can be accomplished with several different algorithms. One effective algorithm, Newton's[4] method, is based on the Taylor expansion. Here we will discuss this method and compare it to another method called the bisection method. Both are members of the family of iterative methods.

10.3.1 Iterative Methods

In Chap. 3, we briefly discussed the idea behind iterative methods. Because the subject is of importance both in computation and econometrics, we illustrate it once more with a similar example. This time we will try to find the cubic root of a number. Suppose we are interested in finding the cubic root of the number, say, 91125. Let us start with an *initial guess* and for the time being we make an off-the-wall guess,

---

[4]Sir Isaac Newton (1643–1727) is a giant in the history of science; indeed, the publication of his *Philosophiae Naturalis Principia Mathematica*, usually referred to as *Principia*, is a turning point in the history of humankind. He invented calculus, discovered important laws of physics, and showed that the universe works on mathematical principles. Yet, he found time to improve the operation of the Royal Mint. He also served as the president of the Royal Society and shaped it to become the leading scientific society in the world. Newton was a loner and secretive. He did not acknowledge the contribution of other scientists and got into a bitter dispute with Leibnitz over the invention of calculus and with Robert Hooke (1635–1703), another pioneer scientist, over the theory of light. There are many good books on the history of science. I suggest John Gribbin's *Science, A History 1543–2001* (2002). On the life of Newton, the reader may be interested in reading *Isaac Newton: The Last Sorcerer* by Michael White (1997). The German mathematician, Gottfried Wilhelm Leibnitz (1646–1716), independently invented calculus. His exposition was easier to understand than Newton's. The two engaged in a bitter dispute over who had priority. Leibnitz organized the Berlin Academy of Sciences and served as its first president. He had other interests, including law and economics, and for a time served as a diplomat.

Table 10.3 Iterative method to find the cubic root of a number in Excel

A	B
91125	20
	=(1/3)*(2*B1+(A$1/(B1^2)))
	=(1/3)*(2*B2+(A$1/(B2^2)))
	⋮

say, $y_0 = 20$. If this guess is a good one, then $91125/(20^2)$ will be very close to our initial guess. But it is not. How can we get closer to the correct number? Let us consider the weighted average of our initial guess and the number $91125/(20^2)$ where the initial guess has the weight of 2.

$$y_1 = \frac{1}{3}\left(2y_0 + \frac{91125}{y_0^2}\right)$$

Our next result is $y_1 = 89.270833$. If we repeat the step above, we have

$$y_2 = \frac{1}{3}\left(2y_1 + \frac{91125}{y_1^2}\right) = 63.325399$$

A few more iterations and we have our number: $\sqrt[3]{91125} = 45$. This procedure can be programmed in Matlab or Excel. In an Excel worksheet make the entries shown in Table 10.3. Highlight the square B2 and drag it down a few rows.

You will get the answer.

A	B
91125	20
	89.270833
	63.325399
	49.791545
	45.446310
	45.004369
	45
	45

A Matlab program accomplishes the same task.

Matlab code

```
% Specify the number
A = 91125
% Initialize y
y = 20
z = 0
```

```
% specify the degree of precision, delta
delta = 0.00001
% Find the cubic root of A
while abs(z-y) > delta
       z = y;
       y = (2*y + A./(y.^2))./3;
 end
```

The above routine illustrates the process of all iterative algorithms, which have three basic ingredients: a starting point, a desired level of accuracy, and a recurrence formula.

$$y_j = \frac{1}{3}\left(2y_{j-1} + \frac{91125}{y_{j-1}^2}\right) \tag{10.45}$$

The speed of computation is greatly enhanced by choosing a starting point closer to the final result (see E.10.11). The desired level of accuracy depends on the purpose of computation. In general, depending on the problem at hand, we should set a limit reflecting the desired precision of the results and terminate the process when the results obtained in two consecutive iterations differ by less than the preset limit Δ. Of course, this limit should not be less than the precision of Matlab or any other software we may be using. Finally, we need a recurrence formula that is convergent, that is, in every step it gets closer to the final answer.

The algorithm is implemented in four steps:

1. Choose a Δ
2. Choose a starting point x_0
3. Calculate x_j based on x_{j-1}
4. Repeat step 3 until $\left|x_j - x_{j-1}\right| < \Delta$

In the next two sections, we flesh out these concepts by discussing two algorithms for finding the roots of an equation.

10.3.2 The Bisection Method

Suppose we are looking for the roots of the equation $f(x) = 0$ where f is a continuous function. Further suppose that within the interval $[a, b]$ the function changes sign such that $f(a)f(b) \leq 0$. Then clearly one root lies in that interval. Calculate the function at the midpoint $m = (a + b)/2$ If the function changes sign between a and m, then the solution lies in the interval $[a, m]$, or else it lies in the interval $[m, b]$. Either way, we divide the interval within which the root lies into two equal segments and repeat the procedure until we are as close as we wish to the root. Because this is an iterative method, we need a starting point and a tolerance level or a stopping rule for ending the iteration. The following Matlab program illustrates the algorithm. First, we specify the function

Matlab code

```
function y = f(x)
y = x.^2 - exp(x)
```

and save it in an `m.file`. The solution based on bisection can be obtained using the following program:

Matlab code

```
% Specify the interval containing the solution
a = -3;
b = 3;
% check that the interval contains the root
if f(a).*f(b) > 0
        disp('The interval does not contain the root
          of the equation.')
        return
end
% specify delta
delta = 0.00001 + eps.*max(abs(a), abs(b))
while abs(a-b) > delta
         m = (a+b)/2;
         if f(a).*f(m) <= 0
             b = m;
         else
             a = m;
        end
end
x = (a+b)/2
```

We could make this program more efficient by cutting down the number of times the function `f` has to be evaluated. This can be accomplished by storing `f(a)`, `f(b)` and `f(m)` once they are evaluated in, say, `fa`, `fb`, and `fm`. Note also that you can change the function in `m.file` as well as `a`, `b`, and the `Delta` in the body of the program.

10.3.3 Newton's Method

Suppose we are interested in finding the roots of the function $y = f(x)$, that is, x^* such that $f(x^*) = 0$. Consider the first two terms of the Taylor expansion of this function

$$P(x) = f(x_0) + (x - x_0) f'(x_0) \tag{10.46}$$

If $P(x)$ is a good approximation for $f(x)$ and if x^* is the root of this function, then

$$P(x^*) = 0 \tag{10.47}$$

and we have

$$f(x_0) + (x^* - x_0)f'(x_0) = 0 \tag{10.48}$$

which implies

$$x^* = x_0 - \frac{f(x_0)}{f'(x_0)} \tag{10.49}$$

But this would be true if either $x_0 = x^*$ or $P(x)$ was exactly equal to $f(x)$. Because $P(x)$ is only an approximation and we may not have been lucky to pick $x_0 = x^*$, our first result will be only an approximation. Let us call it x_1 and ask if we could improve upon it. Certainly x_1 is closer to x^* than x_0. But if we now take x_1 as our initial guess and repeat the process, we will get even closer to x^*. This observation suggests that we should start with some initial guess, x_0 and at every stage replace x_{j-1} by

$$x_j = x_{j-1} - \frac{f(x_{j-1})}{f'(x_{j-1})} \tag{10.50}$$

and stop the process when

$$|x_j - x_{j-1}| < \Delta \tag{10.51}$$

where again Δ is a preset precision level. The method can be illustrated geometrically. In Fig. 10.3 we start with the initial guess x_0. The tangent to the curve at point $(x_0 f(x_0))$ when extended to intersect the x-axis provides us with the next point x_1. The length $|x_1 - x_0|$ is equal to $f(x_0)/f'(x_0)$, but because we are moving in the negative direction, we have $x_1 - x_0 = -f(x_0)/f'(x_0)$. We repeat the same process for point x_1 and gradually approach x^*.

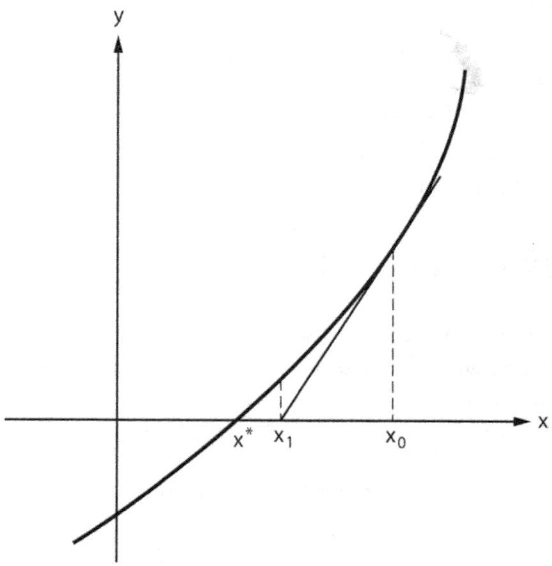

Fig. 10.3 Geometric representation of Newton's method

Example 10.9 In Chap. 3, we showed that the following recurrence formula can be used to compute the square roots of a number:

$$x_j = \frac{1}{2}\left(x_{j-1} + \frac{A}{x_{j-1}}\right)$$

We can show that this formula is a special case of Newton's method. Consider the equation

$$f(x) = x^2 - A = 0$$

whose solution is the square root of A. Since $f' = 2x$, applying Newton's method, we have

$$\begin{aligned} x_j &= x_{j-1} - \frac{f(x_{j-1})}{f'(x_{j-1})} \\ &= x_{j-1} - \frac{x_{j-1}^2 - A}{2x_{j-1}} \\ &= \frac{1}{2}\left(x_{j-1} + \frac{A}{x_{j-1}}\right) \end{aligned}$$

The same can be shown for the recurrence formula of computing the cubic root (see E.10.11).

In order to program Newton's method in Matlab we need to evaluate the derivative of the function. This can be done in three ways. We can use another `m.file` to define the derivative or, alternatively, we can use programs that calculate the derivative of a function. But the simplest way is to use the approximation of derivatives discussed in Chap. 8, that is,

$$f'(x) \approx \frac{f(x+h) - f(x-h)}{2h} \tag{10.52}$$

Matlab code

```
% Specify the initial guess
xi = 0.1;
% specify delta and h
delta = 0.00001 + eps.*f(xi)
h = 0.001
while norm(f(xi)) > delta
% Watch the parentheses as they are important
       xi = xi-f(xi)./((f(xi+h) - f(xi-h))./(2.*h));
 end
```

Again the program can be refined and its parameters reset. Matlab has a ready-made function for finding the roots of a nonlinear equation. It is based on a combination of methods including bisection. Recall that we already have specified the f function. Then we can find the solution as

Matlab code

```
z = fzero(@f,0.1)
```

10.3.4 Exercises

E.10.9 Solve the following equations using both bisection and Newton's methods.

i. $x^3 - 5x + 14 = 0$
ii. $x^3 - 4x^2 + 7x - 8 = 0$
iii. $xe^x - 12 = 0$
iv. $x^4 - 3x^3 + 5x^2 + x - 12$

E.10.10 Use the Matlab `fzero` function to check your results in E.10.9.

E.10.11 Use the Excel routine to find the cube of the following numbers:

i. 456987, ii. 312701, iii. 8123455

After finding the cube root, move your initial guess closer to the final result and see the effect on the number of iterations needed.

E.10.12 Show that the formula in (10.45) for calculating the cubic root of a number is a special case of Newton's method. [*Hint*: Consider the solution of the equation $y^3 - A = 0$ by Newton's method.]

10.4 Taylor Expansion of Functions of Several Variables

The Taylor formula can readily be applied to functions of several variables. Consider the function

$$z = f(x, y) \tag{10.53}$$

To write its Taylor expansion near the point (x_0, y_0), let

$$\delta x = x - x_0$$
$$\delta y = y - y_0$$

Then

$$\begin{aligned} f(x,y) &= f(x_0,y_0) + f_x(x_0,y_0)\delta x + f_y(x_0,y_0)\delta y \\ &+ \frac{1}{2}\left[f_{xx}(x_0,y_0)(\delta x)^2 + 2f_{xy}(x_0,y_0)(\delta x\delta y) + f_{yy}(x_0,y_0)(\delta y)^2\right] \\ &+ \frac{1}{3!}\left[f_{xxx}(x_0,y_0)(\delta x)^3 + 3f_{xxy}(x_0,y_0)(\delta x)^2\delta y \right. \\ &\left. + 3f_{xyy}(x_0,y_0)(\delta y)^2\delta x + f_{yyy}(x_0,y_0)(\delta y)^3\right] + \cdots \end{aligned} \tag{10.54}$$

Example 10.10 Consider the function

$$f(x, y) = 2x^2 - 4xy + 3y^2$$

and the point

$$x_0 = 1, \qquad y_0 = 1$$

because

$$\begin{aligned} f_x &= 4x - 4y \quad f_y = -4x + 6y \\ f_{xx} &= 4 \quad f_{xy} = -4 \quad f_{yy} = 6 \end{aligned}$$

Evaluating the first two derivatives at point (x_0, y_0), we have $f_x(1, 1) = 0$ and $f_y(1, 1) = 2$. The last three derivatives are constant. Thus, the Taylor expansion is

$$\begin{aligned} f(x, y) &= (2 - 4 + 3) + (x - 1)(0) + (y - 1)(2) \\ &\quad + \frac{1}{2}[(x - 1)^2(4) + 2(x - 1)(y - 1)(-4) + (y - 1)^2(6)] \\ &= 2x^2 - 4xy + 3y^2 \end{aligned}$$

Example 10.11 Let us write the Taylor expansion of the function

$$f(x, y) = e^{x+y}$$

near the point

$$x_0 = y_0 = 0$$

The first- and second-order derivatives of the function are

$$f_x = f_y = e^{x+y}, \qquad f_{xx} = f_{xy} = f_{yy} = e^{x+y}$$

Evaluated at point (0, 0), they are all equal to 1. Thus,

$$f(x, y) = 1 + x + y + \frac{1}{2}(x^2 + 2xy + y^2) + \frac{1}{3!}(x^3 + 3x^2y + 3xy^2 + y^3) + R_3$$

Example 10.12 For the function $z = ye^x$ the first three terms of the Taylor series near the point $(x_0 = 0, \; y_0 = 1)$ are

$$z \approx 1 + x + (y - 1) + \frac{1}{2}[x^2 + 2x(y - 1)] + \frac{1}{6}[x^3 + 3x^2(y - 1)]$$

because

$$f_x = ye^x, \qquad f_y = e^x, \qquad f_{xx} = ye^x, \qquad f_{xy} = e^x, \qquad f_{yy} = 0$$

and

$$f_{xxx} = ye^x, \qquad f_{xxy} = e^x, \qquad f_{xyy} = 0, \qquad f_{yyy} = 0$$

Using the notation developed in Chap. 9, we can write the Taylor series in a very compact form. Recall that we defined the gradient of the function

$$y = f(\mathbf{x}) \qquad \mathbf{x} = [x_1, x_2, \ldots, x_k]' \tag{10.55}$$

as

$$\nabla f(\mathbf{x}) = \begin{bmatrix} \partial f / \partial x_1 \\ \vdots \\ \partial f / \partial x_k \end{bmatrix} \tag{10.56}$$

and its Hessian as

$$\nabla^2 f(\mathbf{x}) = \left[\frac{\partial^2 f}{\partial x_i \partial x_j} \right] \qquad i,j = 1, \ldots k \tag{10.57}$$

Using the above notation, we can write the Taylor series of $y = f(\mathbf{x})$ near point $\mathbf{a} = [a_1, a_2, \ldots, a_k]'$ as

$$f(\mathbf{x}) = f(\mathbf{a}) + (\mathbf{x} - \mathbf{a})' \nabla f(\mathbf{x}) + (\mathbf{x} - \mathbf{a})' \nabla^2 f(\mathbf{a}) (\mathbf{x} - \mathbf{a}) + \mathbf{R}_3 \tag{10.58}$$

Note that we stopped at the terms involving second derivatives. The first derivatives are represented by a vector which is a $k \times 1$ column. The Hessian is a matrix and has two dimensions. The set of third-order derivatives have to be presented by a cube of three sides to which a vector of the form $[\mathbf{x} - \mathbf{a}]$ will be attached. The fourth-order derivatives have to be presented in a four-dimensional space that we cannot even visualize. A discussion of symbols to represent such entities is well beyond the scope of this book. But the good news is that for most analytical purposes the first three terms of the Taylor series are all we need. Of course, we can always resort to the form of (10.54) for any number of terms that we may need.

Example 10.13 We can extend the results regarding the mean and variance of a nonlinear function of estimated parameters (Examples 10.5 and 10.6) to the case of a vector of parameters $\boldsymbol{\Theta}$.

Then

$$Ef(\hat{\boldsymbol{\Theta}}) = f(E(\hat{\boldsymbol{\Theta}})) \tag{10.59}$$

and denoting the covariance matrix of $\hat{\boldsymbol{\Theta}}$ by $\mathbf{V}(\hat{\boldsymbol{\Theta}})$ we have

$$\begin{aligned}\text{Var}(f(\hat{\boldsymbol{\Theta}})) &= E[f(\hat{\boldsymbol{\Theta}}) - Ef(\hat{\boldsymbol{\Theta}})][f(\hat{\boldsymbol{\Theta}}) - Ef(\hat{\boldsymbol{\Theta}})]' \\ &= \nabla f(\hat{\boldsymbol{\Theta}})E[\hat{\boldsymbol{\Theta}} - E(\hat{\boldsymbol{\Theta}})][\hat{\boldsymbol{\Theta}} - E(\hat{\boldsymbol{\Theta}})]'(\nabla f(\hat{\boldsymbol{\Theta}}))' \\ &= \nabla f(\hat{\boldsymbol{\Theta}})\mathbf{V}(\hat{\boldsymbol{\Theta}})(\nabla f(\hat{\boldsymbol{\Theta}}))'\end{aligned} \tag{10.60}$$

10.4.1 Exercises

E.10.13 Write the first eight terms of the Taylor series of the following functions:

$$i.\ z = \frac{x-y}{x+y}, \quad ii.\ z = e^{2x-3y}, \quad iii.\ z = \ln(x+y)$$

E.10.14 Write the Taylor expansion of the following functions near point (K_0, L_0).

$$i.\ Q = AK^{\alpha}L^{\beta}$$

$$ii.\ Q = (\alpha K^{\rho} + \beta L^{\rho})^{1/\rho}$$

E.10.15 Suppose we have a model and have estimated $E(y)$ by $\hat{y}$. Further assume that $\text{var}(\hat{y}) = \hat{\sigma}^2$. What would be a reasonable estimate of $E(e^y)$ and its variance?

Chapter 11
Integration

In Chap. 8 we saw that by taking the derivative of the total cost function with respect to the amount of output, we can obtain the marginal cost function. Because an integral is the inverse of the derivative, we may ask if the reverse is true. The answer is, yes, almost. By taking the integral of the marginal cost function we can get the total cost function up to an additive constant (indefinite integral). Whereas the mathematical logic for this will become clear later, the economic logic should be evident to the reader. A knowledge of fixed cost is not contained in the marginal cost function, and, therefore, we will know the total cost function, except for the amount of the fixed cost that we will show by the unknown constant C.

In Chap. 4 we discussed probability distributions and moments of discrete random variables, but we could not do the same for continuous random variables. Suppose there are n random events $A_1, \ldots, A_n$ and we assign to each event the probability $P(A_i) = P_i$, $(i = 1, \ldots, n)$. In order for P_i's to be probabilities, we should have

$$\sum_{i=1}^{n} P_i = 1$$

But when dealing with continuous random variables, which may take any real number between 0 and 1 (e.g., uniform distribution) or on the extended real line (e.g., normal distribution), we cannot use summation and should resort to definite integral. Integration is the counterpart of summation for continuous variables. It allows us to derive probability distributions and moments of continuous random variables. The same device will help us when we deal with continuous time models in economics.

For example, in the continuous time analysis of a compounding interest rate or in discounting future utilities, we need to replace summation with integration. Other applications of integration are in continuous dynamic analysis and in many problems in economics and econometrics that require numerical integration.

There are two types of integrals. One represents the area under a curve[1] with a sign (definite integral). The other is the inverse of a derivative and represents a

[1] Or the volume under a surface in case of a double integral and a 4-, 5-, ... dimensional space in case of triple, ... integrals.

K. Dadkhah, *Foundations of Mathematical and Computational Economics,* 2nd ed.,
DOI 10.1007/978-3-642-13748-8_11,

family of functions (indefinite integral). Whereas the two are related, they are conceptually different. In this chapter we will discuss both and outline their connection and their differences. Certainly it is easier to visualize the definite integral as the area under a curve, but for the ease of exposition we start with the indefinite integral and then present the definite integral. After learning the mathematics of integration, we will show how to conduct numerical integration using Matlab; needless to say, the algorithm can be programmed using other software and languages as well. Then we discuss the very useful subject of differentiation of an integral.

11.1 The Indefinite Integral

Definition 11.1 Consider the function $f(x)$. If we can find a function $F(x)$ (and this may not always exist) such that $dF(x)/dx = f(x)$, the integral of $f(x)$ is defined as

$$\int f(x)dx = F(x) + C \tag{11.1}$$

$F(x)$ is called the indefinite integral, $f(x)$ the integrand, and C the constant of integration.

The qualifier “indefinite" is used because the result of the integration is a family of functions when C is not determined, and a function when the value of C is determined. By contrast we shall see later that a definite integral could be a number. The reader should satisfy herself that the derivative of the RHS is indeed $f(x)$. This integral is referred to as the Riemann integral.[2] If a function is continuous over an interval $[a, b]$, then it has an integral over that interval. Evidently, not all functions have integrals.

Example 11.1 Find the integral of $f(x) = 3ax^2$. Solution:

$$F(x) = \int 3ax^2 dx = ax^3 + C$$

Because taking the derivative is the inverse of finding an integral, we can check the correctness of our results by taking the derivative of the function $F(x) = ax^3 + C$.

$$\begin{aligned}\frac{dF(x)}{dx} &= \frac{d}{dx}ax^3 + \frac{d}{dx}C \\ &= 3ax^2 + 0\end{aligned}$$

---

[2]After the German mathematician Georg Friedrich Berhard Riemann. There are other types of integrals: the Riemann-Stieltjes integral, which we shall discuss briefly later in this chapter, and the Lebesgue integral, a discussion of which is beyond the scope of the present book. We shall refer to the Riemann integral as the integral.

11.1.1 Rules of Integration

In Table 11.1 are the integrals of several simple functions. Check their validity by taking the derivatives of the RHSs and comparing them to the function under the integral sign. There are a number of publications[3] where the reader can find the integrals of a vast array of simple and very complicated functions, although it is quite unlikely that one would encounter more than a handful of these in the first 100 years of life. In addition, computer software such as Mathematica, Maple, and Matlab provide routines for finding integrals of different functions. Still, it is strongly recommended that the reader memorizes these and a few other integrals. In addition to these rules, two properties of the indefinite integral should be mentioned.

Property 11.1 The integral of the sum of two functions is equal to the sum of their integrals, that is,

$$\int [f_1(x) + f_2(x)]dx = \int f_1(x)dx + \int f_2(x)dx \tag{11.2}$$

Example 11.2 Let $f_1\,(x) = 5x$ and $f_2(x) = 9x^2$.
Then

$$\int (5x + 9x^2)dx = \int 5xdx + \int 9x^2dx = \frac{5}{2}x^2 + 3x^3 + C$$

Example 11.3 For the functions $f_1(x) = \sin 2x$ and $f_2(x) = 1/x$, we have

$$\int \left(\sin 2x + \frac{1}{x}\right)dx = \int \sin 2x dx + \int \frac{1}{x}dx = -\frac{1}{2}\cos 2x + \ln x + C$$

Table 11.1 Rules of integration

$\int x^n dx = \dfrac{x^{n+1}}{n+1} + C$	where n is an integer and $n \neq -1$
$\int x^{-1}dx = \int \dfrac{dx}{x} = \ln\|x\| + C$	
$\int e^x dx = e^x + C$	
$\int a^x dx = \dfrac{a^x}{\ln a} + C$	where a is a positive constant
$\int \sin x dx = -\cos x + C$	
$\int \cos x dx = \sin x + C$	
$\int \tan x dx = -\ln\|\cos x\| + C$	

[3]For example, *Table of Integrals, Series and Products* by Gradshteyn and Ryzhik, translated by Alan Jeffrey, 1980.

Property 11.2 A multiplicative constant can be factored out of an integral, that is,

$$\int af(x)dx = a\int f(x)dx \tag{11.3}$$

Example 11.4

$$i. \int 2x^5 dx = 2\int x^5 dx$$

$$ii. \int b\cos\left(x+\frac{\pi}{2}\right)dx = b\int \cos\left(x+\frac{\pi}{2}\right)dx$$

Example 11.5 (Harrod-Domar Growth Model). The Harrod-Domar model of economic growth consists of the following equations[4]:

$$Y(t) = \frac{1}{\alpha}K(t) \tag{11.4}$$

$$\frac{dK}{dt} = I(t) - \rho K(t) \tag{11.5}$$

$$I(t) = i(t)Y(t) \tag{11.6}$$

where Y is output, K capital stock, I investment, t time, α the capital-output ratio, ρ depreciation rate, and i the portion of income devoted to investment. Y, K and I are in constant price monetary units. If we let $i(t) = s$ where s is the marginal propensity to save-assumed to be constant over time—then these three equations represent relaxation of basic short-run Keynesian assumptions: (1) Output is dependent on the productive capacity, and therefore aggregate supply is no more infinitely elastic; (2) capital stock changes as a result of investment; and (3) investment is not autonomous but depends on savings. In this manner, Harrod-Domar adapted the short-run Keynsian model to serve as a long-run growth model. It should be noted that despite its shortcomings and the many criticisms levied at it, the Harrod-Domar model, because of its tractability, for a long time served as a useful tool of macroeconomic planning and policy analysis. Taking the derivative of income, Y, with respect to time and substituting for dK/dt from (11.5) and for I from (11.6), we have

$$\begin{aligned}\frac{dY}{dt} &= \frac{1}{\alpha}\frac{dK}{dt} \\ &= \frac{1}{\alpha}(I - \rho K) \\ &= \frac{1}{\alpha}(iY - \rho\alpha Y)\end{aligned}$$

---

[4]The Harrod and Domar models are similar in their mechanics, but Harrod's model involves expectations that are absent from Domar's model. See "An Essay in Dynamic Theory," by Roy F. Harrod in *Economic Journal* (1939) and "Capital Expansion, Rate of Growth and Employment," by Evsey D. Domar, *Econometrica* (1946).

Thus, the output will grow at the actual rate of

$$\frac{dY/dt}{Y} = \frac{i}{\alpha} - \rho$$

Rearranging, we have

$$\frac{dY}{Y} = \left(\frac{i}{\alpha} - \rho\right) dt$$

Integrating both sides,

$$\ln Y = \int \left(\frac{i}{\alpha} - \rho\right) dt + C$$

or

$$Y = Ae^{\int (\frac{i}{\alpha} - \rho)\, dt} \qquad \text{where} \qquad A = e^C$$

Because i depends on time and we do not know the explicit form of the dependence, we have to leave the integral as it is. If, however, $i(t) = s$, then

$$Y = Ae^{(\frac{s}{\alpha} - \rho)t}$$

$\frac{s}{\alpha} - \rho$ is referred to as the warranted rate of growth.

11.1.2 Change of Variable

Suppose we want to find the following integral:

$$\int f(g(x))g'(x)dx \tag{11.7}$$

Set $u = g(x)$, which implies $du = g'(x)\, dx$, and we have

$$\int f(g(x))g'(x)dx = \int f(u)du \tag{11.8}$$

The change of variable is intended to simplify the integration; therefore, the choice of the substitute variable has to be made with this purpose in mind. We illustrate this with a few examples.

Example 11.6 Find the following integral:

$$z = \int \frac{\cos x}{\sin x} dx$$

If we choose $y = \sin x$ as a substitute variable, we have $dy = \cos x dx$. The integration is simplified to

$$z = \int \frac{dy}{y} = \ln|y| + C = \ln|\sin x| + C$$

Example 11.7

$$z = \int \frac{5x}{x^2+1} dx$$

Make the following change of variable:

$$y = x^2 + 1$$

which implies

$$\text{dy} = \text{2x dx}$$

Now we have

$$\begin{aligned} z &= \frac{5}{2} \int \frac{2xdx}{x^2+1} \\ &= \frac{5}{2} \int \frac{dy}{y} \\ &= \frac{5}{2} \ln y + C \\ &= \frac{5}{2} \ln(x^2+1) + C \end{aligned}$$

Example 11.8 Find the following integral:

$$z = \int \frac{(\ln x)^2}{x} dx$$

Make the substitution $y = \ln x$, which implies dy = (1/x) dx, hence

$$\begin{aligned} z &= \int y^2 dy \\ &= \frac{1}{3} y^3 + C \\ &= \frac{1}{3} (\ln x)^3 + C \end{aligned}$$

11.1.3 Integration by Parts

There are times when the integrand is not easily integrable. In such cases we may resort to integration by parts, that is, finding an equivalent expression that is easier to integrate. Let us first have an example.

Example 11.9 Find the following integral:

$$z = \int x \cos x dx$$

This is not an easy one; at least, we don't find it among our formulas in Table 11.1. Let us replace it with something easier to integrate (don't ask how I know they are equivalent; it will become clear soon).

$$\begin{aligned} z &= \int x \cos x dx \\ &= x \sin x - \int \sin x dx \\ &= x \sin x + \cos x + C \end{aligned}$$

Thus, the integration by parts allowed us to get rid of the multiplicative form $x \cos x$ and have an easier integrand. To verify that indeed the integration is correct, we take the derivative of z:

$$\begin{aligned} \frac{d}{dx}(x \sin x + \cos x + C) &= \sin x + x \cos x - \sin x \\ &= x \cos x \end{aligned}$$

The rationale for the above exercise goes back to Chap. 8. There we had

$$d\,(uv) = vdu + udv \tag{11.9}$$

It follows that

$$\int d(uv) = \int vdu + \int udv \tag{11.10}$$

or

$$\int udv = uv - \int vdu \tag{11.11}$$

Thus, anytime we have a situation where an integrand poses difficulty, we can try to see if it is in the form of udv and replace it with vdu. In the example above we had

$$u = x, \qquad dv = \cos x dx$$

Therefore,

$$du = dx, \qquad v = \sin x$$

Note that there is nothing mechanical here; if we choose our substitution differently, not only will the integration not be easier, it may even be more difficult. For example, instead of what we did above, try

$$u = \cos x \quad \text{and} \quad dv = xdx$$

and see that the integrand is even more intractable. Sometimes we need to use integration by parts twice or more to find the integral. The next two examples illustrate this point.

Example 11.10 Find the following integral:

$$\int xe^x dx$$

Let

$$\begin{aligned} u &= x, & &\Rightarrow & du &= dx, \\ dv &= e^x dx & &\Rightarrow & v &= e^x \end{aligned}$$

Now we have

$$\begin{aligned} \int xe^x dx &= xe^x - \int e^x dx \\ &= xe^x - e^x + C \end{aligned}$$

Example 11.11 Find the following integral:

$$\int x^2 e^x dx$$

Let

$$\begin{aligned} u &= x^2, & du &= 2xdx \\ dv &= e^x dx, & v &= e^x \end{aligned}$$

Now we have

$$\int x^2 e^x dx = x^2 e^x - 2\int xe^x dx$$

The integral on the RHS again requires integration by parts, which we already did in Example 11.11. Thus,

$$\begin{aligned} \int x^2 e^x dx &= x^2 e^x - 2(xe^x - \int e^x dx) + C \\ &= x^2 e^x - 2(xe^x - e^x) + C \\ &= e^x(x^2 - 2x + 2) + C \end{aligned}$$

We can check the correctness of our answer by taking the derivative of $F(x)$:

$$\begin{aligned}\frac{d}{dx}F(x) &= \frac{d}{dx}(e^x(x^2-2x+2)+C)\\ &= e^x(x^2-2x+2)+e^x(2x-2)\\ &= x^2e^x\end{aligned}$$

11.1.4 Exercises

E.11.1 Find the following indefinite integrals.

i. $\int \frac{dx}{x\ln x}$ *ii.* $\int (x^2+x+\sqrt{x})dx$ *iii.* $\int 6x^5dx$

iv. $\int \frac{dx}{\cos^2 3x}$ *v.* $\int \frac{dx}{2x-4}$ *vi.* $\int \frac{xdx}{\sqrt{3x^2+4}}$

vii. $\int \frac{ax^2}{\sqrt{x^3+8}}dx$ *viii.* $\int \frac{\cos xdx}{sin^2x}$

ix. $\int \frac{\ln(2x+a)}{2x+a}dx$ for $a>0$

11.2 The Definite Integral

Consider Fig. 11.1 and let us assume that we want to calculate the area under the solid straight line and above the x-axis, that is, the area of the trapezoid $h_1h_2x_1x_2$. From elementary geometry we know that the area is equal to

$$A = \frac{(y_1+y_2)}{2}\times(x_2-x_1)$$

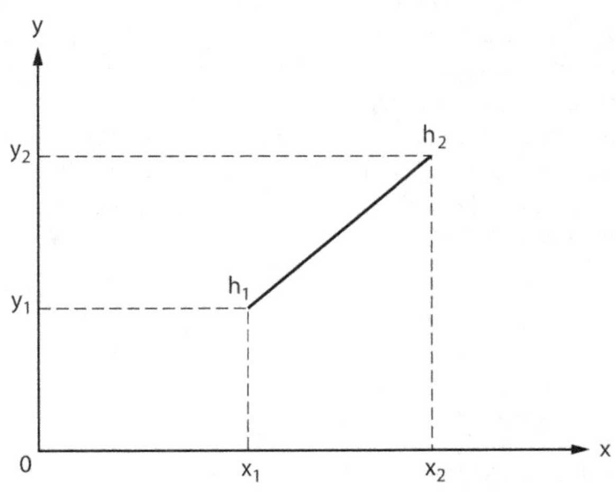

Fig. 11.1 Area under a straight line

The equation for a line is of the form

$$y = a + bx$$

which implies

$$y_1 = a + b\,x_1$$
$$y_2 = a + b\,x_2$$

Therefore, the area is

$$\begin{aligned} A &= \frac{2a + b(x_1 + x_2)}{2}(x_2 - x_1) \\ &= a(x_2 - x_1) + \frac{1}{2}b\left(x_2^2 - x_1^2\right) \\ &= \left(ax_2 + \frac{1}{2}bx_2^2\right) - \left(ax_1 + \frac{1}{2}bx_1^2\right) \\ &= \left[ax + \frac{1}{2}bx^2\right]_{x_1}^{x_2} \end{aligned}$$

where the last equality is a shorthand that indicates the difference between the expression inside the bracket evaluated at x_2 and x_1.

The above is an example of integration and a definite integral. A definite integral is a number, the area of the trapezoid[5] $h_1h_2x_1x_2$. Note that if we take the derivative of the expression inside the brackets, we get the equation of the line we started with, that is, $y = a + bx$. Having the function inside the brackets, we could evaluate it at any two points and obtain the area under that segment of the line. Indeed, we can do this for any curve and not just a straight line. On the other hand, if we concentrate on the function inside the bracket and do not think of any area or points at which to evaluate it, we have the basic idea of the indefinite integral.

You might say that this was easy, I have known how to find the area of a trapezoid since elementary school and there was no point in resorting to the concept of an integral. You are right. We are going to make it more interesting. Consider the function $y = f(x)$ depicted in Fig. 11.2a. Suppose we want to find the area under this curve and above the closed interval $[a, b]$ on the x-axis. We can obtain an estimate of the lower bound of this area by dividing the $[a, b]$ interval into n segments, calculating the areas of the resulting rectangles, and adding them up. Consider the intervals

$$[x_0, x_1], [x_1, x_2], [x_2, x_3], \ldots, [x_{n-1}, x_n],$$

[5] Strictly speaking, it is the area under a curve with a sign because the area under the x-axis is subtracted from the area above it. Moreover, as we will see later, interchanging the limits of integration, that is, changing the place of x_1 and x_2 in the last expression will change the sign of A.

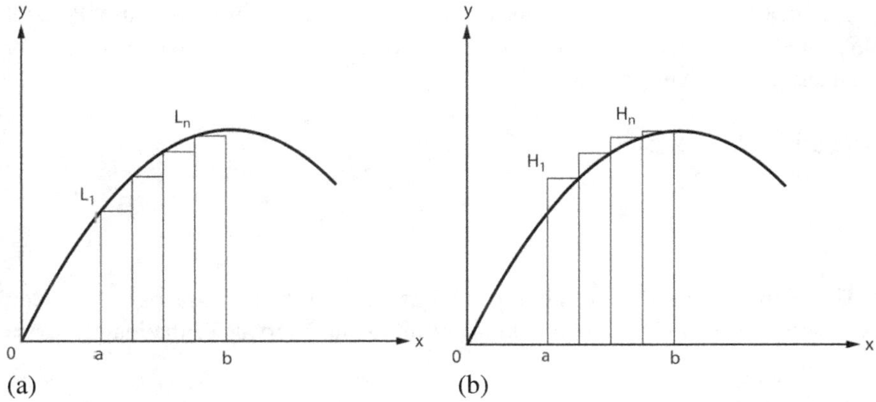

Fig. 11.2 (**a**) Estimating the area under a curve S_n^-; (**b**) Estimating the area under a curve S_n^+

where

$$a = x_0 < x_1 < x_2 \ldots < x_n = b$$

Let

$$\Delta x_i = x_i - x_{i-1} \quad i = 1, \ldots, n$$

Because the height of each interval is $L_i,\ i = 1, \ldots n$, we have our estimate of the lower bound of the area as

$$S_n^- = \sum_{i=1}^{n} L_i \Delta x_i \tag{11.12}$$

Going to Fig. 11.2b, we can repeat the same process to get an upper bound for the area. Because the height of rectangles are $H_i,\ i = 1, \ldots n$, we have

$$S_n^+ = \sum_{i=1}^{n} H_i \Delta x_i \tag{11.13}$$

Now for each interval pick a point u_i such that

$$x_{i-1} < u_i < x_i \quad i = 1, \ldots, n$$

Because for each interval, L_i is the lowest value and H_i the highest value of the function $f(x)$, we have

$$L_i \leq f(u_i) \leq H_i \tag{11.14}$$

It follows that

$$S_n^- = \sum_{i=1}^{n} L_i \Delta x_i \leq \sum_{i=1}^{n} f(u_i) \Delta x_i \leq \sum_{i=1}^{n} H_i \Delta x_i = S_n^+ \tag{11.15}$$

The idea is to increase n and make Δx_i's smaller and smaller, which will result in S_n^- and S_n^+ getting closer and closer to each other. In the limit, the two bounds are equal and we will have the integral.

Example 11.12 Evaluate

$$\int_a^b kxdx.$$

This is the same problem as in Fig. 11.1 except that the new line passes through the origin. Let us divide the interval $[a, b]$ into n subintervals of the length $\Delta x = \dfrac{b-a}{n}$. Then we have

$$\begin{aligned}
x_0 &= a \\
x_1 &= a + \Delta x \\
&\dots \\
x_i &= a + i\Delta x \\
&\dots \\
x_n &= a + n\Delta x = b
\end{aligned}$$

Form the lower sum as

$$S_n^- = \sum_{i=0}^{n-1} k(a + i\Delta x)\Delta x = k\left[na + \left(\sum_{i=0}^{n-1} i\right)\Delta x\right]\Delta x$$

Because

$$\sum_{i=0}^{n-1} i = \frac{n(n-1)}{2}$$

and

$$\Delta x = \frac{b-a}{n}$$

we have

$$\begin{aligned}
S_n^- &= k\left[na + \frac{n(n-1)(b-a)}{2n}\right]\frac{b-a}{n} \\
&= k\left[a + \frac{(n-1)(b-a)}{2n}\right](b-a)
\end{aligned}$$

When we take the limit of S_n^- and $n \to \infty$, we get the integral of the function kx in the interval $[a, b]$:

$$\begin{aligned}\int_a^b kxdx &= \lim_{n\to\infty} S_n^- \\ &= k\left[a + (b-a)\lim_{n\to\infty}\frac{n-1}{2n}\right](b-a) \\ &= k\frac{b^2-a^2}{2}\end{aligned}$$

Example 11.13 Evaluate

$$\int_a^b ke^x dx$$

Following the same procedure as in the previous example, we have

$$S_n^- = \sum_{i=0}^{n-1} ke^{a+i\Delta x}\Delta x = ke^a\left(\sum_{i=0}^{n-1} e^{i\Delta x}\right)\Delta x.$$

Because

$$\sum_{i=0}^{n-1} e^{i\Delta x} = \frac{e^{n\Delta x}-1}{e^{\Delta x}-1}$$

and

$$S_n^- = ke^a\frac{e^{n\Delta x}-1}{e^{\Delta x}-1}\Delta x = ke^a\frac{e^{b-a}-1}{e^{\Delta x}-1}\Delta x = k(e^b-e^a)\frac{\Delta x}{e^{\Delta x}-1}$$

we have

$$\lim_{\Delta x\to 0} S_n^- = k(e^b-e^a)\lim_{\Delta x\to 0}\frac{\Delta x}{e^{\Delta x}-1}$$

By l'Hôpital's rule (see Chap. 8), the limit of the RHS is the same as the limit of the ratio of the derivative of the numerator to the derivative of the denominator with respect to the variable going to the limit. That is,

$$\lim_{\Delta x\to 0}\frac{\Delta x}{e^{\Delta x}-1} = \lim_{\Delta x\to 0}\frac{1}{e^{\Delta x}}$$

Hence, we have

$$\int_a^b ke^x dx = k(e^b-e^a) = ke^x\Big|_a^b$$

As can be seen, the function under the integral sign is the derivative of the right-hand-side function.

11.2.1 Properties of Definite Integrals

Below we state without proof a number of properties of definite integrals that will prove handy.

Theorem 11.1 (Fundamental Theorem of Calculus). *Let $f(x)$ be a continuous function over the interval $[a, b]$ and*

$$F(x) = \int_a^x f(t)dt \tag{11.16}$$

Then

$$\frac{d}{dx}\int_a^x f(t)dt \equiv F'(x) = f(x), \qquad \forall x \in [a, b] \tag{11.17}$$

Moreover, if f is differentiable and if f' is integrable, that is, if the following integral exists

$$\int_a^x f'(t)dt, \qquad \forall x \in [a, b] \tag{11.18}$$

then

$$\int_a^x f'(t)dt = f(x) - f(a) \tag{11.19}$$

Example 11.14 Consider the function $f(x) = 3x^2$ and consider the interval [1, 3], and let

$$F(x) = \int_1^x 3t^2 dt$$

From the rules of integration $F(x) = x^3 - 1$. Thus,

$$\begin{aligned}\frac{d}{dx}\int_1^x f(t)dt &= \frac{d}{dx}[x^3 - 1]\\ &= 3x^2\\ &= f(x)\end{aligned}$$

Moreover, $f'(x) = 6x$ and

$$\int_1^x 6t dt = 3t^2\Big|_1^x = 3x^2 - 3 = f(x) - f(1)$$

Example 11.15 Consider the function $f(x) = e^{2x}$ that is continuous over the extended real line $(-\infty, \infty)$. Let

$$F(x) = \int_{-\infty}^{x} e^{2t} dt$$

Using the rules of integration and recalling that

$$\lim_{x \to -\infty} e^{2x} = 0$$

we have

$$F(x) = \frac{1}{2} e^{2x}$$

It follows that

$$\frac{d}{dx} F(x) = e^{2x} = f(x)$$

Because $f'(x) = 2e^{2x}$ is also continuous over the extended real line, we have

$$\int_{-\infty}^{x} 2e^{2t} dt = e^{2x} = f(x)$$

Property 11.3 Let $F(x)$ be an antiderivative of $f(x)$, that is, $f(x) = dF(x)/dx$. Then

$$\int_{a}^{b} f(x) dx = F(b) - F(a) \tag{11.20}$$

Example 11.16 The antiderivative of $f(x) = 2x$ is $F(x) = x^2$. Thus,

$$\int_{0}^{1} 2x dx = x^2 \Big|_{0}^{1} = (1)^2 - (0)^2 = 1$$

Example 11.17 Let $f(x) = e^x$. Then $F(x) = e^x$, and

$$\int_{a}^{b} e^x dx = e^x \big|_{a}^{b} = e^b - e^a$$

Example 11.18 Consider the function $f(x) = \cos x$. Then

$$\int_{1}^{2} \frac{5}{x} dx = 5 \ln |x| |_{1}^{2} = 3.4657359$$

Example 11.19 For $f(x) = \cos x$ we have

$$\begin{aligned}\int_0^{\frac{\pi}{2}} \cos x dx &= \sin x|_0^{\frac{\pi}{2}} \\ &= \sin\left(\frac{\pi}{2}\right) - \sin(0) \\ &= 1\end{aligned}$$

Property 11.4 If we exchange the place of the lower and upper limits of an integral, its sign will change. That is,

$$\int_a^b f(x)dx = -\int_b^a f(x)dx \tag{11.21}$$

Property 11.5 If the interval over which the function is integrated has length zero, that is, $b = a$, the definite integral is zero:

$$\int_a^a f(x)dx = 0 \tag{11.22}$$

Property 11.6 The integral of the sum of two functions is the sum of their integrals

$$\int_a^b [f_1(x) + f_2(x)]dx = \int_a^b f_1(x)dx + \int_a^b f_2(x)dx \tag{11.23}$$

Property 11.7 A multiplicative constant can be factored out of the integral sign

$$\int_a^b cf(x)dx = c\int_a^b f(x)dx \tag{11.24}$$

Property 11.8 If function $f(x)$ is less than or equal to another function $\phi(x)$ for all points in the interval $[a, b]$, then the definite integral $f(x)$ over that interval is less than or equal to the definite integral of $\phi(x)$ over the same interval. That is, if

$$f(x) \leq \phi(x) \qquad x \in [a, b]$$

then

$$\int_a^b f(x)dx \leq \int_a^b \phi(x)dx \tag{11.25}$$

In particular, if

$$f(x) \geq 0 \qquad x \in [a, b]$$

then

$$\int_a^b f(x)dx \geq 0 \tag{11.26}$$

This property has an application in probability theory. Let $f(x)$ be a density function, which implies that it is nonnegative. Hence its definite integral, called the probability distribution function, is always nonnegative, satisfying the axiom of probability theory.

Property 11.9 (Mean Value Theorem). For the continuous function $f(x)$, and the closed interval $[a, b]$, there exists

$$a \leq u \leq b$$

such that

$$\int_a^b f(x)dx = (b-a)f(u) \tag{11.27}$$

In other words, there is a point u such that the area of a rectangle with the width $b-a$ and the height $f(u)$ is equal to the outcome of the definite integral of $f(x)$ over the interval $[a, b]$.

Property 11.10 The definite integral of a function over the interval $[a, b]$ is equal to the sum of the definite integrals of the same function over two contiguous subintervals $[a, c]$ and $[c, b]$.

$$\begin{gathered}\int_a^b f(x)dx = \int_a^c f(x)dx + \int_c^b f(x)dx \\ a \leq c \leq b\end{gathered} \tag{11.28}$$

Property 11.11 Limits of integration need not be fixed; they can be functions of x. For example,

$$\int_{g(x)}^{h(x)} f(t)dt \tag{11.29}$$

When limits of integration are functions of x, it is better to designate the variable of integration by another symbol, for example, t, to avoid confusion. Note that the variable of integration plays a role similar to that of the counter-variable in summation, and therefore, change of symbol does not signify a change in the function. A special case of the above property occurs when the upper limit of integration is x. We have

$$\begin{aligned}\int_a^x f(t)dt &= F(x) - F(a) \\ &= F(x) + C\end{aligned} \tag{11.30}$$

Now the above integral is a function of x and is not a fixed value. Previously, we noted that an indefinite integral evaluated over an integral $[a, b]$ gives us the definite integral. Now we see that a definite integral with the upper limit equal to x results in an indefinite integral.

11.2.2 Rules of Integration for the Definite Integral

Rules of integration for the definite integral, with some adjustments, are the same as for the indefinite integral presented in Table 11.1 and in Sect. 11.1.1. First calculate the integral and then apply Property 11.3. The following examples illustrate the point.

Example 11.20 Evaluate the following integral:

$$\int_a^b x^n dx \qquad n \neq -1$$

We already know that

$$F(x) = \int x^n dx = \frac{x^{n+1}}{n+1} + C$$

Thus, we have

$$\begin{aligned}\int_a^b x^n dx &= F(b) - F(a) \\ &= \frac{b^{n+1}}{n+1} + C - \frac{a^{n+1}}{n+1} - C \\ &= \frac{b^{n+1} - a^{n+1}}{n+1}\end{aligned}$$

As can be seen, the constant of integration drops out.

Example 11.21 Evaluate the following integral:

$$\int_{\frac{\pi}{2}}^{\frac{3\pi}{2}} \cos x dx$$

Again, we have

$$\begin{aligned}\int_{\frac{\pi}{2}}^{\frac{3\pi}{2}} \cos x dx &= \sin x|_{\frac{\pi}{2}}^{\frac{3\pi}{2}} \\ &= \sin\left(\frac{3\pi}{2}\right) - \sin\left(\frac{\pi}{2}\right) \\ &= -2\end{aligned}$$

11.2.3 Change of Variable

Change of variable in the definite integral is the same as in the indefinite integral with a major difference. Here we need to change the limits of integration as well. We illustrate this with two examples.

Example 11.22 Evaluate the following integral:

$$\int_0^3 \frac{5x}{x^2+1}dx$$

We make the following change of variable:

$$y = x^2 + 1$$

which implies

$$dy = 2xdx$$

Also note that

$$\begin{aligned} x = 0 \quad &\Rightarrow \quad y = 1 \\ x = 3 \quad &\Rightarrow \quad y = 10 \end{aligned}$$

Therefore, we have

$$\begin{aligned} \int_0^3 \frac{5x}{x^2+1}dx &= \frac{5}{2}\int_1^{10} \frac{\ln y}{y}dy \\ &= \frac{5}{2}\ln y\Big|_1^{10} \\ &= \frac{5}{2}\ln(x^2+1)\Big|_0^3 \\ &= 5.7564627 \end{aligned}$$

Similarly when one or both limits of integration are functions of x, we have to make the necessary adjustment(s) as in the next example.

Example 11.23 Evaluate the integral

$$\int_1^{e^x} \frac{(\ln t)^2}{t}dt$$

Let $y = \ln t$ then $dy = dt/t$, and note that $\ln(1) = 0$ and $\ln e^x = x$. We have

$$\int_0^x y^2 dy = \frac{1}{3}y^3\Big|_0^x = \frac{1}{3}(\ln x)^3\Big|_1^{e^x} = \frac{1}{3}x^3$$

11.2.4 Integration by Parts

For a definite integral is the same as for an indefinite integral, that is,

$$\int_a^b udv = uv|_a^b - \int_a^b vdu \tag{11.31}$$

Example 11.24 Find the following integral:

$$\int_1^2 xe^x dx$$

We have

$$\begin{aligned}\int_1^2 xe^x dx &= xe^x\big|_1^2 - \int_1^2 e^x dx\\ &= xe^x\big|_1^2 - e^x\big|_1^2\\ &= 2e^2 - e^2 - e + e\\ &= 7.3890561\end{aligned}$$

Example 11.25

$$\begin{aligned}z &= \int_0^{\frac{\pi}{2}} x\cos x dx\\ &= x\sin x|_0^{\frac{\pi}{2}} - \int_0^{\frac{\pi}{2}} \sin x dx\\ &= x\sin x|_0^{\frac{\pi}{2}} + \cos x|_0^{\frac{\pi}{2}}\\ &= \frac{\pi}{2} - 1\end{aligned}$$

Example 11.26 (Consumers' Lifetime Utility). Most applications of the integral in macroeconomics involve discounting future values to the present. For instance, it is posited that consumers maximize their lifetime utility, U, where

$$U = \int_0^\infty e^{-\rho t} u(c(t))dt \tag{11.32}$$

In (11.32), $c(t)$ is consumption at time t, $u(.)$ the instantaneous utility function, and ρ the discount rate. Thus, the lifetime utility is the integral (sum) of all instantaneous utilities derived from consumption over time discounted to the present. Sometimes the instantaneous utility function is assumed to be of the form

$$u(c(t)) = \frac{c(t)^{1-\theta}}{1-\theta} \tag{11.33}$$

Example 11.27 (Price of a Bond). In Chap. 2 we showed that in case of continuous compounding of interest, the value of A dollars of investment at time 0 would be Ae^{rT} at time T when the interest rate is r. It follows that the present value of H

dollars at time T would be He^{-rT}. In the same chapter, we showed that the value of a bond that pays A dollars per year in perpetuity is A/r. This result has a continuous time counterpart. Let ρ be the rate of interest in continuous time. Then, the value of the bond would be

$$P = A\int_0^{\infty} e^{-\rho t}dt = A\frac{e^{-\rho t}}{-\rho}\bigg|_0^{\infty} = \frac{A}{\rho}$$

Example 11.28 (Consumer and Producer Surplus). Consumer surplus, *CS*, is the area below the demand curve and above the horizontal line P^*. Conceptually, it is the difference between the sum of maximum prices consumers would have been willing to pay for the good or service and the sum of what they paid in the market. Producer surplus, *PS*, is the area above the supply curve and below the horizontal line P^* (see Fig. 11.3a). Again, conceptually, it is the difference between the sum of what producers received in the market and the sum of minimum amounts they would have been willing to settle for when supplying the same good or service, that is, if each unit had been supplied at production cost. Thus,

$$CS = \int_0^{q} (D(t) - P^*)dt$$
$$PS = \int_0^{q} (P^* - S(t))dt$$

where the upper limit of integration is $q = D^{-1}(p^*)$. The sum of consumer and producer surplus is a social welfare function W if we agree to give all individuals equal weights:

$$W = CS + PS = \int_0^{q} (D(t) - S(t))dt$$

As can be seen in Fig. 11.3b, W is maximized when price is equal to the market equilibrium price P^e, and $q = D^{-1}(p^e) = S^{-1}(P^e)$. Let us denote this maximum

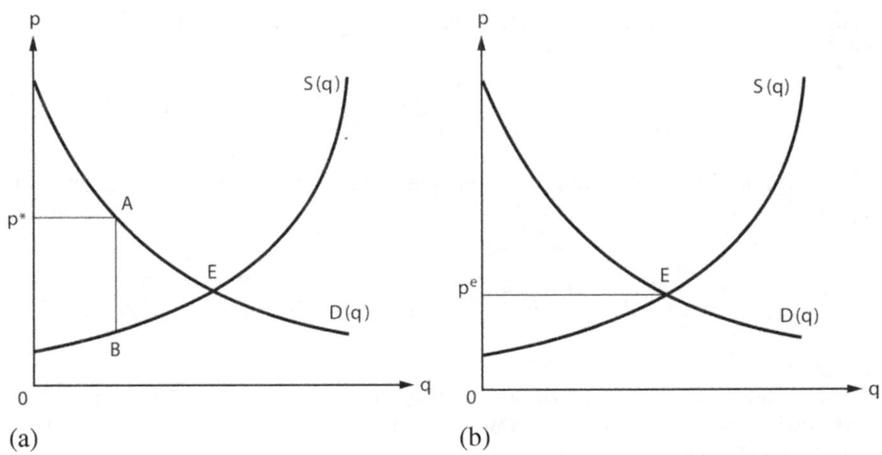

Fig. 11.3 (a) Consumer and producer surplus at disequilibrium price; (b) consumer and producer surplus at equilibrium price

by W_e and any other value of W by W^*;then $W^e - W^*$, that is, the area ABE in Fig. 11.3a, is called the deadweight loss. Finally, because the supply curve in the case of a single producer is the increasing segment of the marginal cost curve, welfare maximization, in the above sense, requires marginal cost pricing.

Example 11.29 (Mean and Variance of Continuous Distributions). Integration is an essential tool of probability theory when dealing with continuous variables. Deriving probability distributions from density functions, showing that indeed a function qualifies as a probability density function, and calculating the mean and variance of a distribution all require integration. Here, we illustrate these applications using uniform and normal distributions. But first let us define a few concepts. Let $X \in (-\infty, \infty)$ be a random variable with density function $f(x)$. Then its probability distribution function $F(x)$, mean $E(X)$, and variance $E[X - E(X)]^2$are defined as[6]

$$P(X \leq x) = F(x) = \int_{-\infty}^{x} f(t)dt \tag{11.34}$$

$$\mu = E(X) = \int_{-\infty}^{\infty} xf(x)dx \tag{11.35}$$

$$\sigma^2 = E[X - E(X)]^2 = \int_{-\infty}^{\infty} (x - \mu)^2 f(x)dx \tag{11.36}$$

Example 11.30 (Mean and Variance of the Uniform Distribution). The random variable $X \in [a, b]$ whose density function is

$$f(x) = \begin{cases} \dfrac{1}{b-a} & \text{if} \quad a \leq x \leq b \\ 0 & \text{otherwise.} \end{cases}$$

is uniformly distributed. A special case of the uniform distribution is when $X \in [0, 1]$. Then $f(x) = 1$. Because $f(x) \geq 0$, in order to show that $f(x)$ is indeed a density function, we have only to show that its integral is equal to 1.

$$\int_a^b f(x)dx = \int_a^b \frac{1}{b-a}dx = \left.\frac{x}{b-a}\right|_a^b = 1.$$

The probability distribution function, the mean, and variance of the uniform distribution are

$$F(x) = \int_a^x \frac{1}{b-a}dt = \left.\frac{t}{b-a}\right|_a^x = \frac{x-a}{b-a}$$

---

[6]In this definition and in what follows we use X as the random variable and x as its value or realization. For instance, $P(X \leq x) = p$ means that the probability of random variable X being less than or equal to the number x is p.

Again note that, if $X \in [0, 1]$, then $F(x) = x$.

$$\mu = \int_a^b x\frac{1}{b-a}dx = \left.\frac{x^2}{2(b-a)}\right|_a^b = \frac{a+b}{2}$$

$$\sigma^2 = \int_a^b (x-\mu)^2\frac{1}{b-a}dx$$

Let

$$y = x - \mu, \qquad dy = dx$$

$$\sigma^2 = \frac{1}{b-a}\int_{\frac{a-b}{2}}^{\frac{b-a}{2}} y^2 dy = \left.\frac{1}{3(b-a)}y^3\right|_{\frac{a-b}{2}}^{\frac{b-a}{2}} = \frac{(b-a)^2}{12}$$

Example 11.31 (Mean of the Normal Distribution). The normal distribution has the density function

$$f(x) = \frac{1}{\sqrt{2\pi}\sigma}\exp\left\{-\frac{1}{2}\left(\frac{x-\mu}{\sigma}\right)^2\right\}$$

Hence the probability distribution function will be

$$F(x) = \int_{-\infty}^{x}\frac{1}{\sqrt{2\pi}\sigma}\exp\left\{-\frac{1}{2}\left(\frac{t-\mu}{\sigma}\right)^2\right\}dt$$

The above integral does not have a closed form and its values for different x have to be calculated numerically. To show that indeed $F(x)$ is a probability distribution, we note that

$$f(x) \geq 0 \quad \forall x$$

implies

$$F(x) = \int_{-\infty}^{x} f(t)dt \geq 0 \quad \forall x$$

Next we have to show that $F(\infty) = 1$. This is a bit tricky. First, make the following change of variable:

$$z = \frac{t-\mu}{\sqrt{2}\sigma}$$

which implies

$$dz = \frac{dt}{\sqrt{2}\sigma}$$

Thus,

$$\int_{-\infty}^{\infty} \frac{1}{\sqrt{2\pi}\sigma} \exp\left\{-\frac{1}{2}\left(\frac{t-\mu}{\sigma}\right)^2\right\} dt = \frac{1}{\sqrt{\pi}} \int_{-\infty}^{\infty} e^{-z^2} dz$$

The last integral is equal to $\sqrt{\pi}$. On this point you have to trust me. Showing this result is not difficult, but it takes us off our track. Once you accept it, it follows that $F(\infty) = 1$ and the normal distribution is indeed a probability distribution function. Next we show that

$$\begin{aligned}
E(X) &= \int_{-\infty}^{\infty} x \frac{1}{\sqrt{2\pi}\sigma} \exp\left\{-\frac{1}{2}\left(\frac{x-\mu}{\sigma}\right)^2\right\} dx \\
&= \int_{-\infty}^{\infty} x \frac{\mu + \sqrt{2}\sigma z}{\sqrt{\pi}} e^{-z^2} dz \\
&= \frac{\mu}{\sqrt{\pi}} \int_{-\infty}^{\infty} e^{-z^2} dz + \frac{\sqrt{2}\sigma}{\sqrt{\pi}} \int_{-\infty}^{\infty} z e^{-z^2} dz \\
&= \mu - \frac{\sqrt{2}\sigma}{2\sqrt{\pi}} e^{-z^2} \Big|_{-\infty}^{\infty} \\
&= \mu
\end{aligned}$$

11.2.5 Riemann-Stieltjes Integral

So far we have been concerned with integration with respect to changes in the variable x, that is, dx. We can generalize the notion of an integral by evaluating it when there is a change in an increasing function of x, that is, $\alpha(x)$. Thus, the Riemann-Stieltjes integral is a generalization of the Riemann integral. We have

$$z = \int_a^b f(x) d\alpha(x) \tag{11.37}$$

An example of the application of Riemann-Stieltjes integral is in probability theory. Suppose the random variable x has the probability distribution function $F(x)$ over the interval $[a, b]$. Then

$$E(x) = \int_a^b x dF(x)$$

Indeed, the expectation of any function of x, say, $g(x)$, can be written

$$E(g(x)) = \int_a^b g(x)dF(x)$$

Example 11.32 Evaluate the integral

$$\int_0^2 x^3 d(x^2)$$

Because x^2 is continuous and $d(x^2) = 2xdx$, we have

$$\int_0^2 x^3 d(x^2) = \int_0^2 2x^4 dx = \frac{2}{5}x^5|_0^2 = 12.8$$

You might ask: "What is the big idea?" This seems to be another case of counting legs of sheep and dividing by four to find the number of sheep. You are right, but not totally. As long as we are dealing with continuously differentiable functions, the Riemann-Stieltjes integral reduces to the Riemann integral. Thus, if $g(x)$ is continuously differentiable, we have

$$\int_a^b f(x)dg(x) = \int_a^b f(x)g'(x)dx \tag{11.38}$$

But there are times that $g'(x)$ does not exist and in such cases the Riemann-Stieltjes integral is a generalization of the Riemann integral.

Example 11.33 Evaluate the integral

$$\int_0^{10} xd(x-[x])$$

where $[x]$ denotes the integer part of x and therefore $x-[x]$ is the part of x after the decimal point. Thus, if $x = 5.367, [x] = 5$, and $x - [x] = 0.367$,

$$\begin{aligned}\int_0^{10} xd(x-[x]) &= \int_0^{10} xdx - \int_0^{10} xd[x] \\ &= \frac{x^2}{2}\bigg|_0^{10} - \sum_{i=1}^{10} i \\ &= 50 - 55 \\ &= -5\end{aligned}$$

As can be seen, because of the discontinuity of $[x]$, the Riemann integral does not exist, but the more general Riemann-Stieltjes integral does.

11.2.6 Exercises

E.11.2 Evaluate the following definite integrals:

$$i. \int_0^2 x^3 dx \qquad ii. \int_0^1 2e^x dx \qquad iii. \int_{\frac{\pi}{2}}^{\pi} \cos x dx$$

$$iv. \int_{-\frac{\pi}{2}}^{0} \sin x d \qquad xv. \int_0^3 \frac{dx}{1+x} \qquad vi. \int_1^e \frac{dx}{x}$$

$$vii. \int_{-5}^{5} (x^3 + 3x) dx \qquad viii. \int_1^5 \frac{dx}{2x-1} \qquad ix. \int_0^{\frac{\pi}{2}} \cos^2 x dx$$

E.11.3 Show that σ^2 is the variance of normal distribution.

11.3 Computer and Numerical Integration

11.3.1 Computer Integration

As in the case of derivatives, you can use a computer program to find both the definite and indefinite integral of a function. Below we discuss the Maple commands for this purpose. In order to take an integral, you may define the function first and then take the integral, or simply specify the function within the integral command.

Maple code

```
# Specify the function
f := 5*x^2 - 3*x + 14 ;
# Find the indefinite integral
int(f, x);
# The result will be
```

$$\frac{5}{3}x^3 - \frac{3}{2}x^2 + 14x$$

```
# Note that the constant of integration does not
# appear in the answer. More important, the command
# is case sensitive. For example, if you type
Int(f, x);
# then instead of the indefinite integral above,
# you get
```

$$\int 5x^2 - 3x + 14dx$$

```
# To find the definite integral of the function
# from x = 1 to x = 3, use the command
int(f, x=1..3);
# the result will be
```

$$\frac{178}{3}$$

```
# Again if you use the command Int(f, x=1..3);
# the result would be
```

$$\int_1^3 5x^2 - 3x + 14dx$$

```
# You need not specify the function outside the
# Integration command. The following commands will
# result, respectively, in the indefinite and
# definite integral of the function specified above.
int(5*x^2 - 3*x + 14 , x);
int(5*x^2 - 3*x + 14 , x=1..3);
```

11.4 Numerical Integration

On many occasions we need to numerically evaluate a definite integral because the integral cannot be written in a closed form or the problem may not lend itself to analytical solution. For example, the following function can be written in closed form and evaluated:

$$\int_a^b xdx = \left.\frac{x^2}{2}\right|_a^b = \frac{b^2 - a^2}{2}$$

But the following integral has no closed form and has to be evaluated numerically:

$$\int_0^2 e^{-t^2} dt$$

Recall that a closed form is a simple formula that does not involve integrals or infinite sums. For example,

$$\int_0^u e^{-t^2} dt = \sum_{j=0}^{\infty} \frac{(-1)^j u^{2j+1}}{j!(2j+1)}$$

and, therefore, the integral does not have a closed form. On the other hand,

$$\int_0^{\infty} e^{-t^2} dt$$

has a closed form because it is equal to $\sqrt{\pi}/2$. Similarly, in Bayesian statistics and econometrics some problems involving multivariate integration may not be analytically tractable and one has to resort to numerical integration.

We shall introduce the reader to numerical integration by first discussing the *trapezoid method* that has the advantage of being simple and intuitively appealing. The trapezoid method is a special case of Newton-Cotes,[7] which is the most popular method of numerical integration. The method is based on the Lagrange interpolation formula and, therefore, we shall discuss it before presenting the Newton-Cotes method. We leave it to the reader to show that indeed the trapezoid method is a special case of the Newton-Cotes method (see E.11.6). We shall also discuss *the Simpson's rule* and show that it, too, is a special case of Newton-Cotes method.

11.4.1 The Trapezoid Method

This is perhaps the simplest method of numerical integration and is based on approximating the area under the curve by the sum of the areas of trapezoids that can be fitted under it (see Fig. 11.2a). Our objective is to find the integral

$$\int_a^b f(x)dx$$

Let us divide the interval of integration $[a, b]$ into n equal subintervals

$$[x_0, x_1], [x_1, x_2], \ldots, [x_{n-1}, x_n]$$

and define

$$h = x_{i+1} - x_i = \frac{b-a}{n}$$

that is, the width of each trapezoid under the curve is h. The area of the trapezoid is

$$\frac{h}{2}\,[f(x_{i+1}) + f(x_i)]$$

Summing over all intervals, we have

$$\begin{aligned}\int_a^b f(x)dx &\approx \sum_{i=0}^{n} \frac{h}{2}\,[f(x_{i+1}) + f(x_i)]\\ &= h\left[\frac{f(x_0)+f(x_n)}{2} + f(x_1) + f(x_2) + \ldots + f(x_{n-1})\right]\end{aligned}$$

[7] The British mathematician Roger Cotes (1682–1716) was a professor of astronomy in Cambridge and made contributions to the study of logarithms, Newton's method of interpolation, and numerical integration. He also edited the second edition of Newton's *Principia*. Cotes died young and Newton who was 40 years his senior said "if he had lived we might have known something."

It is clear that the precision of the result increases with the number of subintervals. To illustrate the effect of fineness of interval length on the precision of the result, below we will first write a routine in Matlab that evaluates the above expression using different subdivisions. The function we will evaluate is the standard normal distribution and the interval of integration starts at the mean and extends to two standard deviations above it. Recall that the mean of this distribution is equal to zero and its standard deviation one.

$$\int_0^2 \frac{1}{\sqrt{2\pi}} e^{-\frac{1}{2}t^2}\, dt$$

Matlab code

```
% Define a function. In this case normal density
% function.
F=inline('(1/sqrt(2*pi))*exp(-0.5*x.^2)');
% Specify the limits of integration
a=0;
b=2;
% Define a counter to keep track of number of
% intervals
k=0;
% We are going to compute the integral with
% 4, 6, ..., 22 subintervals. That is, 2*j.
for j=2:11
        k = k + 1;
        x = 0;
        % Compute the length of each subinterval
        H = (b-a)/(2*j);
        % calculate [f(x0)+f(xn)/2]
        S(k) = 0.5*(F(a) + F(b));
        % Add f(x1),f(x2),...,f(xn-1)
        for i=1:(2*j-1)
                x = x+h;
                S(k) = S(k) + F(x);
        end
        % Multiply the sum by the length of
          intervals h.
        S(k)=S(k)*h;
end
% Print the result

S
```

If you run the preceding Matlab program, you get the results in Table 11.2. Because we know from the table of normal distribution that a reasonable

Table 11.2 Numerical integration with increasing subintervals

Number of subintervals	Area under the curve
4	0.4750
6	0.4763
8	0.4767
10	0.4769
12	0.4770
14	0.4771
16	0.4771
18	0.4771
20	0.4772
22	0.4772

approximation is 0.4772, it is clear that with about 20 subintervals, we get the desired result. Of course the above is for illustration; otherwise Matlab has two built-in functions that evaluate the integral. We will discuss them at the end of this section.

11.4.2 The Lagrange Interpolation Formula

Suppose we have $n + 1$ points

$$(x_0, y_0),\ (x_1, y_1),\ \ldots,\ (x_n, y_n)$$

Can we find a polynomial of degree n such that it passes through all the points above? One answer to this question is *the Lagrange interpolation formula.* Consider the polynomial

$$\begin{aligned} P_n(x) = {} & C_0(x - x_1)(x - x_2)\ldots(x - x_n) \\ & + C_1(x - x_0)(x - x_2)\ldots(x - x_n) \\ & + \quad \ldots \\ & + C_n(x - x_0)(x - x_1)\ldots(x - x_{n-1}) \end{aligned} \tag{11.39}$$

Letting $x = x_0$ makes all terms except the first one equal to zero. This allows us to determine C_0 as

$$C_0 = \frac{y_0}{(x_0 - x_1)(x_0 - x_2)\ldots(x_0 - x_n)}$$

Similarly,

$$C_1 = \frac{y_1}{(x_1 - x_0)(x_1 - x_2)\ldots(x_1 - x_n)}$$

and

$$C_n = \frac{y_n}{(x_n - x_0)(x_n - x_2)\dots(x_n - x_{n-1})}$$

Thus, we can write the polynomial as

$$\begin{aligned}P_n(x) = &\frac{(x - x_1)(x - x_2)\dots(x - x_n)}{(x_0 - x_1)(x_0 - x_2)\dots(x_0 - x_n)} y_0 \\ &+ \frac{(x - x_0)(x - x_2)\dots(x - x_n)}{(x_1 - x_0)(x_1 - x_2)\dots(x_1 - x_n)} y_2 \\ &+ \quad \dots \\ &+ \frac{(x - x_1)(x - x_2)\dots(x - x_{n-1})}{(x_n - x_0)(x_n - x_1)\dots(x_n - x_{n-1})} y_n\end{aligned}$$

or in a more compact form

$$\begin{aligned}P_n(x) &= \sum_{k=0}^{n} y_k \prod_{\substack{j=0 \\ j \neq k}}^{n} \frac{x - x_j}{x_k - x_j} \\ &= \sum_{k=0}^{n} L_k(x) y_k\end{aligned} \tag{11.40}$$

where

$$L_k(x) = \prod_{\substack{j=0 \\ j \neq k}}^{n} \frac{x - x_j}{x_k - x_j}, \qquad k = 0, \dots, n$$

Example 11.34 Consider the two points ($x_0 = 1,\ y_0 = 3$) and ($x_1 = 2,\ y_1 = 5$). Let us fit a first degree polynomial, that is, a linear function to these points. We have

$$C_0 = \frac{3}{-1} \qquad C_1 = \frac{5}{1}$$

and the polynomial is

$$\begin{aligned}P_2(x) =& -3\,(x - 2) + 5\,(x - 1) \\ =& 2x + 1\end{aligned}$$

Example 11.35 Consider the three points

$$\begin{aligned}x_0 &= 1 & y_0 &= 4 \\ x_1 &= 2 & y_1 &= 6 \\ x_2 &= -1 & y_2 &= 18\end{aligned}$$

For the coefficients of a second-degree polynomial that passes through these points, we have

$$C_0 = -\frac{4}{2}, \qquad C_1 = \frac{6}{3}, \qquad C_2 = \frac{18}{6}$$

and the polynomial is

$$\begin{aligned} P_3(x) &= -2(x-2)(x+1) + 2(x-1)(x+1) + 3(x-1)(x-2) \\ &= 3x^2 - 7x + 8 \end{aligned}$$

11.4.3 Newton-Cotes Method

To numerically evaluate the integral

$$\int_a^b f(x)dx$$

we divide the interval $[a, b]$ into m subintervals:

$$[\alpha_0, \alpha_1],\ [\alpha_1, \alpha_2], \ldots, [\alpha_{m-1}, \alpha_m]$$

In each subinterval $[\alpha_j, \alpha_{j+1}]$ select $n + 1$ equidistance points

$$(x_0, f(x_0)),\ \ldots, (x_n, f(x_n))$$

Consider the approximation to the value of the integral over the j-th subinterval

$$\int_{x_0}^{x_n} f(x)dx \approx \int_{x_0}^{x_n} P_n(x)dx \tag{11.41}$$

Substituting for $P_n(x)$ from (11.40),

$$\begin{aligned} \int_{x_0}^{x_n} f(x)dx &\approx \int_{x_0}^{x_n} \sum_{k=0}^{n} L_k(x) f(x_k)dx \\ &= \sum_{k=0}^{n} f(x_k) \int_{x_0}^{x_n} L_k(x)dx \\ &= \sum_{k=0}^{n} w_k f(x_k) \end{aligned} \tag{11.42}$$

where $w_k = \int_{x_0}^{x_n} L_k(x)dx$.

The formula in (11.42) gives the value of the integral over one of the subintervals. In order to obtain the value of the integral over $[a, b]$, we repeat the same process for all subintervals and add up the results. Let the value of the integral over the j-th subinterval be denoted by s_j. Then

$$\int_a^b f(x)dx \approx \sum_{j=1}^{m} s_j \tag{11.43}$$

11.4.4 Simpson's Method

If we choose $P_n(x)$ to be of degree three, we will have Simpson's method.[8] Here we are approximating the area under the curve with a series of parabolas instead of a series of trapezoids. First, we divide the interval $[a, b]$ into $2m$ subintervals of length

$$h = \frac{b-a}{2m}$$

Thus,

$$\begin{aligned} x_0 &= a \\ x_1 &= x_0 + h \\ &\dots \\ x_{2m} &= x_0 + 2mh = b \end{aligned}$$

The reason we have an even number of subintervals is that we approximate the function $f(x)$ in two contiguous subintervals by a parabola. Let us consider the first three points, $(x_0, f(x_0))$, $(x_1, f(x_1))$, and $(x_2, f(x_2))$. Because they are equally distanced, we can write

$$x_0 = x_1 - h, \qquad x_2 = x_1 + h$$

Then

$$\begin{aligned} P_n(x) = &\frac{(x-x_1)(x-x_1-h)}{-h(2h)} f(x_0) + \frac{(x-x_1+h)(x-x_1-h)}{h(-h)} f(x_1) \\ &+ \frac{(x-x_1+h)(x-x_1)}{h(2h)} f(x_2) \end{aligned} \tag{11.44}$$

Let

$$z = x - x_1$$

[8] Thomas Simpson (1710–1761) worked on many areas of mathematics including calculus, numerical methods, astronomy, and probability theory. He is the author of several high-quality textbooks. Simpson was a self-taught mathematician and for a time he lectured in London coffee houses. This may seem strange today, but coffee houses were known as Penny Universities where customers could listen to lectures on mathematics, art, law, and other subjects while drinking coffee. See *A History of the World in Six Glasses* by Tom Standage (2005). Simpson's rule discussed in the text is due to Newton not Simpson. Justice, however, has prevailed as the Newton method discussed in Chap. 10 is due to Simpson.

Then

$$P_n(x) = \frac{1}{2h^2}[z(z-h)f(x_0) - 2(z+h)(z-h)f(x_1) + z(z+h)f(x_2)]$$

Thus,

$$\int_{x_0}^{x_2} f(x)dx \approx \int_{x_0}^{x_2} P_n(x)dx = \int_{-h}^{h} P_n(z)dz \tag{11.45}$$

We now evaluate the last integral:

$$\int_{-h}^{h} P_n(z)dz = \int_{-h}^{h} \frac{1}{2h^2}[z(z-h)f(x_0) - 2(z+h)(z-h)f(x_1) + z(z+h)f(x_2)]dz$$

$$= \frac{1}{2h^2}\left[\frac{f(x_0) - 2f(x_1) + 2f(x_2)}{3}z^3 + \frac{(f(x_2) - f(x_0))hz^2}{2} + 2h^2 f(x_1)z\right]\Bigg|_{-h}^{h}$$

$$= \frac{h}{3}[f(x_0) + 4f(x_1) + f(x_2)]$$

Similarly, for the next three points we have

$$\int_{x_2}^{x_4} f(x)dx \approx \frac{h}{3}[f(x_2) + 4f(x_3) + f(x_4)]$$

and for the entire interval $[a, b]$

$$\begin{aligned}\int_a^b f(x)dx \approx & \frac{b-a}{6m}\{f(x_0) + f(x_{2m}) + 4[f(x_1) + f(x_3) + \ldots + f(x_{2m-1})] \\ & + 2[f(x_2) + f(x_4) + \ldots + f(x_{2m-2}]\}\end{aligned} \tag{11.46}$$

Of course, you do not need to program Simpson's method. Matlab already has a function for integration.

Example 11.36 Find the following integral:

$$\int_0^{\frac{\pi}{2}} \cos x dx$$

Matlab code

```
% Define the function
F = inline('cos(x)')
% Find the integral
q = quad(F,0,pi/2)
```

Matlab has another procedure based on the Gaussian quadrature method. We next use this function to calculate the standardized normal table.

Matlab code

```
% Define the function
F=inline('(1/sqrt(2*pi))*exp(-0.5*x.^2)')
% Find the integral for x=0 to 3
for j=1:31
    for i=1:10
            % Compute the value of x
            h=(j-1).*0.1+(i-1).*.01;
            % Compute the integral
            B(j,i)=quadl(F,0,h);
    end
end
% Matlab will give you a Warning. This does not
% affect your result. To understand its meaning look
% up Matlab help under Integration, Numerical, quadl.
% Print the table
B
```

After printing the table, compare it to the normal table at the end of your statistics or econometrics book. The function computes the area under the curve to the right of the mean.

11.4.5 Exercises

E.11.4 For each of the following two sets of points find a second-degree polynomial that passes through them.

i. x	$f(x)$	*ii. x*	$f(x)$
1	6	1	0
2	7	−1	−10
3	10	2	−1

E.11.5 For each of the following two sets of points find a third-degree polynomial that passes through them

i. x	$f(x)$	*ii. x*	$f(x)$
1	2	1	0
−1	14	−1	16
2	8	2	13
−4	2	−2	21

E.11.6 Show that the trapezoid rule is a special case of the Newton-Cotes method.

E.11.7 Use Matlab's `quadl` function to numerically evaluate integrals in E. 11.2.

11.5 Special Functions

Two functions play prominent roles in probability theory and econometrics, neither of which has a closed form. We discuss them here and show how Matlab can be used to calculate them.

Definition 11.2 (Gamma Function). The gamma function has one parameter α and has the form

$$\Gamma(\alpha) = \int_0^\infty e^{-t} t^{\alpha-1} dt \qquad \alpha > 0 \tag{11.47}$$

α need not be real, but if it is complex, then its real part must be positive. The gamma function has some interesting properties:

i. $\Gamma(\alpha+1) = \alpha\,\Gamma(\alpha)$

ii. If the argument of the function is an integer, that is, if $\alpha = n$, then

$$\Gamma(n+1) = n! \qquad n = 0, 1, 2, \ldots \tag{11.48}$$

iii.

$$\Gamma\left(\frac{1}{2}\right) = \sqrt{\pi} \tag{11.49}$$

The gamma function can be evaluated using a Matlab function

Matlab code

```
gamma(a)
```

Definition 11.3 (Beta Function). The beta function has the form

$$B(\alpha, \beta) = \int_0^1 t^{\alpha-1}(1-t)^{\beta-1} dt \qquad \alpha,\ \beta > 0 \tag{11.50}$$

The beta function is related to the gamma function

$$B(\alpha, \beta) = \frac{\Gamma(\alpha)\Gamma(\beta)}{\Gamma(\alpha+\beta)} \tag{11.51}$$

Matlab can be used to evaluate the beta function

Matlab code

```
beta(a,b)
```

11.5.1 Exercises

E.11.8 Prove the properties of the gamma function listed in Definition 11.2.

E.11.9 The density of the t-distribution is

$$f(t) = \frac{1}{\sqrt{\nu}B(\frac{1}{2}, \frac{1}{2}\nu)} \left(1 + \frac{t^2}{\nu}\right)^{-\frac{1}{2}(\nu+1)}$$

where ν is the degrees of freedom. Write a program in Matlab that reads the value of t and ν and calculates

$$P(T > t) = 1 - P(T \leq t)$$

Note that the t-distribution is symmetric around zero.

11.6 The Derivative of an Integral

In certain instances one has to find the derivative of an integral. In particular if the objective function of an optimization problem is in the form of an integral, we need to calculate such derivatives. Below we state the rules of such differentiations and illustrate them with examples.

Property 11.12 The derivative of a definite integral with fixed limits with respect to the variable of integration is zero because

$$\frac{d}{dx} \int_a^b f(x)dx = \frac{d}{dx}[F(b) - F(a)] = 0 \tag{11.52}$$

Property11.13 The derivative of an indefinite integral with respect to the variable of integration is the integrand. This property follows from the definition of indefinite integral:

$$\frac{d}{dx} \int f(x)dx = \frac{d}{dx}[F(x) + C] = f(x) \tag{11.53}$$

Property 11.14 The derivative of a definite integral with a fixed lower limit and variable upper limit is equal to the integrand evaluated at the point of upper limit times the derivative of the upper limit. That is,

$$\begin{aligned} \frac{d}{dx} \int_a^{h(x)} f(t)dt &= \frac{d}{dx}[F(h(x)) - F(a)] \\ &= f(h(x))h'(x) \end{aligned} \tag{11.54}$$

Example 11.37

$$\frac{d}{dx}\int_a^{2x}\frac{dt}{t}=\frac{1}{x}$$

Let us check this result:

$$\frac{d}{dx}\int_a^{2x}\frac{dt}{t}=\frac{d}{dx}\left[\ln t|_a^{2x}\right]=\frac{d}{dx}(\ln 2x-\ln a)=\frac{1}{x}$$

Example 11.38 Check the following result first by using Property 11.14 and then by actually carrying out the integration and taking the derivative.

$$\frac{d}{dx}\int_a^{\ln x}t^2e^tdt=(\ln x)^2$$

Consider the function $f(x, y)$. Suppose we want to integrate this function with respect to x and then take its derivative with respect to y. If the limits of integration do not depend on x, then the result of integration would be a function of y alone. Therefore, we can first integrate with respect to x and then differentiate with respect to y. There are, however, cases where the integral does not exist. In such cases we have the following results.

Property 11.15 In the case where both limits of integration are fixed,

$$\frac{d}{dy}\int_a^b f(x,y)dx=\int_a^b\frac{\partial f(x,y)}{\partial y}dx \tag{11.55}$$

Example 11.39 Find

$$\frac{d}{dy}\int_a^b x\ln y dx$$

Using Property 11.15, we have

$$\int_a^b\frac{\partial(x\ln y)}{\partial y}=\int_a^b\frac{x}{y}dx=\frac{x^2}{2y}|_a^b=\frac{b^2-a^2}{2y}$$

We can verify this result by first carrying out the integration with respect to x and then taking the derivative with respect to y:

$$\int_a^b x\ln y dx=\ln y\int_a^b xdx=(\ln y)\frac{b^2-a^2}{2}$$

Then

$$\left(\frac{d}{dy}\ln y\right)\frac{b^2-a^2}{2}=\frac{b^2-a^2}{2y}$$

Property 11.16 In the case where both limits of integration are functions of y,

$$\begin{aligned}\frac{d}{dy}\int_{g(y)}^{h(y)} f(x,y)dx &= \int_{g(y)}^{h(y)}\frac{\partial f(x,y)}{\partial y}dx\\ &+f(h(y),y)\frac{dh(y)}{dy}\\ &-f(g(y),y)\frac{dg(y)}{dy}\end{aligned} \tag{11.56}$$

Example 11.40 Let us evaluate the following expression in two ways. First, by using Property 11.16 and, second, in the old-fashioned way of integrating and then differentiating.

$$\frac{d}{dy}\int_{y}^{3y+1} xy^2dx$$

Using Property 11.16, we have

$$\begin{aligned}\frac{d}{dy}\int_{y}^{3y+1} xy^2dx &= \int_{y}^{3y+1} 2xydx + 3(3y+1)y^2 - y^3\\ &= x^2y\Big|_{y}^{3y+1} + 9y^3 + 3y^2 - y^3\\ &= 16y^3 + 9y^2 + y\end{aligned}$$

On the other hand, we can integrate the function to obtain

$$\begin{aligned}\frac{d}{dy}\int_{y}^{3y+1} xy^2dx &= \frac{1}{2}x^2y^2\Big|_{y}^{3y+1}\\ &= \frac{1}{2}[(3y+1)^2y^2 - y^4]\end{aligned}$$

Taking the derivative of the RHS, we get

$$\frac{d}{dy}\frac{1}{2}[(3y+1)^2y^2-y^4] = 16y^3+9y^2+y$$

11.6.1 Exercises

E.11.10 Evaluate the following expressions.

i. $\frac{d}{dx}\int_0^{3x}(t^3 - 2t^2 + 5t + 17)dt$

ii. $\frac{d}{dx}\int_1^{x}\frac{dt}{2t-1}$

iii. $\frac{d}{dx}\int_1^{e^x}5\frac{dt}{t}$

iv. $\frac{d}{dy}\int_0^{y}xe^y dx$

v. $\frac{d}{dy}\int_{-y}^{y}x\ln y dx$

vi. $\frac{d}{dy}\int_y^{y^2+y}\frac{x^2}{y}dx$

Part IV
Optimization

Chapter 12
Static Optimization

A theory or model is the tool by which we organize our thought about a phenomenon and has to have the ability to explain or forecast. Economic processes are the outcome of the interaction of decisions made by many economic agents. It follows that any economic theory has to be based on some model of decision making by economic agents, be it individual, household, firm, or government. Preferably the behavioral assumptions underlying such a model are applicable to a variety of agents and do not vary in an ad hoc manner, because a science worthy of the name cannot consist of a bunch of unrelated models, each of which is applicable to only a special case. Indeed, it is quite easy to find a rationalization for any event or phenomenon after the fact.

The work of marginalists in the nineteenth and early twentieth centuries laid the foundation of a unified economic theory. Antoine Augustin Cournot studied the profit maximizing firm, Johann Heinrich von Thünen applied it to cost minimization by firms, and Hermann Heinrich Gossen studied utility maximizing consumers. In the latter part of the nineteenth century the first generation of professional economists, Léon Walras, William Stanley Jevons, and Carl Menger, developed these ideas into a comprehensive theory of the economy. Many others including Alfred Marshall contributed to this endeavor.

The neoclassical theory found its most eloquent statement in John Hicks's *Value and Capital* and Paul Samuelson's *Foundations of Economic Analysis*. The behavioral assumption underlying the neoclassical theory is that all decision makers in the economy try to maximize an objective function subject to the constraints put on them by resources, technology, regulations, and market forces. Individuals and households maximize their utility subject to their budget constraint, firms try to minimize their cost for each level of output, and then given the market price try to maximize their profit. The objective of the theory is to explain the behavior of individuals, households, and firms, and through their interactions the behavior of the economy as a whole.

In the second half of the twentieth century the neoclassical model was extended to analyze areas once deemed outside the realm of economics. These include decisions as to the number of children a family would produce, allocation of time between work, entertainment, and spiritual activities, behavior of criminals, and the political process.

K. Dadkhah, *Foundations of Mathematical and Computational Economics,* 2nd ed.,
DOI 10.1007/978-3-642-13748-8_12, © Springer-Verlag Berlin Heidelberg 2011

There is little doubt that in the real world decisions are made differently than what the theory assumes. Thus, if the idea is to study the way businesses behave or to advise them on how to improve their performance, one has to appeal to decision sciences, statistical decision theory, and behavioral economics. But if the goal is to understand the behavior of an industry or market, one can ask if there is another theory that is tractable and yields as many insights and testable hypotheses as the neoclassical theory. There has been talk of theories based on *bounded rationality* and *satisficing* instead of optimizing behavior, but it is hard to find a body of theory based on these assumptions that produces hypotheses and conclusions that are materially different from neoclassical predictions. Thus, pending an unforeseen breakthrough in economics, optimization theory remains the cornerstone of economic analysis.

Static optimization is concerned with finding the optimum point of a function. In contrast, dynamic optimization (Chap. 14) tries to find a function of time that maximizes (minimizes) an objective functional. In economic application, static optimization applies when we are making a decision for a point in time, abstracting from what might happen in the future. Alternatively we can say that the decision is made on the assumption that no significant changes will occur in the environment and parameters of our decision problem. Thus, suppose we know the price and marginal cost of our product. Then the question is, how many units should we produce? In the real world consumers' tastes, technology, and laws change over time and so do the price and marginal cost of our product. Furthermore, our information about the present and future demand for our product are estimates and, therefore, subject to forecast error. Finally, we should take into account the reaction of our competitors, who may surprise us with their innovation or cost-cutting strategies. But for the time being our goal is modest, static optimization within a deterministic model.

Neoclassical theory is only one place where optimization theory proves quite useful. Economists have many more reasons to make sure that they are well versed in the subject. In statistics, econometrics, and decision analysis, optimization plays a pivotal role. In this chapter we deal with static optimization when no constraints limit our choice variables. The next chapter is devoted to constrained optimization, when we try to maximize an objective function while our decision variables have to satisfy one or more constraints.

12.1 Maxima and Minima of Functions of One Variable

One has little trouble in intuitively understanding the concept of the maximum and minimum of a function or even to distinguish between a local and the global maximum or minimum of a function. The maximum is the highest point and the minimum the lowest. If a point is the highest within a subset of the domain of a function, then it is a local maximum. On the other hand, if it is the highest over the entire domain of the function, that is, the highest of the high points, then it is the global maximum. The same applies to the case of local and global minima. Some functions have only one maximum (minimum) and the local and global maxima (minima) coincide. Other functions may have one global and several local maxima (minima). We are interested in maxima and minima that are finite.

Definition 12.1 The real-valued function $f(x)$ whose domain is $D \in \Re$ has its local maximum at the point $x^* \in D$ if

$$f(x^*) \geq f(x) \qquad \forall x \neq x^* \tag{12.1}$$

where

$$|x - x^*| < \delta \quad \text{and} \quad \delta > 0$$

In the above definition the real number δ defines a neighborhood within which x^* is the maximum. If (12.1) holds for all $x \in D$, then $f(x^*)$ is the global maximum. We can define local and global minima in a similar fashion except that in this case $x^* \in D$ is a local minimum[1] if

$$f(x^*) \leq f(x) \qquad \forall x \neq x^* \tag{12.2}$$

where

$$|x - x^*| < \delta, \quad \delta > 0$$

and is the global minimum if the relationship holds for all $x \in D$.

Figure 12.1 depicts both local and global maxima as well as the local minimum of a function. It is easy to determine the maximum or minimum of a function on its graph just as it is easy to see the summit of a mountain and the deepest point

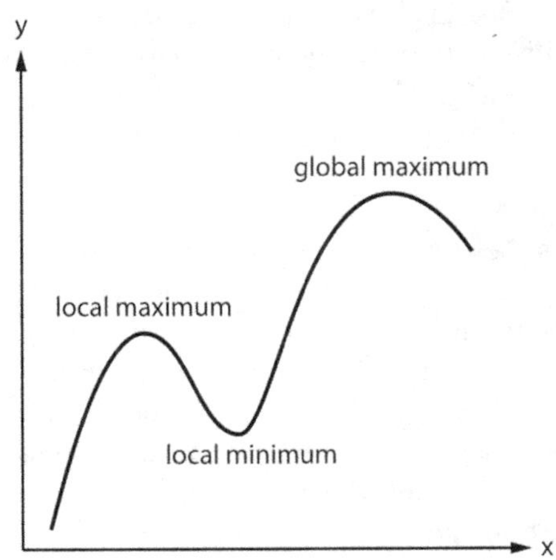

Fig. 12.1 Local and global maxima and minima

[1] If $f(x) < f(x^*)$, then $f(x^*)$ is referred to as the strict maximum and if $f(x) > f(x^*)$, then it is the strict minimum.

of a valley. Carrying on with the analogy a bit further, we can say that as long as you are moving upward you have not reached the summit, and as soon as you start downward, you have passed it. A function is increasing as long as its derivative is positive and decreases when the derivative turns negative. It follows that maximum is reached at the point of transition between a positive and negative derivative, that is, at the point when the derivative is zero. Thus, a necessary condition for a maximum is that the first derivative be zero. But this is not sufficient, because at a trough, too, the derivative would be zero. Note, however, that a summit is reached when we ascend and then descend. That requires the derivative to be declining, going from positive to zero to negative. In other words, the derivative of the derivative, the second derivative, is negative. For the minimum the opposite holds and the second derivative should be positive.

These ideas are summarized in the following theorems. They are stated for the case of maxima but, *mutatis mutandis*, they hold for minima. Before getting formal, however, let us take a casual tour of the argument.

Let $(x^*, f(x^*))$ be the maximum point of a function (the same argument with appropriate changes would apply to the point of minimum). Using Taylor expansion we can write

$$f(x) = f(x^*) + (x - x^*)f'(x^*) + \frac{1}{2}(x - x^*)^2 f''(\zeta) \tag{12.3}$$

where x can be any point on either side of x^* such that $|x - x^*| < \varepsilon$ and ζ is a point between x and x^*.

If x^* is to be the maximum, we should have

$$f(x) - f(x^*) = (x - x^*)f'(x^*) + \frac{1}{2}(x - x^*)^2 f''(\zeta) \leq 0 \tag{12.4}$$

for all x within ϵ distance of x^*.

But this is not guaranteed because $(x - x^*)f'(x^*)$ could be both positive and negative unless we require $f'(x^*)$ to be equal to zero, thus eliminating the possibility of $(x - x^*)f'(x^*) > 0$ spoiling the game. Thus,

$$f'(x^*) = 0 \tag{12.5}$$

is the necessary condition for x^* being a maximum. Now note that $(x - x^*)^2$ is always positive, therefore, for the inequality in (12.4) to hold we need $f''(\zeta) < 0$. Furthermore, $f''(x)$ cannot change sign, otherwise we do not have a maximum. Recall that $f'(x)$ has to be decreasing, be positive, become zero, and then negative. If $f''(x)$ changes sign, then the pattern is disrupted. Of course we may have $f''(x^*) = 0$, but then we will have to restate the same condition for the fourth derivative. Thus, within the neighborhood under consideration, including at the point of maximum we should have

$$f''(x^*) < 0 \tag{12.6}$$

which is our sufficient condition.

Theorem 12.1 *Let $f(x)$ be a real-valued function that is at least twice continuously differentiable over its domain D. The necessary condition for $f(x^*)$ to be a maximum is*

$$f'(x^*) = 0 \tag{12.7}$$

Proof We need to show that if $f(x) \le f(x^*)$ for all $x \in D$ such that $|x - x^*| \le \delta$, then $f'(x^*) = 0$.We show that if $f'(x^*) \ne 0$, then it contradicts the assumption of the theorem. By the mean value theorem,

$$f(x) = f(x^*) + (x - x^*)f'(\xi) \tag{12.8}$$

where ξ is a point between x and x^*. Let

$$\begin{aligned} \theta &= x - x^* \\ \xi &= x^* + \lambda\theta, \qquad \lambda \in (0, 1) \end{aligned}$$

Then

$$f(x^* + \theta) = f(x^*) + \theta f'(x^* + \lambda\theta) \tag{12.9}$$

If $f'(x^*) > 0$, then by continuity there will be a $\theta > 0$ and $|\theta| < \varepsilon$ such that

$$f'(x^* + \lambda\theta) > 0$$

for all λ's between 0 and 1. But that contradicts the assumption of the theorem because $\theta f'(x^* + \lambda\theta) > 0$ implies that $f(x^* + \theta) > f(x^*)$. Alternatively, if $f'(x^*) < 0$, then we can find a $\theta < 0$ and $|\theta| < \varepsilon$ such that

$$f'(x^* + \lambda\theta) < 0$$

and again we have a contradiction because $\theta f'(x^* + \lambda\theta) > 0$. The only alternative left is $f'(x^*) = 0$. Note that this proof equally applies to the case of a minimum.

Intuitively, the theorem says that if $f'(x^*) \ne 0$, then it is either positive or negative. Because $x - x^*$ could be either positive or negative, there are points within the neighborhood $|x - x^*| < \delta$ where $x - x^*$ and $f'(x^*)$ have the same sign resulting in $(x - x^*)f'(x^*) > 0$, which violates the assumption that $f(x^*) > f(x)$ everywhere in the neighborhood.

Theorem 12.2 *Let $f(x)$ be a real-valued, at least twice differentiable function over its domain D. Further assume that $f''(x) \ne 0$.[2] Then the sufficient condition for $f(x^*)$to be a maximum is*

$$f''(x^*) < 0 \tag{12.10}$$

---

[2]The case where the second and higher derivatives are zero is taken up later.

Proof The Taylor expansion of the function near point x^* with the Lagrange remainder is

$$f(x) = f(x^*) + (x - x^*)f'(x^*) + \frac{1}{2}(x - x^*)^2 f''(\xi) \tag{12.11}$$

where $|x - x^*| < \delta$ and ξ is a point between x and x^*.

If x^* is a maximum, then $f'(x) = 0$, and we can write

$$f(x) = f(x^*) + \frac{1}{2}\theta^2 f''(x^* + \lambda\theta) \tag{12.12}$$

Now suppose $f''(x^*) > 0$. Then by continuity we can find a $|\theta| < \varepsilon$ such that $f''(x^* + \lambda\theta) > 0$ for all λ between 0 and 1, which would contradict the assumption that $f(x^*)$ is a maximum, therefore, $f''(x^*) \leq 0$. Because we ruled out $f''(x) = 0$, it follows that $f''(x^*) < 0$. The proof of the theorem for the case of a minimum follows the same line of argument.

Example 12.1 Find the maximum of the function

$$y = -2x^2 + 3$$

Setting the first derivative equal to zero we have

$$-4x = 0$$

which results in

$$x = 0, \qquad y = 3$$

This is indeed a maximum because the second derivative is negative

$$y'' = -4 < 0$$

Example 12.2 Find the minimum of the function

$$y = 5x^2 - 10x + 8$$

We have

$$10x - 10 = 0$$

with the minimum point being

$$x = 1, \qquad y = 3$$

This is a minimum because

$$\frac{d^2y}{dx^2} = 10 > 0$$

Example 12.3 Find the extrema[3] of the function

$$f(x) = x^3 - 9x^2 + 15x + 3$$

Setting the first derivative equal to zero

$$3x^2 - 18x + 15 = 0$$

we have

$$x_1 = 1, \qquad f(x_1) = 10$$

and

$$x_2 = 5, \qquad f(x_2) = -22$$

because the second derivative of the function is

$$f''(x) = 6x - 18$$

Evaluating it at the extrema points, we find

$$f''(1) = -12, \qquad f''(5) = 12$$

Thus, the point $x = 1$, $f(x) = 10$ is a local maximum and the point $x = 5$, $f(x) = -22$ a local minimum. The reader is urged to graph the functions in Examples 12.1, 12.2, and 12.3 and ascertain that we indeed have found the maxima and minima of the functions.

Optimization of functions of one variable finds many applications in economics, as the next few examples illustrate.

Example 12.4 Consider the short-run decision problem of a competitive firm whose capital stock is fixed at $K = K_0$. The production function of the firm is represented by

$$Q = F(K_0, L)$$

where Q and L are, respectively, output and labor.

$$\frac{dQ}{dL} = F_L > 0 \qquad \text{and} \qquad \frac{d^2Q}{dL^2} = F_{LL} < 0$$

That is, the marginal product of labor is positive, but decreasing. Because the firm is a price taker, the prices of its product, P, the rental cost of capital, R, and the

---

[3]Extremum is a generic word for any point on the function where the first derivative is equal to zero and includes maximum, minimum, and as will be seen later, the inflection point.

wage rate, W, are all given and fixed. A profit maximizing firm faces the decision of how much to produce and how many workers to hire. Let π denote the profit. Then

$$\pi = PF(K_0, L) - RK_0 - WL$$

Maximization of profit requires setting the derivative of profit with respect to labor input equal to zero:

$$\frac{d\pi}{dL} = PF_L - W = 0$$

which results in

$$PF_L = W \qquad \Rightarrow \qquad F_L = \frac{W}{P}$$

That is, the firm expands its production and hires additional labor to the point where the value of marginal product of labor equals the wage rate. Alternatively, we could say that production will expand until the marginal product of labor is equal to the real wage. We can ascertain that this indeed is the maximum profit by examining the second derivative

$$\frac{d^2\pi}{dL^2} = PF_{LL} < 0$$

Example 12.5 A famous proposition in microeconomics involves the equality of average and marginal cost at the point of minimum average cost. Intuitively the argument is that average cost will be decreasing as long as it is above marginal cost and will be increasing as long as it is below marginal cost. It follows that the two will be equal when average cost is neither increasing nor decreasing. Because average cost can increase indefinitely, the point of equality must be the minimum.

To show this rigorously, we need to be more specific about our assumptions. Let $C = C(Q)$ denote the total cost function that depends on output. It is assumed that the marginal cost $C'(Q)$ is first decreasing and then increasing, which implies that on the increasing portion of the marginal cost function $C''(Q) > 0$. Minimizing the average cost, we get

$$\frac{d}{dQ}\frac{C(Q)}{Q} = \frac{QC'(Q) - C(Q)}{Q^2} = 0$$

which leads to

$$C'(Q^*) = \frac{C(Q^*)}{Q^*}$$

where Q^* denotes the output level at which the average cost is minimized. That this point corresponds to the minimum average cost is confirmed by noting that

$$\frac{d^2}{dQ^2}\frac{C(Q)}{Q} = \frac{Q^2[C'(Q) + QC''(Q) - C'(Q)] - 2Q[QC'(Q) - C(Q)]}{Q^4}$$

Evaluating it at the point of $Q = Q^*$, we have

$$\frac{d^2}{dQ^2}\frac{C(Q)}{Q}\bigg|_{Q=Q^*} = \frac{C''(Q^*)}{Q} > 0$$

Example 12.6 A competitive firm maximizes its profit, that is, the difference between its revenue and cost. Thus, the problem is

$$\max \quad \pi = PQ - C(Q)$$

The necessary and sufficient conditions are

$$P = C'(Q) \quad \text{and} \quad -C''(Q) < 0$$

The first condition states that the firm will expand its output to the point where marginal cost equals the market determined price. The second condition states that such a point entails maximum profit if marginal cost is on the rise. Because marginal cost is assumed to be first declining and then increasing, the meaning of the sufficient condition is that the maximum profit happens on the rising portion of marginal cost function.

12.1.1 Inflection Point

In proving Theorem 12.2 we assumed that the second derivative was not equal to zero. We now consider the possibility that the second- and higher-order derivatives may be zero. First, let us assume that not only the second derivative, but also the third derivative is zero, and the fourth-order derivative is nonzero. Then we can rewrite (12.4) as

$$f(x) = f(x^*) + (x - x^*)f'(x^*) + \frac{1}{4!}(x - x^*)^4 f^{(4)}(\xi) \tag{12.13}$$

We can apply the same argument for the proof of Theorem 12.2 to (12.13). Thus, if the second and third derivatives are zero, we look to the fourth derivative. If it is negative, we have a maximum, and if it is positive, we have a minimum. Indeed, the same argument applies if the fourth, fifth, ... derivatives are zero and the first nonzero derivative is even numbered.

Example 12.7 Consider the function

$$y = (x - 2)^4 + 7$$

Fig. 12.2 Inflection point

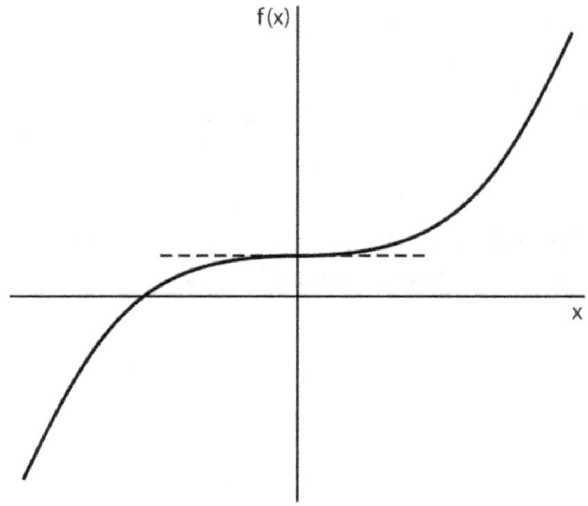

Setting the first derivative equal to zero, we get $x = 2$ and $y = 7$. However, at the extremum point, y'' and y''' are also zero. For the fourth derivative we have

$$y^{(4)} = 24 > 0$$

Therefore, we have a minimum. The reader should verify the conclusion by graphing the function.

The case where the first nonzero derivative is odd numbered is a different story. The third derivative is the derivative of $f''(x)$, therefore, a nonzero third derivative tells us that either the second derivative is increasing ($f'''(x) > 0$) or decreasing ($f'''(x) < 0$). Because $f''(x)$ passes through zero, it follows that either $f''(x)$ goes from negative to positive or from positive to negative, implying that the first derivative goes from decreasing to increasing or vice versa. Thus, the function itself has an *inflection point* (see Fig. 12.2). The same is true if the third, fourth, ... derivatives are zero, and the first nonzero derivative is odd numbered.

Example 12.8 Let $y = x^3 + 3$. Then

$$y' = 3x^2$$

Setting it equal to zero yields $x = 0$, $y = 0$, then

$$y'' = 6x$$

which is equal to zero at the point $x = 0$. Turning to the third derivative, we get

$$y''' = 6 \neq 0$$

Thus, we have an inflection point. This function is shown in Fig. 12.2.

Inflection points are not confined to the points of extremum. We could have points of inflection without the first derivative being zero.

Example 12.9 The function $y = e^{-x^2}$ has two points of inflection. The first derivative is

$$y' = -2xe^{-x^2}$$

and the second derivative

$$y'' = -2e^{-x^2} + 4x^2e^{-x^2} = 2(2x^2 - 1)e^{-x^2}$$

Setting the second derivative equal to zero and confining ourselves to finite points, we have

$$x_1 = -\frac{\sqrt{2}}{2}, \quad y_1 = e^{-\frac{1}{2}}$$

and

$$x_2 = \frac{\sqrt{2}}{2}, \quad y_2 = e^{-\frac{1}{2}}$$

12.1.2 Exercises

E.12.1 Find the extrema of the following functions and determine their nature.

i. $y = x^2 - 2x + 5,$ *ii.* $y = \frac{x^3}{3} - 3x^2 + 2x + 1,$

iii. $y = 7 - (x - 4)^{2/3},$ *iv.* $y = \frac{x}{\ln x},$

v. $y = -2x^4 + 6x^2,$ *vi.* $y = \frac{x^2 - 3x + 2}{x^2 + 3x + 2},$

vii. $y = \frac{(x-2)(3-x)}{x^2},$ *viii.* $y = \frac{1}{1 + x^2},$

ix. $y = x^4,$ *x.* $y = \frac{a^2}{x} + \frac{b^2}{a - x}$

E.12.2 Suppose a firm's technology can be represented by the Cobb-Douglas production function $Q = AK_0^{\alpha}L^{1-\alpha}$, where K_0 is fixed. The price of firm's product is P, the wage rate W, and the rental price of capital R. Find the optimum conditions for the firm's production decision.

E.12.3 Consider the case of a monopoly firm whose cost and price functions are

$$C = C(Q)$$
$$P = P(Q)$$

Find the conditions for profit maximization of the firm.

12.2 Unconstrained Optima of Functions of Several Variables

The definition of optimum points for functions of several variables is very similar to the case of functions of one variable

Definition 12.2 The real-valued function $f : \mathbf{D} \rightarrow \Re$, where $\mathbf{D} \subset \Re^n$ has its local maximum at the point $\mathbf{x}^* \in \mathbf{D}$ if

$$f(\mathbf{x}^*) \geq f(\mathbf{x}) \qquad \forall \mathbf{x} \neq \mathbf{x}^* \tag{12.14}$$

where

$$d(\mathbf{x}, \mathbf{x}^*) < \delta \quad \text{and} \quad \delta > 0$$

In Definition 12.2 the real number δ defines a neighborhood within which $\mathbf{x}^*$ is the maximum. If (12.14) holds for all $\mathbf{x} \in \mathbf{D}$, then $f(\mathbf{x}^*)$ is the global maximum. If it holds with strict inequality, then the maximum is a strict maximum. We can define the local and global minimum in a similar fashion except that in this case $\mathbf{x}^* \in \mathbf{D}$ is the local minimum if

$$f(\mathbf{x}^*) \leq f(\mathbf{x}), \qquad \forall \mathbf{x} \neq \mathbf{x}^* \tag{12.15}$$

where

$$d(\mathbf{x}, \mathbf{x}^*) < \delta, \qquad \delta > 0$$

and $\mathbf{x}^*$ is the global minimum if the relationship holds for all $\mathbf{x} \in \mathbf{D}$. The theorems establishing the necessary and sufficient conditions for the maximum and minimum of functions of several variables are stated below without proof, as the proofs are very similar to the case of functions of one variable.

Theorem 12.3 *Let the function f in Definition 12.2 be at least twice continuously differentiable over its domain* $\mathbf{D}$. *The necessary condition for* $f(\mathbf{x}^*)$ *to be a maximum is*

$$\nabla f(\mathbf{x}^*) = 0 \tag{12.16}$$

Theorem 12.4 *Let f be as in Theorem 12.3. Then the sufficient condition for* $f(\mathbf{x}^*)$ *to be a maximum is that its Hessian matrix (the matrix of its second-order derivatives) be negative definite.*

$$\mathbf{z}'\nabla^2 f(\mathbf{x}^*)\mathbf{z} = \mathbf{z}'\mathbf{H}(\mathbf{x}^*)\mathbf{z} < 0 \tag{12.17}$$

for all vectors $\mathbf{z} \neq \mathbf{0}$. *For a minimum, the Hessian has to be positive definite.*

Example 12.10 Find the extremum point of the function

$$z = (x-1)^2 + (y-3)^2 - 5$$

We have

$$\frac{\partial z}{\partial x} = 2(x-1)$$
$$\frac{\partial z}{\partial y} = 2(y-3)$$

Setting the partial derivatives equal to zero, we have

$$x = 1, \qquad y = 3, \qquad z = -5$$

Second-order partial derivatives are

$$\frac{\partial^2 z}{\partial x^2} = 2, \qquad \frac{\partial^2 z}{\partial y^2} = 2, \qquad \frac{\partial^2 z}{\partial x \partial y} = \frac{\partial^2 z}{\partial y \partial x} = 0$$

Therefore, the Hessian is

$$\begin{bmatrix} 2 & 0 \\ 0 & 2 \end{bmatrix}$$

Because both principal minors are positive, the matrix is positive definite and we have a minimum.

Example 12.11 For the function

$$z = -2x^2 + xy - y^2 + 3x + y + 6$$

we have

$$-4\mathrm{x} + \mathrm{y} + 3 = 0$$
$$\mathrm{x} - 2\mathrm{y} + 1 = 0$$

Thus,

$$x = 1, \qquad y = 1, \qquad z = 8$$

and the Hessian is

$$\begin{bmatrix} -4 & -1 \\ -1 & -2 \end{bmatrix}$$

which is negative definite, and we have a maximum.

Example 12.12 We can visualize the maximum and minimum of a function of two variables, for example

$$z = x\exp(-x^2 - y^2)$$

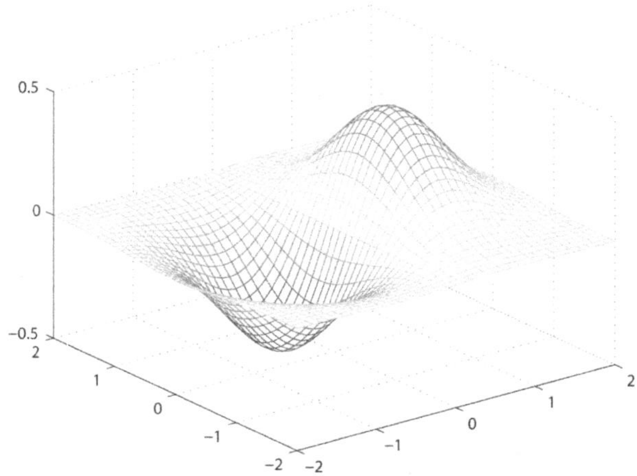

Fig. 12.3 Maximum and minimum of the function $z = x\exp(-x^2 - y^2)$

which is depicted in Fig. 12.3. We cannot, however, visualize the extrema of functions of more than two variables. Finding the maximum and minimum of the function above is left to the reader as an exercise (see E.12.4 *vi*).

Example 12.13 Let us revisit Example 12.4 and this time consider the long-run problem when the firm can decide on its capital stock as well as its labor force. Then the problem of profit maximization will be

$$\max \pi = PF(K, L) - RK - WL$$

where, as before, P is the market determined price of the firm's product, K is the services of capital, L, the labor force, R, the rental price of capital, and W, the wage rate. Furthermore, it is assumed that

$$F_K = \frac{\partial F}{\partial K} > 0, \qquad F_L = \frac{\partial F}{\partial L} > 0$$

where F_K is marginal product of capital and F_L, marginal product of labor. Also

$$F_{KK} = \frac{\partial^2 F}{\partial K^2} < 0, \qquad F_{LL} = \frac{\partial^2 F}{\partial L^2} < 0$$

$$F_{KL} = \frac{\partial^2 F}{\partial K \partial L} = F_{LK} = \frac{\partial^2 F}{\partial L \partial K} > 0$$

First-order conditions require

$$PF_K - R = 0$$
$$PF_L - W = 0$$

The above equations imply that each factor should be paid the value of its marginal product. Alternatively, it means that at the point of maximum profit, real wage equals the marginal product of labor, and real rental of capital equals the marginal product of capital. Moreover, in order to have a maximum, we require the matrix

$$\begin{bmatrix} F_{KK} & F_{KL} \\ F_{LK} & F_{LL} \end{bmatrix}$$

to be negative definite, which means we should have

$$F_{KK}F_{LL} - F_{KL}^2 > 0$$

Example 12.14 In elementary econometrics, we encounter the problem of estimating the parameters of the linear equation

$$y_i = \alpha + \beta x_i + u_i$$

where y is the dependent variable, x, the explanatory variable, α and β unknown, but constant parameters of the model, and u the error term. Suppose we have n observations on y and x. How can we estimate our parameters? One way is the method of least squares. It is based on minimizing the sum of squared residuals. The i-th residual is defined as

$$e_i = y_i - \hat{\alpha} - \hat{\beta}x_i$$

Squaring and summing over all n observations,

$$\sum_{i=1}^{n} e_i^2 = \sum_{i=1}^{n} (y_i - \hat{\alpha} - \hat{\beta}x_i)^2$$

Note that now y's and x's are data and variables to be determined are $\hat{\alpha}$ and $\hat{\beta}$. Taking partial derivatives with respect to $\hat{\alpha}$ and $\hat{\beta}$, we get

$$\frac{\partial \sum e_i^2}{\partial \hat{\alpha}} = -2\sum_{i=1}^{n} (y_i - \hat{\alpha} - \hat{\beta}x_i)$$

and

$$\frac{\partial \sum e_i^2}{\partial \hat{\beta}} = -2\sum_{i=1}^{n} x_i(y_i - \hat{\alpha} - \hat{\beta}x_i)$$

Setting them equal to zero results in the *normal equations* of least squares

$$\sum_{i=1}^{n} y_i = n\hat{\alpha} + \hat{\beta}\sum_{i=1}^{n} x_i$$

$$\sum_{i=1}^{n} x_i y_i = \hat{\alpha}\sum_{i=1}^{n} x_i + \hat{\beta}\sum_{i=1}^{n} x_i^2$$

Solving the equations, we have

$$\hat{\beta} = \frac{\sum x_i(y_i - \bar{y})}{\sum x_i^2}$$

$$\hat{\alpha} = \bar{y} - \hat{\beta}\bar{x}$$

where $\bar{y} = \sum y_i/n$ and $\bar{x} = \sum x_i/n$ are sample means of y and x, respectively. We can make sure, that this is indeed a point of minimum by calculating the Hessian matrix:

$$\mathbf{H} = 2\begin{bmatrix} n & \sum x_i \\ \sum x_i & \sum x_i^2 \end{bmatrix}$$

This matrix is positive definite because its principal minors are positive

$$n > 0$$

and

$$n\sum x_i^2 - \left(\sum x_i\right)^2 = n\sum (x_i - \bar{x})^2 > 0$$

Example 12.15 We can generalize the results of Example 12.14 to the case of k variables. Let the regression model be

$$\mathbf{y} = \mathbf{X}\boldsymbol{\beta} + \mathbf{u}$$

Letting $\mathbf{e} = \mathbf{y} - \mathbf{X}\hat{\boldsymbol{\beta}}$, the sum of squared residuals would be

$$\mathbf{e}'\mathbf{e} = (\mathbf{y} - \mathbf{X}\hat{\boldsymbol{\beta}})'(\mathbf{y} - \mathbf{X}\hat{\boldsymbol{\beta}})$$

The first-order conditions for minimizing $\mathbf{e}'\mathbf{e}$ are

$$2\mathbf{X}'\mathbf{X}\hat{\boldsymbol{\beta}} - 2\mathbf{X}'\mathbf{y} = \mathbf{0}$$

which result in the least squares estimator of $\boldsymbol{\beta}$

$$\hat{\boldsymbol{\beta}} = (\mathbf{X}'\mathbf{X})^{-1}\mathbf{X}'\mathbf{y}$$

and the Hessian matrix is $2\mathbf{X}'\mathbf{X}$ which is positive definite.

12.2.1 Convex and Concave Functions

In our definitions we made a distinction between local and global optima. But our necessary and sufficient conditions referred simply to maxima and minima. In this section we shall talk about conditions under which a local maximum or minimum will be the global maximum or minimum. Whether a local optima is also a global optima depends on the shape of the objective function, and we shall look for classes of functions with the property that local and global optima coincide. The long and short of the story is that if we have a convex function (to be defined below), then the local minimum is also the global minimum. On the other hand if we have a concave function (again, to be defined below), then the local maximum is the global maximum.

Definition 12.3 The function $f : \mathbf{D} \rightarrow \Re$, where $\mathbf{D} \in \Re^n$ is convex if for any two points $\mathbf{x}_1,\ \mathbf{x}_2 \in \mathbf{D}$ and any real number $\lambda \in [0, 1]$

$$f(\lambda \mathbf{x}_1 + (1 - \lambda)\mathbf{x}_2) \leq \lambda f(\mathbf{x}_1) + (1 - \lambda)f(\mathbf{x}_2) \tag{12.18}$$

If in (12.18) the inequality is strict and $\lambda \in (0, 1)$, the function is strictly convex.

Definition 12.4 If f is convex, then -f is a concave function. Similarly, if a function is strictly convex, then -f is strictly concave.

These ideas for the case of $D \subset \Re$ are depicted in Figs. 12.4 and 12.5. Observe that a convex function carves out a convex set above the curve, whereas a concave function carves out a convex set below the curve.

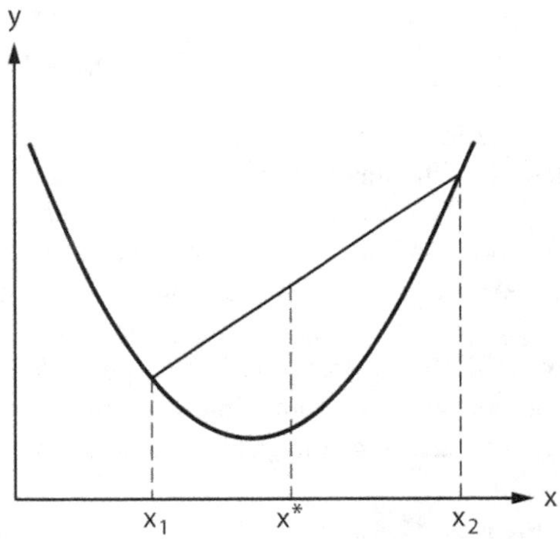

Fig. 12.4 Convex function

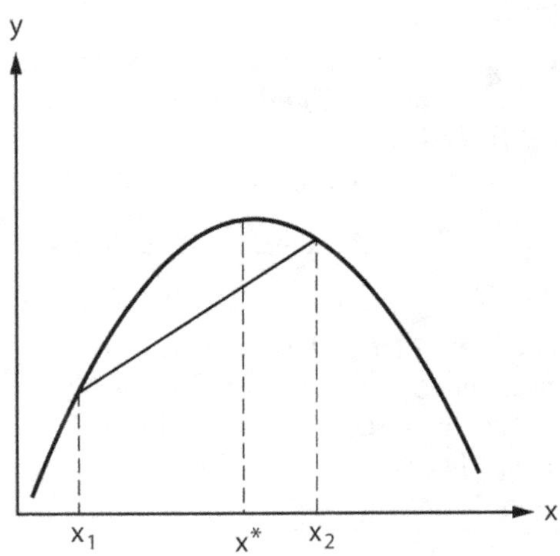

Fig. 12.5 Concave function

Example 12.16 The function $f(x) = x^2$ is convex. For two arbitrary points x_1 and x_2, we have

$$\begin{aligned}[\lambda x_1 + (1-\lambda)x_2]^2 - \lambda x_1^2 - (1-\lambda)x_2^2 \\ &= (\lambda^2 - \lambda)x_1^2 + [(1-\lambda)^2 - (1-\lambda)]x_2^2 + 2\lambda(1-\lambda)x_1x_2 \\ &= -\lambda(1-\lambda)(x_1 - x_2)^2 \leq 0\end{aligned}$$

Numerically, consider two points $x_1 = -1$ and $x_2 = 3$. Let $\lambda = 0.4$ and consider the point

$$\lambda x_1 + (1-\lambda)x_2 = -0.4 + 1.8 = 1.4$$

Then

$$1.4^2 < 0.4f(x_1) + 0.6f(x_2) = 5.8$$

Example 12.17 On the other hand, the function $f(x) = -x^2$ is concave. If we take the same points as in Example 12.15 and, as before, choose $\lambda = 0.4$, we have

$$-(1.4)^2 > 0.4f(x_1) + 0.6f(x_2) = -5.8$$

The importance of convex and concave functions stems from the fact that they facilitate the analysis of optimal points. But before presenting our results, and in order to be mathematically kosher, let us rule out the improper convex and concave functions. A convex (concave) function is said to be improper if it is either identical to ∞ or somewhere in its domain takes the values $-\infty$ or ∞. Ruling out such functions, we will be dealing with proper convex and concave functions, although we shall drop the designation proper and simply refer to these functions as convex and concave. The following theorems, stated without proof, contain the important results regarding convex and concave functions.

Theorem 12.5 *Let* $f : \mathbf{D} \rightarrow \Re$ *be a convex and differentiable function over its convex domain where* $\mathbf{D} \subset \Re^n$. *Then all its partial derivatives are continuous in* $\mathbf{D}$.

Theorem 12.6 *Let f be as in Theorem 12.5 with continuous second derivatives. f is convex (concave) if and only if its Hessian is positive (negative) semidefinite for all* $\mathbf{x} \in \mathbf{D}$. *That is,*

$$\begin{aligned} f \text{ is convex if } \mathbf{z}'\nabla f(\mathbf{x})\mathbf{z} \geq 0 \\ f \text{ is concave if } \mathbf{z}'\nabla f(\mathbf{x})\mathbf{z} \leq 0 \end{aligned} \quad \forall\, \mathbf{x} \in \mathbf{D}, \forall \mathbf{z} \in \Re^n \tag{12.19}$$

Theorem 12.7 *If f is a convex (concave) function, then every local minimum (maximum) of it is the global minimum (maximum).*

The importance of convexity and concavity, especially in theoretical work, should be clear by now. We can ensure the existence of a global minimum or maximum by requiring some functions to be convex or concave. For example, we require indifference curves to be convex and production frontier to be concave.

12.2.2 Quasi-convex and Quasi-concave Functions

Whereas the properties of convex and conave functions facilitate our analysis, sometimes they are too restrictive. For one, they do not allow for flat portions in the curve or surface of a function. In economic analysis, for technological reasons or institutional requirements, a function may be nonincreasing or nondecreasing over a segment of its domain. Similarly, in statistical and econometric estimation, one encounters likelihood functions that are flat for a range of parameter values. Second, to be convex or concave, a function has to be continuous. Again, there are instances when we may want to allow discontinuity in the function to be optimized or at least we may not be able to rule out discontinuity. To accommodate such eventualities, we define quasi-convex and quasi-concave functions. Figure 12.6 shows a quasi-concave function of one variable with a flat segment and a point of discontinuity.

Definition 12.5 The real-valued function $f : \mathbf{D} \to \Re$ where $\mathbf{D} \subset \Re^n$ is quasi-convex if

$$f(\lambda \mathbf{x}_1 + (1-\lambda)\mathbf{x}_2) \leq \max[f(\mathbf{x}_1), f(\mathbf{x}_2)], \qquad \mathbf{x}_1, \mathbf{x}_2 \in \mathbf{D} \tag{12.20}$$

In other words, if $f(\mathbf{x}_2) \geq f(\mathbf{x}_1)$, then $f(\mathbf{x}_2)$ is greater than or equal to f at any point that is a convex combination of $\mathbf{x}_1$ and $\mathbf{x}_2$. If we look at any cross section of this function (i.e., by holding some of its arguments constant), then the function will be monotone or unimodal. Thus, starting at any point, if f increases in a direction, then it will remain nondecreasing in that direction.

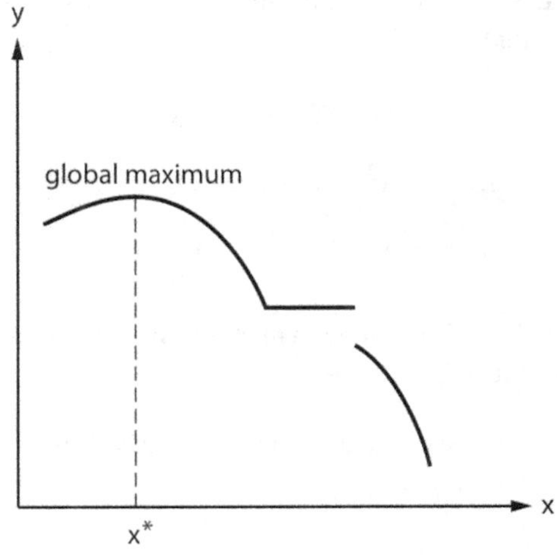

Fig. 12.6 Quasi-concave function

Definition 12.6 The real-valued function $f : \mathbf{D} \rightarrow \Re$ where $\mathbf{D} \subset \Re^n$ is quasi-concave if

$$f(\lambda \mathbf{x}_1 + (1-\lambda)\mathbf{x}_2) \geq \min[f(\mathbf{x}_1), f(\mathbf{x}_2)], \qquad \mathbf{x}_1, \mathbf{x}_2 \in \mathbf{D} \tag{12.21}$$

It follows that if f is quasi-convex, then $-f$ is quasi-concave. For quasi-concave functions the local maximum is not necessarily the global maximum, and the same is true for quasi-convex functions and the minimum. This is the price we pay for relaxing the assumption regarding the behavior of the objective function. But if we define our functions on a convex set, that is, if the domain of the function is a convex set, then a strict local maximum of a quasi-concave function will also be the global maximum. Again the same would be the case for a quasi-convex function and a strict local minimum. Alternatively, if the functions are strictly quasi-convex or strictly quasi-concave, that is, when we rule out a flat segment or surface for the function, we have the following theorems.

Theorem 12.8 *Let $f : D \rightarrow \Re$ where $\mathbf{D}$ is a convex set, be a strictly quasi-concave function, that is,*

$$f(\lambda \mathbf{x}_1 + (1-\lambda)\mathbf{x}_2) > \min[f(\mathbf{x}_1), f(\mathbf{x}_2)], \qquad \mathbf{x}_1, \mathbf{x}_2 \in \mathbf{D} \tag{12.22}$$

Then if $\mathbf{x}^* \in \mathbf{D}$ is a local maximum, it is also global maximum.

Indeed, we can note that a strictly quasi-concave function defined over a convex set will have only one maximum. The above results, mutatis mutandis, *apply to quasi-convex functions.*

12.2.3 Exercises

E.12.4 Find the extremum points of the following functions and determine their nature.

i.	$z = (x-2)^2 + (y-5)^2 - 3$	*ii.*	$z = x^2 - 3xy + y^2 + 5x - 2y + 2$
iii.	$z = x^3 + y^3 - 3xy$	*iv.*	$z = x^2 + xy + y^2 + \dfrac{2}{x} + \dfrac{2}{y}$
v.	$z = x^2 y^3 (1 - x - y)$	*vi.*	$z = x \exp(-x^2 - y^2)$

E.12.5 Determine which of the functions in E.12.4 are convex and which are concave.

E.12.6 Show that a convex (concave) function is also quasi-convex (quasi-concave) function.

12.3 Numerical Optimization

In applied work we frequently need to find numerical solutions to optimization problems. Even in theoretical work, an analytical solution may be difficult or impossible to find, and one has to resort to approximation. More importantly, a number of estimation techniques, such as least squares and maximum likelihood, require numerical optimization. Whereas the methods we discussed in previous sections are powerful in finding optimal solutions, they are not appropriate for numerical optimization. For one, they require the computer to calculate partial derivatives and then solve a system of possibly nonlinear equations. In contrast numerical methods start from an initial guess and move toward the optimal solution. Suppose we are interested in minimizing[4] the function $f(\mathbf{x})$. Let the initial guess be $\mathbf{x}_0$. Then the procedure would be to move toward the optimal point $\mathbf{x}^*$ by revising our initial guess in accordance with

$$\mathbf{x}_k = \mathbf{x}_{k-1} + \lambda_k \mathbf{d}_k \tag{12.23}$$

where λ_k is a scalar and $\mathbf{d}_k$ is a vector that determines the direction of our movement. Below we will first discuss a method for finding $\mathbf{d}_k$ and then a technique for finding a reasonably good λ_k. We continue the discussion with other methods of finding $\mathbf{d}_k$.

12.3.1 Steepest Descent

The idea here is to take the step that results in the fastest movement toward the minimum. Such a direction is provided by the gradient of the function. Thus, we set

$$\mathbf{d}_k = -\nabla f(\mathbf{x}_k) \tag{12.24}$$

which implies

$$\mathbf{x}_k = \mathbf{x}_{k-1} - \lambda_k \nabla f(\mathbf{x}_k) \tag{12.25}$$

To see the rationale behind this choice, consider the function $f(x) = x^2$ that attains its minimum at $x_0 = 0$. Let the starting point be $x_0 = -5$. It is clear that a move toward the minimum requires an increase in x. The derivative of the function at $x_0 = -5$ is $2x_0 = -10$ and therefore the direction of steepest descent is positive. On the other hand, at the point of $x_0 = 5$, the direction of steepest descent will be negative. As another example consider the function $f(x, y) = x^2 + y^2$ that attains its minimum at $x = y = 0$. The direction of steepest descent is

$$\begin{bmatrix} -2x \\ -2y \end{bmatrix}$$

---

[4]Procedures for maximizing a function are similar to minimization discussed in this section. Note that minimizing $-f(\mathbf{x})$ results in maximizing $f(\mathbf{x})$.

Thus, at any point $x_k,\ y_k$, the direction of the steepest descent requires the variable that has a positive (negative) value to decrease (increase).

The algorithm for the steepest descent is:

1. Specify $\mathbf{x}_0$ and the convergence criterion Δ.
2. Compute $\mathbf{d}_k = -\nabla f(\mathbf{x}_k)$.
3. If $\|\mathbf{d}_k\| \leq \Delta$ stop.
4. Find λ_k that minimizes $f(\mathbf{x}_k + \lambda_k \mathbf{d}_k)$ (see 12.3.2 below).
5. Compute $\mathbf{x}_k = \mathbf{x}_{k-1} + \lambda_k \mathbf{d}_k$ and go back to step 2.

Other optimization methods differ from the steepest descent in the choice of $\mathbf{d}_k$. But before discussing other methods, let us find out how we can choose λ_k.

12.3.2 Golden Section Method

The idea is to find λ_k that minimizes $f(\mathbf{x}_k + \lambda_k \mathbf{d}_k)$. But, as the problem is auxiliary to our main mission, we want to accomplish this cheaply, that is, with minimum effort. First, note that this is a unidimensional optimization problem, and because we are looking at a small segment of the function, it is reasonable to assume that there is only one minimum. Thus, we can define our problem as finding the minimum of a function of one variable $g(z)$ where $z \in [z_1, z_2]$ (see Fig. 12.6). Let $b = 0.618034$ which is the reciprocal of the golden ratio $(1 + \sqrt{5})/2$. The following is the golden section algorithm.

1. Let
$$z_3 = b\, z_1 + (1 - b)\, z_2$$
$$z_4 = b z_2 + (1 - b)\, z_1$$
2. If $f(z_3) > f(z_4)$, then the minimum is in $[z_3, z_2]$. Replace z_1 with z_3 and go to step 1; else if $f(z_4) > f(z_3)$, then the minimum is in $[z_1, z_4]$. Replace z_2 with z_4 and go to step 1 (see Fig. 12.7).

12.3.3 Newton Method

In this method

$$\mathbf{d}_k = -[\nabla^2 f(\mathbf{x})]^{-1} \nabla f(\mathbf{x}) \tag{12.26}$$

The rationale behind this choice is the Taylor formula. Recall that we can write the Taylor expansion of $\nabla f(\mathbf{x})$ as

$$\nabla f(\mathbf{x}) \approx \nabla f(\mathbf{x}_0) + (\mathbf{x} - \mathbf{x}_0)\nabla^2 f(\mathbf{x}_0) \tag{12.27}$$

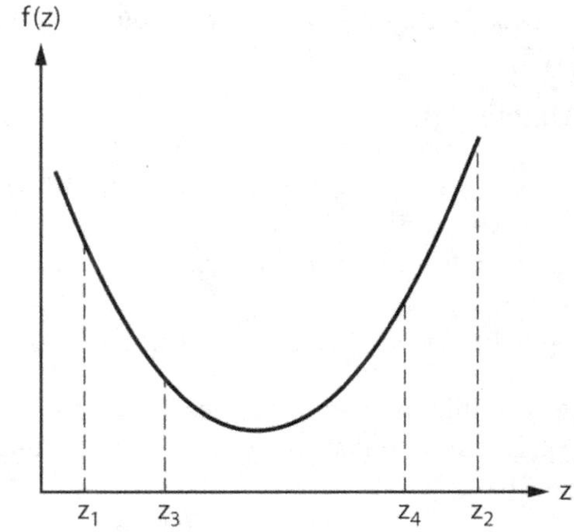

Fig. 12.7 The golden section method

At the point of optimum, $\nabla f(\mathbf{x}) = \mathbf{0}$, therefore, a reasonable iterative method would be

$$\nabla f(\mathbf{x}) \approx \nabla f(\mathbf{x}_0) + (\mathbf{x} - \mathbf{x}_0)\nabla^2 f(\mathbf{x}_0) \tag{12.28}$$

The problem with this algorithm is the calculation of the inverse of the Hessian matrix. Algorithms known as quasi-Newton differ in their suggestion as how to approximate $[\nabla^2 f(\mathbf{x}_0)]^{-1}$. Here we mention the Davidson-Fletcher-Powell algorithm.

1. Specify x_0, the convergence criterion Δ, and $\mathbf{H}_1 = \mathbf{I}$, where $\mathbf{I}$ is the identity matrix.
2. Compute $\mathbf{d}_k = -\mathbf{H}_k \nabla f(\mathbf{x}_k)$.
3. Find λ_k that minimizes $f(\mathbf{x}_k + \lambda_k \mathbf{d}_k)$.
4. Compute $\mathbf{x}_{k+1} = \mathbf{x}_k + \lambda_k \mathbf{d}_k$.
5. If $\|\nabla f(\mathbf{x}_{k+1})\| \leq \Delta$ stop.
6. Compute $\mathbf{q}_k = \nabla f(\mathbf{x}_{k+1}) - \nabla f(\mathbf{x}_k)$, $\mathbf{p}_{k+1} = -\mathbf{H}_{k+1}\nabla f(\mathbf{x}_{k+1})$, and

$$\mathbf{H}_{k+1} = \mathbf{H}_k + \lambda_k \frac{\mathbf{p}'_k \mathbf{p}_k}{\mathbf{p}_k \mathbf{q}_k} - \frac{\mathbf{H}_k \mathbf{q}_k \mathbf{q}'_k \mathbf{H}_k}{\mathbf{q}'_k \mathbf{H}_k \mathbf{q}_k}$$

Go back to step 2.

12.3.4 Matlab Functions

Matlab has two procedures for finding the minimum of a function. `fminbnd` finds the minimum of a function of one variable. Suppose you have defined the following function in an `M.file` called `clara`:

$$f(x) = 2x^3 - 7x + 13$$

Moreover, suppose that we would like to conduct the search over the interval $[1,5]$.

Matlab code

```
% To find x at which the function attains its minimum
x = fminbnd(@clara,1,5)
% To find both x and the minimum value of the
% function
[x, fval] = fminbnd(@clara,1,5)
```

The minimum of functions of several variables is found using `fminsearch`. Again, suppose that we have defined the function of interest

$$f(\mathbf{x}) = 7(x_1 - x_2)^2 + 6x_3^2 + 3(1 - x_2)^2$$

in an `M.file` called `nigel`. The function `fminsearch` requires a starting point. Let $[x_1^0, x_2^0, x_3^0] = [-1, 2, 0.5]$.

Matlab code

```
% To find x at which the function attains its minimum
x = fminsearch(@nigel, [-1, 2, 0.5])
% To find both x and the minimum value of the
% function
[x, fval] = fminsearch(@nigel, [-1, 2, 0.5])
```

12.3.5 Exercises

E.12.7 Implement the golden section method on Matlab.

E.12.8 Use the Matlab function `fminbnd` to solve problems in E.12.1.

E.12.9 Use the `fminsearch` to solve problems in E.12.4.

Chapter 13
Constrained Optimization

In a scene in *The Godfather*, the late Marlon Brando, playing Don Vito Corleone, tells his son, "Well, this wasn't enough time, Michael. It wasn't enough time." There is never enough time, nor is there ever enough money. In personal life, in the affairs of a company or university, and in the government budget, there are never enough resources, be it time, money, or energy. Economics has always been concerned with optimal allocation of scarce resources to competing and unbounded wants. We can imagine that if resources were infinite or wants were limited, there would be no economic problems. Thus, in deciding the family consumption, the hiring for a university, the number and types of courses to offer in a discipline, the amount of R&D expenditures in the company, the allocation of the federal budget among national defense, education, and welfare programs, we face the same problem of allocating scarce resources to achieve the best result possible. But if the behavior of economic decision makers is determined by choosing the best allocation subject to budget constraints, then the starting point of economic theory of households and firms ought to be constrained optimization.

13.1 Optimization with Equality Constraints

Constraints come in two different forms of equality and inequality. We have a theory that can handle a combination of both types of constraints. Whereas starting with a unified general theorem is mathematically appealing, we will lose the opportunity to gain an intuitive understanding of the issues involved. Thus, in accordance with our policy to have a kinder, gentler math book, we start with equality constraints, then discuss inequality constraints, and finally present the Karush-Kuhn-Tucker (KKT) theorem that handles both types of constraints. The constrained optimization problem is formally defined as

$$\begin{aligned} &\max f(\mathbf{x}), \qquad \mathbf{x} \in \Re^n \\ &\quad \text{subject to} \quad g_i(\mathbf{x}) = 0, \quad i = 1, \ldots, m \end{aligned} \tag{13.1}$$

K. Dadkhah, *Foundations of Mathematical and Computational Economics,* 2nd ed.,
DOI 10.1007/978-3-642-13748-8_13,

or

$$\begin{aligned} &\min f(\mathbf{x}), \qquad \mathbf{x} \in \Re^n \\ &\text{subject to} \quad g_i(\mathbf{x}) = 0, \quad i = 1, \ldots, m \end{aligned} \tag{13.2}$$

where $f : \mathbf{D} \to \Re$ is a twice continuously differentiable function, and $\mathbf{D} \subset \Re^n$.

Example 13.1

$$\begin{aligned} \max\ z &= xy \\ \text{s.t.}\ x + y &= 12 \end{aligned}$$

An old-fashioned way of solving this problem is to start with the constraint and write y in terms of x. By substituting the result in the objective function, we will have a function of one variable to maximize. Because

$$y = 12 - x$$

we can change our problem to

$$\max z = x(12 - x) = -x^2 + 12x$$

Taking the first derivative and setting it equal to zero

$$\frac{dz}{dx} = -2x + 12 = 0$$

we get

$$x = 6, \quad y = 6, \quad z = 36$$

This is a maximum because the second derivative is negative.

$$\frac{d^2z}{dx^2} = -2$$

The above solution runs into two problems. First, it is not always easy to solve for y in terms of x or, in general, to solve for as many variables as there are constraints and substitute them in the objective function. Second, by eliminating the constraint, we can only indirectly study its effect on the optimal solution. In other words, the value of relaxing the constraint just a bit is not part of the solution to the problem. This is an important issue, especially in economics, because a constraint means scarce resources and, as will be seen later, Lagrange multipliers are marginal valuations or shadow prices of these resources. Hence, the widespread use of the Lagrange method in economics.

We will discuss the Lagrange method twice. First, when there is only one constraint, which is by far the most recurring problem in economics. Then we will describe the general method with m constraints. The essence of the Lagrange method is the observation that we can replace the objective function in problems in (13.1) and (13.2), when $m = 1$ with

$$L(\mathbf{x}, \lambda) = f(\mathbf{x}) + \lambda g(\mathbf{x}) \tag{13.3}$$

L s called the Lagrangian and λ is called the Lagrange multiplier. For the new problem we have the following theorem that we state without proof. Whereas proof of the theorem is not difficult, it hardly adds anything for those who are interested in its applications. Instead, we will provide an intuitive justification for the theorem.

Theorem 13.1 *Let $f : \mathbf{D} \to \Re$ and $g(x)$ be twice continuously differentiable function. Then the necessary conditions for f to be optimum subject to $g(\mathbf{x}) = 0$ are*

$$\begin{aligned} \nabla L(\mathbf{x}^*, \lambda^*) &= \mathbf{0} \\ g(\mathbf{x}^*) &= 0 \end{aligned} \tag{13.4}$$

What Theorem 13.1 says is that by taking the first-order derivatives of L with respect to x's and λ and setting them equal to zero, we will reach an optimum. Thus, to find the optimum point, we need to solve the following system of equations:

$$\begin{aligned} \frac{\partial f(\mathbf{x}^*, \lambda^*)}{\partial x_1} + \lambda^* \frac{\partial g(\mathbf{x}^*, \lambda^*)}{\partial x_1} &= 0 \\ &\vdots \\ \frac{\partial f(\mathbf{x}^*, \lambda^*)}{\partial x_n} + \lambda^* \frac{\partial g(\mathbf{x}^*, \lambda^*)}{\partial x_n} &= 0 \\ g(\mathbf{x}^*) &= 0 \end{aligned} \tag{13.5}$$

Example 13.2 Let us solve the problem in Example 13.1 using the Lagrange method. We have

$$L = xy + \lambda(x + y - 12)$$

Taking derivatives and setting them equal to zero result in the following equations:

$$\begin{aligned} y + \lambda &= 0 \\ x + \lambda &= 0 \\ x + y - 12 &= 0 \end{aligned}$$

Solving the equations, the optimal point is reached at

$$x = y = 6, \qquad xy = 36$$

and

$$\lambda = -6$$

The equations in (13.5) define the necessary condition for L to reach its optimum, that is, its total differential to be equal to zero.

$$dL = \sum_{i=1}^{n} \frac{\partial f(\mathbf{x}^*)}{\partial x_i} dx_i + \lambda^* \sum_{i=1}^{n} \frac{\partial g(\mathbf{x}^*)}{\partial x_i} dx_i + g(\mathbf{x}^*) d\lambda = 0 \tag{13.6}$$

To see why the above equations result in the constrained optimum of f, note that such an optimum requires, as a necessary condition, $df = 0$ when evaluated at the point $\mathbf{x}^*$ subject to $g(\mathbf{x}^*) = 0$. The fulfillment of the constraint makes the last term in (13.6) equal to zero. Furthermore, because there is only one constraint when $\mathbf{x}$ has n elements, the constraint defines a large set of values that are consistent with it. As a matter of fact, because of continuity there are an infinite number of points that satisfy the constraint. Let us denote this set by $\mathbf{D}_g \subset \mathbf{D}$. Moving within this set requires

$$dg(\mathbf{x})|_{\mathbf{x}\in\mathbf{D}_g} = \left. \sum_{i=1}^{n} \frac{\partial g(\mathbf{x}^*)}{\partial x_i} dx_i \right|_{\mathbf{x}\in\mathbf{D}_g} = 0 \tag{13.7}$$

This leaves us with

$$\sum_{i=1}^{n} \frac{\partial f(\mathbf{x}^*)}{\partial x_i} dx_i = 0$$

which is the optimal condition for $f(\mathbf{x})$, but now the optimum is found not in $\mathbf{D}$ but in $\mathbf{D_g}$.

Example 13.3 The most important example of constrained maximization occurs in consumer theory. An individual or household maximizes its utility $U(x_1, x_2)$ subject to the budget constraint $y = p_1x_1 + p_2x_2$, where p_1 and p_2 are prices of the first and second commodities, respectively, and y is the household's income. We are assuming two consumer goods, but extension to n goods is straightforward. The Lagrangian is

$$L = U(x_1, x_2) + \lambda(p_1x_1 + p_2x_2 - y) \tag{13.8}$$

The first-order conditions are

$$\begin{aligned} U_1 + \lambda p_1 &= 0 \\ U_2 + \lambda p_2 &= 0 \\ p_1x_1 + p_2x_2 &= y \end{aligned} \tag{13.9}$$

Solving the first two equations, we get

$$\frac{U_1}{U_2} = \frac{p_1}{p_2} \tag{13.10}$$

In other words, the ratio of marginal utilities should be the same as the ratio of respective prices. Because the slope of the budget line is $-p_1/p_2$ and the slope of

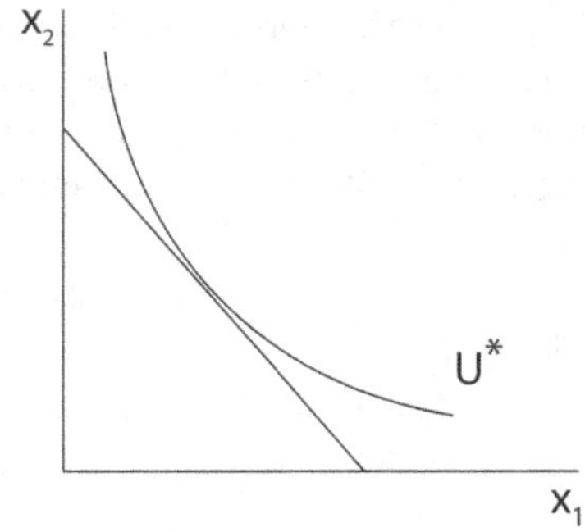

Fig. 13.1 Tangency of indifference curve and the budget constraint

the indifference curves $-U_1/U_2$, (13.10) states that the maximum utility subject to the budget constraint is reached when the two curves have the same slope, meaning that they are tangents (Fig. 13.1). The last equation in (13.9) requires the total expenditures to be equal to the available resources, y.

Example 13.4 Let the technology of a firm be represented by

$$Q = AK^{\alpha}L^{1-\alpha} \tag{13.11}$$

where, in familiar notation, Q, K, and L are, respectively, output, capital, and labor. Let us consider the problem of minimizing cost subject to producing a given level of output $\bar{Q}$. The problem is

$$\begin{aligned} \min \quad & rK + wL \\ \text{s.t.} \quad & AK^{\alpha}L^{1-\alpha} - \bar{Q} = 0 \end{aligned} \tag{13.12}$$

where r is the rental cost of capital and w the wage rate. The Lagrangian is

$$L = (rK + wL) + \lambda(AK^{\alpha}L^{1-\alpha} - \bar{Q}) \tag{13.13}$$

The optimality conditions are

$$\begin{aligned} & r + \lambda\alpha AK^{\alpha-1}L^{1-\alpha} = 0 \\ & w + \lambda(1-\alpha)AK^{\alpha}L^{-\alpha} = 0 \\ & AK^{\alpha}L^{1-\alpha} = \bar{Q} \end{aligned}$$

Solving these equations, we get

$$K = \frac{\bar{Q}}{A}\left(\frac{\alpha}{1-\alpha}\frac{w}{r}\right)^{1-\alpha}, \qquad L = \frac{\bar{Q}}{A}\left(\frac{1-\alpha}{\alpha}\frac{r}{w}\right)^{\alpha}$$

The case of more than one constraint does not differ substantively from the above, except that we have to incorporate the additional constraints. The sufficient conditions, however, are a bit tricky, and the reader is advised to pay particular attention. Again we will form the Lagrangian

$$L(\mathbf{x}, \lambda) = f(\mathbf{x}) + \sum_{i=1}^{m} \lambda_i g(\mathbf{x}) \tag{13.14}$$

Let

$$\boldsymbol{\lambda} = [\lambda_1 \ \ldots \ \lambda_m]'$$

The necessary condition for optimum requires

$$\nabla L(\mathbf{x}^*, \ \boldsymbol{\lambda}^*) = \mathbf{0} \tag{13.15}$$

which implies

$$\begin{aligned} \frac{\partial f(\mathbf{x}^*, \boldsymbol{\lambda}^*)}{\partial x_1} + \sum_{i=1}^{m} \lambda_i^* \frac{\partial g_i(\mathbf{x}^*, \boldsymbol{\lambda}^*)}{\partial x_1} &= 0 \\ &\vdots \\ \frac{\partial f(\mathbf{x}^*, \boldsymbol{\lambda}^*)}{\partial x_n} + \sum_{i=1}^{m} \lambda_i^* \frac{\partial g_i(\mathbf{x}^*, \boldsymbol{\lambda}^*)}{\partial x_n} &= 0 \\ g_1(\mathbf{x}^*) &= 0 \\ &\vdots \\ g_m(\mathbf{x}^*) &= 0 \end{aligned} \tag{13.16}$$

Example 13.5 Frequently in econometrics we face the problem of estimating the parameters of a model subject to constraints either because such constraints are inherent in the theory or because we would like to test such restrictions. As an example of the former, suppose we want to estimate a Cobb-Douglas production function with constant returns to scale. That is,

$$Q = AK^{\alpha}L^{\beta}$$

subject to the constraint

$$\alpha + \beta = 1$$

Of course we can incorporate the restriction by estimating

$$\ln Q = \ln A + \alpha \ln K + (1 - \alpha) \ln L$$

The same is true if we would like to test zero restrictions on the effects of specific variables. We can estimate the model once with the variable in the equation and once excluding it. Then a test could be conducted using, say, the likelihood ratio. But there are instances when such simple tricks don't work and we need constrained least squares. The problem is to estimate the parameters of the model

$$\mathbf{y} = \mathbf{X}\boldsymbol{\beta} + \mathbf{u} \tag{13.17}$$

subject to restrictions

$$\mathbf{R}\boldsymbol{\beta} = \mathbf{c} \tag{13.18}$$

For example, if $\boldsymbol{\beta}$ has three components and we want $\beta_1 = \beta_3$ and $\beta_2 + \beta_3 = 1$, the restrictions would be

$$\begin{bmatrix} 1 & 0 & -1 \\ 0 & 1 & 1 \end{bmatrix} \begin{bmatrix} \beta_1 \\ \beta_2 \\ \beta_3 \end{bmatrix} = \begin{bmatrix} 0 \\ 1 \end{bmatrix}$$

To solve the problem, we form the Lagrangian

$$L = (\mathbf{y} - \mathbf{X}\hat{\boldsymbol{\beta}}_r)'(\mathbf{y} - \mathbf{X}\hat{\boldsymbol{\beta}}_r) + \boldsymbol{\lambda}(\mathbf{R}\hat{\boldsymbol{\beta}}_r - \mathbf{c}) \tag{13.19}$$

First-order conditions for a minimum are

$$\begin{aligned} -2\mathbf{X}'\mathbf{y} + 2\mathbf{X}'\mathbf{X}\hat{\boldsymbol{\beta}}_r + \mathbf{R}'\boldsymbol{\lambda} &= 0 \\ -\mathbf{c} + \mathbf{R}\hat{\boldsymbol{\beta}}_r &= 0 \end{aligned} \tag{13.20}$$

Premultiplying both sides of the first equation in (13.17) by $\mathbf{R}(\mathbf{X}'\mathbf{X})^{-1}$, we can solve for $\boldsymbol{\lambda}$:

$$\boldsymbol{\lambda} = -2[\mathbf{R}(\mathbf{X}'\mathbf{X})^{-1}\mathbf{R}']^{-1}[\mathbf{c} - \mathbf{R}(\mathbf{X}'\mathbf{X})^{-1}\mathbf{y}] \tag{13.21}$$

Substituting $\boldsymbol{\lambda}$ from (13.21) back into the first equation of (13.20) and premultiplying both sides by $(\mathbf{X}'\mathbf{X})^{-1}$, we have the restricted least squares estimator

$$\begin{aligned} \hat{\boldsymbol{\beta}}_{\mathbf{r}} &= (\mathbf{X}'\mathbf{X})^{-1}\mathbf{X}'\mathbf{y} + (\mathbf{X}'\mathbf{X})^{-1}\mathbf{R}'[\mathbf{R}(\mathbf{X}'\mathbf{X})^{-1}\mathbf{R}']^{-1}[\mathbf{c} - \mathbf{R}(\mathbf{X}'\mathbf{X})^{-1}\mathbf{y}] \\ &= \hat{\boldsymbol{\beta}} + (\mathbf{X}'\mathbf{X})^{-1}\mathbf{R}'[\mathbf{R}(\mathbf{X}'\mathbf{X})^{-1}\mathbf{R}']^{-1}[\mathbf{c} - \mathbf{R}\hat{\boldsymbol{\beta}}] \end{aligned} \tag{13.22}$$

The interesting point to observe is that the truer the restrictions imposed, that is, the more support the restrictions get from the data, the closer will be the restricted least squares to ordinary least squares estimator and, therefore, the more difficult to reject the restrictions.

13.1.1 The Nature of Constrained Optima and the Significance of λ

We stated that maximizing $f(\mathbf{x})$ subject to a constraint is equivalent to maximizing the Lagrangian. Let us do an experiment: Take any of the functions with numerical coefficients in this section, form their Lagrangian, and try to maximize them using a computer routine for maximization, for example, `fminbnd` or `fminsearch`. You will get a strange result. The computer will set $\lambda = 0$ and you will get an error message because the maximum is not finite. The same will happen for minimization. This anomalous result is obtained because the routine will try to maximize the objective function with respect to all of its arguments including λ. But that is not what the Lagrange method is about. In fact, the Lagrangian is maximized with respect to $\mathbf{x}$ and minimized with respect to λ. Therefore, we have a saddle point solution. This is best understood by saying

$$L(\mathbf{x}, \lambda^*) \leq L(\mathbf{x}^*, \lambda^*) \leq L(\mathbf{x}^*, \lambda) \tag{13.23}$$

But, what is this λ? Its meaning is best understood in the context of Example 13.3. Referring to (13.8) we have

$$\frac{\partial L}{\partial y} = -\lambda \tag{13.24}$$

For any infinitesimal relaxation of the constraint the objective function is increased by that amount times $-\lambda$. Loosely speaking, a one unit increase in the amount of resources, y, has a reward of $-\lambda$. Thus, $-\lambda$ is the opportunity cost or price of the resource, which was scarce and formed a constraint. In optimizing our resource allocation, we were minimizing the cost of resources we used. Thus, we solved two problems as if one was the twin or dual of the other. Indeed, cost minimization and optimal allocation of resources are dual problems and the solution of one is tantamount to the solution of the other.

13.1.2 Exercises

E.13.1 Solve the following constrained optimization problems. In all these problems we have the additional constraints $x \geq 0$ and $y \geq 0$.

i.	$\min z = (x-2)^2 + 2(y-5)^2 - 7$	s.t.	$x + y = 12$
ii.	$\max z = (x-2)^2 + 2(y-5)^2 - 7$	s.t.	$x + y = 12$
iii.	$\min z = x^2 - 3xy + y^2 + 5x - 2y + 2$	s.t.	$x + y = 44$
iv.	$\min z = x^3 + y^3 - 3xy$	s.t.	$x + y = 3$
v.	$\max z = \ln(x) + \ln(y)$	s.t.	$x + y = 15$
vi.	$\max z = 2\ln(x) + \ln(y)$	s.t.	$x + y = 26$

E.13.2 Solve the following constrained optimization problems. In these problems we have the additional constraints $x_1,\ x_2,\ x_3 \geq 0$.

i.
$$\begin{aligned} \max\quad & \ln(x_1) + 0.9\ln(x_2) + 0.81\ln(x_3) \\ \text{s.t.}\quad & x_1 + x_2 + x_3 = 125 \end{aligned}$$

ii.
$$\begin{aligned} \max\quad & x_1 + x_2 + x_3 \\ \text{s.t.}\quad & x_1 + 2x_2 + 5x_3 = 150 \end{aligned}$$

E.13.3 Solve the minimum cost function when production technology does not exhibit constant returns to scale.

$$\begin{aligned} \min\quad & rK + wL \\ \text{s.t.}\quad & AK^{\alpha}L^{\beta} - \bar{Q} = 0 \end{aligned}$$

13.2 Value Function

The assumption of optimizing behavior on the part of economic players is intended to lead to testable hypotheses and an explanation of economic behavior. In particular, we may be interested in the effects of a change in relative prices or the budget constraint on the behavior of the consumer or the effects of a change in input or product prices on the behavior of the firm. The optimization by the consumer or firm leads to a function where any configuration of the parameters of the problem uniquely determine the behavior of the economic player. Thus, we can investigate the effects of a change in any parameter on the variables of interest. The emphasis is on the uniqueness of connections between parameters such as prices, which are outside the control of the firm or consumer on the one hand, and quantities such as the amount of output or consumption of a particular good that are under the control of the decision makers. Without a unique relationship between the two sets of variables, no conclusion can be reached regarding the effects of the former on the latter.

The function that uniquely connects parameters and decision variables in an optimization problem is called the *value function.* We shall explore this function. But the interesting point is that the optimization problem itself has a twin called the *dual problem.* The dual problem also leads to a value function of its own. In some problems, using the value function from the dual problem leads to a function whose elements are observable while the same is not true for the value function of the *primal problem.*

Consider the optimization problem

$$\begin{aligned} \max_{\mathbf{x}}\quad & f(\mathbf{x}, \boldsymbol{\theta}) \\ \text{s.t.}\quad & g(\mathbf{x}, \boldsymbol{\theta}) = 0 \\ & \mathbf{x} \geq \mathbf{0} \end{aligned} \tag{13.25}$$

where $\boldsymbol{\theta} = (\theta_1, \ldots, \theta_k)'$ are parameters of the model. The solution of the problem in (13.25) will depend on the parameters and can be written as $\mathbf{x} = \mathbf{x}(\boldsymbol{\theta})$. The value function is defined as

$$V(\boldsymbol{\theta}) \equiv \max_{\mathbf{x}}[f(\mathbf{x}, \boldsymbol{\theta}) \mid g(\mathbf{x}, \boldsymbol{\theta}) = 0, \mathbf{x} \geq \mathbf{0}] \tag{13.26}$$

Thus,

$$V(\boldsymbol{\theta}) \equiv f(\mathbf{x}(\boldsymbol{\theta}), \boldsymbol{\theta}) \tag{13.27}$$

Example 13.6 The consumer problem is

$$\begin{aligned} \max_{\mathbf{x}} \quad & U(\mathbf{x}) \\ \text{s.t.} \quad & \mathbf{p}'\mathbf{x} = y \\ & \mathbf{x} \geq \mathbf{0} \end{aligned}$$

where $\mathbf{x} = (x_1, \ldots, x_n)'$ is the vector of the quantities of each good or service consumed, $\mathbf{p} = (p_1, \ldots, p_n)'$ is the vector of prices of the same goods and services, and y is the consumer's income. The solution of this problem will be of the form $\mathbf{x} = \mathbf{x}(\mathbf{p}, y)$, which is the system of the *Marshallian demand functions*. For example,

$$x_j = x_j(\mathbf{p}, y) \tag{13.28}$$

is the Marshallian demand function for the j-th commodity. Substituting the solutions back into the objective function, $V(\mathbf{p}, y) = U(\mathbf{x}(\mathbf{p}, y))$, we get the *indirect utility function*.

Example 13.7 Consider the utility function

$$U(x_1, x_2) = x_1 x_2$$

and the budget constraint

$$p_1 x_1 + p_2 x_2 = y$$

The consumer problem is

$$\begin{aligned} & \max \quad x_1 x_2 \\ & \text{s.t.} \quad p_1 x_1 + p_2 x_2 = y \\ & x_1, x_2 \geq 0 \end{aligned}$$

The problem can be reformulated as

$$\max \quad \mathcal{L} = x_1 x_2 + \lambda(p_1 x_1 + p_2 x_2 - y)$$

The optimality conditions are

$$x_2 + \lambda p_1 = 0$$
$$x_1 + \lambda p_2 = 0$$
$$p_1x_1 + p_2x_2 = y$$

solving the equations we get

$$x_1 = \frac{1}{2p_1}y$$
$$x_2 = \frac{1}{2p_2}y$$
$$\lambda = -\frac{1}{2p_1p_2}y$$

The first two equations are Marshallian demand functions for the first and second commodities. Substituting these results in the utility function, gives us the indirect utility function.

$$V(p_1, p_2, y) = \frac{1}{4p_1p_2}y^2$$

Example 13.8 Let us assume that the utility function of the consumer has constant elasticity of substitution (CES), which for the case of two goods is

$$U(x_1, x_2) = (x_1^\rho + x_2^\rho)^{1/\rho}$$

Then the utility maximization problem will be

$$\begin{aligned} \max\ & \left(x_1^\rho + x_2^\rho\right)^{1/\rho} \\ \text{s.t.}\ \ & p_1x_1 + p_2x_2 = y \\ & x_1, x_2 \geq 0 \end{aligned}$$

or

$$\max\ \ \mathcal{L} = \left(x_1^\rho + x_2^\rho\right)^{1/\rho} + \lambda(p_1x_1 + p_2x_2 - y)$$

The first order conditions are

$$x_1^{\rho-1}\left(x_1^\rho + x_2^\rho\right)^{(1/\rho)-1} + \lambda p_1 = 0$$
$$x_2^{\rho-1}\left(x_1^\rho + x_2^\rho\right)^{(1/\rho)-1} + \lambda p_2 = 0$$
$$p_1x_1 + p_2x_2 = y$$

Solving these equations we get

$$\begin{aligned}
x_1 &= \frac{p_1^{1/(1-\rho)} y}{p_1^{\rho/(\rho-1)} + p_2^{\rho/(\rho-1)}} \\
&= \frac{p_1^{\beta-1} y}{p_1^{\beta} + p_2^{\beta}} \\
x_2 &= \frac{p_2^{1/(1-\rho)} y}{p_1^{\rho/(\rho-1)} + p_2^{\rho/(\rho-1)}} \\
&= \frac{p_2^{\beta-1} y}{p_1^{\beta} + p_2^{\beta}}
\end{aligned}$$

where in obvious notation $\beta = \rho/(\rho - 1)$. Now we can find the indirect utility function by plugging the optimal values of x_1 and x_2 in the utility function. Thus,

$$V(p_1, p_2, y) = \left[\left(\frac{p_1^{\beta-1} y}{p_1^{\beta} + p_2^{\beta}} \right)^{\rho} + \left(\frac{p_2^{\beta-1} y}{p_1^{\beta} + p_2^{\beta}} \right)^{\rho} \right]^{1/\rho}$$

An important proposition regarding the value function is the *Envelop Theorem*, which is quite useful in comparative static analysis.

Theorem 13.2 *Let $\mathcal{L}$ be the Lagrangian, $\mathbf{x}(\boldsymbol{\theta})$ and $\boldsymbol{\lambda}(\boldsymbol{\theta})$ the solutions, and $V(\boldsymbol{\theta})$ the value function in (13.26). Further assume that $\mathbf{x} >> \mathbf{0}$, which means $\mathbf{x}$ is much greater than zero.*[1] *Then*

$$\frac{\partial V(\boldsymbol{\theta})}{\partial \theta_j} = \left. \frac{\partial \mathcal{L}}{\partial \theta_j} \right|_{\mathbf{x}(\boldsymbol{\theta}),\, \boldsymbol{\lambda}(\boldsymbol{\theta})} \tag{13.29}$$

We will not prove this theorem but illustrate it with Example 13.7. Taking the derivative of the value function with respect to its parameters p_1, p_2, and y, we have

$$\begin{aligned}
\frac{\partial V}{\partial p_1} &= -\frac{1}{4 p_1 p_2} y^2 \\
\frac{\partial V}{\partial p_2} &= -\frac{1}{4 p_1 p_2} y^2 \\
\frac{\partial V}{\partial y} &= \frac{1}{2 p_1 p_2} y
\end{aligned}$$

[1]This assumption is made so that we rule out the possibility of a corner solution.

Taking the derivative of the Lagrangian, we have

$$\frac{\partial \mathcal{L}}{\partial p_1} = \lambda x_1$$
$$\frac{\partial \mathcal{L}}{\partial p_2} = \lambda x_2$$
$$\frac{\partial V}{\partial y} = -\lambda$$

Substituting the optimal values of x_1, x_2 and λ in the above equations, completes our illustration. For example, for the first equation,

$$\begin{aligned}\frac{\partial \mathcal{L}}{\partial p_1} &= -\frac{1}{2p_1p_2}y \cdot \frac{1}{2p_1}y \\ &= -\frac{1}{4p_1^2p_2}y^2\end{aligned}$$

Showing the other two equalities are left to the reader. Another interesting characteristic of the value function is the *Roy's identity.*

Theorem 13.3 (Roy's Identity). *If* $x_j = x_j(\mathbf{p}, y)$ *is the Marshallian demand function, then*

$$x_j(\mathbf{p}, y) = -\frac{\frac{\partial V(\mathbf{p}, y)}{\partial p_j}}{\frac{\partial V(\mathbf{p}, y)}{\partial y}} \tag{13.30}$$

Again, we shall illustrate this proposition with Example 13.7. Note that

$$\frac{\partial V(\mathbf{p}, y)}{\partial p_1} = -\frac{1}{4p_1^2p_2}y^2 \tag{13.31}$$

and

$$\frac{\partial V(\mathbf{p}, y)}{\partial y} = \frac{1}{2p_1p_2}y \tag{13.32}$$

substituting (13.31) *and* (13.32) in (13.30)*, we get the demand function for commodity 1.*

An alternative way of solving the consumer problem is to minimize the expenditures, subject to a given level of utility.

$$\begin{aligned}&\min_{\mathbf{x}} \ \mathbf{p}'x \\ &\text{s.t.} \quad U(\mathbf{x}) = U^* \\ &\mathbf{x} \geq \mathbf{0}\end{aligned}$$

The value function resulting from this problem is called the expenditure function *and depends on prices and the preselected level of utility* U^**.*

$$e = e(\mathbf{p}, U^*)$$

The demand functions for different goods will also depend on prices and utility and they are called the Hicksian demand functions. *We shall illustrate these concepts with Example 13.7.*

Example 13.9 The Lagrangian function is

$$\mathcal{L} = p_1x_1 + p_2x_2 + \mu(x_1x_2 - U^*)$$

First-order conditions are

$$\begin{aligned} p_1 + \mu x_2 &= 0 \\ p_2 + \mu x_1 &= 0 \\ x_1x_2 &= U^* \end{aligned}$$

Again we can solve for x_1, x_2, and μ:

$$x_1^2 = \frac{p_2}{p_1}U^*, \qquad x_2^2 = \frac{p_1}{p_2}U^*$$

Recalling that only positive values of commodities are allowed, we have the Hicksian demand functions for commodities 1 and 2 as

$$x_1 = \left(\frac{p_2}{p_1}U^*\right)^{\frac{1}{2}}$$

$$x_2 = \left(\frac{p_1}{p_2}U^*\right)^{\frac{1}{2}}$$

and

$$\mu = -\left(\frac{p_1p_2}{U^*}\right)^{\frac{1}{2}}$$

Substituting the optimal values of x_1 and x_2 into the objective function, we get the expenditure function.

$$\begin{aligned} e(\mathbf{p}, U^*) &= p_1\left(\frac{p_2}{p_1}U^*\right)^{\frac{1}{2}} + p_2\left(\frac{p_1}{p_2}U^*\right)^{\frac{1}{2}} \\ &= 2\left(p_1p_2U^*\right)^{\frac{1}{2}} \end{aligned}$$

Note that the Hicksian demand functions depend on unobservable U^*

whereas the Marshallian functions can be estimated using observations on income and prices. But there is a connection between the Marshallian and Hicksian demand functions. Let us denote the latter by $h_j(\mathbf{p}, U^*)$. Then

$$h_j(\mathbf{p}, U^*) = x_j(\mathbf{p}, e(\mathbf{p}, U^*)) \tag{13.33}$$

Example 13.10 (Slutsky[2] Equation). A benefit of having the expenditure function is that we can derive the Slutsky equation in a different and perhaps more straightforward way. Consider (13.33) and let us differentiate it with respect to p_j. Then

$$\frac{\partial h_j(\mathbf{P}, U^*)}{\partial p_j} = \frac{\partial x_j(\mathbf{P}, y)}{\partial p_j} + \frac{\partial x_j(\mathbf{P}, y)}{\partial y}\frac{\partial e(\mathbf{p}, U^*)}{\partial p_j} \tag{13.34}$$

The last term on the RHS is the partial derivative of the expenditure function with respect to p_j. Because the expenditure function is the optimal value of $\mathbf{p}'x$. its derivative with respect to p_j is equal to the optimal value of x_j. Making the substitution and rearranging terms, we have the Slutsky equation decomposing the effect of a price change into substitution and income effects:

$$\frac{\partial x_j(\mathbf{P}, y)}{\partial p_j} = \frac{\partial h_j(\mathbf{P}, U^*)}{\partial p_j} - \frac{\partial x_j(\mathbf{P}, y)}{\partial y}x_j \tag{13.35}$$

Example 13.11 Going back to Example 13.4, we can substitute the optimal values of K and L into the cost function to get

$$C = \frac{\bar{Q}}{A}\left[r\left(\frac{1-\alpha}{\alpha}\frac{r}{w}\right)^{\alpha-1} + w\left(\frac{1-\alpha}{\alpha}\frac{r}{w}\right)^{\alpha}\right] \tag{13.36}$$

The above equation allows us to illustrate Shephard's lemma.

Theorem 13.4 (Shephard's Lemma). *Let* $\mathbf{q}$ *be the vector of input prices and* $C(\mathbf{q}, Q)$, *the cost function. Suppose* C *is differentiable. Then the demand function for the* j*-th input* $x_j(\mathbf{q}, Q)$ *can be obtained as*

$$x_j(\mathbf{q}, Q) = \frac{\partial C(\mathbf{q}, Q)}{\partial q_j} \tag{13.37}$$

---

[2]Evgeny Evgenievich Slutsky (1880–1948), Russian mathematician, economist, and statistician, published this result in a highly mathematical article "Sulla teoria del bilancio del consumatore," in *Giornali degli Economisti* in 1915. Because of the war, it received little attention. In the 1930s, economists R. G. D. Allen, John Hicks, and Henry Schultz, working on consumer theory, rediscovered it.

Note that from (13.36) we have

$$\frac{\partial C}{\partial r} = \frac{\bar{Q}}{A}\left(\frac{1-\alpha}{\alpha}\frac{r}{w}\right)^{\alpha-1}$$

and

$$\frac{\partial C}{\partial w} = \frac{\bar{Q}}{A}\left(\frac{1-\alpha}{\alpha}\frac{r}{w}\right)^{\alpha}$$

which are demand functions for K and L.

13.2.1 Exercises

E.13.4 Show that the indirect utility function is homogeneous of degree zero. Show this first for the functions in Examples 13.7 and 13.8 and then for a general consumer problem.

E.13.5 Show that the expenditure function is homogeneous of degree one in prices. Show this first for the function in Examples 13.9 and then for a general consumer problem.

E.13.6 Verify the Slutsky equation using utility functions in Examples 13.7 and 13.8.

E.13.7 Verify Shephard's lemma when the firm's technology is represented by

$$Q = AK^{\alpha}L^{\beta}$$

E.13.8 Derive the indirect utility and expenditure functions for the consumer problem in part *i.* of E.13.2.

13.3 Second-Order Conditions and Comparative Static

So far we have been discussing the first-order or the necessary conditions for the solution of an optimization problem. We still need to distinguish between maximum and minimum. The distinction is made through the second-order or the sufficient conditions that are presented in the following two theorems.

Theorem 13.5 *Let* $\mathbf{x}^*$ *be the optimum point of f subject to* $g(\mathbf{x}^*) = 0$. *It will be a*

$$\begin{array}{lll} \text{maximum if} & \nabla^2_{\mathbf{x}} L(\mathbf{x}^*, \lambda^*) & \text{is negative definite} \\ \text{minimum if} & \nabla^2_{\mathbf{x}} L(\mathbf{x}^*, \lambda^*) & \text{is positive definite} \end{array} \tag{13.38}$$

subject to $g(\mathbf{x}^*) = 0$. *The constraint implies*

$$\mathbf{z}'\nabla g(\mathbf{x}^*) = 0 \tag{13.39}$$

for all vectors $\mathbf{z}$.

Theorem 13.6 *The conditions of Theorem 13.5 for a minimum are realized if the* bordered Hessians

$$|\mathbf{H_p}| = \begin{vmatrix} \frac{\partial^2 L(\mathbf{x}^*, \lambda^*)}{\partial x_1^2} & \cdots & \frac{\partial^2 L(\mathbf{x}^*, \lambda^*)}{\partial x_1 \partial x_p} & \frac{\partial g(\mathbf{x}^*)}{\partial x_1} \\ \vdots & \vdots & \vdots & \vdots \\ \frac{\partial^2 L(\mathbf{x}^*, \lambda^*)}{\partial x_p \partial x_1} & \cdots & \frac{\partial^2 L(\mathbf{x}^*, \lambda^*)}{\partial x_p^2} & \frac{\partial g(\mathbf{x}^*)}{\partial x_p} \\ \frac{\partial g(\mathbf{x}^*)}{\partial x_1} & \cdots & \frac{\partial g(\mathbf{x}^*)}{\partial x_p} & 0 \end{vmatrix} < 0 \tag{13.40}$$

for $p = 2, \ldots, n$. *For a maximum we should have*

$$(-1)^p \left|\mathbf{H_p}\right| > 0 \tag{13.41}$$

for $p = 2, \ldots, n$.

Example 13.12 Recall the first-order conditions of the optimization problem in Example 13.2:

$$\begin{aligned} y + \lambda &= 0 \\ x + \lambda &= 0 \\ x + y - 12 &= 0 \end{aligned}$$

The solution is a maximum because

$$\begin{vmatrix} 0 & 1 & 1 \\ 1 & 0 & 1 \\ 1 & 1 & 0 \end{vmatrix} = 2 > 0$$

Note that for a maximum the determinant has to be positive, because $p = 2$ and $(-1)^p > 0$.

Example 13.13 In Example 13.3, the first-order conditions were

$$\begin{aligned} U_1 + \lambda p_1 &= 0 \\ U_2 + \lambda p_2 &= 0 \\ p_1 x_1 + p_2 x_2 &= y \end{aligned} \tag{13.42}$$

The bordered Hessian

$$|\mathbf{H}_2| = \begin{vmatrix} U_{11} & U_{12} & p_1 \\ U_{21} & U_{22} & p_2 \\ p_1 & p_2 & 0 \end{vmatrix} \tag{13.43}$$

has to be positive and, indeed, it is. Because marginal utilities are declining, but marginal utility of one commodity increases if the consumption of the other is increased,

$$U_{11} < 0 \qquad U_{22} < 0 \qquad U_{12} = U_{21} > 0$$

and

$$|\mathbf{H}_2| = p_1 p_2 (U_{12} + U_{21}) - p_1^2 U_{22} - p_2^2 U_{11} > 0 \tag{13.44}$$

The first-order conditions allow us to solve, at least in principle, for the optimal values of each good or service as well as λ in terms of the parameters of the model. The parameters of the consumer choice model consist of prices of goods and the income of consumer. Thus, for the case of two goods we can write

$$\begin{aligned} x_1^* &= x_1(p_1, p_2, y) \\ x_2^* &= x_2(p_1, p_2, y) \\ \lambda^* &= \lambda(p_1, p_2, y) \end{aligned}$$

Were we able to find the exact form of the above functions, we could have conducted a comparative static analysis gauging the effects of changes in prices and income on the amount of each good consumed. But even without an explicit solution we are able to perform such an analysis.

Example 13.14 Let us take the total differentials of all three equations in (13.42):

$$\begin{aligned} U_{11}dx_1^* + U_{12}dx_2^* + p_1 d\lambda^* &= -\lambda^* dp_1 \\ U_{21}dx_1^* + U_{22}dx_2^* + p_2 d\lambda^* &= -\lambda^* dp_2 \\ p_1 dx_1^* + p_2 dx_2^* \quad &= dy - x_1^* dp_1 - x_2^* dp_2 \end{aligned} \tag{13.45}$$

Now we take up the parameters one at a time and consider their effect on the three decision variables.[3] First, we set $dp_2 = dy = 0$ and divide all equations by dp_1. Because we are holding p_2 and y constant, all derivatives will be partial. Writing the result in matrix form, we have

[3]In this problem parameters play the same role as exogenous variables in Chap. 9 and variables of interest or decision variables play the role of endogenous variables.

$$\begin{bmatrix} U_{11} & U_{12} & p_1 \\ U_{21} & U_{22} & p_2 \\ p_1 & p_2 & 0 \end{bmatrix} \begin{bmatrix} \partial x_1^*/\partial p_1 \\ \partial x_2^*/\partial p_1 \\ \partial \lambda^*/\partial p_1 \end{bmatrix} = \begin{bmatrix} -\lambda^* \\ 0 \\ -x_1 \end{bmatrix} \tag{13.46}$$

Using one of the methods of Chap. 6, we can solve for the unknowns. In particular, let us solve for the effect of p_1 on x_1^* :

$$\begin{aligned} \frac{\partial x_1^*}{\partial p_1} &= \frac{-x_1^* p_2 U_{12} + x_1^* p_1 U_{22} + p_2^2 \lambda^*}{|\mathbf{H}_2|} \\ &= -x_1^* \frac{p_2 U_{12} - p_1 U_{22}}{|\mathbf{H}_2|} + \frac{p_2^2 \lambda^*}{|\mathbf{H}_2|} \end{aligned} \tag{13.47}$$

If we let $dp_1 = dp_2 = 0$, we get

$$\begin{bmatrix} U_{11} & U_{12} & p_1 \\ U_{21} & U_{22} & p_2 \\ p_1 & p_2 & 0 \end{bmatrix} \begin{bmatrix} \partial x_1^*/\partial y \\ \partial x_2^*/\partial y \\ \partial \lambda^*/\partial y \end{bmatrix} = \begin{bmatrix} 0 \\ 0 \\ 1 \end{bmatrix} \tag{13.48}$$

and we can compute the effect of income on x_1 as

$$\frac{\partial x_1^*}{\partial y} = \frac{p_2 U_{12} - p_1 U_{22}}{|\mathbf{H}_2|} \tag{13.49}$$

Furthermore, going back to the last equation in (13.45), recall that we set $dp_2 = dy = 0$. Instead, let us make $dy = x_1^* dp_1$ while keeping $dp_2 = 0$. In other words, let us compensate the consumer for any change in income due to a change in p_1.

Under this assumption, if we solve the equations in (13.45), we get

$$\left.\frac{\partial x_1^*}{\partial p_1}\right|_{dy = x_1^* dp_1} = \left.\frac{\partial x_1^*}{\partial p_1}\right|_{\text{compensated}} = \frac{p_2^2 \lambda^*}{|\mathbf{H}_2|} \tag{13.50}$$

Substituting (13.49) and (13.50) in (13.47) we have

$$\frac{\partial x_1^*}{\partial p_1} = -x_1^* \left.\frac{\partial x_1^*}{\partial y}\right|_{dp_1 = 0} + \left.\frac{\partial x_1^*}{\partial p_1}\right|_{\text{compensated}} \tag{13.51}$$

which is the Slutsky equation. We derived this equation from the expenditure function in previous sections, but this alternative way of deriving the Slutsky equation allows us to better understand its meaning and significance. What can we say about the sign of $\partial x_1^*/\partial p_1$?

Intuitively, the formula states that an increase (decrease) in the price of a good in your consumption basket affects your demand for that good through two channels: income effect and substitution effect. First, the price increase (decrease) acts like a reduction (expansion) in the amount of resources at your disposal. As such it will affect your demand for all goods including the one under consideration. If the sign of the first term is negative, that is, when an increase in real income, with relative prices held constant, increases the demand for a good, then that good is called a *normal good.* Otherwise it is called an *inferior good.* The second term is definitely negative (why?) and it states that if the change in your income is neutralized—by compensating you for the loss or by taking away your extra resources—then an increase in the price of a good reduces its demand. Putting the two effects together, we can state that for a normal good, $\partial x_1^*/\partial p_1 < 0$. The effect of a change in p_2 on x_2^*, as well as the effects of p_1 on x_2^* and p_2 on x_1^* can be similarly analyzed. Indeed, the Slutsky equation is more general. If there are n goods, we can write

$$\frac{\partial x_i^*}{\partial p_j} = -x_j^* \left.\frac{\partial x_i^*}{\partial y}\right|_{dp_j=0} + \left.\frac{\partial x_i^*}{\partial p_j}\right|_{\text{compensated}} \quad \forall i,j \tag{13.52}$$

So far we have been dealing with optimization problems that had only one constraint. Here we turn to sufficient conditions for problems with m constraints. It turns out that the sufficient conditions depend on the number of constraints.

Theorem 13.7 *If the first-order conditions are satisfied, we will have a minimum if*

$$(-1)^m \left|\mathbf{H_p}\right| > 0 \tag{13.53}$$

for $p = m+1, \ldots, n$. *Where*

$$\left|\mathbf{H_p}\right| = \begin{vmatrix} \frac{\partial^2 L(\mathbf{x}^*,\lambda^*)}{\partial x_1^2} & \cdots & \frac{\partial^2 L(\mathbf{x}^*,\lambda^*)}{\partial x_1 \partial x_p} & \frac{\partial g_1(\mathbf{x}^*)}{\partial x_1} & \cdots & \frac{\partial g_m(\mathbf{x}^*)}{\partial x_1} \\ \vdots & & \vdots & \vdots & & \vdots \\ \frac{\partial^2 L(\mathbf{x}^*,\lambda^*)}{\partial x_p \partial x_1} & \cdots & \frac{\partial^2 L(\mathbf{x}^*,\lambda^*)}{\partial x_p^2} & \frac{\partial g_1(\mathbf{x}^*)}{\partial x_p} & \cdots & \frac{\partial g_m(\mathbf{x}^*)}{\partial x_p} \\ \frac{\partial g_1(\mathbf{x}^*)}{\partial x_1} & \cdots & \frac{\partial g_1(\mathbf{x}^*)}{\partial x_p} & 0 & \cdots & 0 \\ \vdots & & \vdots & \vdots & & \vdots \\ \frac{\partial g_m(\mathbf{x}^*)}{\partial x_1} & \cdots & \frac{\partial g_m(\mathbf{x}^*)}{\partial x_p} & 0 & \cdots & 0 \end{vmatrix}$$

The condition for a maximum is the same as above except that $(-1)^m$ *should be replaced with* $(-1)^p$. *That is, if*

$$(-1)^p \left|\mathbf{H_p}\right| > 0 \tag{13.54}$$

for $p = m + 1, \ldots, n$.

Example 13.15

$$\begin{aligned} \max f(x, y, z) &= xyz \\ s.t. \quad y + 2x &= 15 \\ 2z + y &= 7 \end{aligned}$$

We can still solve this problem the old-fashioned way by substituting for y and z to maximize

$$f(x, y, z) = x(15 - 2x)(x - 4)$$

The solution is

$$x = 6, \quad y = 3, \quad z = 2$$

Of course, the function has a minimum, too at

$$x \approx 1.67, \qquad y \approx 11.66, \qquad z \approx -2.33$$

The reader is urged to solve the problem and check the results. Next, we solve this problem using Lagrange multipliers:

$$\max xyz + \lambda_1(y + 2x - 15) + \lambda_2(2z + y - 7)$$

Setting the derivatives equal to zero, the first-order conditions are

$$\begin{aligned} yz + 2\lambda_1 &= 0 \\ xz + \lambda_1 + \lambda_2 &= 0 \\ xy + 2\lambda_2 &= 0 \\ y + 2x &= 15 \\ 2z + y &= 7 \end{aligned}$$

Solving the above equations, we get two sets of solutions

$$\begin{aligned} &i. \quad x = 6, \qquad y = 3, \qquad z = 2, \\ &\qquad \lambda_1 = -3, \qquad \lambda_2 = -9 \end{aligned}$$

$$\begin{aligned} &ii. \quad x \approx 1.67, \quad y \approx 11.66, \quad z \approx -2.33, \\ &\qquad \lambda_1 \approx 13.58, \quad \lambda_2 \approx -6.80 \end{aligned}$$

The second-order condition for a maximum requires the bordered Hessian to be negative:

$$|\mathbf{H}_3| = \begin{vmatrix} 0 & z & y & 2 & 0 \\ z & 0 & x & 1 & 1 \\ y & x & 0 & 0 & 2 \\ 2 & 1 & 0 & 0 & 0 \\ 0 & 1 & 2 & 0 & 0 \end{vmatrix} = -16x + 8y - 16z$$

Evaluated at the point of solution in (i), the bordered Hessian is negative, therefore, we have a maximum. On the other hand, evaluated at the point of (ii), $|\mathbf{H}_3|$ is positive, signifying a minimum.

13.3.1 Exercises

E.13.9 Check the second-order conditions for problems E.13.1, 13.2, and E.13.3.

E.13.10 Show that the constrained least squares estimator of (13.24) indeed minimizes the sum of squared residuals.

13.4 Inequality Constraints and Karush-Kuhn-Tucker Conditions

There are many instances when because of technical necessity, legal restrictions, policy imperatives, and other reasons, inequality constraints are imposed on an optimization problem. A simple example is the fact that many economic variables, including prices and quantities, are positive or at least nonnegative. Wages cannot go below the minimum prescribed by law, no one can work more than 52 weeks a year, and many production processes cannot be maintained unless a certain minimum production is scheduled. As far as resources are concerned, they pose an upper limit, but there is no imperative to utilize any resource to the fullest. For example, in solving the consumer problem, we assumed that the individual or the household exhausts their income on the goods available. There is no reason why any family should spend exactly the amount of their income. A family can save for future consumption or borrow and live beyond its present means.

Inequality constraints complicate the optimization problem and the Lagrangian method has to be modified. Karush-Kuhn-Tucker (KKT) conditions provide the necessary conditions for the optima with inequality and equality constraints. Because we are dealing with inequality constraints, they are not necessarily binding. In the language of resource allocation, we may not exhaust a resource and may have some slack. If a constraint is not binding, then it cannot have an effect on the optimum and its Lagrange multiplier will be zero. But if a constraint is binding, then it turns into an equality. It follows that the product of a constraint and its Lagrange multiplier has to be zero.

Note also that if a Lagrange multiplier is different from zero, then the optimality conditions deviate from the unconstrained optimum. The amount and sign of deviation depend on the Lagrange multiplier. It follows that the multiplier should have opposite signs in the case of a minimum and maximum. Before stating these conditions formally, let us get an intuitive understanding through a simple example involving one variable and one constraint. Consider the following problem:

$$\begin{aligned} &\max f(x) \\ &\text{s.t.} \quad x \geq x_0 \end{aligned} \tag{13.55}$$

Three possibilities can be envisaged. First, the unconstrained optimum is greater than the constraint $x^*>x_0$ in which case the constraint is nonbinding, the Lagrangian multiplier is zero, and $f'(x^*) = 0$. This situation is depicted in Fig. 13.2a. The second

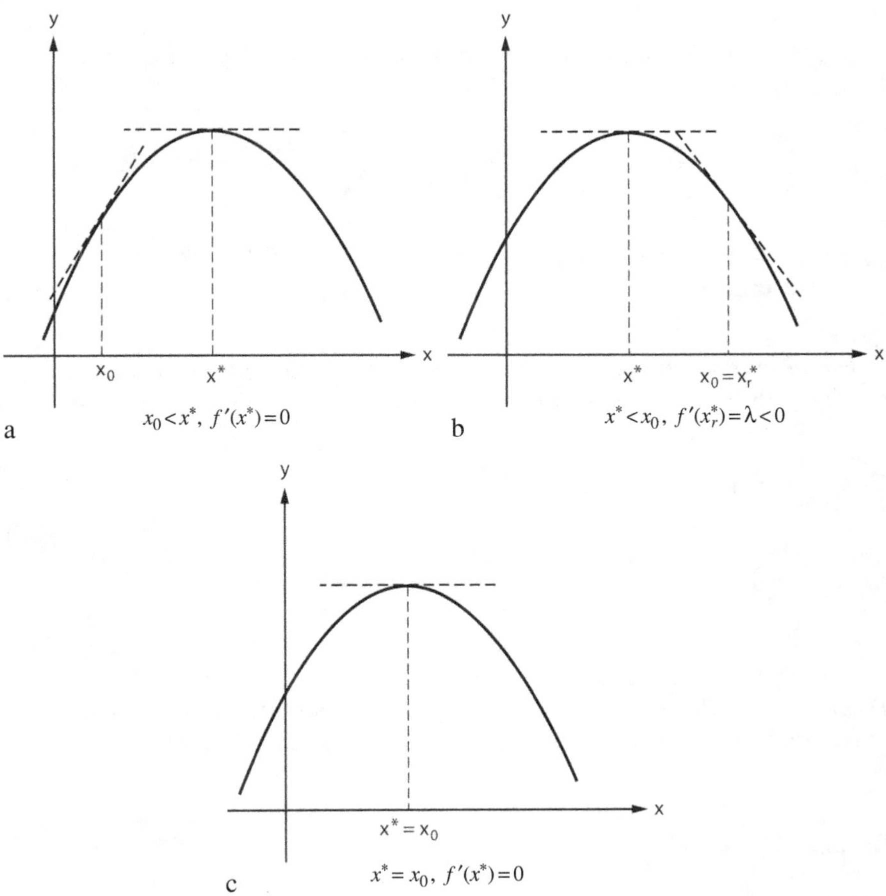

Fig. 13.2 Three possible configurations in a simple optimization problem with inequality constraint

possibility occurs when $x_0 > x^*$. Although $f'(x^*) = 0$, this point is excluded because it violates the constraint. Hence the solution is at $x_r^* = x_0$ where $f'(x_r^*) = \lambda < 0$ (see Fig. 13.2b). Finally, we have the special case where the unconstrained optimum point and the constraint coincide, $x^* = x_0$. This situation is depicted in Fig.13.2c. In practice this is unlikely to happen, but it poses no problem for our analysis. Because the constraint is binding, $\lambda \leq 0$, but because we are at the unconstrained optimum, the value of relaxing the constraint a bit is zero, hence $\lambda = 0$ and $f'(x^*) = 0$.

All these possibilities are subsumed under the Karush-Kuhn-Tucker conditions.

Theorem 13.8 *Consider the problem*

$$\begin{aligned} &\min f(\mathbf{x}) \\ &\text{s.t.} \quad g_i(\mathbf{x}) \leq 0 \qquad i = 1, \dots, m \\ &\qquad\; h_j(\mathbf{x}) = 0 \qquad j = 1, \dots, J \end{aligned} \tag{13.56}$$

If $\mathbf{x}^*$ *is a solution to this problem, then there exists a set of multipliers* λ_i^*, $i = 1, \dots, m$ *and* μ_j^*, $j = 1, \dots, J$, *such that*

$$\begin{aligned} &\nabla f(\mathbf{x}^*) + \sum_{i=1}^{m} \lambda_i^* \nabla g_i(\mathbf{x}^*) + \sum_{j=1}^{J} \mu_j^* \nabla h_j(\mathbf{x}^*) = 0 \\ &\lambda_i^* \geq 0, \qquad g_i(\mathbf{x}^*) \leq 0, \qquad \lambda_i^* g_i(\mathbf{x}^*) = 0 \qquad i = 1, \dots, m \\ &h_j(\mathbf{x}^*) = 0 \qquad j = 1, \dots, J \end{aligned} \tag{13.57}$$

For the maximization problem, we can either minimize $-f(\mathbf{x})$ *and keep the KKT conditions as in* (13.57) *or alternatively, we can write them as*

$$\begin{aligned} &\nabla f(\mathbf{x}^*) + \sum_{i=1}^{m} \lambda_i^* \nabla g_i(\mathbf{x}^*) + \sum_{j=1}^{J} \mu_j^* \nabla h_j(\mathbf{x}^*) = 0 \\ &\lambda_i^* \leq 0, \qquad g_i(\mathbf{x}^*) \leq 0, \qquad \lambda_i^* g_i(\mathbf{x}^*) = 0 \qquad i = 1, \dots, m \\ &h_j(\mathbf{x}^*) = 0 \qquad j = 1, \dots, J \end{aligned} \tag{13.58}$$

Note that if there are no inequality constraints, KKT conditions specialize to Theorem 13.1. On the other hand, we may have no equality constraints, only inequalities, in which case the part involving μ_j*'s and equality constraints is omitted.*

Example 13.16 Find

$$\begin{aligned} \max \quad & f(x) = -(x-3)^2 \\ \text{s.t.} \quad & x \geq 5 \end{aligned}$$

The Lagrangian will be

$$L = -(x-3)^2 + \lambda(5-x)$$

The optimality conditions are

$$\begin{aligned} &-2x + 6 - \lambda = 0, \\ &\lambda \le 0, \qquad \lambda(5-x) = 0 \end{aligned}$$

Starting with the last equation, either $x = 5$ or $\lambda = 0$. If $x = 5$, then $\lambda = -4$ and all conditions are satisfied. Thus, $x = 5$ is the point of constrained maximum. But, if we set $\lambda = 0$, the first equation implies $x = 3$, which violates the constraint. Now suppose that we change the constraint to

$$\begin{aligned} \max \quad & f(x) = -(x-3)^2 \\ \text{s.t.} \quad & x \ge 1 \end{aligned}$$

The KKT conditions are

$$\begin{aligned} &-2x + 6 - \lambda = 0 \\ &\lambda \le 0, \qquad \lambda(1-x) = 0 \end{aligned}$$

Starting with the last equation, if we set $x = 1$, we get $\lambda = 4 > 0$, which violates the second condition. Hence, we should set $\lambda = 0$, which results in $x = 3$ where all conditions are satisfied.

Karush-Kuhn-Tucker conditions pertain to local optima, but under certain conditions a local optima will also be the global optima.

Theorem 13.9 *If*

1. *f is convex (concave),*
2. *g_i, $i = 1, \ldots, m$ are concave,*
3. *h_j, $j = 1, \ldots, J$ are linear.*

Then the local minimum (maximum) satisfying the conditions of Theorem 13.8 is also the global minimum (maximum). Alternatively, if

1. *f is convex (concave),*
2. *g_i, $i = 1, \ldots, m$ and h_j, $j = 1, \ldots, J$ define a convex set.*

Then the local minimum (maximum) satisfying the conditions of Theorem 13.8 is also the global minimum (maximum). Indeed, we can somewhat relax the above conditions. For this purpose we need to define a pseudo-convex function.

Definition 13.1 A differentiable function $f : \mathbf{D} \rightarrow \Re$ is called pseudo-concave if for all $\mathbf{x}_1, \mathbf{x}_2 \in \mathbf{D}$

$$\nabla f(\mathbf{x}_1)'(\mathbf{x}_2 - \mathbf{x}_1) \leq 0 \tag{13.59}$$

implies that

$$f(\mathbf{x}_2) \leq f(\mathbf{x}_1) \tag{13.60}$$

If f is pseudo-convex, then $-f$ is pseudo-concave.

Theorem 13.10 *The local minimum (maximum) obeying KKT conditions is the global minimum (maximum) if*

1. *f is pseudo-convex (pseudo-concave),*
2. *g_i, $i = 1, \ldots, m$ are quasi-concave,*
3. *h_j, $j = 1, \ldots, J$ are quasi-concave or quasi-convex.*

13.4.1 Duality

Every optimization problem has a twin or dual. Under certain conditions, the solution of one entails the solution of the other. Intuitively, suppose we are allocating the available quantities of three resources, labor, capital, and energy, say, to two activities (outputs). Given the price of outputs, the *primal problem* is to find the amount of each resource to be allocated to each activity so as to maximize an objective function. On the other hand, we may pose a *dual problem* asking how we could price our resources to minimize the cost of producing at least a given level of each output. The importance of duality in economics is to understand the underlying meaning of an optimization problem, in particular, the role of Lagrange multipliers as optimal prices of resources.

Example 13.17 Consider the following *linear programming*[4] problem involving two products and three resources. We shall denote the amount of each output by x_j, $j = 1, 2$ and the amount of each resource by r_i, $i = 1, 2, 3$. The following table shows how much of each resource is needed for every unit of each output.

[4]It is called linear programming because both the objective function and constraints are linear. Russian mathematician Leonid Vitalyevich Kantorovich (1912–1986) was the first to formulate the linear programming problem in response to a request by a Soviet enterprise wanting to optimize the use of its resources. He solved the problem using Lagrange multipliers. His work received scant attention at the time, but he later received the Nobel Prize in economics for his contribution to this subject. The simplex method for solving linear programming problem was discovered by the American mathematician and statistician George Dantzig (1914–2005). In 1984, Narendra Karmarkar discovered a different method for solving such problems.

	Resources		
Outputs	1	2	3
1	6	3	3
2	3	3	6

Let us suppose that 50 units of the first resource, 40 of the second, and 75 of the third are available. Finally, assume that the first product brings \$15 of profit while the second brings \$12 of profit. Our problem is to maximize the total profit subject to resource constraints.

$$\begin{aligned} \max \quad & \pi = 15x_1 + 12x_2 \\ \text{s.t.} \quad & \begin{bmatrix} 6 & 3 \\ 3 & 3 \\ 3 & 6 \end{bmatrix} \begin{bmatrix} x_1 \\ x_2 \end{bmatrix} \le \begin{bmatrix} 50 \\ 40 \\ 75 \end{bmatrix} \\ & x_1,\ x_2 \ge 0 \end{aligned}$$

The dual of this problem is

$$\begin{aligned} \min \quad & c = 50y_1 + 40y_2 + 75y_3 \\ \text{s.t.} \quad & \begin{bmatrix} 6 & 3 & 3 \\ 3 & 3 & 6 \end{bmatrix} \begin{bmatrix} y_1 \\ y_2 \\ y_3 \end{bmatrix} \ge \begin{bmatrix} 15 \\ 12 \end{bmatrix} \\ & y_1,\ y_2,\ y_3 \ge 0 \end{aligned}$$

More generally, we can write a linear programming problem and its dual as

$$\begin{aligned} \max \quad & \pi = \mathbf{p}'x \\ \text{s.t.} \quad & \mathbf{Ax} \le \mathbf{r} \\ & \mathbf{x} \ge \mathbf{0} \end{aligned} \tag{13.61}$$

and

$$\begin{aligned} \min \quad & c = \mathbf{r}'y \\ \text{s.t.} \quad & \mathbf{A}'y \ge \mathbf{p} \\ & \mathbf{y} \ge \mathbf{0} \end{aligned} \tag{13.62}$$

where $\mathbf{p}$ is the vector of output prices and $\mathbf{y}$ the vector of input (resource) prices.

The issue of duality is not confined to linear models. For a nonlinear primal problem with one equality and one inequality constraint,

$$\begin{aligned} \max \quad & f(\mathbf{x}) \\ \text{s.t.} \quad & g(\mathbf{x}) \leq \mathbf{0} \\ & h(\mathbf{x}) = \mathbf{0} \\ & \mathbf{x} \geq \mathbf{0} \end{aligned} \tag{13.63}$$

The dual problem is

$$\begin{aligned} \min \quad & \Phi(\lambda, \mu) \\ \text{s.t.} \quad & \mu \geq 0 \end{aligned} \tag{13.64}$$

where

$$\Phi(\lambda, \mu) = \inf_{\mathbf{x}} [f(\mathbf{x}) + \mu g(\mathbf{x}) + \lambda h(\mathbf{x})] \tag{13.65}$$

Example 13.18 Let us consider the following maximization problem and this time use numerical values for prices and income:

$$\begin{aligned} \max \quad & 2x_1^2 + x_2^2 \\ \text{s.t.} \quad & 3x_1 + x_2 \leq 110 \\ & x_1,\ x_2 \geq 0 \end{aligned}$$

The Lagrangian is

$$L = 2x_1^2 + x_2^2 + \lambda(3x_1 + x_2 - 110)$$

The first-order conditions imply

$$\begin{aligned} x_1 &= -\frac{3}{4}\lambda \\ x_2 &= -\frac{1}{2}\lambda \end{aligned}$$

Combining the above equations with the constraint, we get

$$x_1 = 30, \quad x_2 = 20, \quad \lambda = -40$$

Now, we can form the Lagrangian of the dual problem:

$$\Phi(\lambda) = 2\left(-\frac{3}{4}\lambda\right)^2 + \left(-\frac{1}{2}\lambda\right)^2 + \lambda\left(-3\frac{3}{4}\lambda - \frac{1}{2}\lambda - 110\right)$$

or

$$\Phi(\lambda) = -\frac{11}{8}\lambda^2 - 110\lambda$$

Minimizing this function with respect to λ will give the same answer as the primal problem.

Under certain conditions, the optimal values of objective functions in the primal and dual problems are equal. In particular, if all constraints are differentiable and convex and we can find a vector $\mathbf{x}$ such that

$$g_j(\mathbf{x}) < 0, \qquad \forall j$$

then the optimal values of the objective functions for primal and dual problems are equal. This is known as the *Slater constraint qualification.* It can be modified by saying that equality constraints should be *affine*[5] and inequality constraints convex. If the optimal values of the two objective functions are not equal, then there exists a *duality gap.*

13.4.2 Exercises

E.13.11 Solve the following optimization problems:

i.

$$\begin{aligned} &\max xy \\ \text{s.t.} \quad & x + y = 12 \\ & x \geq 7 \end{aligned}$$

ii.

$$\begin{aligned} &\max xy \\ \text{s.t.} \quad & x + y = 12 \\ & x \geq 5 \end{aligned}$$

E.13.12 Solve the following problem:

$$\begin{aligned} &\max f(x,y,z) = xyz \\ s.t. \quad & y + 2x = 15 \\ & 2z + y = 7 \\ & y \geq 5, \quad x, z \geq 0 \end{aligned}$$

[5] An affine equation is linear but it is not homogeneous; that is, it has a constant term or intercept.

E.13.13 Solve the following problem:

$$\begin{aligned} \max f(x,y,z) &= xyz \\ s.t. \quad y + 2x &= 15 \\ 2z + y &\leq 7 \\ y \geq 5, \qquad & x, z \geq 0 \end{aligned}$$

E.13.14 Show that a convex (concave) function is also a pseudo-convex (pseudo-concave) function.

E.13.15 Write the dual problems of the primal optimization problems in E.13.11, E.13.12, and E.13.13.

Chapter 14
Dynamic Optimization

14.1 Dynamic Analysis in Economics

In the real world time passes and, if you think about it, you will agree with me that the only truly exogenous variable in economic models is time. Dynamic analysis takes time into consideration in an essential way. In this chapter we shall discuss dynamic optimization and, in the next three chapters, the modeling of economic behavior over time. In Chaps. 12 and 13, we studied static optimization to find the maximum or minimum point of a function with or without constraints on the variables involved. In this chapter we are interested in optimization over time.

Examples of optimization over time abound in economics. A consumer can spend his entire current income on consumption or save some for the future that will earn him future income and enhance his future consumption possibilities. On the other hand, he can borrow against his future earnings and expand his current budget at the expense of future consumption when the debt has to be paid back. The question, then, is how can a consumer balance consumption, savings, and borrowing to achieve the highest lifetime utility?

As another example, we can consider an oil producer, be it a small firm in Texas or as large an operation as the entire oil industry of Saudi Arabia. Profit from an oil well depends on the total output of the well, oil prices over time, and the rate of extraction. How can the extraction rate be regulated so as to result in the highest profit over the lifetime of the well?

A worker can spend her entire lifetime working and earning money. On the other hand, she can set aside time to learn new skills and enhance her human capital, which would cut into her current income because there is less time to work and because of education expenses. How does a worker maximize her lifetime earnings or utility by allocating her time between work and education?

All of the above problems can be formulated as a dynamic optimization or control problem. Let us have some specific examples.

Example 14.1 (Ramsey Problem). The objective is to maximize discounted social utility that depends on per capita consumption over the infinite horizon.

$$\max U = \int_0^{\infty} u(c)e^{-\theta t}dt \tag{14.1}$$

K. Dadkhah, *Foundations of Mathematical and Computational Economics,* 2nd ed.,
DOI 10.1007/978-3-642-13748-8_14, © Springer-Verlag Berlin Heidelberg 2011

where $c(t)$ is the per capita consumption, u the utility function, and θ is the discount factor. The technology is represented by a production function that depends on labor and capital and is homogeneous of degree one:

$$Y = F(K,L) = \frac{1}{L}F\left(\frac{K}{L},1\right) = \frac{1}{L}f(k) \tag{14.2}$$

where Y is output, K, capita, L, labor, and $k = K/L$. Output is divided between consumption and capital formation.

$$Y = C + \frac{dK}{dt} \tag{14.3}$$

or in per capita form

$$f(k) = c + \frac{1}{L}\frac{dK}{dt} = c + \frac{dk}{dt} + k\frac{dL/dt}{L} \tag{14.4}$$

Denoting the rate of growth of labor by n, we have

$$\frac{dk}{dt} = f(k) - c - nk \tag{14.5}$$

and the control problem can be written as

$$\begin{aligned} \max\ U &= \int_0^\infty u(c)e^{-\theta t}dt \\ \text{s.t.}\quad \frac{dk}{dt} &= f(k) - c - nk \\ c,\ k &\geq 0,\ \forall t \end{aligned} \tag{14.6}$$

Example 14.2 Human Capital Accumulation.[1] Consider a worker who enters the labor market at time $t = 1$ with human capital stock k_1 and retires at $t = T$. Her earnings depend on the number of hours worked, h_t, and the wage rate, w_t, which is proportional to the worker's human capital k_t. That is,

$$w_t = \alpha k_t \tag{14.7}$$

For simplicity, we assume $\alpha = 1$. Given the discount factor $\beta = 1/(1+r)$ where r is the interest rate, the worker's objective is to maximize the present value of her lifetime earnings.

$$\max_{\{k_{t+1}\}_{t=1}^{T-1}} \sum_{t=1}^{T} \beta^t h_t k_t \tag{14.8}$$

---

[1]This problem is from *Recursive Methods in Economic Dynamics* by Nancy Stokey and Robert Lucas (1989).

The worker can use all the time at her disposal to work and earn income in which case her human capital diminishes or can devote part or all of her time to enhancing the capital stock which would diminish earnings. In general, we can write

$$h_t = \phi\left(\frac{k_{t+1}}{k_t}\right), \qquad \phi' < 0 \tag{14.9}$$

Moreover the worker's human capital can grow at most at the rate λ, for which she has to devote all her time to education causing h_t to be equal to zero. On the other hand, as the result of working full time, $h_t = 1$, the worker's human capital would diminish at most at the rate δ. Thus, we have

$$\begin{gathered} (1-\delta)k_t \leq k_{t+1} \leq (1+\lambda)k_t \\ 0 = \phi(1+\lambda) \leq \phi\left(\frac{k_{t+1}}{k_t}\right) \leq \phi(1-\delta) = 1, \quad \forall t \end{gathered} \tag{14.10}$$

Combining all the elements above, we have the following control problem:

$$\begin{aligned} &\max_{\{k_{t+1}\}_{t=1}^{T-1}} \sum_{t=1}^{T} \beta^t k_t \phi\left(\frac{k_{t+1}}{k_t}\right) \\ &\text{s.t.} \quad (1-\delta)k_t \leq k_{t+1} \leq (1+\lambda)k_t, \quad \forall t \\ &\qquad\quad k_1 > 0, \text{ given} \end{aligned} \tag{14.11}$$

14.2 The Control Problem

The aim of a dynamic optimization problem is to find the maximum or minimum of an objective functional (see Sect. 14.1.1 for the meaning of the term functional), which is either an integral (continuous time) or sum of future terms (discrete time). The objective functional depends on the variable of interest called the *state variable*. In the Ramsey problem the *state variable* is the per capita capital, k_t, and in the accumulation of human capital, the state variable is the number of hours worked, h_t.

Because we are interested in optimization over time, a crucial component of the problem is the equation of motion, which describes the evolution of the state variable over time. In Example 14.1, the equation of motion is (14.5) and in Example 14.2, it is (14.9). The solution to a control problem is a path for the state variable, which results in the maximum or minimum of the objective functional. Such a path depends on the equation of motion and, therefore, we need an instrument, called a *control variable*, to make sure that the motion of the state variable does not deviate from the optimal path. In Example 14.1 the control variable is the per capita consumption, $c(t)$, and in Example 14.2, it is the worker's human capital, k_t. In some problems (see Sect. 14.3 on the calculus of variations), the control variable is the derivative of the state variable and therefore, we do not have an equation of motion. To these elements of a dynamic optimization problem, we could add initial and terminal conditions, that is, restrictions on state variables at the beginning and endpoint of the path. In addition, in some problems the admissible values of state and control variables may be restricted to a particular set.

It should be pointed out that in a deterministic control problem, it is assumed that future variables are either known or they can be determined by the decision maker. Usually in stochastic dynamic optimization, all random variables are replaced by their expected values, and the expected value of the objective functional is maximized or minimized. As a result, the control problem deals with a future that is known in advance. The nature of control problems is different from real-world problems where the decision maker has to respond to evolving situations.

The real-life problems involve uncertainty about the future that entails evaluating risk against return. Moreover, the decision maker may face an evolving situation where other decision makers' moves affect the outcome of his or her decisions. In such cases one may only be able to adopt a strategy and respond to short-term changes with moves consistent with the adopted strategy. Among games, backgammon epitomizes such situations. Not only does a player not know what will be the outcome of the next throw of the dice, but he also cannot be sure of the opponent's reaction. Thus, a player, at best, can devise a strategy and revise it as events unfold. The real-world situation in politics, business, diplomacy, and war resembles backgammon. Dynamic optimization is a step closer to decision making in the real world because it incorporates time. But it does not emulate real-world situations. On the other hand, even in devising a strategy that has to be revised at every stage, one needs to use dynamic optimization for formulating such a strategy, be it a business venture, diplomatic negotiation, or economic policy.

At this point, the reader may ask, why do we need special methods for dynamic optimization? Why can't we simply take the best possible route when dealing with a continuous time problem? And if we have a multistep problem, why can't we simply take the best decision at every point? As the examples in this chapter will show, an important lesson of dynamic optimization is that a piecemeal approach to such problems will result in suboptimal outcomes. In optimizing over time and in multistage decision-making problems, one has to seek the global optimum.

Suppose that we have a bead on a string that connects two points (see Fig. 14.1). If we allow the bead to slide down the string under the force of gravity, and assuming zero friction, what is the fastest path between the high and low points? This is the famous Brachistochrone[2] problem. It may seem that the fastest route is the straight line. It is not. A curved line allowing the bead to first pick up speed would result in a shorter travel time. The suboptimality of one-step-at-a-time decisions for multistage problems is illustrated in the traveling salesman problem of Sect. 14.4.

Three methods for solving control problems will be discussed in this book: the calculus of variations, which is the oldest method for tackling such problems; dynamic programming, which is usually applied to problems where time is treated

---

[2]It consists of two Greek words meaning the shortest time.

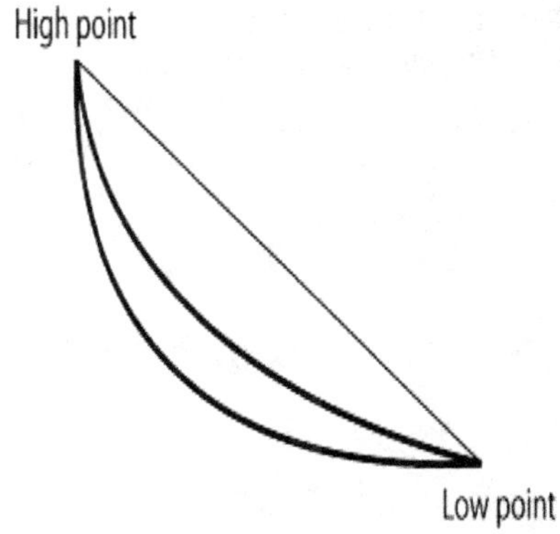

Fig. 14.1 The Brachistochrone problem

as a discrete variable; and the maximum principle, which has been the favorite and most useful method for economic analysis. Before presenting these methods, however, we need to discuss the general setup of a dynamic optimization problem and clarify a few useful concepts.

14.2.1 The Functional and Its Derivative

In Example 14.1, the lifetime utility is a *functional*. For each path of consumption, lifetime utility assumes a particular value. In other words, a path or a function is mapped to a unique number. Recall that a function maps every point in its domain to a unique point in its range, where the domain may be the set of real or complex numbers. Now suppose that we define the domain as a set of functions and define a mapping from each function in our domain to a unique point in the set of real numbers. Such mapping is called a functional. For instance, consider a family of functions

$$\phi(\beta), \quad \beta \in \Re \tag{14.12}$$

and define

$$v = V[\phi], \quad v \in \Re \tag{14.13}$$

Then $V[\phi]$ is a functional. Notice the difference in notation: We write a function as $y = \phi(x)$, but we denote a functional as $v = V[\phi]$ to avoid confusion. Alternatively, we can look at Fig. 14.2 where each curve represents one path. In order to choose among the set of available paths, we attach a number to each one—in the example of a bead on a string, the number is the time it takes for the bead to reach the end of line. The mapping between those paths and the time elapsed is a functional.

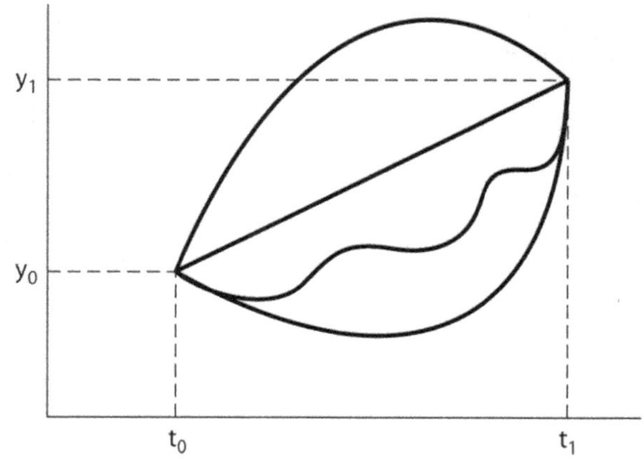

Fig. 14.2 Different paths between points (t_0, y_0) and (t_1, y_1)

Example 14.3 Consider definite integrals of functions that depend on one or more parameters. For example, the following is a functional:

$$v[\beta] = \int_0^1 e^{-\beta t} dt$$

because for every value of β, we have a different function $e^{-\beta}$ and for every such function we have a different value for $v[\beta]$. Similarly,

$$v[\sigma] = \int_{-\infty}^{\infty} (x-\mu)^2 \frac{1}{\sigma\sqrt{2\pi}} e^{-\frac{1}{2}\left(\frac{x-\mu}{\sigma}\right)^2} dx$$

is a functional. For every value of σ, we have a different normal distribution and each results in a different value of the functional.

In order to maximize the functional, $v = V[\phi]$, we need to search among the admissible functions ϕ and choose the one that maximizes v. In Chap. 12, we were searching for the maximum point on a function. This was accomplished by taking the derivatives of the function with respect to its arguments and setting them equal to zero. Needless to say, we cannot take the derivative with respect to a function. Recall that the derivative is the limit of $(f(x+\Delta x) - f(x))/\Delta x$ as Δx tends to zero. But we cannot say that the change in a function tends to zero.

Suppose we are interested in finding the maximum of a functional that depends on a family of functions that we generically represent as $y(t)$. We define the family of functions in a way that each member is identified by the value of a single parameter:

$$y(t) = y^*(t) + \varepsilon h(t) \tag{14.14}$$

For every value of ϵ we get a different function. Thus, by varying the value of ϵ we create a neighborhood around the function $y^*(t)$ (see Fig. 14.3), similar to the neighborhood around the extremum of a function. Within this neighborhood we can search for the function that results in the extremal for the functional, thus arriving at

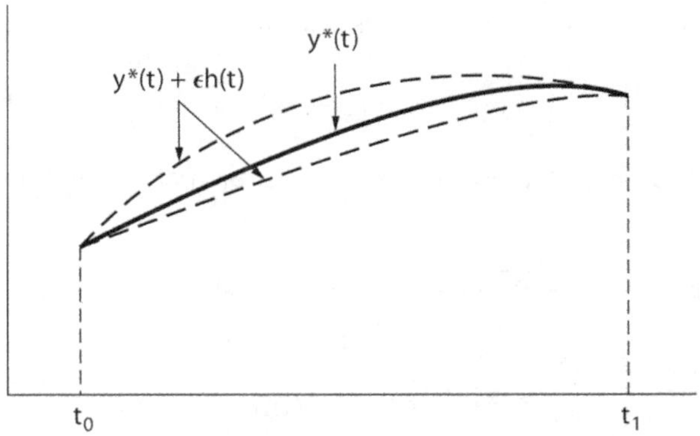

Fig. 14.3 Creating a neighborhood around the function $y^*(t)$

the conditions that the optimal function should satisfy. The technique is called the *variational method*. Let us illustrate this variational method with an example from static optimization. We are not implying that dynamic optimization is the same as static optimization; this example is for illustrative purposes only.

Example 14.4 Consider the problem of maximizing lifetime utility:

$$\begin{aligned} \max U &= \sum_{t=0}^{T} (1+\delta)^{-t} u(c_t) \\ \text{s.t.} \quad & \sum_{t=0}^{T} (1+r)^{-t} (c_t - y_t) = A_0 \end{aligned} \tag{14.15}$$

where c_t and y_t are, respectively, consumption and income at time t, δ is the constant rate of time preference, r, the constant interest rate, T, the lifetime of the consumer, and A_0, initial wealth.

The problem can be solved by forming the Lagrangian

$$\max L = \sum_{t=0}^{T} (1+\delta)^{-t} u(c_t) + \lambda \left[\sum_{t=0}^{T} (1+r)^{-t} (c_t - y_t) - A_0 \right] \tag{14.16}$$

and setting partial derivatives of L with respect to c_t, $t = 0, \ldots, T$ and with respect to λ equal to zero. We arrive at the following optimality conditions:

$$\begin{aligned} u'(c_{t+1}) &= \frac{1+\delta}{1+r} u'(c_t), \quad t = 0, \ldots, T-1 \\ \sum_{t=0}^{T} & (1+r)^{-t} (c_t - y_t) = A_0 \end{aligned} \tag{14.17}$$

We can solve this problem with the variational method and arrive at the same results. Let c_t^*, $t = 0, \ldots, T$ denote the optimal sequence of consumptions. Let us change two elements of this sequence:

$$c_s = c_s^* + x, \qquad c_{s+1} = c_{s+1}^* - (1+r)x \tag{14.18}$$

Thus, our lifetime utility maximization problem will become

$$\begin{aligned} \max_x V(x) = &\sum_{t=0}^{s-1} (1+\delta)^{-t} u(c_t^*) + (1+\delta)^{-s} u(c_s^* + x) \\ &+ (1+\delta)^{-s-1} u(c_{s+1}^* - (1+r)x) + \sum_{t=s+2}^{T} (1+\delta)^{-t} u(c_t^*) \end{aligned} \tag{14.19}$$

First note that (14.19) satisfies the budget constraint. Next, minimize the variation around the optimal path by taking the derivative of V with respect to x and set it equal to zero. We get

$$(1+\delta)^{-s} u'(c_s^* + x) - (1+r)(1+\delta)^{-s-1} u'(c_{s+1}^* - (1+r)x) = 0 \tag{14.20}$$

Evaluating (14.20) at point $x = 0$, we have

$$u'(c_{t+1}) = \frac{1+\delta}{1+r} u'(c_t) \tag{14.21}$$

Because this process can be repeated for all pairs of time periods, we can obtain all of the conditions in (14.17).

14.3 Calculus of Variations

The calculus of variations[3] is the oldest method of dynamic optimization. In its simplest form, the problem of the calculus of variations can be written as

---

[3]The calculus of variations has its origin in the challenge that the Swiss mathematician John Bernoulli (1667–1748) issued to mathematicians in 1696. He asked them to solve the Brachistochrone problem. Prominent mathematicians of the day, Newton, Leibnitz, l'Hôpital, as well as John and his elder brother James Bernoulli (1654–1705) solved the problem. But it was James's solution and his call for the solution of the more general problem of the calculus of variations that led the way in the development of the subject. Leonhard Euler (1707–1783), one of the great mathematicians and the most prolific of them all, came up with the general solution embodied in the differential equation bearing his name. It was Joseph Lagrange (1736–1813) who devised the method of variation of a function, considering $y(t) = y^*(t) + \varepsilon h(t)$, which Euler promptly adopted for finding the generalized solution of the problem of the calculus of variations; hence sometimes the equation is referred to as the Euler-Lagrange equation. It should also be noted that the genesis of the problem of the calculus of variations can be found in the work of Newton and even Galileo.

$$\max \ J = \int_{t_0}^{t_1} f(y(t), y'(t), t)dt \\ y(t_0) = y_0, \qquad y(t_1) = y_1 \tag{14.22}$$

Example 14.5 Consider the problem of the shortest distance between two points, $[t_0, y_0]$ and $[t_n, y_n]$, in a plane. We can divide the interval $[t_0, t_n]$ into n equal subintervals

$$[t_{i-1}, t_i], \qquad i = 1, \ldots, n$$

with

$$\Delta t_i = t_i - t_{i-1} = \frac{t_n - t_0}{n} = \Delta t, \qquad i = 1, \ldots, n$$

Now the distance (the Euclidean norm) between (t_{i-1}, y_{i-1}) and (t_i, y_i) is $\sqrt{(\Delta y_i)^2 + (\Delta t)^2}$, and for the whole interval,

$$J(n) = \sum_{i=1}^{n} \sqrt{(\Delta y_i)^2 + (\Delta t)^2} \tag{14.23}$$

We can rewrite (14.23) as

$$J(n) = \sum_{i=1}^{n} \sqrt{\left(\frac{\Delta y_i}{\Delta t}\right)^2 + 1}\ \Delta t \tag{14.24}$$

Letting $n \to \infty$, we have

$$J = \lim_{n \to \infty} J(n) = \int_{t_0}^{t_1} \sqrt{y'^2 + 1}\ dt \tag{14.25}$$

The problem is to minimize J.

Example 14.6 A company has to make the delivery at time T of y_T units of its product. Suppose the marginal cost of production is proportional to the rate of production, $c_1 y'$, and the cost of inventory is proportional to the level of inventory. Thus, the total cost function can be written as

$$C(t) = c_1[y'(t)]^2 + c_2 y(t) \tag{14.26}$$

The problem facing the company is

$$\min \ J[y] = \int_0^T [c_1[y'(t)]^2 + c_2 y(t)]dt \tag{14.27}$$

14.3.1 The Euler Equation

In order to solve the problem of the calculus of variations we resort to the variational method. Suppose the optimal path for $y(t)$ is $y^*(t)$. We consider deviations from this path and note that if indeed $y^*(t)$ is the path that maximizes J, then for all other paths we should have

$$\Delta J = J[y^*] - J[y] = \int_{t_0}^{t_1} f(y^*, y'^*, t)dt - \int_{t_0}^{t_1} f(y, y', t)dt \geq 0 \tag{14.28}$$

For the path that minimizes J the inequality sign is changed to $\leq$. Let us write all possible paths in relation to the optimal path $y^*(t)$. In other words, let us create a neighborhood around $y^*(t)$:

$$y(t) = y^*(t) + \varepsilon h(t) \tag{14.29}$$

Note that because we have fixed the initial and terminal points of the optimal path, we require all paths to start and end at the same points (see Fig. 14.3). Thus,

$$h(t_0) = h(t_1) = 0 \tag{14.30}$$

Now the objective functional can be written as

$$J[\varepsilon] = \int_{t_0}^{t_1} f(y^* + \varepsilon h(t), y^{*\prime} + \varepsilon h'(t), t) \tag{14.31}$$

and our problem is to maximize $J[\varepsilon]$ with respect to ϵ and evaluate the result at $\varepsilon = 0$, that is, at $y(t) = y^*(t)$ to obtain the properties of the optimal path:

$$\frac{\partial J}{\partial \varepsilon} = \int_{t_0}^{t_1} \left[f_y h(t) + f_{y'} h'(t) \right] dt = 0 \tag{14.32}$$

The second term on the RHS of (14.32) could be integrated by part:

$$\begin{aligned} \int_{t_0}^{t_1} f_{y'} h' dt &= f_{y'} h(t) \Big|_{t_0}^{t_1} - \int_{t_0}^{t_1} \left(\frac{d}{dt} f_{y'} \right) h(t)dt \\ &= - \int_{t_0}^{t_1} \frac{d}{dt} f_{y'} h(t)dt \end{aligned} \tag{14.33}$$

where the last equality is based on (14.30). Substituting (14.33) in (14.32), we have

$$\int_{t_0}^{t_1} \left[f_y - \frac{d}{dt} f_{y'} \right] h(t)dt = 0 \tag{14.34}$$

which would hold for all $h(t)$ if and only if

$$f_y - \frac{d}{dt} f_{y'} = 0 \tag{14.35}$$

This is the *Euler equation* and the necessary condition for solving the problem of the calculus of variations.

Example 14.7 Consider the problem of finding the extremal of the following functional:

$$J = \int_0^1 (4ty + y'^2)\, dt,$$
$$y(0) = \frac{2}{3}, \quad y(1) = 2$$

The Euler equation is

$$4t - \frac{d}{dt}(2y') = 0, \quad \Rightarrow \quad y'' = 2t$$

Thus,

$$y' = t^2 + c_1, \quad \text{and} \quad y = \frac{1}{3}t^3 + c_1 t + c_2$$

Using the initial and terminal conditions, we have

$$c_2 = \frac{2}{3}, \qquad c_1 = 1$$

Combining the results, we have

$$y = \frac{1}{3}t^3 + t + \frac{2}{3}$$

Example 14.8 In Example 14.5, our problem was

$$\min J = \int_{t_0}^{t_1} \sqrt{y'^2 + 1}\, dt$$
$$y(t_0) = y_0, \quad y(t_1) = y_1$$

Applying the Euler equation to the above problem, we have

$$-\frac{d}{dt}\left[\frac{y'}{\sqrt{y'^2 + 1}}\right] = 0$$

which implies

$$\frac{y'}{\sqrt{y'^2 + 1}} = \text{constant}$$

which in turn requires $y\prime$ to be constant. But if the derivative of the function y is constant, then

$$y = a + bt$$

In other words, the shortest distance between two points is a straight line. You don't look surprised! We can determine the free parameters a and b using the initial and terminal conditions. Thus,

$$b = \frac{y_n - y_0}{t_n - t_0}, \quad a = \frac{t_n y_0 - t_0 y_n}{t_n - t_0}$$

Example 14.9 In Example 14.6, our problem is

$$\begin{aligned} \min C &= \int_0^T (c_1 y'^2 + c_2 y) dt \\ y(0) = 0,& \quad y(T) = y_T \end{aligned}$$

Applying the Euler formula, we get

$$c_2 - \frac{d}{dt}(2c_1 y') = c_2 - 2c_1 y'' = 0$$

Thus, we have

$$y'' = \frac{c_2}{2c_1}$$

Integrating both sides twice, we have

$$y = \frac{c_2}{4c_1} t^2 + \theta_1 t + \theta_2$$

where θ_1 and θ_2 are constants of integrations. Using the initial and terminal conditions

$$\theta_2 = 0, \quad \text{and} \quad \frac{c_2}{4c_1} T^2 + \theta_1 T = y_T$$

Solving for θ_1, we can write the production schedule as

$$y = \frac{c_2}{4c_1} t^2 + \left(\frac{y_T}{T} - \frac{c_2}{4c_1} T \right) t, \qquad 0 \le t \le T$$

14.3.2 Second-Order Conditions

In static optimization we had second-order conditions that distinguished between maxima and minima. The corresponding condition for the calculus of variations is called the *Legendre condition*.[4] The proof of the Legendre theorem is too involved to be presented here. We simply state, without proof, that a necessary condition for the functional

$$J = \int_{t_0}^{t_1} f(y(t), y'(t), t)dt \tag{14.36}$$

to have a minimum is

$$f_{y'y'} \geq 0, \qquad t_0 \leq t \leq t_1 \tag{14.37}$$

Similarly for a maximum, the necessary condition is

$$f_{y'y'} \leq 0, \qquad t_0 \leq t \leq t_1 \tag{14.38}$$

Two points deserve attention. First, the minimum condition (14.37) and the maximum condition (14.38) should hold for all points on the function $y(t)$. Second, whereas in static optimization the second-order condition was a sufficient condition, the Legendre condition is necessary, but not sufficient.

Example 14.10 In Example 14.7, we had

$$f_{y'} = 2y'$$

Therefore,

$$f_{y'y'} = 2 \geq 0$$

which is the necessary condition for a minimum.

Example 14.11 In Example 14.8, we had

$$f_{y'} = \frac{y'}{\sqrt{y'^2 + 1}}$$

Taking the second derivative with respect to y',

---

[4]For the French mathematician Adrien-Marie Legendre (1752–1833).

we get

$$
\begin{aligned}
f_{y'y'} &= \frac{(y'^2+1)^{\frac{1}{2}} - y'^2(y'^2+1)^{\frac{-1}{2}}}{y'^2+1} \\
&= \frac{y'^2+1-y'^2}{(y'^2+1)\sqrt{y'^2+1}} \\
&= \frac{1}{(y'^2+1)\sqrt{y'^2+1}} \geq 0
\end{aligned}
$$

which is the necessary condition for a minimum. Thus, we indeed have found the minimum distance between two points.

Example 14.12 In Example 14.9, we had

$$f_{y'} = 2c_1 y'$$

Therefore,

$$f_{y'y'} = 2c_1 \geq 0$$

Thus, the path we found represents the minimum cost production schedule.

14.3.3 Generalizing the Calculus of Variations

The calculus of variations can be generalized in several directions. These subjects, however, are beyond the scope of the present chapter, and we simply mention them here.[5] These include the *transversality condition*, generalizing the functional to depend on several state variables, and introducing constraints. So far we have assumed that the terminal condition is in the form of $y(t_1) = y_1$, that is, we know exactly where the curve ends. Suppose instead we only know that the curve should land on a particular line. Then the terminal value of the function $y^*(t_1)$ has to conform to the restriction and at the same time obey the optimality condition. The equation that incorporates these conditions for the terminal value of $y^*(t_1)$ is called the transversality condition.

The problem of the calculus of variations in (14.22) can be generalized to include many state variables. We can consider a vector of state variables

$$
\begin{aligned}
\mathbf{y}(t) &= (y_1(t), y_2(t), \ldots, y_k(t)) \\
\mathbf{y}'(t) &= (y'_1(t), y'_2(t), \ldots, y'_k(t))
\end{aligned}
\tag{14.39}
$$

[5] The interested reader may want to consult more specialized texts, for example, Gelfand and Fomin (1963), Intriligator (1971), Bryson and Ho (1975), Kamien and Schwartz (1991), and Chiang (1992).

Then the k dimensional problem can be written as

$$\max\ J = \int_{t_0}^{t_1} f(\mathbf{y(t)}, \mathbf{y}'(\mathbf{t}), t)dt$$
$$\mathbf{y}(t_0) = \mathbf{y_0}, \qquad \mathbf{y}(t_1) = \mathbf{y_1} \tag{14.40}$$

Finally, once we have several state variables we may impose a constraint on them. For example, we may impose the constraint

$$\int_{t_0}^{t_1} g(\mathbf{y}, \mathbf{y}', t)dt = c \tag{14.41}$$

Then the constraint could be incorporated into the objective functional using *costate variables*, which are the counterparts of Lagrange multipliers in dynamic optimization.

14.3.4 Exercises

E.14.1 Write the Euler equation for the following problems:

$$i.\quad \int_0^1 (ty' + y'^2)dt, \qquad ii.\quad \int_0^1 (y^2 + 4yy' + 3y'^2)dt$$

$$iii.\quad \int_0^1 (t - y)^2 dt, \qquad iv.\quad \int_0^1 (y^2 + y'^2 + 2ye^t)dt$$

E.14.2 Check the Legendre condition for all functionals in E.14.1.

E.14.3 Show that the following optimization problem has no solution. [*Hint*: Show that the function that satisfies the Euler equation does not meet other conditions.]

$$\min J = \int_0^1 yy'tdt$$
$$y(0) = 0, \quad y(1) = 1$$

14.4 Dynamic Programming

Optimization models involving time as a discrete variable are dynamic programming problems.[6] To illustrate the technique, it is best to start with a simple problem that has become a classic. Suppose a salesman has to travel from point A to point

[6]Dynamic programming is the brainchild of the brilliant American mathematician Richard Bellman (1920–1984).

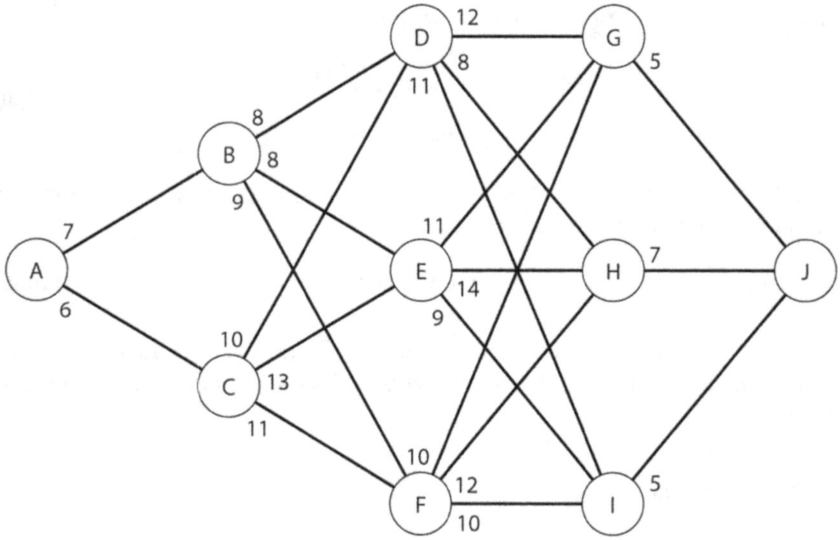

Fig. 14.4 The traveling salesman problem

J, as in Fig. 14.4. He has to travel through four stages and in each stage he has a number of options. The cost of travel on each route is shown as a number near the line connecting one point to another. Which route should he choose to minimize his cost?

Before embarking on a systematic solution, observe two important points. First, whereas starting at point A and choosing the least expensive route at every stage seems an intuitively appealing solution, it is not necessarily the optimal solution. In our example, such a route is *ACDHJ* at a cost of 31. As we shall see below, we can do better. Second, evaluating all possible routes is not an appealing option. In our example we have two options in stage 1 and three options in stages 2 and 3, resulting in a total of 18 possible routes. It is clear that in a larger problem, the number of routes would be far greater. For example, if there are 11 stages, and in each of the first 10 stages we have 9 options, then there would be close to 3.5 billion routes to evaluate. It seems that we need a more efficient way of finding the optimal path. Let us denote the cost between every two points by $f(X_t,\ X_{t+1})$ where t denotes the stage. For example,

$$f(A,B) = 7, \quad f(C,F) = 11, \quad \text{and} \quad f(E,I) = 9$$

The method we shall employ here, which will be generalized later, is a recursive one. We start at the endpoint n and work our way backward to point 0. At each stage i we choose the best option for the path from point i to point n and assume that the decisions for all previous stages up to that point have been optimal. In our example we start at point J and consider all our options. They are

	J
G	5
H	7
I	5

It is clear that if in the penultimate stage we are at G or I, then the optimal path to the last stage would cost 5, whereas if we are at H, then the last stage would cost 7. Now suppose that we are at a particular point, say, G at stage $t = 3$. Let us denote the optimal value of the travel cost for the remaining stage by $v(G)$. Thus, we have

$$v(G) = 5, \quad v(H) = 7, \quad \text{and} \quad v(I) = 5$$

Next, we move one step backward and consider the cost of moving from stage 3 to point J. We have

$f(X_2, X_3)$				$f(X_1, X_2) + v(X_3)$			
	G	*H*	*I*	*GJ*	*HJ*	*IJ*	$\min[f+v]$
D	12	8	11	17	15	16	15
E	11	14	9	16	21	14	14
F	10	12	10	15	19	15	15

From the vantage point of stage 2, therefore, optimum paths and costs are

Path	Cost
DHJ	15
EIJ	14
FGJ or *FIJ*	15

We need only concern ourselves with the four paths listed in the table above. Thus, we eliminate from consideration all other paths. Stepping back to stage 1, we repeat the exercise

$f(X_2, X_3)$				$f(X_1, X_2) + v(X_3)$			
	D	*E*	*F*	*DHJ*	*EIJ*	*FGJ* or *FIJ*	$\min[f+v]$
B	8	8	9	23	22	24	22
C	10	13	11	25	27	26	25

At this stage for optimal paths and costs we have

Path	Cost
DEIJ	22
CDHJ	25

Again we retain the two optimal paths and discard the other four. There is one more step to take.

$f(X_0, X_1)$			$f(X_0, X_1) + v(X_1)$		
	B	*C*	*BEIJ*	*CDHJ*	$\min[f + v]$
A	7	6	29	31	29

Thus, the optimal path is *ABEIJ* and the cost is 29. Note that we did not evaluate the cost for all possible paths as some routes were eliminated along the way. Yet we had to evaluate all the admissible options at every stage. It is not difficult to see that increasing the number of stages and available options in each stage will cause a tremendous increase in the number of required calculations, making the task of finding the optimal route time consuming, costly, and at times impossible. This is the famous *curse of dimensionality*, which precludes the use of dynamic programming in many real-world numerical problems.

In economics, however, we are more interested in the general construct of dynamic programming and its solution. The general setup is to minimize or maximize the sum of $f(x_t, x_{t+1}),\ t = 0, \ldots, n$, where x_t is the value of the variable of interest at time t to be decided by the optimizing agent. The problem has a starting and an end point and the value of x has to be chosen from an admissible set. The general format of the problem when seeking a minimum, therefore, can be written as

$$\begin{aligned} &\min_{\{x_{t+1}\}_{t=0}^{n-1}} \sum_{t=0}^{n} \beta^t f(x_t, x_{t+1}) \\ &\text{s.t.}\quad x_0 \in X_0, \qquad x_{t+1} \in \Gamma(x_t), \quad t = 0, 1, 2, \ldots \end{aligned} \tag{14.42}$$

and for a maximum problem as[7]

$$\begin{aligned} &\max_{\{x_{t+1}\}_{t=0}^{n-1}} \sum_{t=0}^{n} \beta^t f(x_t, x_{t+1}) \\ &\text{s.t.}\quad x_0 \in X_0, \qquad x_{t+1} \in \Gamma(x_t), \quad t = 0, 1, 2, \ldots \end{aligned} \tag{14.43}$$

[7] A more general formulation recognizes that *min* or *max* of an objective function may be unattainable; therefore, they are replaced by *inf* and *sup*, respectively.

Based on the example of traveling salesman, we can define the optimal function at time t as

$$v(x_t) = \min_{z \in \Gamma(x_t)} [f(x_t, z) + \beta v(z)] \tag{14.44}$$

where $v\,(x_t)$ is the optimal value of the functional at time t. The variable z ranges over all possible values of x_{t+1}, which denotes the optimal value of x at stage t+1. The same could be written for the case of maximization:

$$v(x_t) = \max_{z \in \Gamma(x_t)} [f(x_t, z) + \beta v(z)] \tag{14.45}$$

Thus, the solutions to (14.42) and (14.43) require working recursively the formulas in (14.44) and (14.45), respectively.

Example 14.13 (Human Capital Accumulation). Now it can be recognized that the problem of human capital accumulation in Example 14.2 is one of dynamic programming. Thus, the solution to the problem

$$\max_{\{k_{t+1}\}_{t=1}^{T-1}} \sum_{t=1}^{T} \beta^t k_t \phi\left(\frac{k_{t+1}}{k_t}\right) \tag{14.46}$$
$$\text{s.t. } (1-\delta)k_t \leq k_{t+1} \leq (1+\lambda)k_t, \quad \forall t, \quad k_1 > 0, \text{ given}$$

is

$$v(k_t) = \max_{(1-\delta)k_t \leq z \leq (1+\lambda)k_t} [k_t \phi(z/k_t) + \beta v(z)] \tag{14.47}$$

Example 14.14 (Job Search). Consider a worker who lives n periods and in each period faces two alternatives: She is offered a job that pays x income that she can take or she can stay unemployed in that period and search for a possibly better job. Income, y, from jobs among which the worker is searching, is randomly distributed with density function $\varphi(y)$. Therefore, the expected income for the next period is

$$E(y) = \int_Y y\phi(y)dy \tag{14.48}$$

where Y is the set of all possible values of y. There is no point in staying unemployed and searching in the last stage of one's working life. Therefore, the worker accepts the job offered and her income is x. But in one period before the last, there are two options: to accept the job and have an income of $x + \delta x$ or search for a job with an expected income of $\delta E(y)$ where $\delta$is the discount rate. The recursion formula is

$$v_{n-1}(x) = \max[(1+\delta)x, \delta E(y)] \tag{14.49}$$

If we go back k periods, her choices are to accept the job in which case her earnings for the remainder of her working life will be

$$(1 + \delta + \delta^2 + \ldots + \delta^{k-1})x = x\frac{1 - \delta^k}{1 - \delta} \tag{14.50}$$

Or stay unemployed and search with the prospect being

$$\delta E[v_{n-k+1}(y)] \tag{14.51}$$

Thus, the recursion formula is

$$v_{n-k}(x) = \max\left\{x\frac{1 - \delta^k}{1 - \delta}, \delta E[v_{n-k+1}(y)]\right\} \tag{14.52}$$

14.4.1 Exercises

E.14.4 Figure 14.5 shows the routes from A to J together with the costs of each segment of the road. Using the method outlined in Sect. 14.4, find the least expensive route from A to J.

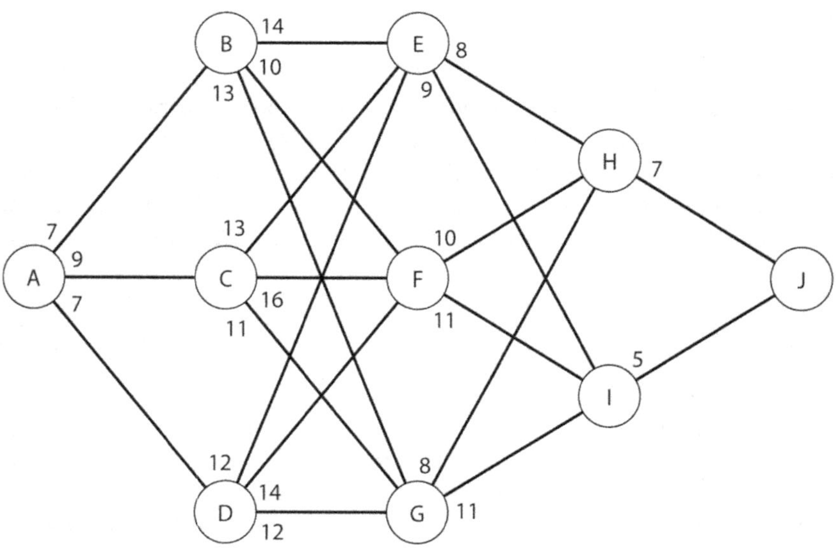

Fig. 14.5 The traveling salesman problem

E.14.5 In Example 14.14 find the reservation wage of the worker, which is the wage rate that makes the worker indifferent between accepting the job and staying unemployed and continuing the search in the next period.

14.5 The Maximum Principle

The control problem[8] in its simplest form can be written as

$$\begin{aligned} &\max \int_{t_0}^{t_1} f(t, y, u)dt \\ &\text{s.t. } y' = g(y, u, t) \\ &y(t_0) = y_0, \quad y(t_1) = y_1 \end{aligned} \tag{14.53}$$

where y is the state variable, u, the control variable and, t, time. By letting $u = y'$ we get the calculus of variations problem. In other words, calculus of variations is a special case of the control problem where the control variable is the derivative of the state variable. The problem in (14.53) is one of constrained maximization and we use the method of Lagrange multipliers. Thus, our problem becomes

$$\max \ L = \int_{t_0}^{t_1} \left\{ f(t, y, u)dt + \lambda(t)[g(y, u, t) - y'] \right\} dt \tag{14.54}$$

We have to maximize this function by choosing $\lambda(t)$, $y(t)$, and $u(t)$. Because these are all functions, we should resort to the same device we used in solving the problem of the calculus of variations, that is, by defining a family of functions or creating a neighborhood for each of them. Because this would make for a cluttered equation with many variables, we proceed one function at a time. First let

$$\lambda(t) = \lambda^*(t) + \varepsilon p(t) \tag{14.55}$$

where $\lambda^*(t)$ is the path of λ that would maximize L. Substituting (14.55) in (14.54), we have

$$L[\varepsilon] = \int_{t_0}^{t_1} \left\{ f(t, y, u)dt + (\lambda^*(t) + \varepsilon p(t))[g(y, u, t) - y'] \right\} dt \tag{14.56}$$

Taking the derivative of $L[\varepsilon]$ with respect to ε and setting it equal to zero, we get

$$\int_{t_0}^{t_1} p(t)[g(y, u, t) - y']\, dt = 0 \tag{14.57}$$

[8] Russian mathematician Lev Semenovich Pontryagin (1908–1988) and his colleagues proposed the method of maximum principle.

(12.57) will hold for all functions $p(t)$ if and only if

$$y' = g\,(y,\ u,\ t) \tag{14.58}$$

In other words, the constraint or the equation of motion is a necessary condition for maximizing L. Before tackling the other two functions, note that λ is a function of time and we can integrate the term $\lambda y'$ by part

$$\int_{t_0}^{t_1} -\lambda y' dt = -\lambda y|_{t_0}^{t_1} + \lambda' y \tag{14.59}$$

Substituting (14.59) in (14.54), we have

$$\begin{aligned}\max \quad L = & \int_{t_0}^{t_1} \{f(t, y, u)dt + \lambda g(y, u, t) + y\lambda'\}dt \\ & -[y(t_1)\lambda(t_1) - y(t_0)\lambda(t_0)]\end{aligned} \tag{14.60}$$

Now let us define the *Hamiltonian* as

$$H(t, y, u, \lambda) = f(t, y, u) + \lambda g(y, u, t) \tag{14.61}$$

Then our problem becomes

$$\max \quad L = \int_{t_0}^{t_1} \{H(t, y, u, \lambda) + y\lambda'\}dt - [y(t_1)\lambda(t_1) - y(t_0)\lambda(t_0)] \tag{14.62}$$

But again we are dealing not with a function but a functional. Thus, as before, we define

$$y(t) = y^*(t) + \varepsilon h(t) \tag{14.63}$$

Substituting (14.63) in (14.62) we have

$$\begin{aligned}L[\varepsilon] = & \int_{t_0}^{t_1} \left\{H\left(t, y^*(t) + \varepsilon h(t), u, \lambda\right) + \left(y^*(t) + \varepsilon h(t)\right) \lambda'\right\} dt \\ & -[y(t_1)\lambda(t_1) - y(t_0)\lambda(t_0)]\end{aligned} \tag{14.64}$$

Taking the derivative of $L[\varepsilon]$ with respect to ε, setting it equal to zero, and evaluating the resulting function at $\varepsilon = 0$, we have

$$\int_{t_0}^{t_1} \left[\frac{\partial H}{\partial y} + \lambda'\right] h(t)dt = 0 \tag{14.65}$$

which again would hold for all $h(t)$ if and only if

$$\frac{\partial H}{\partial y} + \lambda' = 0 \quad \Rightarrow \quad \frac{\partial H}{\partial y} = -\lambda' \tag{14.66}$$

Finally, repeating for u what we did for y, we write

$$u(t) = u^*(t) + \varepsilon q(t) \tag{14.67}$$

Substituting (14.67) in (14.62) results in

$$\begin{aligned} L[\varepsilon] = &\int_{t_0}^{t_1} \left\{H(t, y^*(t), u^*(t) + \varepsilon q(t), \lambda) + y^*(t)\lambda'\right\} dt \\ &-[y(t_1)\lambda(t_1) - y(t_0)\lambda(t_0)] \end{aligned} \tag{14.68}$$

Taking the derivative of $L[\varepsilon]$ with respect to ε and setting it equal to zero,

$$\int_{t_0}^{t_1} \frac{\partial H}{\partial u} q(t) dt = 0 \tag{14.69}$$

which would hold for all $q(t)$ if

$$\frac{\partial H}{\partial u} = 0 \tag{14.70}$$

Putting all the conditions together, we have

$$\begin{aligned} \frac{\partial H}{\partial y} &= -\lambda' \\ \frac{\partial H}{\partial u} &= 0 \\ y' &= g(y, u, t) \end{aligned} \tag{14.71}$$

In addition, the initial and terminal conditions have to be satisfied. Note that we assumed that both the initial and terminal values of the state variable $y(t)$ are fixed. If, however, the terminal condition is left free, that is, $y(t_1)$ is not fixed, then we need the transversality condition

$$\lambda(t_1) = 0 \tag{14.72}$$

Example 14.15

$$\begin{aligned} &\max \int_0^4 -(3t^2 + 2u^2) dt \\ &y' = 4u, \\ &y(0) = 8, \quad y(4) = 28 \end{aligned}$$

The Hamiltonian is

$$H(t, y, u, \lambda) = -3t^2 - 2u^2 + 4\lambda u$$

and the optimality conditions are

$$\frac{\partial H}{\partial u} = -4u + 4\lambda = 0, \quad \Rightarrow \quad u = \lambda$$
$$\frac{\partial H}{\partial y} = 0 = \lambda' \quad \Rightarrow \quad \lambda = c_1$$

It follows that $u = c_1$ and because $y' = 4u$, we have

$$y = 4c_1 t + c_2$$

Using the initial and terminal conditions, we can determine the values of c_1 and c_2:

$$c_2 = 8, \qquad 4c_1 = 5$$

and the solution is

$$y = 8 + 5t$$

Example 14.16

$$\max\ J = \int_0^2 (2y + u)dt$$
$$y' = 5 - u^2, \quad y(0) = 1$$

The Hamiltonian is

$$H = 2y + u + \lambda(5 - u^2)$$

and the optimality conditions are

$$H_u = 1 - 2\lambda u = 0$$
$$-H_y = \lambda' = -2$$

First we can solve for $\lambda(t)$

$$\lambda = c_1 - 2t$$

Using the transversality condition in (14.73) we can determine c_1. Thus,

$$\lambda = 4 - 2t$$

and

$$u(t) = \frac{1}{2\lambda(t)} = \frac{1}{4(2 - t)}$$

On the other hand,

$$y' = 5 - u^2 = 5 - \frac{1}{16(2 - t)^2}$$

Integrating both sides, the solution is

$$y(t) = 5t - \frac{1}{16(2-t)} + c$$

Using the initial condition $y(0) = 1$, we obtain the specific solution as

$$y(t) = 5t - \frac{1}{16(2-t)} + \frac{33}{32}$$

Example 14.17 (Ramsey Problem). Our discussion of the Ramsey problem at the beginning of this chapter ended with the following control problem:

$$\begin{aligned} \max\ U &= \int_0^{\infty} u(c)e^{-\theta t}dt \\ \text{s.t.}\ \frac{dk}{dt} &= f(k) - c - nk \\ c,\ k &\geq 0,\ \forall t \end{aligned} \tag{14.73}$$

where, k, is the state variable and, c, the control variable. The Hamiltonian is

$$H = u(c)e^{-\theta t} + \mu[f(k) - c - nk] \tag{14.74}$$

The optimality conditions are

$$\begin{aligned} \frac{\partial H}{\partial c} &= u'(c)e^{-\theta t} - \mu = 0, \\ \frac{\partial H}{\partial k} &= \mu[f'(k) - n] = -\mu' \end{aligned} \tag{14.75}$$

To make life a bit easier, let $\lambda = \mu e^{\theta t}$. Then the optimality conditions will be

$$\lambda e^{-\theta t} = u'(c)e^{-\theta t} \quad \Rightarrow \quad u'(c) = \lambda \tag{14.76}$$

and

$$\mu' = \lambda' e^{-\theta t} - \theta\lambda e^{-\theta t} = \lambda e^{-\theta t}[n - f'(k)] \tag{14.77}$$

The last equation can be rewritten as

$$\lambda' = \lambda[n + \theta - f'(k)] \tag{14.78}$$

Substituting (14.76) in (14.78) and noting that

$$\lambda' = u''(c)\frac{dc}{dt} \tag{14.79}$$

the solution to our problem is

$$\frac{u''(c)}{u'(c)}\frac{dc}{dt} = n + \theta - f'(k) \tag{14.80}$$

As a specific example, if $u(c) = \ln c$

$$\frac{1}{c}\frac{dc}{dt} = f'(k) - n - \theta \tag{14.81}$$

In plain English, the optimal plan requires the rate of growth of per capita consumption to be equal to the marginal product of per capita capital less the rate of growth of population, less the rate of time preference. The transversality condition requires some explanation. Recall from (14.62) that the term $y(t_1)\lambda(t_1)$ appears in the objective functional. We noted, however, that for problems in which the state variable has a free terminal value, the transversality condition is $\lambda(t_1) = 0$. In the Ramsey problem we do not have a fixed terminal value for the state variable and, in addition, the time horizon is infinite. In such a problem the transversality condition is

$$\lim_{t_1\to\infty} y(t_1)\lambda(t_1) = 0 \tag{14.82}$$

In this way we can find a finite solution to the problem. In the Ramsey problem, the equivalent of (14.82) is

$$\lim_{t\to\infty} k(t)\mu(t) = \lim_{t\to\infty} k(t)\lambda(t)e^{-\theta t} = 0 \tag{14.83}$$

Example 14.18 (Ramsey Model in an Open Economy). Our previous example dealt with a closed economy in that no allowances were made for international trade and investment. For an open economy the per capita income is defined as

$$f(k) - nk = c + i + x$$

where x is per capita net export—the difference between per capita export and import. If we let b to denote the amount of capital U.S. nationals hold abroad less the amount of capital foreigners hold in the United States, then the net per capita income from abroad is rb where r is the international rate of interest. Note that b and, as a result, rb can be negative if the claim of the rest of the world on the United States exceeds the claims of U.S. citizens on the rest of the world. The current account would be equal to net export plus the net income from abroad. Because the current account would be the change in b, we can write

$$\begin{aligned}\frac{db}{dt} &= x + rb\\ &= f(k) - nk - c - i + rb\end{aligned}$$

Now we can reformulate the Ramsey problem as

$$\begin{aligned}
\max U = & \int_0^{\infty} u(c)e^{-\theta t}dt \\
\text{s.t.} \quad & \frac{dk}{dt} = i \\
& \frac{db}{dt} = f(k) - nk - c - i + rb \\
& c,\ k,\ i \geq 0,\ \forall t
\end{aligned}$$

We leave it to the reader to form the Hamiltonian and solve this problem (see E.14.9).

Example 14.19 (Tobin's q). Consider a firm whose profit Π is a function of its capital stock K:

$$\Pi = \Pi(K) \tag{14.84}$$

Change in capital stock is equal to investment I less depreciation δK :

$$\frac{dK}{dt} = I - \delta K \tag{14.85}$$

A profit maximizing firm will expand its capital stock until the marginal profit is equal to the cost of obtaining an additional unit of capital. The cost of an additional unit of capital is equal to investment cost plus adjustment cost:

$$C(I) = I + \alpha(I) \tag{14.86}$$

We have set the price of the capital good equal to one, which causes no loss of generality because the model has only one good. The adjustment cost $\alpha(I)$ can be justified on the ground that installing additional machinery, expanding the buildings, and using new equipment require planning, installation, training, and usually disrupt the operation of a firm. The profit maximization problem can be formulated as

$$\begin{aligned}
\max \quad & \int_0^{\infty} [\Pi(K) - I - \alpha(I)]e^{-rt}dt \\
\text{s.t.} \quad & \frac{dK}{dt} = I - \delta K \\
& I,\ K \geq 0,\ \forall t
\end{aligned} \tag{14.87}$$

where r is the real rate of interest. The Hamiltonian for this problem is

$$H = [\Pi(K) - I - \alpha(I)]e^{-rt} + \lambda(I - \delta K)$$

The first optimality condition is

$$\frac{\partial H}{\partial K} = e^{-rt}\Pi'(K) - \lambda\delta = -\lambda'$$

Let

$$\lambda = qe^{-rt}$$

which implies

$$\lambda' = -rqe^{-rt} + q'e^{-rt}$$

Therefore, the first optimality condition can be written as

$$q' - (r + \delta)q = -\Pi'(K) \tag{14.88}$$

This is a differential equation whose solution we shall discuss in Chap. 15. Here we simply state that the solution is

$$q(t) = \int_{\zeta=t}^{T} e^{-(r+\delta)(\zeta-t)}\Pi'(K)d\zeta + e^{-(r+\delta)(T-\zeta)}q(T) \tag{14.89}$$

You can check that indeed the above is the solution of our differential equation by computing *dq/dt* and plugging q and q' in (14.88). The transversality condition requires that

$$\lim_{t\to\infty} e^{-rt}q(t)K(t) = 0 \tag{14.90}$$

This requires that as $T \to \infty$, the second term in (14.89) go to zero. Therefore, we can write

$$q(t) = \int_{\zeta=t}^{\infty} e^{-(r+\delta)(\zeta-t)}\Pi'(K)d\zeta$$

Because the value of any asset equals the present value of the income stream it generates, then q is the value of an additional unit of capital. Returning to our optimization problem, the second condition requires

$$\frac{\partial H}{\partial I} = (-1 - \alpha'(I))e^{-rt} + \lambda = 0$$

Or substituting for λ,

$$q = 1 + \alpha'(I)$$

that is, q is also equal to marginal cost (including adjustment cost) of one unit of additional capital. Because adjustment cost increases with the size of investment I, it follows that $\alpha'(I) > 0$ and $q > 1$. This is the q theory of investment introduced by James Tobin (Nobel Laureate, 1981). The rate of investment, according to the theory, is a function of the ratio of market value of new additional investment goods to its cost of replacement (average q). We derived the marginal q that is the ratio of the market value of additional units of capital to its replacement cost (recall that we set the price of a unit of capital good at 1).

14.5.1 Necessary and Sufficient Conditions

In addition to the necessary conditions derived above, a maximum should also satisfy the condition

$$H_{uu} = \frac{\partial^2 H}{\partial u^2} \le 0 \tag{14.91}$$

and a minimum should satisfy

$$H_{uu} = \frac{\partial^2 H}{\partial u^2} \ge 0 \tag{14.92}$$

It is important to keep in mind that the above conditions are necessary and not sufficient conditions. Based on a theorem due to the Nobel prize-winning economist Kenneth Arrow, it turns out that with certain restrictions on the Hamiltonian, the necessary conditions for the solution of the control problem are also sufficient conditions. Consider the Hamiltonian in which u is replaced by its optimal value u^*

$$H^0(t,y,u^*,\lambda) = f(t,y,u^*) + \lambda g(t,y,u^*) \tag{14.93}$$

Note that other arguments are not replaced by their optimal values. Now if H^0 is concave in the variables y and t at all points in the interval $[t_0,\ t_1]$, then the necessary conditions of optimality in (14.71) are also sufficient conditions.

14.5.2 Exercises

E.14.6 Write the first-order optimality conditions for the following control problems:

i. $\max \int_0^{10} (yu - 2y^2 - 5u^2)dt$

$y' = y + 3u, \qquad y(0) = 6$

ii. $\max \int_0^{10} \sqrt{3y + u^2 dt}$

$y' = 2u, \qquad y(0) = 1$

E.14.7 Check the second-order necessary condition for problems in E.14.6 and for Examples 14.15 and 14.16.

E.14.8 Redo the Ramsey problem with the Cobb-Douglas production function

$$Y = AK^{\alpha}L^{1-\alpha}$$

E.14.9 Solve the Ramsey problem for an open economy in Example 14.18.

Part V
Differential and Difference Equations

Chapter 15
Differential Equations

The mathematical device that epitomizes the birth of modern sciences is the differential equation.[1] Newton's second law states that the acceleration of a particle is inversely proportional to its mass and is directly related to the force applied to it. Thus,

$$\frac{d^2x}{dt^2} = \frac{1}{m}F(x) \tag{15.1}$$

where m is mass of the particle, x is its location, and F denotes the force field. Differential equations model the dynamics of a system and show how the variables of interest evolve over time.

Economic life is a dynamic process and it seems natural to model economic phenomena using differential equations. An objection may be raised here that in economics all variables are measured at discrete time intervals. Therefore, models that treat time as a continuous variable may not be suitable for economic analysis. We can offer two counterarguments.

First, in economic life, time is a continuous variable; it is the measurement convention that is artificial. At every moment numerous decisions are made, many transactions are concluded, and many production processes never stop. It is always possible to model an economic process using continuous time and differential equations and then approximate it with the discrete time model for estimation and simulation. Second, mathematical theory of differential equations is too rich and powerful to forego it in economic analysis.

15.1 Examples of Continuous Time Dynamic Economic Models

Equilibrium analysis in economics produces a benchmark against which to gauge actual events and the workings of an economy. Comparative static analysis compares two equilibrium points in order to determine the effects of a change in exogenous variables or parameters of a model on endogenous variables. Yet we know that in the

---

[1] Of course today the equation that symbolizes science is Einstein's $E = mc^2$.

K. Dadkhah, *Foundations of Mathematical and Computational Economics,* 2nd ed., DOI 10.1007/978-3-642-13748-8_15, © Springer-Verlag Berlin Heidelberg 2011

real world the economy is always in the process of adjustment and, therefore, out of equilibrium. It is as if the economy is chasing an equilibrium and before reaching it, the position of the equilibrium has changed. To model the movement from one equilibrium to another and to study the path of different variables when the system is out of equilibrium, we need dynamic models.

Example 15.1 The purchasing power parity (PPP) model of exchange rates determination in its absolute version postulates that the exchange rate is equal to the ratio of the price indices in the two countries involved. Because any two countries have different base years for their price indices and the indices may contain nontradable goods and services, we write the model as

$$S^e = a\frac{P^d}{P^f} \tag{15.2}$$

where S is the exchange rate, that is, the price of foreign currency in terms of domestic currency, say the number of dollars (domestic currency) exchanged for one Swiss franc (foreign currency). P^d and P^f are, respectively, domestic and foreign price indices. Let ρ_d and ρ_f be domestic and foreign inflation rates, respectively. Then we can write

$$\frac{P^d}{P^f} = \frac{P_0^d}{P_0^f} e^{(\pi_d - \pi_f)t} = P_0 e^{\pi t} \tag{15.3}$$

where P_0^d and P_0^f are, respectively, the domestic and foreign price indices at time $t = 0$, $P_0 = P_0^d/P_0^f$, and $\pi = \pi_d - \pi_f$. Combining (15.2) and (15.3) we can write

$$s^e = \ln S^e = \ln a + \ln P_0 + \pi t \tag{15.4}$$

Research on international finance has shown that the PPP is an equilibrium concept, not a causal relationship. If that is true, then for the theory to make sense we need to have an adjustment process or an error correction mechanism. A continuous time error correction model would be

$$\frac{ds}{dt} = \gamma(s^e - s) \tag{15.5}$$

or

$$\frac{ds}{dt} + \gamma s = \beta_0 + \beta_1 t \tag{15.6}$$

where $\beta_0 = \gamma(\ln a + \ln P_0)$ and $\beta_1 = \gamma\pi$. This is a first-order linear differential equation.

Example 15.2 Excess demand—that is, the difference between demand and supply of a good or service—would cause the price of that good or service to rise:

$$\frac{dP}{dt} = \gamma(D - S) \tag{15.7}$$

Let

$$\begin{aligned} D &= D(P, Y) \\ S &= S(P) \\ Y &= Y(t) \end{aligned}$$

Then

$$\frac{dP}{dt} = \gamma(D(P, Y) - S(P)) = \gamma E(P, t) \tag{15.8}$$

which is a nonhomogeneous differential equation.

Example 15.3 The most famous model in economics using differential equations is *Solow's growth model*. Solow assumed that output depends on labor and capital and the production function to be homogeneous of degree one. Thus,

$$Q = F(K, L) = LF\left(\frac{K}{L}, 1\right) = Lf(k) \tag{15.9}$$

Because

$$F_K > 0, \quad F_{KK} < 0$$

we have

$$f' > 0, \quad f'' < 0$$

where Q, K, and L are, respectively, output, capital and labor, and $k = K/L$. Furthermore, he assumed that a portion s of income is saved and invested, which immediately turns into additional capital:

$$\frac{dK}{dt} = I = S = sQ = sLf(k) \tag{15.10}$$

where the marginal propensity to save is assumed to be positive and less than one: $0 < s < 1$. Labor is assumed to grow at the constant rate n, thus,

$$\frac{dL}{dt} = nL \tag{15.11}$$

Note that

$$\frac{dk}{dt} = \frac{d}{dt}\left(\frac{K}{L}\right) = \frac{1}{L^2}\left[L\frac{dK}{dt} - K\frac{dL}{dt}\right] \tag{15.12}$$

Substituting for dK/dt and dL/dt, we get Solow's famous first-order differential equation:

$$\frac{dk}{dt} = sf(k) - nk \tag{15.13}$$

We will analyze this equation later in this chapter.

15.2 An Overview

To model a dynamic process with continuous time we need differential equations that connect the change in a variable with its position at any moment in time. In other words, differential equations involve both a variable and its derivatives.

Definition 15.1 The equation

$$F\left(t, y, \frac{dy}{dt}, \frac{d^2y}{dt^2}, \ldots, \frac{d^ny}{dt^n}\right) = 0 \tag{15.14}$$

is called an *ordinary differential equation* of order n.

It is called ordinary because it involves only one independent variable, t. If, in addition, it involves another variable say, x, then it is a *partial differential equation.*

Example 15.4 The equation

$$\frac{d^2y}{dt^2} - 3ty + 6 = 0 \tag{15.15}$$

is an ordinary differential equation of order two, Whereas

$$t\frac{\partial y}{\partial t} = x\frac{\partial y}{\partial x} \tag{15.16}$$

is a partial differential equation. If the equation has a linear form such as

$$\frac{d^2y}{dt^2} + b_1\frac{dy}{dt} + b_2y = g(t) \tag{15.17}$$

it is called a linear differential equation. Nonlinear differential equations involve nonlinear terms of y and its derivatives.

Example 15.5 The following are nonlinear differential equations:

$$\frac{d^2y}{dt^2} + \alpha y\frac{dy}{dt} + \beta\frac{dy}{dt} + y = 0$$

$$\frac{d^2y}{dt^2} + a\left(\frac{dy}{dt}\right)^2 - by = 0$$

$$\frac{d^2y}{dt^2} + 2\frac{dy}{dt} - 3\ln y = 0$$

If the equation in (15.14) does not explicitly depend on t, then the equation is called homogeneous; else it is called nonhomogeneous. For example, the equation in (15.17) is nonhomogeneous but if we replace $g(t)$ on the RHS in (15.17) with zero, then it will become a second-order linear homogeneous equation. If in a linear differential equation the coefficients of y and its derivatives are constant, then we are dealing with a *linear differential equation with constant coefficients*.

By the solution of a differential equation we mean a function of the form $y = \phi(t)$ such that the substitution of $\phi(t)$ and its derivatives in the original equation will turn it into an identity, in other words, a function $\phi(t)$ that satisfies the differential equation.

Example 15.6 The function

$$y = Ae^{-\rho t}$$

where A is an arbitrary constant, is the solution to the differential equation

$$\frac{dy}{dt} + \rho y = 0$$

This can be checked. Because $dy/dt = -\rho Ae^{-\rho t}$, we have

$$-\rho Ae^{-\rho t} + \rho Ae^{-\rho t} = 0$$

Example 15.7 The function

$$y = A_1 e^{-t} + A_2 e^{2t}$$

is the solution to the equation

$$\frac{d^2y}{dt^2} - \frac{dy}{dt} - 2y = 0$$

We can check this by noting that

$$\frac{dy}{dt} = -A_1 e^{-t} + 2A_2 e^{2t}$$
$$\frac{d^2y}{dt^2} = A_1 e^{-t} + 4A_2 e^{2t}$$

Substituting in the original equation, we get

$$A_1 e^{-t} + 4A_2 e^{2t} + A_1 e^{-t} - 2A_2 e^{2t} - 2A_1 e^{-t} - 2A_2 e^{2t} = 0$$

The reader can verify that both e^{-t} and e^{2t} are solutions to the differential equation. Thus, $y = A_1 e^{-t} + A_2 e^{2t}$ is a linear combination of the two solutions.

Based on the way we found the solution of second-order homogenous equations, we can infer that a third-order linear homogeneous equation has three solutions, a fourth-order equation, four solutions, and so on. In the same vein, the solution of a third-order equation involves three arbitrary constants, the solution of a fourth-order equation, four arbitrary constants, and so on.

To understand the nature of a differential equation and its solution, let us ponder its geometric meaning. Consider the differential equation

$$\frac{dy}{dt} = f(t, y) \tag{15.18}$$

and its solution

$$y = \phi(t) \tag{15.19}$$

At any point on the t-axis, say $t_1, f(t_1, y)$is the equation of tangent to the curve $\phi(t)$. Because the differential equation provides us with such an equation for all the points along the curve, it enables us to trace out the curve itself. As Fig. 15.1 shows, such a curve is not unique because we know only the slope of the curve and not its exact location. But if we know the location of one point on the curve, we can obtain a *specific solution* and pinpoint the curve.

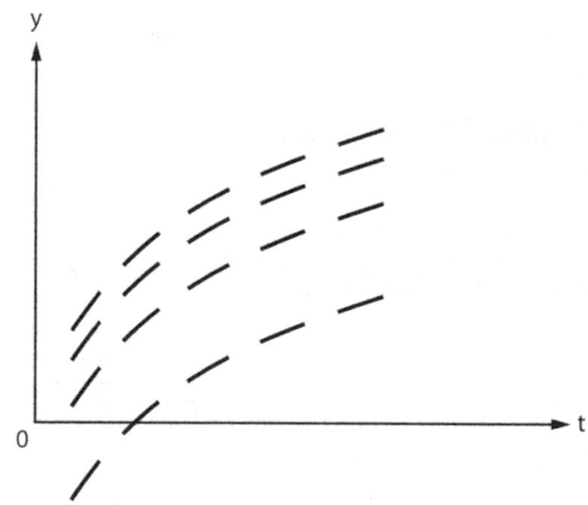

Fig. 15.1 A differential equation traces out a family of functions

15.2.1 Initial Value Problem

A differential equation with as many initial conditions as needed to get one specific solution is referred to as an *initial value problem*. Thus,

$$\frac{dy}{dt} = f(t, y), \qquad y(t_0) = y_0 \tag{15.20}$$

is an initial value problem. Among all the curves that are tangent to $f(t, y)$ for all values of t, only one passes through the point (t_0, y_0) (Fig. 15.2). This is called

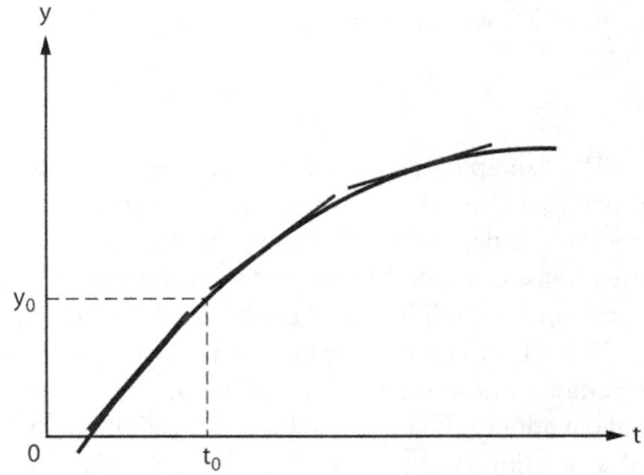

Fig. 15.2 Solution of an initial value problem with $y(t_0) = y_0$

the specific solution, which depicts the trajectory of the variable of interest y over time.

Example 15.8 To solve the initial value problem

$$\frac{dy}{dt} + \rho y = 0, \qquad y(0) = 2$$

we recall that the solution to the homogeneous equation was

$$y = Ae^{-\rho t}$$

Letting $t = 0$ and $y = 2$, we have

$$2 = Ae^{-\rho 0}, \qquad \Rightarrow \qquad A = 2$$

and the specific solution is

$$y = 2e^{-\rho t}$$

Example 15.9 Consider the initial value problem

$$\frac{d^2y}{dt^2} - \frac{dy}{dt} - 2y = 0$$

$$y(0) = 4, \qquad y'(0) = 0.5$$

Recalling the general solution of the equation from Example 15.7, we can write

$$\begin{aligned} y(0) &= A_1 + A_2 = 4 \\ y'(0) &= -A_1 + 2A_2 = 0.5 \end{aligned}$$

from which we obtain $A_1 = 2.5$ and $A_2 = 1.5$, and the specific solution as

$$y = 2.5e^{-t} + 1.5e^{2t}$$

The examples above make it clear that to obtain the specific solution of a first-order equation, we need one initial condition because there is only one arbitrary constant to determine. The solution of a second-order equation involves two arbitrary constants and, therefore, requires two initial conditions. In general, we need as many initial conditions as there are arbitrary constants to obtain a specific solution.

Not all differential equations can be solved. Indeed, the number of differential equations whose solutions are not known is far greater than the number of differential equations with known solutions. In addition, in economics, oftentimes the shape of a function is not known (see Example 15.3). One way to qualitatively analyze such equations is to draw the *phase diagram* of the equation. In particular, a phase diagram enables us to analyze dynamic economic models where the shape of functions involved is unspecified. An alternative is to compute the numerical solution of the equation, which is suitable when coefficients of an equation can be estimated or calibrated from the available data.

Three important issues regarding the solutions of differential equations are *existence*, *uniqueness*, and *stability*. We take up these issues in the next two subsections.

15.2.2 Existence and Uniqueness of Solutions

Does a differential equation have a solution, or more properly, under what conditions does it have a solution? To answer this question we need the following definition.

Definition 15.2 (Lipschitz Condition). A function $f(t, y)$ is said to satisfy the Lipschitz condition in its domain D if for some $L \geq 0$

$$|f(t_1, y) - f(t_2, y)| \leq L\,|t_1 - t_2|\,, \qquad \forall t_1, t_2 \in D \tag{15.21}$$

f is said to satisfy the *one-sided Lipschitz condition* if for some $L \geq 0$

$$t_2 > t_1 \quad \Rightarrow \quad f(t_2, y) - f(t_1, y) \leq L(t_2 - t_1) \tag{15.22}$$

Intuitively, the Lipschitz condition rules out the functions that have a vertical segment, because the condition guarantees that the ratio

$$\frac{f(t_2, y) - f(t_1, y)}{t_2 - t_1} \tag{15.23}$$

is finite. Therefore, at no point the slope of the curve or a line connecting two points of the function (in case of discontinuity) can become infinite. This condition is automatically satisfied for continuously differentiable functions in a bounded closed convex domain. Recall the mean value theorem: For a continuously differentiable function f on an interval $[a, b]$, we can write

$$f(b) - f(a) = (b - a)f'(\xi), \qquad a < \xi < b \tag{15.24}$$

Letting $L = \sup_D f'$ shows that the Lipschitz condition is fulfilled. Needless to say, the reverse is not true. A function that satisfies the Lipschitz condition is not necessarily continuously differentiable. Indeed, it may have discontinuity. Now we state, without proof, an important result regarding differential equations.

Theorem 15.1 *If the function $f(t, y)$ satisfies the Lipschitz condition, then the differential equation*

$$\frac{dy}{dt} = f(t, y) \tag{15.25}$$

has a unique solution.[2]

15.2.3 Equilibrium and Stability

A differential equation describes the motion of an object or variable. Thus, the equilibrium is the point of the rest of the object or variable, that is, when $dy/dt = 0$ or alternatively when $f(t, y) = 0$. We may not always be able to explicitly find the point of equilibrium but we are interested in knowing its properties. In particular, we are interested in knowing if the equilibrium is stable or not. Stability involves the question of whether deviations from the equilibrium point remain small and the system shows a tendency to return to equilibrium.

Formally, if every solution remains bounded and thus deviations from equilibrium remain bounded, the solution is called *stable*. If every solution tends to zero as $t \to \infty$, that is, if the system returns to equilibrium, it is called *strictly stable*.

Example 15.10 The equilibrium point of the equation in Example 15.7 is not stable because

$$\lim_{t\to\infty} A_1 e^{-t} = 0$$

$$\lim_{t\to\infty} A_2 e^{2t} \to \infty$$

[2]This is not the only existence and uniqueness theorem regarding differential equations, but for a deeper discussion of the subject the reader is referred to specialized books.

Example 15.11 The solution of the equation

$$\frac{d^2y}{dt}+5.5\frac{dy}{dt}+2.5y=0$$

is

$$y=A_1e^{-5t}+A_2e^{-0.5t}$$

Because

$$\lim_{t\to\infty}A_1e^{-5t}=0, \qquad \text{and} \qquad \lim_{t\to\infty}A_2e^{-0.5t}=0$$

the solution is strictly stable.

15.3 First-Order Linear Differential Equations

Consider the following first-order linear homogeneous differential equation:

$$\frac{dy}{dt}+ay=0 \tag{15.26}$$

By a solution we mean an equation of the form $y=\phi(t)$ that turns (15.26) into an identity. That is,

$$\phi'(t)+a\phi(t)=0 \tag{15.27}$$

Rewriting the equation in (15.26) in the following form makes the solution apparent:

$$\frac{\frac{dy}{dt}}{y}=-a \tag{15.28}$$

The equation in (15.28) states that the instantaneous rate of growth of y is $-a$. It follows that

$$y=Ae^{-at} \tag{15.29}$$

Despite the fact that the solution is self-evident, let us check that indeed it is a solution to (15.26). Because $\frac{dy}{dt}=-aAe^{-at}$, we have

$$\frac{dy}{dt}+ay=-aAe^{-at}+aAe^{-at}=0 \tag{15.30}$$

Example 15.12 The solution to the differential equation

$$\frac{dy}{dt}-2y=0$$

is

$$y = Ae^{2t}$$

As can be seen, instead of one equation, we have a family of equations. For every value of A we have a different equation. The reason is that the differential equation in (15.26) specifies the rate of change of y. Thus, the position of y after the lapse of time t will depend on where it was at the outset. If a denotes the speed of a car and we started in Boston going at 60 miles an hour in a southerly direction, three hours later we are close to New York City. But if we started in Houston, Texas, we are close to San Antonio where the Alamo is located.

Example 15.13 Suppose y is the population of a country with the growth rate of 1.1% a year. The differential equation expressing the dynamic of population is

$$\frac{dy}{dt} - 0.011y = 0$$

Now if the country is China and starts with a population of 1.35 billion, then in 30 years its population will be

$$1.35e^{0.011\times 30} = 1.878 \text{ billion}$$

But the same equation applied to the United States with a population of 310 million will result in

$$310e^{0.011\times 30} = 431.300 \text{ million}$$

To pinpoint the exact trajectory of y, we need an *initial condition*, the location of y at one particular value of t. Because y is a function of t, we can denote its value at time t by $y(t)$. Let

$$y(0) = y_0 \tag{15.31}$$

Then

$$y = y_0e^{-at} \tag{15.32}$$

Example 15.14 In Example 15.12, let $y(0) = 5$. Then the specific solution will be

$$y = 5e^{2t}$$

Example 15.15 The solution to the initial value problem

$$\frac{dy}{dt} + 3y = 0, \qquad y(0) = 7$$

is

$$y = 7e^{-3t}$$

Note that $y(0) = 7e^{-3(0)} = 7$, and we can verify that the solution satisfies the differential equation.

$$\frac{dy}{dt} + 3y = -21e^{-3t} + 21e^{-3t} = 0$$

As discussed above, an important issue in the study of differential equations is the behavior of the resulting function. In particular, does the function have an equilibrium where it will come to rest? And if this equilibrium point is disturbed, would the function return to it or would it forever diverge from it? In other words, is the equilibrium stable?

$$\lim_{t\to\infty} y = y_0 \lim_{t\to\infty} e^{-at} \tag{15.33}$$

Therefore the behavior of the function depends on the sign of a.

$$\lim_{t\to\infty} e^{-at} = \begin{cases} 0 \\ 1 \\ \infty \end{cases} \text{if} \quad a \begin{cases} > 0 \\ 0 \\ < 0 \end{cases} \tag{15.34}$$

We have a stable equilibrium if $a > 0$ and an unstable one if $a < 0$. In the case of $a = 0$, $dy/dt = 0$, and y is a constant function.

Example 15.16 The solution to the equation in Example 15.14 is not stable because $\lim_{t\to\infty} 5e^{2t} \to \infty$. On the other hand, the equation in Example 15.15 has a stable equilibrium because $\lim_{t\to\infty} 7e^{-3t} = 0$.

15.3.1 Variable Coefficient Equations

The method of solving first-order linear homogeneous equations described above can be extended to an equation with variable coefficients. Consider the equation

$$\frac{dy}{dt} + a(t)y = 0 \tag{15.35}$$

We can write it as

$$\frac{dy}{y} = -a(t)dt \tag{15.36}$$

Thus,

$$y = Ae^{-\int a(t)dt} \tag{15.37}$$

The fact that (15.37) is a solution to (15.35) can be ascertained. First note that

$$\frac{dy}{dt} = -a(t)Ae^{-\int a(t)dt}$$

and

$$\frac{dy}{dt} + a(t)y = -a(t)Ae^{-\int a(t)dt} + a(t)Ae^{-\int a(t)dt} = 0$$

Example 15.17 The solution of the differential equation

$$\frac{dy}{dt} + 2ty = 0$$

is

$$y = Ae^{-\int 2tdt} = Ae^{-t^2}$$

The reader is urged to verify the above result.

Example 15.18 The solution to the differential equation

$$\frac{dy}{dt} + (3t^2 - \cos t)y = 0$$

is

$$y = A^{-\int (3t^2 - \cos t)dt} = Ae^{-t^3 + \sin t}$$

15.3.2 Particular Integral, the Method of Undetermined Coefficients

The solution of nonhomogeneous differential equations of the form

$$\frac{dy}{dt} + \rho y = g(t) \tag{15.38}$$

is the sum of two parts: the *complementary function*, y_c, and the *particular integral*, y_p. The complementary function is the solution to the homogeneous part of the equation that we have already discussed. The particular integral deals with the nonhomogeneous part. In this subsection we discuss the method of undetermined coefficients and in the next, the alternative method of separable equations.

The method of undetermined coefficients starts by assuming that the particular integral y_p is of the same form as the nonhomogeneous part of the equation, that is, $g(t)$. For example, if $g(t)$ is a linear function of t, then we start by assuming $y_p = \alpha + \beta t$, and if $g(t)$ is a trigonometric function, we start with $y_p = \alpha \cos wt + \beta \sin wt$. By substituting y_p and its derivative in the differential equation and comparing the coefficients to those of $g(t)$, we determine the specific values of the parameters of y_p.

Example 15.19 Consider the nonhomogeneous equation

$$y' - 2y = 4t, \qquad y(0) = 2$$

The solution to the homogeneous part is $y_c = Ae^{2t}$. To find the particular integral—the solution to the nonhomogeneous part—we start by assuming

$$y_p = \alpha + \beta t$$

Substituting y_p and its derivative in the equation, we have

$$\beta - 2\alpha - 2\beta t = 4t$$

Equating the coefficients on both sides of the equation, we have

$$\begin{aligned} -2\beta &= 4 \qquad \Rightarrow \qquad \beta = -2 \\ \beta - 2\alpha &= 0 \qquad \Rightarrow \qquad \alpha = -1 \end{aligned}$$

The complete solution is

$$y = Ae^{2t} - 1 - 2t$$

and the specific solution is

$$y = 3e^{2t} - 1 - 2t$$

Let us check the result. First, note that the initial condition is satisfied because $y(0) = 2$. Next, we have $y' = 6e^{2t} - 2$. Substituting y and y' in the differential equation, we have

$$6e^{2t} - 2 - 2(3e^{2t} - 1 - 2t) = 4t$$

Example 15.20 Consider the initial value problem:

$$y' - 2ty = 4t, \qquad y(0) = 2$$

To solve the homogeneous part of the equation, we rewrite it as

$$\frac{dy}{dt} - 2ty = 0, \qquad \Rightarrow \qquad \frac{dy}{y} = 2tdt$$

Integrating both sides

$$\ln|y| = t^2 + C \qquad \Rightarrow \qquad y_c = Ae^{t^2}$$

where $A = e^C$. Again, we assume y_p to be of the form

$$y_p = \alpha + \beta t$$

Substituting y_p and its derivative in the differential equation, we have

$$\beta - 2\alpha t - 2\beta t^2 = 4t$$

and, therefore,

$$\alpha = -2, \quad \beta = 0$$

Thus, the complete solution is

$$y = 4e^{t^2} - 2$$

Example 15.21 The solution to the homogeneous part of the differential equation

$$y' + 5y = 8\sin 3t$$

is

$$y_c = Ae^{-5t}$$

For the particular integral, we start with

$$y_p = \alpha\cos 3t + \beta\sin 3t$$

To determine α and β, we substitute y_p and its derivative in the differential equation:

$$-3\alpha\sin 3t + 3\beta\cos 3t + 5\alpha\cos 3t + 5\beta\sin 3t = 8\sin 3t$$

It is easy to see that y_p satisfies the differential equation if

$$-3\alpha + 5\beta = 8$$
$$3\beta + 5\alpha = 0$$

Solving the for α and β, we obtain the particular integral

$$y_p = -\frac{12}{17}\cos 3t + \frac{20}{17}\sin 3t$$

and the complete solution is

$$y = y_c + y_p = Ae^{-5t} - \frac{12}{17}\cos 3t + \frac{20}{17}\sin 3t$$

Example 15.22 Solve the initial value problem

$$y' + 2y = 5e^{-7t}, \qquad y(0) = 3$$

We have $y_c = Ae^{-2t}$ and for y_p we start with

$$y_p = \alpha e^{-7t}$$

Substituting y_p and its derivative in the differential equation

$$-7\alpha e^{-7t} + 2\alpha e^{-7t} = 5e^{-7t}$$

we obtain $\alpha = -1$. The complete solution is

$$y = Ae^{-2t} - e^{-7t}$$

From the initial condition, we obtain

$$3 = A - 1, \qquad \Rightarrow \qquad A = 4$$

and the specific solution is

$$y = 4e^{-2t} - e^{-7t}$$

It is important to remember that in solving initial value problems one has to find the complete solution before trying to find the arbitrary constant. Had we tried to find A using the complementary function, we would have gotten A=3, which would have been the wrong answer. You can verify that setting A=3 would contradict the initial condition $y(0) = 3$.

15.3.3 Separable Equations

This is the second method for solving a nonhomogeneous differential equation. It appears a bit involved, but it is easier than it looks. If you find it difficult to follow, study the examples to see how the general formula is applied. Consider the initial value problem

$$y' + a(t)y = g(t), \qquad y(0) = y_0 \tag{15.39}$$

Define the related system

$$x' - a(t)x = 0, \qquad x(0) = 1 \tag{15.40}$$

We already know that the solution to this second problem is

$$x = \exp\left(\int_0^t a(s)ds\right) \tag{15.41}$$

Now consider the function xy and its total derivative with respect to t:

$$\begin{aligned} \frac{d}{dt}(xy) &= x'y + y'x \\ &= a(t)xy + xg(t) - a(t)xy \\ &= xg(t) \end{aligned} \tag{15.42}$$

It follows that

$$xy = \int_0^t xg(s)ds + C \tag{15.43}$$

Because $x(0)\, y(0) = 1 \times y_0$, we have $C = y_0$. Thus,

$$y = \frac{1}{x}\int_0^t xg(s)ds + \frac{y_0}{x} \tag{15.44}$$

Substituting for x in (15.44), the solution is

$$y = \exp\left(-\int_0^t a(s)ds\right)\left[y_0 + \int_0^t g(\xi)\left(\exp\left(\int_0^t a(s)ds\right)d\xi\right)\right] \tag{15.45}$$

Although it does not look appetizing, it is still easy to apply.

Example 15.23 Consider the initial value problem

$$y' - 2ty = 4t, \qquad y(0) = 2$$

Applying (15.45) to this problem we have

$$\begin{aligned} y &= e^{t^2}\left[y_0 + \int 4te^{-t^2}dt\right] \\ &= e^{t^2}\left[2 - 2e^{-t^2}\right] \\ &= 2e^{t^2} - 2 \end{aligned}$$

We can check that this is indeed the solution. First, note that at $t = 0$, $y = 2$. Next, we have

$$y' = 4e^{t^2}$$

Thus,

$$y' - 2ty = 4te^{t^2} - 2t\left[2e^{t^2} - 2\right] = 4t$$

Example 15.24 Applying (15.45) to the differential equation

$$y' - 2y = \sin t, \qquad y(0) = y_0$$

we have

$$y = e^{2t}\left(y_0 + \int e^{-2t}\sin t dt\right)$$

The integral in the above equation is a bit complicated. Looking up the tables of integrals or using Maple, the solution is

$$\begin{aligned} y &= e^{2t}\left[y_0 - \frac{1}{5}(2\sin t + \cos t)e^{-2t}\right] \\ &= y_0e^{2t} - \frac{2}{5}\sin t - \frac{1}{5}\cos t \end{aligned}$$

Solving this problem with the method of undetermined coefficients gets us the same answer.

15.3.4 Exact Differential Equations

Taking the differential of the function $u(t, y) = C$ we can write

$$\frac{\partial u(t,y)}{\partial t}dt + \frac{\partial u(t,y)}{\partial y}dy = 0 \tag{15.46}$$

Let

$$M(t,y) = \frac{\partial u(t,y)}{\partial t}, \qquad \text{and} \qquad N(t,y) = \frac{\partial u(t,y)}{\partial y}$$

Then

$$M(t,\ y)dt + N(t,\ y)dy = 0 \tag{15.47}$$

or

$$\frac{dy}{dt} = -\frac{M(t,y)}{N(t,y)} \tag{15.48}$$

The differential equation in (15.47) is called an *exact differential equation* and, obviously, its solution is

$$u\,(t,y) = C \tag{15.49}$$

Example 15.25 The equation

$$t\frac{dy}{dt} + y = 0, \qquad t > 0$$

is an exact differential equation with the solution $ty = c$ where c is a constant. First, note that its differential is equal to our original equation. Moreover, we can check that, it is indeed a solution.

$$\frac{dy}{dt} = -\frac{y}{t}$$

and we have

$$-t\frac{y}{t} + y = 0$$

The question how do we determine if a differential equation is exact is answered by the following theorem.

Theorem 15.2 *The differential equation in (15.47), where $M(t,\ y)$ and $N(t,\ y)$ are continuous and continuously differentiable functions, is exact iff*

$$\frac{\partial M(t,y)}{\partial y} = \frac{\partial N(t,y)}{\partial t}$$

To prove the necessary part of the proposition, suppose that indeed

$$M(t,y) = \frac{\partial u(t,y)}{\partial t}, \qquad \text{and} \qquad N(t,y) = \frac{\partial u(t,y)}{\partial y} \tag{15.50}$$

Then differentiating $M(t, \ y)$ with respect to y *and $N(t, \ y)$) with respect to t results in*

$$\frac{\partial M(y,t)}{\partial y}=\frac{\partial^2 u(t,y)}{\partial t\partial y}=\frac{\partial^2 u(t,y)}{\partial y\partial t}=\frac{\partial N(t,y)}{\partial t} \tag{15.51}$$

which shows that the equation is an exact differential equation. The proof of the sufficient condition is by construction. Because

$$\frac{\partial u(t,y)}{\partial t}=M(t,y) \tag{15.52}$$

we can integrate the RHS of (15.52) to find u:

$$u(t,y)=\int_{t_0}^{t} M(s,y)ds+\phi(y) \tag{15.53}$$

The reason for the inclusion of the second term on the RHS is that in integrating we assume y *to remain constant. Therefore, the arbitrary constant of integration can possibly depend on* y*. On the other hand,*

$$\frac{\partial u}{\partial y}=\int_{t_0}^{t}\frac{\partial M(s,y)}{\partial y}ds+\phi'(y)=N(t,y) \tag{15.54}$$

Because $\partial M(t,y)/\partial y=\partial N(t,y)/\partial t$, we can write

$$\begin{aligned}\int_{t_0}^{t}\frac{\partial N(s,y)}{\partial t}ds+\phi'(y)&=N(s,y)|_{t_0}^{t}+\phi'(y)\\&=N(t,y)-N(t_0,y)+\phi'(y)\end{aligned} \tag{15.55}$$

Because the LHS of (15.55) and the middle expression in (15.54) are equal, the upshot is that $\phi'(y)=N(t_0,y)$, which, in turn, implies

$$\phi(y)=\int_{y_0}^{y}N(t_0,z)dz+C_1 \tag{15.56}$$

Substituting (15.56) in (15.53)

$$u(t,y)=\int_{t_0}^{t}M(s,y)ds+\int_{y_0}^{y}N(t_0,z)dz+C_1 \tag{15.57}$$

Because $u(y, \ t)$ is equal to an arbitrary constant, we have

$$\int_{t_0}^{t}M(s,y)ds+\int_{y_0}^{y}N(t_0,z)dz=C \tag{15.58}$$

Example 15.26 Solve the differential equation

$$y\,dt + t\,dy = 0$$

This is an exact differential equation because $M = y$ and $N = t$ and we have

$$\frac{\partial M}{\partial y} = 1 = \frac{\partial N}{\partial t}$$

We solve the problem with the following steps:

$$1. \quad u = \int y dt + \phi(y) = yt + \phi(y)$$

$$2. \quad \frac{\partial u}{\partial y} = t + \phi'(y) = t$$

which results in

$$\phi'(y) = 0, \quad \Rightarrow \quad \phi(y) = C_1$$

$$3. \quad u = yt + C_1$$

Because $u = C$ and C is an arbitrary constant, the solution is

$$yt = C$$

Example 15.27 Consider the differential equation

$$\frac{2t}{y^3}\,dt + \frac{y^2 - 3t^2}{y^4}\,dy = 0$$

This is an exact equation because

$$\frac{\partial M}{\partial y} = \frac{-6t}{y^4} = \frac{\partial N}{\partial t}$$

Again we take the following steps to solve the problem

$$1. \quad u = \int \frac{2t}{y^3} dt + \phi(y) = \frac{t^2}{y^3} + \phi(y)$$

$$2. \quad \frac{\partial u}{\partial y} = \frac{-3t^2}{y^4} + \phi'(y) = \frac{y^2 - 3t^2}{y^4}$$

which implies

$$\phi'(y) = \frac{1}{y^2}, \quad \Rightarrow \quad \phi(y) = -\frac{1}{y} + C_1$$

3.
$$u = \frac{t^2}{y^3} - \frac{1}{y} + C_1$$

Again we can write the solution as

$$\frac{t^2}{y^3} - \frac{1}{y} = C$$

Example 15.28 Verify that the following differential equation is exact:

$$(2t + y)\,dt + (6y + t)\,dy = 0$$

It is, because

$$\frac{\partial(2t + y)}{\partial y} = 1 = \frac{\partial(6y + t)}{\partial t}$$

Furthermore,

$$u = t^2 + yt + \phi(y)$$

and

$$\phi(y) = 3y^2$$

Thus, the solution is

$$t^2 + yt + 3y^2 = C$$

15.3.5 Integrating Factor

There are equations of the form

$$M\,(t, y)\,dt + N\,(t, y)\,dy = 0 \tag{15.59}$$

that are not exact, but it is possible to find a function $\mu(t, y)$ such that

$$\mu(t, y)M(t, y)dt + \mu(t, y)N(t, y)dy = 0 \tag{15.60}$$

is an exact differential equation. In such a case we should have

$$\frac{\partial(\mu M)}{\partial y} = \frac{\partial(\mu N)}{\partial t} \tag{15.61}$$

Carrying out the differentiation, we have

$$\mu \frac{\partial M}{\partial y} + M \frac{\partial \mu}{\partial y} = \mu \frac{\partial N}{\partial t} + N \frac{\partial \mu}{\partial t} \tag{15.62}$$

Rearranging,

$$M \frac{\partial \mu}{\partial y} - N \frac{\partial \mu}{\partial t} = \mu \left[\frac{\partial N}{\partial t} - \frac{\partial M}{\partial y} \right] \tag{15.63}$$

Dividing through by μ and recalling that

$$\begin{aligned} \frac{\partial \mu / \partial y}{\mu} &= \frac{\partial \ln \mu}{\partial y} \\ \frac{\partial \mu / \partial t}{\mu} &= \frac{\partial \ln \mu}{\partial t} \end{aligned} \tag{15.64}$$

we have

$$M \frac{\partial \ln \mu}{\partial y} - N \frac{\partial \ln \mu}{\partial t} = \frac{\partial N}{\partial t} - \frac{\partial M}{\partial y} \tag{15.65}$$

We are really in a proper mess now. We have traded an ordinary differential equation (15.59) for a partial differential equation (15.65). But there is hope. If μ depends only on either t or y, we may be able to solve the equation.

If μ does only depend on y, then

$$\frac{d \ln \mu}{dy} = \frac{\dfrac{\partial N}{\partial t} - \dfrac{\partial M}{\partial y}}{M} \tag{15.66}$$

On the other hand, if If μ does only depend on t, then

$$\frac{d \ln \mu}{dt} = \frac{\dfrac{\partial M}{\partial y} - \dfrac{\partial N}{\partial t}}{N} \tag{15.67}$$

The next two examples illustrate the use of integrating factor.

Example 15.29 The differential equation

$$ydt - tdy = 0$$

is not exact because $M = y$ and $N = -t$ and

$$1 = \frac{\partial M}{\partial y} \neq \frac{\partial N}{\partial t} = -1$$

but

$$\frac{d\ln\mu}{dy} = \frac{\dfrac{\partial N}{\partial t} - \dfrac{\partial M}{\partial y}}{M} = \frac{-2}{y}$$

depends only on y. It follows that $\mu = 1/y^2$ and the solution to the equation is:

$$\frac{t}{y} = C$$

Example 15.30 Let us solve the differential equation

$$-\,tdy + (y + ty^2)dt = 0$$

First, let us check if the equation is exact. Because $M = y + ty^2$ and $N = -t$, we have

$$\frac{\partial M}{\partial y} = 1 + 2ty, \qquad \frac{\partial N}{\partial t} = -1$$

which shows

$$\frac{\partial M}{\partial y} \neq \frac{\partial N}{\partial t}$$

Next let us check if we can find an integrating factor.

$$\frac{\dfrac{\partial N}{\partial t} - \dfrac{\partial M}{\partial y}}{M} = \frac{-1 - 1 - 2ty}{y + ty^2} = -\frac{2}{y}$$

Because μ depends on y alone, we have

$$\frac{d\ln\mu}{dy} = -\frac{2}{y} \quad \Rightarrow \quad \ln\mu = -2\ln y \quad \Rightarrow \quad \mu = \frac{1}{y^2}$$

Multiplying both sides of the original equation by μ, we have the exact differential equation

$$\left(\frac{1}{y} + t\right) dt - \frac{t}{y^2} dy = 0$$

whose solution is

$$\frac{t}{y} + \frac{t^2}{2} = C$$

15.3.6 Exercises

E.15.1 Solve the following differential equations.

i. $\frac{dy}{dt} + 4y = 0$ *ii.* $\frac{dy}{dt} - 0.4y = 0$ *iii.* $\frac{dy}{dt} - 2y = 0$

iv. $\frac{dy}{dt} + 3y = \cos t$ *v.* $\frac{dy}{dt} - 4y = 2t$ *vi.* $\frac{dy}{dt} = 2ty$

vii. $\frac{dy}{dt} + y = e^t$ *viii.* $y' - y = e^{2t}$ *ix.* $y' + t\cos t = \frac{1}{2}\sin 2t$

x. $y' + e^t y = e^t$ *xi.* $(y-2)dt - 2t^2 dy = 0$ *xii.* $ydt - tdy = 0$

E.15.2 Solve the following initial value problems.

i. $\frac{dy}{dt} + 0.4y = 0,\quad y(0) = 3$ *ii.* $\frac{dy}{dt} + 0.4y = 2t,\quad y(0) = 4$

iii. $\frac{dy}{dt} = 2ty,\quad y(0) = 1$ *iv.* $\frac{dy}{dt} = \frac{t}{y},\quad y(1) = 1$

v. $\frac{dy}{dt} = e^{-y},\quad y(0) = 2.5$ *vi.* $\frac{dy}{dt} - 2y = \sin t,\quad y(0) = y_0$

vii. $y' + 5\frac{y}{t} = 6t,\quad y(1) = 3$ *viii.* $t^3 y' + 2y = t^3 + 2t,\quad y(1) = e + 1$

ix. $y' + y = t,\quad y(0) = 5$

E.15.3 In Solow's model assume that the production function is $Q = AK^{\alpha}L^{1-\alpha}$. Derive the differential equation and solve it. [*Hint*: A differential equation of the form

$$\frac{dy}{dt} + b_1 y^{\alpha} + b_2 y = 0$$

is called *the Bernoulli equation*. A simple trick facilitates its solution. Divide the equation by y^{α} and let $z = y^{1-\alpha}$. Substituting z and its derivative into the Bernoulli equation results in a first-order linear nonhomogeneous equation.]

15.4 Phase Diagram

There are many differential equations for which we cannot find an explicit solution. In addition, in many economic applications the functional form of the equation is unspecified. Still, we could analyze the behavior of such equations using a *phase diagram*. Let us graph the equation

$$\frac{dy}{dt} = f(y) \tag{15.68}$$

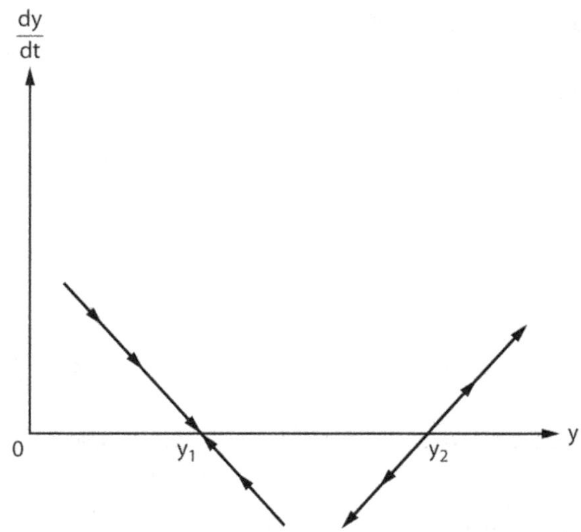

Fig. 15.3 Stable equilibrium at y_1 and unstable equilibrium at y_2

with *dy*/*dt* on the vertical axis and *y* on the horizontal axis. As long as $f(y) > 0$, we have $dy/dt > 0$ and, therefore, *y* is increasing. This is shown with rightward pointing arrows above the horizontal axis in Fig. 15.3. On the other hand, when $f(y) < 0$—that is, in the area below the horizontal axis—we have $dy/dt < 0$ and *y* is decreasing. This is shown by leftward pointing arrows. Equilibrium is reached when $dy/dt = 0$, that is, when the function $f(y)$ intersects with the horizontal axis. In Fig. 15.3, both y_1 and y_2 are points of equilibrium.

An equilibrium is called stable if when there is a slight disturbance and *y* is moved away from its equilibrium value, the system corrects itself and the equilibrium is restored. In Fig. 15.3 y_1 is a stable equilibrium. It can be seen that when $y < y_1$, $dy/dt > 0$, and *y* moves toward y_1 as shown by the arrows. If $y > y_1$, then $dy/dt < 0$ and *y* decreases, as shown by leftward pointing arrows.

On the other hand, the point y_2 is an unstable equilibrium. Once the equilibrium is disturbed, regardless of the direction of the disturbance, *y* moves away from equilibrium, as shown by the arrows pointing away from the equilibrium in both directions.

There may be more than one equilibrium point (Fig. 15.4). In such cases, two adjacent equilibrium points cannot both be stable. In Fig. 15.4, y_2 is a stable equilibrium, but y_1 is unstable.

Example 15.31 In Example 15.3 we discussed Solow's growth model and derived the differential equation

$$\frac{dk}{dt} = sf(k) - nk \tag{15.69}$$

For some forms of the production function $f(k)$, we are able to solve the equation (see Exercise E.15.3). But because the functions *F* and *f* are not specified, this equation cannot be solved. Still, we can study its behavior using a phase diagram.

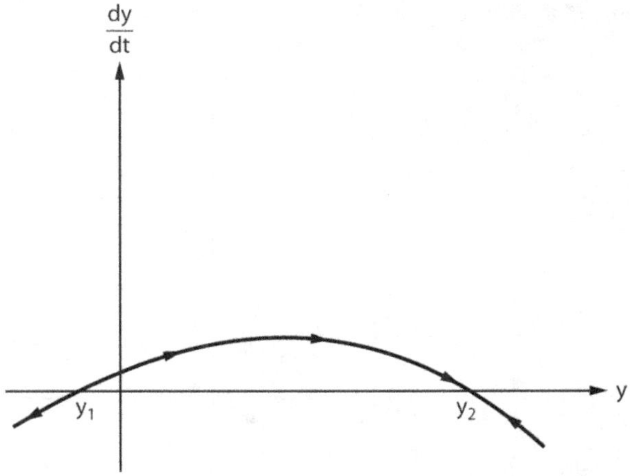

Fig. 15.4 Unstable equilibrium at y_1 and stable equilibrium at y_2

Note that because F is homogeneous of degree one, $F(0,\ L) = F(K,0) = f(0) = 0$. Thus, one equilibrium point is at $k = 0$. It is conceivable that for all $k > 0$ the curve $sf(k)$ remains below the line nk, in which case $dk/dt < 0$ for all values of k and we will have one equilibrium point that will be stable. Ruling this possibility out, we note that by (15.9) $f(k)$ and $sf(k)$ are positive and increasing, but at a decreasing rate because $f'(k) > 0$ and $f''(k) < 0$. Therefore, there must be another point k^* where $sf(k^*) = nk^*$(see Figs. 15.5 and 15.6). In that case we have two equilibrium points: one unstable, at $k = 0$, and the other stable, at $k = k^*$.

Because L is growing at the rate of n, a constant k implies that K is also growing at the rate n. Because we have constant returns to scale in production, the growth rate of output is also n.

$$\left.\frac{dQ/dt}{Q}\right|_{k=k^*} = \left.\frac{f(k)(dL/dt) + Lf'(k)(dk/dt)}{Lf(k)}\right|_{k=k^*} = \frac{dL/dt}{L} = n \tag{15.70}$$

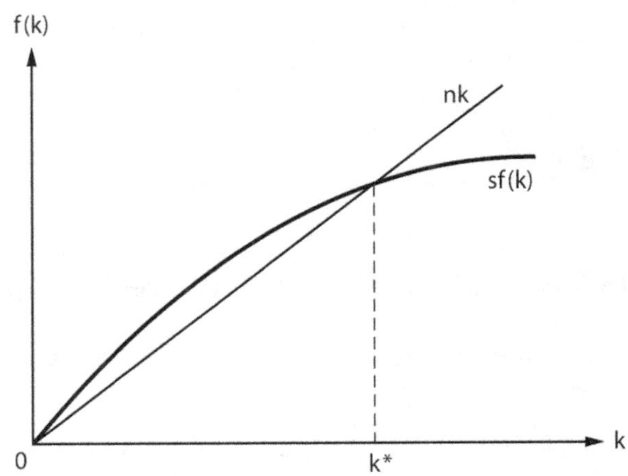

Fig. 15.5 Equilibria of Solow's model

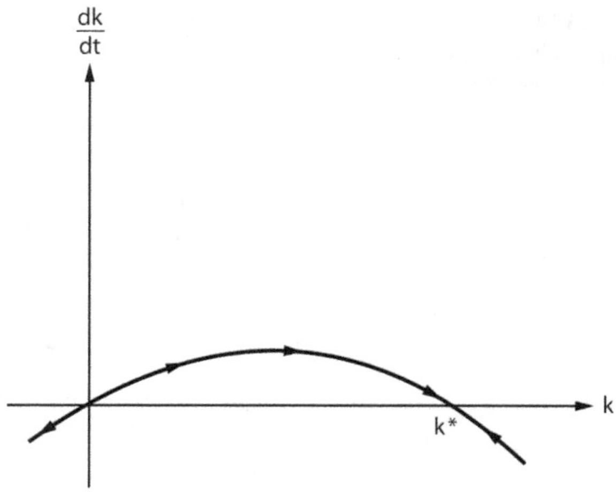

Fig. 15.6 Phase diagram of Solow's model

It follows that savings $sLf(k)$ is also growing at the same rate. When all variables are growing at the same rate and are on their long-run trajectory, the system is in its *steady state*. The fact that $k = k^*$ is a stable equilibrium means that, should the capital output ratio k differ from the equilibrium k^*, then economic forces cause a move toward equilibrium. This points to the conclusion that, in the long run, economies that use the same technology or are at the same stage of technological progress, should converge and grow at the same rate.

15.4.1 Exercises

E.15.4 Draw the phase diagram of the following differential equations, find their stationary points, and determine if they are stable or unstable.

i. $\dfrac{dy}{dt} = y^3 - 4y$ *ii.* $\dfrac{dy}{dt} = y^2 - 3y + 2$

iii. $\dfrac{dy}{dt} = 4 - 2y + \ln y$ *iv.* $\dfrac{dy}{dt} = 2e^{3y} - 5y + 1$

15.5 Second-Order Linear Differential Equations

A second-order linear differential equation is of the form

$$\frac{d^2y}{dt^2} + b_1\frac{dy}{dt} + b_2y = g(t) \tag{15.71}$$

Again, if $g(t) = 0$, we have a second-order homogeneous equation. In line with what we learned from solving a first-order differential equation, we may start with the solution

$$y = Ae^{rt} \tag{15.72}$$

for the homogeneous part of the equation. Then we have

$$r^2 Ae^{rt} + b_1 rAe^{rt} + b_2 Ae^{rt} = 0 \tag{15.73}$$

Dividing through by $Ae^{rt} \neq 0$, we get the *characteristic equation*

$$r^2 + b_1 r + b_2 = 0 \tag{15.74}$$

This is is a second-order equation and, therefore, will have two solutions. Let us denote them by r_1 and r_2 and recall that

$$r_1 + r_2 = -b_1, \qquad \text{and} \qquad r_1 r_2 = b_2 \tag{15.75}$$

Because r_1 and r_2 are solutions of a quadratic equation, we may have two real and distinct solutions, a repeated root solution, or a complex solution. We take up each case in turn.

15.5.1 Two Distinct Real Roots

In this case both $y = e^{r_1 t}$ and $y = e^{r_2 t}$ are solutions of the equation. But we need not choose between them as a linear combination of them is also a solution. Thus, the solution would be

$$y = A_1 e^{r_1 t} + A_2 e^{r_2 t} \tag{15.76}$$

which can easily be checked. The first and second derivatives are

$$\begin{aligned} \frac{dy}{dt} &= A_1 r_1 e^{r_1 t} + A_2 r_2 e^{r_2 t} \\ \frac{d^2 y}{dt^2} &= A_1 r_1^2 e^{r_1 t} + A_2 r_2^2 e^{r_2 t} \end{aligned}$$

Substituting in (15.71) when $g(t) = 0$, we have

$$\begin{aligned} A_1 r_1^2 e^{r_1 t} + A_2 r_2^2 e^{r_2 t} + b_1 \left(A_1 r_1 e^{r_1 t} + A_2 r_2 e^{r_2 t}\right) + b_2 \left(A_1 e^{r_1 t} + A_2 e^{r_2 t}\right) \\ = A_1 e^{r_1 t} \left(r_1^2 + b_1 r_1 + b_2\right) + A_2 e^{r_2 t} \left(r_2^2 + b_1 r_2 + b_2\right) = 0 \end{aligned}$$

The last equality is based on the fact that r_1 and r_2 are indeed solutions of the equation $r^2 + b_1 r + b_2 = 0$ and therefore, the expressions in parentheses are identically equal to zero.

Example 15.32 For the following differential equation

$$y'' - \frac{3}{2}y' - y = 0$$

the characteristic equation is

$$r^2 - \frac{3}{2}r - 1 = 0$$

with the roots

$$r_1 = 2, \qquad r_2 = -\frac{1}{2}$$

Thus, the solution is

$$y = A_1 e^{2t} + A_2 e^{-\frac{1}{2}t}$$

At this point we may ask, how do we determine if two (or more) solutions of a differential equation are independent and therefore, can they be combined as in (15.76)? To answer this question, we first define the *Wronskian*[3] of two (or more) equations.

Definition 15.3 The Wronskian of functions $f_1(x), f_2(x), \ldots, f_n(x)$ is the determinant

$$W = \begin{vmatrix} f_1 & f_2 & f_3 & \cdots & f_n \\ f'_1 & f'_2 & f'_3 & \cdots & f'_n \\ f''_1 & f''_2 & f''_3 & \cdots & f''_n \\ \vdots & \vdots & \vdots & \cdots & \vdots \\ f_1^{(n-1)} & f_2^{(n-1)} & f_3^{(n-1)} & \cdots & f_n^{(n-1)} \end{vmatrix} \tag{15.77}$$

In particular, for the case of two solutions we have

$$W = \begin{vmatrix} A_1 e^{r_1 t} & A_2 e^{r_2 t} \\ A_1 r_1 e^{r_1 t} & A_2 r_2 e^{r_2 t} \end{vmatrix} = A_1 A_2 e^{(r_1 + r_2)t}(r_2 - r_1) \tag{15.78}$$

Two or more functions are independent if their Wronskian is not zero. Evidently, in (15.78) $W \neq 0$ if $r_1 \neq r_2$, that is, as long as the two roots are distinct.

[3] After the Polish mathematician Hoëné Wronski (1778–1853).

Example 15.33 Consider the equation

$$y'' + 7y' + 10y = 0$$

with the initial conditions

$$y(0) = 10, \qquad y'(0) = -29$$

whose solution is

$$y = 3e^{-5t} + 7e^{-2t}$$

Computing the Wronskian, we have

$$W = \begin{vmatrix} 3e^{-5t} & 7e^{-2t} \\ -15e^{-5t} & -14e^{-2t} \end{vmatrix} = -63e^{-7t} \neq 0$$

We now turn to the question of stability of the equilibrium point of the equation. Because

$$\lim_{t\to\infty} y = \lim_{t\to\infty} A_1 e^{r_1 t} + \lim_{t\to\infty} A_2 e^{r_2 t} \tag{15.79}$$

it follows that, the equilibrium is stable iff $r_1 < 0$ and $r_2 < 0$, because in that case the RHS of (15.79) will be zero when $t \to \infty$. Thus, the solution of the equation in Example 15.32 is unstable whereas the solution of the equation in Example 15.33 is stable.

15.5.2 Repeated Root

In this case, the differential equation is of the form

$$\frac{d^2y}{dt^2} + b\frac{dy}{dt} + \frac{b^2}{4}y = 0 \tag{15.80}$$

and we have $r_1 = r_2 = r = -b/2$. Therefore, we do not have two independent solutions, and we cannot form the solution of the homogenous equation as

$$y = A_1 e^{rt} + A_2 e^{rt} = (A_1 + A_2)e^{rt} \tag{15.81}$$

because A_1 and A_2 and, therefore, $A_1 + A_2$ are arbitrary constants and one solution will get lost. On the other hand, we cannot allow the solution of a second-order equation to degenerate into the solution of a first-order equation. Here we form the second solution as

$$A_2 t e^{rt} \tag{15.82}$$

and the complete solution is

$$y = A_1 e^{rt} + A_2 t e^{rt} \tag{15.83}$$

The logic behind this choice will become clear in Chap. 17. Here, we show that the two solutions are independent and (15.83) is indeed a solution to (15.80). First, note that

$$\frac{dy}{dt} = rA_1 e^{rt} + rA_2 t e^{rt} + A_2 e^{rt} \tag{15.84}$$

and

$$\frac{d^2y}{dt^2} = r^2 A_1 e^{rt} + r^2 A_2 t e^{rt} + 2rA_2 e^{rt} \tag{15.85}$$

The Wronskian of the functions is

$$\begin{vmatrix} A_1 e^{rt} & A_2 t e^{rt} \\ rA_1 e^{rt} & rA_2 t e^{rt} + A_2 e^{rt} \end{vmatrix} = A_1 A_2 e^{2rt} \neq 0 \tag{15.86}$$

Therefore, the two functions are independent. Moreover, substituting (15.83) and its derivatives in (15.80), and recalling that $r = -b/2$, we have

$$\begin{aligned} &r^2 A_1 e^{rt} + r^2 A_2 t e^{rt} + 2rA_2 e^{rt} + b(rA_1 e^{rt} + rA_2 t e^{rt} + A_2 e^{rt}) \\ &\quad + \frac{b^2}{4}(A_1 e^{rt} + A_2 t e^{rt}) \\ &= \frac{b^2}{4} A_1 e^{rt} + \frac{b^2}{4} A_2 t e^{rt} - bA_2 e^{rt} - \frac{b^2}{2} A_1 e^{rt} - \frac{b^2}{2} A_2 t e^{rt} \\ &\quad + bA_2 e^{rt} + \frac{b^2}{4} A_1 e^{rt} + \frac{b^2}{4} A_2 t e^{rt} = 0 \end{aligned} \tag{15.87}$$

Example 15.34 To find the solution of differential equation

$$\frac{d^2y}{dt^2} - 3\frac{dy}{dt} + \frac{9}{4}y = 0$$

we solve the characteristic equation

$$r^2 - 3r + \frac{9}{4} = 0 \qquad \Rightarrow \qquad r = \frac{3}{2}$$

and the solution is

$$y = A_1 e^{1.5t} + A_2 t e^{1.5t}$$

If we add the initial conditions $y(0) = 5$ and $y'(0) = 9$, we have

$$A_1 = 5, \qquad \text{and} \qquad 1.5A_1 + A_2 = 9, \qquad \Rightarrow \qquad A_2 = 1.5$$

and the specific solution is

$$y = 5e^{1.5t} + 1.5te^{1.5t}$$

Example 15.35 Find the solution of the initial value problem

$$\frac{d^2y}{dt^2} + 2\sqrt{2}\,\frac{dy}{dt} + 2y = 0, \qquad y(0) = \sqrt{2},\ \ y'(0) = 3$$

The characteristic equation is

$$r^2 + 2\sqrt{2}r + 2 = 0 \qquad \Rightarrow \qquad r = -\sqrt{2}$$

The solution is

$$y = A_1e^{-\sqrt{2}t} + A_2te^{-\sqrt{2}t}$$

Utilizing the initial conditions, we have

$$A_1 = \sqrt{2}$$

and

$$\sqrt{2}A_1 + A_2 = 3, \qquad \Rightarrow \qquad A_2 = 1$$

and the specific solution is

$$y = \sqrt{2}e^{-\sqrt{2}t} + te^{-\sqrt{2}t}$$

The condition for the stability of the equilibrium when we have a repeated root is similar to the case of two distinct roots, except that now we have only one root and it should be less than one. It is not difficult to see that one solution converges to zero if $r < 0$, because

$$\lim_{t\to\infty} Ae^{rt} \to 0, \qquad \text{if} \qquad r < 0 \tag{15.88}$$

For the second solution we can write

$$\lim_{t\to\infty} Ate^{rt} = \lim_{t\to\infty} \frac{At}{e^{-rt}} \tag{15.89}$$

Using l'Hôpital's rule we get

$$\lim_{t\to\infty} \frac{At}{e^{-rt}} = \lim_{t\to\infty} \frac{A}{-re^{-rt}} = 0, \qquad \text{if} \qquad r < 0 \tag{15.90}$$

15.5.3 Complex Roots

In this case the roots will be complex conjugates, that is,

$$r_1 = \rho + i\theta, \qquad r_2 = \rho - i\theta \tag{15.91}$$

Therefore the solution is

$$y = A_1 e^{(\rho+i\theta)t} + A_2 e^{(\rho-i\theta)t} \tag{15.92}$$

Recall that

$$e^{\pm i\theta t} = \cos\theta t \pm i \sin\theta t \tag{15.93}$$

and

$$\begin{aligned} y &= e^{\rho t}[A_1(\cos\theta t + i\sin\theta t) + A_2(\cos\theta t - i\sin\theta t)] \\ &= e^{\rho t}[(A_1 + A_2)\cos\theta t + (A_1 - A_2) i \sin\theta t] \\ &= e^{\rho t}[B_1 \cos\theta t + B_2 \sin\theta t] \end{aligned} \tag{15.94}$$

Note that because y is real, $A_1 - A_2$ has to be complex, hence B_2 is real. Indeed, A_1 and A_2 are complex conjugates (see Exercise E.15.8).

Example 15.36 Consider the equation

$$y'' + 6y' + 10y = 0$$

The auxiliary equation

$$r^2 + 6r + 10 = 0$$

yields the roots

$$r_1 = -3 + i, \qquad r_2 = -3 - i$$

Thus, the solution is

$$y = e^{-3t}(B_1 \cos t + B_2 \sin t)$$

Example 15.37 Solve the initial value problem

$$y'' - 2y' + 17y = 0, \qquad y(0) = 2, \quad y'(0) = -6$$

The characteristic equation

$$r^2 - 2r + 17 = 0$$

yields the roots

$$r_1 = 1 + 4i, \qquad r_2 = 1 - 4i$$

Thus, the solution is

$$y = e^t(B_1 \cos 4t + B_2 \sin 4t)$$

Applying the initial conditions, we get

$$B_1 = 2, \qquad B_2 = -2$$

and we obtain the specific solution as

$$y = e^t(2\cos 4t - 2\sin 4t)$$

Example 15.38 To analyze the stability of the equilibrium in the case of complex roots, we consider the two parts of the solution in turn. First, note that

$$B_1 \cos \theta t + B_2 \sin \theta t$$

is a circular or trigonometric function. Therefore, it will fluctuate over time. Left to itself, it will have the same amplitude and will stay within certain bounds. In particular recall that $\cos \theta t$ and $\sin \theta t$ oscillate between 1 and –1. The second part, $e^{\rho t}$, determines the amplitude and, therefore, the extent of oscillations. If $\rho > 0$, the fluctuations amplify as time goes by. But if $\rho < 0$, then y gradually approaches zero. The case of $\rho = 0$ corresponds to constant amplitudes oscillations. Thus,

$$\rho \begin{cases} < 0, & \text{damped oscillations} \\ = 0, & \text{constant oscillations} \\ > 0, & \text{explosive oscillations} \end{cases} \tag{15.95}$$

The solution in Example 15.36 exhibits damped oscillation whereas the solution in Example 15.37 is explosive. We can summarize what we have learned so far about the stability of the equilibrium of second-order homogeneous differential equations (real and distinct roots, repeated roots, and complex roots) in the following theorem.

Theorem 15.3 *A second-order differential equation is stable if the roots of its characteristic equation are real and negative or complex and their real part is negative.*

15.5.4 Particular Integral

The solution of second-order nonhomogeneous equations follows the same procedure we used to find the particular integral of the first-order equations. We illustrate this with a few examples.

Example 15.39 Consider the differential equation

$$y'' - \frac{3}{2}y' - y = \frac{1}{2}t$$

The complementary function is

$$y = A_1 e^{2t} + A_2 e^{-0.5t}$$

To find the particular integral, we start with

$$y_p = \alpha + \beta t$$

Then $y' = \beta$ and $y'' = 0$. Substituting them in the differential equation

$$-\frac{3}{2}\beta - \alpha - \beta t = \frac{1}{2}t$$

Equating the coefficients on the left and right sides of the equation we get $\alpha = -3/4$ and $\beta = -1/2$. Thus, the complete solution is

$$y = A_1 e^{2t} + A_2 e^{-0.5t} - \frac{3}{4} - \frac{1}{2}t$$

Let us now add the initial conditions $y(0) = 8$, $y'(0) = 15$. The specific solution is

$$y = 8e^{2t} + 1.5e^{-0.5t} - \frac{3}{4} - \frac{1}{2}t$$

Again note that the arbitrary constants are determined using the complete solution.

Example 15.40 To find the particular integral of the differential equation

$$y'' + 6y' + 10y = 2\cos t$$

we start with

$$y_p = \alpha \cos t + \beta \sin t$$

Substituting the particular integral and its first and second derivatives in the differential equation, we have

$$-\alpha \cos t - \beta \sin t - 6\alpha \sin t + 6\beta \cos t + 10\alpha \cos t + 10\beta \sin t = 2\cos t$$

By equating the coefficients on both sides of the equation, we obtain the particular integral as

$$y_p = \frac{2}{13}\cos t + \frac{4}{39}\sin t$$

and the complete solution is

$$y = e^{-3t}(B_1 \cos t + B_2 \sin t) + \frac{2}{13} \cos t + \frac{4}{39} \sin t$$

Example 15.41 Consider an investment model where investors balance the costs and benefits of an investment today as opposed to the next period.[4] There are two industries, one producing capital goods and the other consumer goods. All capital goods are bought by those who produce consumer goods. The price of capital goods and consumer goods are, respectively, q and p. We shall assume that p is constant and can be treated as a parameter:

$$\frac{dq}{dt} + pf_K = (\delta + \phi)q \tag{15.96}$$

The LHS is the benefit of investing today, which consists of the change in the price of capital goods and the value of the marginal product of capital. The RHS is composed of the interest foregone due to investment today and the depreciation of capital, where δ is the rate of interest, and ϕ the depreciation rate. Let us assume that the marginal product of capital starts at a very high level but decreases in proportion to the amount of capital:

$$f_K = \theta_0 - \theta_1 K \tag{15.97}$$

Thus,

$$\frac{dq}{dt} = (\delta + \phi)q - p(\theta_0 - \theta_1 K) \tag{15.98}$$

Suppose that the demand for an investment good depends on the ratio of the price of capital goods and consumer goods:

$$I = \frac{dK}{dt} + \phi K = \alpha \frac{q}{p} \tag{15.99}$$

or

$$\frac{dK}{dt} = \alpha \frac{q}{p} - \phi K \tag{15.100}$$

Now we have two differential equations. In Chap. 17 we will learn how to solve and analyze a system of differential equations. But now we approach the problem differently and will merge the two first-order differential equations into

[4]David, K. H., Begg, *The Rational Expectations Revolution in Macroeconomics, Theories and Evidence*, (1982), Chap. 7.

one second-order differential equation. Let us take the derivative of both sides of (15.100):

$$\frac{d^2K}{dt^2} = \frac{\alpha}{p}\frac{dq}{dt} - \phi\frac{dK}{dt} \tag{15.101}$$

Substituting for dq/dt in (15.99),

$$\frac{d^2K}{dt^2} = \frac{\alpha}{p}(\delta + \phi)q - \alpha(\theta_0 - \theta_1 K) - \phi\frac{dK}{dt} \tag{15.102}$$

Because

$$q = \frac{p}{\alpha}\left(\frac{dK}{dt} + \phi K\right) \tag{15.103}$$

we have

$$\frac{d^2K}{dt^2} - \delta\frac{dK}{dt} - (\phi^2 + \delta\phi + \alpha\theta_1)K = -\alpha\theta_0 \tag{15.104}$$

which is a second-order differential equation.

15.5.5 Exercises

E.15.5 Solve the following homogeneous differential equations.

i. $y'' - y' - 12y = 0$ *ii.* $y'' - 2y' + y = 0$

iii. $y'' + y' + \frac{1}{4}y = 0$ *iv.* $y'' - 12y' + 9y = 0$

v. $y'' + \frac{7}{2}y' - 15y = 0$, *vi.* $y'' + y' + 6y = 0$

vii. $y'' - 2y' + 9y = 0$

E.15.6 Solve the following nonhomogeneous equations.

i. $y'' - y' - 12y = 5$ *ii.* $y'' - 2y' + y = 2t$

iii. $y'' + y' + \frac{1}{4}y = e^{-t}$ *iv.* $y'' - 12y' + 9y = \sin t$

v. $y'' + \frac{7}{2}y' - 15y = \cos t$ *vi.* $y'' + y' + 6y = 5 + 3t$

vii. $y'' - 2y' + 9y = 2$

E.15.7 Solve the following initial value problems.

i. $y'' - y' - 12y = 0$	$y(0) = 2,\ y'(0) = 3$
ii. $y'' - 2y' + y = 0$	$y(0) = 7,\ y'(0) = 4$
iii. $y'' + \frac{7}{2}y' - 15y = \cos t$	$y(0) = 1,\ y'(0) = 2$
iv. $y'' + y' + 6y = 5 + 3t$	$y(0) = 3,\ y'(0) = 6$
v. $y'' + y' + 6y = 0$	$y(0) = 1,\ y'(0) = 1$

E.15.8 Show that in (15.94) A_1 and A_2 are complex conjugates.

15.6 Computer Solution of Differential Equations

Computer programs such as Maple and Mathematica include functions for solving differential equations. In this section with the help of a few examples we shall illustrate the use of `dsolve` function in Maple for solving differential equations.

Consider the first order differential equation

$$\frac{dy}{dt} = a + by$$

You input the equation in Maple as

$$ode := diff(y(t), t) = 2 + 3 \cdot y(t)$$

You get

$$\frac{\mathrm{d}}{\mathrm{d}t} y(t) = a + b\, y(t)$$

To get the solution, you input

dsolve(ode)

Maple returns

$$y(t) = -\frac{a}{b} + \mathrm{e}^{b\,t}\_C1$$

Note that C1 is the constant of integration.

Now consider the second order equation:

$$\frac{d^2y}{dt^2} = 5 + 3y$$

We can input it as:

$$ode := diff(y(t), t, t) = 5 + 3 \cdot y(t)$$

and solve it

dsolve(*ode*)

Which will result in

$$y(t) = \mathrm{e}^{\sqrt{3}\,t}\_C2 + \mathrm{e}^{-\sqrt{3}}\,t\_C1 - \frac{5}{3}$$

Note that we now have two constants of integration. As a final example consider the following equation:

$ode := diff(y(t), t, t) = 2 + 7 \cdot t + 5 \cdot y(t)$

$$\frac{\mathrm{d}^2}{\mathrm{d}t^2}y(t) = 2 + 7\,t + 5\,y(t)$$

dsolve(*ode*)

$$y(t) = \mathrm{e}^{\sqrt{5}\,t}\_C2 + \mathrm{e}^{-\sqrt{5}\,t}\_C1 - \frac{2}{5} - \frac{7}{5}t$$

Come to think of it, life was not this easy when I went to school.

15.7 Numerical Analysis of Differential Equations

Many differential equations cannot be solved with our present mathematical knowledge. As we saw, phase diagrams afford us a way for qualitatively analyzing differential equations and gaining insight into their behavior. But if we are ready to specify the shape of the functions involved and assign numerical values to the parameters of the equation, then we can use numerical algorithms to simulate differential equations and study their exact behavior.

15.7.1 The Euler Method

This is the easiest and the most intuitively appealing algorithm for the numerical solution of differential equations. Consider the following initial value problem:

$$\frac{dy}{dt} = f(t, y), \qquad y(t_0) = y_0 \tag{15.105}$$

Recall that

$$\frac{dy}{dt} = \lim_{t \to \infty} \frac{y(t_2) - y(t_1)}{t_2 - t_1}$$

We can write

$$y(t_2) \approx y(t_1) + (t_2 - t_1) f(y(t_1), t_1) \tag{15.106}$$

Let $[t_0, t_n]$ be the interval in the domain of the function over which we would like to study the behavior of the equation. We divide this interval into n equal subintervals of the length h:

$$[t_0, t_1], \quad [t_2, t_3], \quad \ldots, \quad [t_{n-1}, t_n]$$

Then we can recursively calculate the values of y starting with y_0 in the following way:

$$\begin{aligned} y_1 &= y_0 + hf(y_0, t_0) \\ &\vdots \\ y_n &= y_{n-1} + hf(t_{n-1}, y_{n-1}) \end{aligned} \tag{15.107}$$

To gauge the accuracy of Euler's formula, consider the Taylor expansion of $y(t_n)$ near y_n,

$$y(t_n) = y_n + (t_n - t_{n-1})y'(t_{n-1}) + \frac{(t_n - t_{n-1})^2}{2} y''(\tau) \tag{15.108}$$

where $\tau \in [t_{n-1}, t_n]$. The error of calculation for y_n, therefore, is

$$y(t_n) - y_n = \frac{(t_n - t_{n-1})^2}{2} y''(\tau) \tag{15.109}$$

Recall that $t_n - t_{n-1} = h$ and let

$$M \geq y''(\tau), \qquad \tau \in [t_0, t_n] \tag{15.110}$$

In other words, M is the upper limit of the second derivative of the function in the $[t_0, t_n]$ interval. Thus, the maximum error of calculation for any value of y would be $Mh^2/2$. This is referred to as *local truncation error*. Under certain circumstances, the sum of such errors is equal to the global error of computation. Thus, we can take the sum of absolute local errors, that is, $nMh^2/2$, as an estimate of the global error. This allows us to determine the number of intervals based on the total error we are ready to tolerate. Let the acceptable error be d. Recalling that $h = (t_n - t_0)/n$, we have

$$\frac{nMh^2}{2} = \frac{M(t_n - t_0)^2}{2n} \leq d$$

which means

$$n \geq \frac{M(t_n - t_0)^2}{2d} \tag{15.111}$$

Thus, we can reduce the error of computation to an acceptable level by increasing the number of intervals n. Euler's formula can be programmed in Matlab. Let us assume that you have already programmed the function $f(y, t)$ in an `m.file` named `horatio.m`. To be specific, suppose we would like to solve the equation

$$\frac{dy}{dt} = -2y + 0.4t, \qquad y(0) = 1 \tag{15.112}$$

for the interval [0, 3]. The reason for choosing this equation, which is an easy one to solve, is that the solution obtained from numerical analysis can be checked against the exact solution. The `horatio.m` file will look like

Matlab code

```
function z = horatio(t,y)
z = -2*y + 0.4*t;
```

Because we know the solution to (15.112), we can find the largest value of d^2y/dt^2 over the interval of interest, which turns out to be 4. The main program will call the function `horatio` and solves the problem.

Matlab code

```
% Set the parameters of the problem
t0 = 0;
tn = 3;
M = 4;
D = 0.0001;
n = ceil(((tn-t0).^2.*M)./(2.*d))+1;
H = (tn-t0)./(n-1);
% The number of y and t values is one more than the
% number of intervals. That is why we first added 1
% to n and then calculated n-1 intervals. Note also
% that Matlab starts its counter from 1 not 0.
% Now define a vector to hold y values.
y = zeros(n,1);
% Define n equally spaced values of t.
t = linspace(t0,tn,n);
% Set the initial value of y.
y(1) = 1;
% Recursively use the Euler method.
for i=1:n-1
        fval = feval(@horatio,t(i),y(i));
        y(i+1) = y(i) + h.*fval;
end
```

15.7.2 Runge-Kutta Methods

The essence of the Runge-Kutta methods is to reevaluate the function f in the interval t_n, t_{n+1}. Thus, depending on the number of points at which the function is reevaluated, we have the second-order, third-order, ..., Runge-Kutta method. Let us consider the second-order method. Here we evaluate the function at points

$$\begin{aligned} k_1 &= f(t_n, y_n) \\ k_2 &= f(t_n + \alpha h, y_n + \beta h k_1) \end{aligned} \tag{15.113}$$

and y_{n+1} is computed as

$$y_{n+1} = y_n + h\,(a k_1 + b k_2) \tag{15.114}$$

By comparing (15.114) to the Taylor expansion in (15.108), it is also clear that f is approximated by a weighted average of k_1 and k_2. To make the method operational we have to determine four parameters α, β, a and b. First, note that

$$\begin{aligned} y' &= f\,(y,\ t) \\ y'' &= f_y y' + f_t = f_y f + f_t \end{aligned} \tag{15.115}$$

Therefore,

$$y(t_{n+1}) = y(t_n) + hf + \frac{h^2}{2}(f_y f + f_t) + O(h^3) \tag{15.116}$$

On the other hand, (15.112) can be written as

$$\begin{aligned} y_{n+1} &= y_n + ahf(t_n, y_n) + bhf(t_n + \alpha h, y_n + \beta h k_1) \\ &= y_n + ahf(t_n, y_n) + bhf(t_n, y_n) + \alpha b h^2 f_t \\ &\quad + \beta b h^2 f_y f + O(h^3) \\ &= y_n + (a + b)hf(t_n, y_n) + bh^2(\alpha f_t + \beta f_y f) + O(h^3) \end{aligned} \tag{15.117}$$

Comparing (15.117) and (15.116), it follows that for the method to be consistent and the error to be $O(h^3)$, we should have

$$\begin{aligned} a + b &= 1 \\ 2b\alpha = 2b\beta &= 1 \end{aligned} \tag{15.118}$$

One solution to the above system would be $\alpha = \beta = 1$ and $a = b = 1/2$, which results in the *improved Euler method.*

$$\begin{aligned} z(k+1) &= y(k) + hf(t_k, y(k)) \\ y(k+1) &= y(k) + \frac{h}{2}\left[f(t_{k+1}, z(k+1)) + f(t_k, y(k))\right] \end{aligned} \tag{15.119}$$

We can use the improved Euler formula by modifying our previous program.

Matlab code

```
for i=1:n-1
     fval=feval(@horatio,t(i),y(i));
     z = y(i) + h.*fval;
     fval=feval(@horatio,t(i),z);
     y(i+1) = y(i) + (h./2).*(fval+z);
end
```

The fourth-order Runge-Kutta method is the most popular. In this method

$$y_{n+1} = y_n + \frac{1}{6}h(k_1 + 2k_2 + 2k_3 + k_4) \tag{15.120}$$

where

$$\begin{aligned} k_1 &= f(y_n, t_n) \\ k_2 &= f\left(y_n + \frac{1}{2}h, y_n + \frac{1}{2}hk_1\right) \\ k_3 &= f\left(y_n + \frac{1}{2}h, y_n + \frac{1}{2}hk_2\right) \\ k_4 &= f(y_n + h, y_n + hk_3) \end{aligned} \tag{15.121}$$

This method is available as `ode45` function in Matlab. Again suppose that we have already defined the function in the file `horatio.m`.

Matlab code

```
[T, Y] = ode45(@horatio,[0 3],1);
% [0 3] is the range of t and 1 is the initial value
% of y.
% Call T and Y
[T, Y]
```

15.7.3 Exercises

E.15.9 Use the Euler method to solve the differential equations in E.15.1

E.15.10 Use the `ode45` function in Matlab to solve the differential equations in E.15.1

Chapter 16
Difference Equations

Change is the essence of economic life. Production, income, prices, money in circulation, exchange rates, and other economic variables are increasing or decreasing all the time. Jobs are created and destroyed, some activities and products disappear, and new ones replace them. On a longer horizon, economic institutions undergo gradual evolution and sometimes abrupt changes. Yet it is rare that we observe a clean break with the past as if yesterday did not exist. The past exerts an influence on the present.

Dynamic economic models reflect one aspect of this dynamism by positing that the behavior of a variable is determined, among other things, by its own past values. For example, consumption of a family or a nation is determined not only by its current income but also by its past consumption. Similarly, the inflation rate is affected by its past levels as well as changes in the money supply.

There are several sources for this influence of the past on the present. First, there may be a gestation period. It takes time for an investment to turn into capital, and it takes time for a product to become known to the public. Second, there may be inertia in attaining new equilibrium levels; an immediate adjustment to new circumstances could be costly. It is not easy to shut down all offshore oil operations because demand and price have declined or build a new platform because prices are on the rise. Third, decisions imply action in the future and, therefore, are based on the expectation of the future behavior of relevant variables. Decisions to buy or sell a stock are based on the expectation of its future values, future interest rates, and the future earnings and dividends of the company. When the expectations of the future are formed based on the past behavior of a set of variables, decisions and behavior are affected by the past. Finally, optimization over time takes into account the values of different variables at different times. The optimal solutions and decisions based on that, therefore, tie together the values of variables at different time periods. Differential and difference equations are mathematical tools to model this dynamic phenomenon of economic life. If we are concerned with more than one variable, we need a system of differential or difference equations (see the next chapter). In the previous chapter we dealt with differential equations. Difference equations are usually described as the discrete version of differential equations and have received scant attention from mathematicians. This is not surprising given the immense task of solving differential equations and their wide applicability in physics. In

K. Dadkhah, *Foundations of Mathematical and Computational Economics,* 2nd ed.,
DOI 10.1007/978-3-642-13748-8_16, © Springer-Verlag Berlin Heidelberg 2011

economics a good deal of theoretical work is conducted within the framework of differential equations. But economic variables are measured in discrete time intervals and, in applied work, estimation, forecasting, simulation, and policy evaluation, we rely on difference equations.

16.1 An Overview

The first order of business is terminology and definitions. An ordinary linear difference equation of order n with constant coefficients is defined as

$$y_t = b_1 y_{t-1} + b_2 y_{t-2} + \dots b_n y_{t-n} + \gamma x_t \tag{16.1}$$

It is called ordinary because it involves only one independent variable t. If another variable were also involved, it would be called a *partial difference equation*. It is linear as it does not involve any nonlinear functions of y. And it is a constant coefficient because $b_i,\ i = 1, \dots, n$ are constant and do not vary with t. In this book we are concerned only with ordinary linear difference equations; therefore, the term difference equation will be used in place of the longer designation, and whenever there is no fear of ambiguity, the term will be shortened to equation.

If $\gamma = 0$, the equation is homogeneous and when $\gamma \neq 0$, it is called nonhomogeneous. x_t could take many different forms including an explicit function of time $x_t = f(t)$, or a member of a deterministic or stochastic series where $x_t \in \{x_t\}_{t=-\infty}^{\infty}$. $f(t)$, in turn, could be a polynomial in t, or a trigonometric, or exponential function of time.

Example 16.1 The following are examples of difference equations:

$$\begin{aligned}
y_t &= \frac{2}{3} y_{t-1} \\
y_t &= \frac{2}{3} y_{t-1} + 2 \\
y_t &= 5 y_{t-1} - \frac{1}{2} y_{t-2} \\
y_t &= 0.85 y_{t-1} + 0.35 y_{t-2} + 3 + 2t \\
y_t &= 0.9 y_{t-1} + 2x_t + 0.5 x_{t-1} + 1
\end{aligned}$$

Definition 16.1 By a solution of a difference equation we mean a function of the form $y_t = g(t)$ that satisfies the equation. That is, the substitution of y with the solution makes the difference equation an identity.

Example 16.2

$$y_t = A\left(\frac{2}{3}\right)^t$$

is a solution to the equation $y_t = \frac{2}{3}y_{t-1}$ (we will shortly learn how to find such solutions). We have

$$A\left(\frac{2}{3}\right)^t = \frac{2}{3}A\left(\frac{2}{3}\right)^{t-1}$$

Example 16.3 Let

$$y_t = \frac{2}{3}y_{t-1} + 2$$

Then

$$y_t = A\left(\frac{2}{3}\right)^t + 6$$

is a solution because

$$A\left(\frac{2}{3}\right)^t + 6 = \frac{2}{3}A\left(\frac{2}{3}\right)^{t-1} + \frac{2}{3} \times 6 + 2$$

A characteristic of such solutions is that if both $A_1\lambda_1^t$ and $A_2\lambda_2^t$ with $\lambda_1 \neq \lambda_2$ are both solutions of a difference equation, then

$$y_t = A_1\lambda_1^t + A_2\lambda_2^t \tag{16.2}$$

is also a solution. This is particularly important in the case of second- and higher-order difference equations. We illustrate this with an example.

Example 16.4 Check that

$$y_t = A_1 3^t \qquad \text{and} \qquad y_t = A_2\left(\frac{1}{2}\right)^t$$

are both solutions to the equation

$$y_t = 3.5y_{t-1} - 1.5y_{t-2}$$

But so is

$$y_t = A_1 3^t + A_2\left(\frac{1}{2}\right)^t$$

As can be seen, such solutions are not unique because they involve the free parameter A (in the case of a first-order equation) and parameters A_1 and A_2 (in the case of a second-order equation).[1] In other words, they define a family of solutions. In order to have a specific solution we need to know the value of y in as many points as there are free parameters. These values are referred to as *initial conditions.*

Example 16.5 In Example 16.2, let $y_0 = 5$. Then

$$y_0 = A\left(\frac{2}{3}\right)^0 = A = 5$$

and the specific solution is

$$y_t = 5\left(\frac{2}{3}\right)^t$$

Example 16.6 In the case of Example 16.4, we need two points. Thus, let

$$y_0 = 7, \qquad y_1 = 6$$

Then

$$\begin{aligned} A_1 + A_2 &= 7 \\ 3A_1 + 0.5\,A_2 &= 6 \end{aligned}$$

Verify that the following is a specific solution:

$$y_t = 3^t + 6\left(\frac{1}{2}\right)^t$$

Once we have found a specific solution for a linear difference equation with initial conditions, the solution is *unique.* To get an intuitive understanding of the uniqueness of the solution, consider an n-th order difference equation of the form

$$y_t = b_1 y_{t-1} + \ldots + b_n y_{t-n} + f(t) \tag{16.3}$$

with the initial conditions

$$y_j = y_j^*, \qquad j = 0, \ldots, n-1 \tag{16.4}$$

Given the values of b's and the function f, we substitute the values of y_j^*'s and t in the equation and compute y_n^*. This value of y is unique. Next we substitute the values of y_1^* to y_n^* in the equation and compute y_{n+1}^*. Again this value is unique. We can continue in this fashion and trace the path of y_t for as long as we desire. The path is unique because each and every value of y is uniquely determined. Therefore,

[1] In general, the solution of an n-th order linear difference equation involves n free parameters.

the solution of the difference equation with initial conditions that depict this path has to be unique as well. Incidentally, this argument shows how a difference equation is numerically simulated, a practice prevalent in present-day dynamic economic analysis.

The solutions of stochastic difference equations are unique in an expected value or conditional expected value sense. That is, they are unique if we are interested in the expected value $E(y_t)$ or when we have observed the value of stochastic terms and therefore, our solution is conditional on the realization of the random variables.

An important issue is to determine the point at which the difference equation may come to rest, that is, once at that point, there is no tendency in the variable y to change. Such points are referred as the *equilibrium points*. A dynamic system may not have such a point. Instead, it may have a steady-state path on which it can stay forever.

A related question we are interested in is if for some reason, the equilibrium is disturbed or the variable has left the steady state, would it return to equilibrium or steady state? This is the question of the *stability* of the equation. If an equation converges to a particular value or a path, it is called stable; else it is unstable. The issue is of significance for economic analysis in that every economic equilibrium is constantly disturbed by external forces. Every economy is constantly subjected to shocks from within and without the system. Examples include inventions and innovations in production processes such as the introduction of new products and new technology, changes in consumer tastes, political changes both at home and abroad; and economic changes in other countries. These changes are not predictable and influence the economy in unforeseen ways; hence, the stochastic nature of economic variables and relationships. Consumers, producers, and government officials are cognizant of these shocks and have to make decisions based on their expectations and with imperfect information.

Since the 1960s and the seminal work of John Muth, Robert Lucas, and Thomas Sargent on models of rational expectations, random components have explicitly been introduced into economic models. The work of Finn Kydland and Edward Prescott and others on real business cycle theory has also been instrumental in making stochastic difference equations an integral part of economic theory. What are the short- and long-term effects of these shocks? Will the economy return to its long-run growth path? Or is the long-term trajectory of the economy itself shaped by the shocks and innovations? These are some of the questions whose answers require dynamic models and the use of difference equations.

16.2 Examples of Discrete Dynamic Economic Models

That economic life is a dynamic process is self-evident. To capture this dynamism in a model which is based on plausible behavior on the part of economic actors is a challenge for economists. In this section we will discuss a few such models and will encounter more in later sections, including models reflecting more recent directions in economic theory.

Everyday experience indicates that obtaining information, processing it, making decisions, and implementing them take time. Therefore, it is not stretching things too far if we simply start by noting that a variable y at time t depends on x at time t–1. For example, some Keynesian models of business cycles (see Example 16.29) assume that consumption depends on income in the previous period:

$$C_t = \beta Y_{t-1}$$

or investment depends on changes in consumption:

$$I_t = \gamma(C_t - C_{t-1})$$

On the other hand, some models start from plausible assumptions regarding economic behavior that lead to dynamic economic models. We discuss three such models here.

16.2.1 Adaptive Expectations

Commenting on Keynes's *General Theory*, Schumpeter pointed out that, if expectations are assumed to be exogenous to the model, they would resemble a *deux ex machina* that would save the theorist when the model cannot explain a phenomenon. Therefore, if expectations are introduced into a theory, they should be endogenous to the model. In other words, the mechanism or the model for their formation should be specified. Economists have proposed two models of expectations formation: adaptive expectations and rational expectations.

The *adaptive expectations* model was suggested by Cagan in 1946 in relation to money and hyperinflation. It posits that people revise their expectations in light of their recent experience. If x^e is the expectation of the variable x, then adaptive expectation model is

$$x^e_{t+1} - x^e_t = \lambda(x_t - x^e_t) \qquad 0 < \lambda < 1 \tag{16.5}$$

Note that this model implies that expectations formed at time t for one period ahead is a weighted average of the expectation of the same variable formed one period earlier for time t and the actual realization of that variable.

$$x^e_{t+1} = \lambda x_t + (1 - \lambda)x^e_t \tag{16.6}$$

Because the same relationship holds for one period earlier, we have

$$x^e_t = \lambda x_{t-1} + (1 - \lambda)x^e_{t-1} \tag{16.7}$$

Substituting (16.7) in (16.6) we have

$$x^e_{t+1} = \lambda x_t + \lambda(1 - \lambda)x_{t-1} + (1 - \lambda)^2 x^e_{t-1} \tag{16.8}$$

Continuing in this fashion, we get

$$x_{t+1}^{e} = \lambda \sum_{j=0}^{\infty} (1-\lambda)^{j} x_{t-j} \tag{16.9}$$

Now suppose a variable y depends on the expectation of x. For example, we can think of x^e to be permanent income and y consumption. Then

$$y_t = \beta_0 + \beta_1 x_{t+1}^{e} \tag{16.10}$$

Substituting for x_{t+1}^{e} we have

$$y_t = \beta_0 + \beta_1 \lambda \sum_{j=0}^{\infty} (1-\lambda)^{j} x_{t-j} \tag{16.11}$$

Subtracting

$$(1-\lambda) y_{t-1} = (1-\lambda)\beta_0 + \beta_1 \lambda \sum_{j=1}^{\infty} (1-\lambda)^{j} x_{t-j} \tag{16.12}$$

from both sides of (16.10) and rearranging terms. We have

$$y_t = \beta_0 \lambda + (1-\lambda) y_{t-1} + \beta_1 \lambda x_t \tag{16.13}$$

Thus, the magnitude of the variable y at any time period t is determined by its values in the past, and we have a first-order nonhomogeneous difference equation. It is a nonhomogeneous equation because in addition to the dynamic of the process itself, in every period the exogenous variables x_t affects the variable y_t.

Economic relationships hardly, if ever, can be captured by deterministic equations like (16.10) and (16.13). In both theoretical and applied work we usually deal with the stochastic version of a dynamic relationship. For example, (16.10) could include a random component, that is

$$y_t = \beta_0 + \beta_1 x_{t+1}^{e} + \varepsilon_t \tag{16.14}$$

where ε_t is white noise. In this case the final equation would be

$$y_t = \beta_0 \lambda + (1-\lambda) y_{t-1} + \beta_1 \lambda x_t + \varepsilon_t - (1-\lambda)\varepsilon_{t-1} \tag{16.15}$$

16.2.2 Partial Adjustment

This model takes into account the fact that sometimes adjustment to an optimal or planned level of output, investment, consumption, and many other variables may be costly or altogether impossible. Therefore, there may be a partial move toward the desired level of a variable. Suppose the desired or optimal level of variable y denoted by y^* depends on variable x:

$$y_t^* = \alpha_0 + \alpha_1 x_t + \nu_t \tag{16.16}$$

where μ is white noise. But the move toward the optimal level is gradual,

$$y_t - y_{t-1} = \gamma(y_t^* - y_{t-1}) \qquad 0 < \gamma < 1 \tag{16.17}$$

Therefore,

$$y_t^* = \frac{1}{\gamma} y_t - \frac{1-\gamma}{\gamma} y_{t-1} \tag{16.18}$$

Substituting (16.18) in (16.17), we have

$$y_t = \gamma\alpha_0 + (1-\gamma)y_{t-1} + \gamma\alpha_1 x_t + \gamma\nu_t \tag{16.19}$$

Again we have a first-order nonhomogeneous difference equation.

16.2.3 Hall's Consumption Function

In 1978 Robert E. Hall literally shook the economics profession by presenting results that showed aggregate consumption follows a random walk. He started with the lifetime utility function

$$\sum_{j=t}^{T} \frac{U(C_j)}{(1+\delta)^{j-t}} \tag{16.20}$$

and the budget constraint

$$\sum_{j=t}^{T} \frac{C_j - W_j}{(1+r)^{j-t}} = A_t \tag{16.21}$$

and

$\delta =$ rate of time preference
$T =$ the length of economic life
$C_t =$ consumption in year t
$U =$ utility function
$r =$ real rate of interest
$W_t =$ earnings at time t
$A_t =$ Assets excluding human capital

Furthermore he assumed r to be constant over time and $r \geq \delta$. Maximizing the conditional expectation of the utility function subject to budget constraint, he arrived at

$$E_t U'(C_{t+1}) = \frac{1+\delta}{1+r} U'(C_t) \tag{16.22}$$

where E_t denotes conditional expectation—conditional on all information available at time t. Using different utility functions, he concluded that (16.22) can be approximated by the simple function

$$C_t = \lambda C_{t-1} + \varepsilon_t \tag{16.23}$$

which is a first-order difference equation.

By now it should be obvious that in order to analyze and understand economic dynamics, make forecasts, and evaluate different policies within a dynamic context, we need to understand the behavior of deterministic and stochastic difference equations. This in turn requires us to find and analyze the nature of solutions of difference equations.

16.2.4 Exercises

E.16.1 Verify the solution of each of the following equations.

Equation	Solution
$y_t = -3y_{t-1} + 4$	$y_t = A(-3)^t + 1$
$y_t = y_{t-1} + 6$	$y_t = A + 6t$
$y_t = 7y_{t-1} - 10y_{t-2}$	$y_t = A_1(2^t) + A_2(5^t)$
$y_t = 0.1y_{t-1} + 0.3y_{t-2},\ y_0 = 1,\ y_1 = 1.7$	$y_t = 2(0.6^t) - (-0.5)^t$
$y_t = 2y_{t-1} - y_{t-2},\ y_0 = 2,\ y_1 = 5$	$y_t = 2 + 3t$

E.16.2 Show that the solution to the maximization problem

$$\max\ E_t \sum_{j=t}^{T} \frac{U(C_j)}{(1+\delta)^{j-t}}$$

$$\text{subject to} \quad \sum_{j=t}^{T} \frac{C_j - W_j}{(1+r)^{j-t}} = A_t$$

is

$$E_t U'(C_{t+1}) = \frac{1+\delta}{1+r} U'(C_t)$$

E.16.3 In Exercise 16.2, show that we can write

$$U'(C_{t+1}) = \frac{1+\delta}{1+r}U'(C_t) + \varepsilon_{t+1}$$

where ε_{t+1} is white noise.

E.16.4 In Exercise 16.2, let

$$U(C_t) = -\frac{1}{2}(\bar{C} - C_t)^2$$

where $\bar{C}$ is a constant. Show that the solution to the utility maximization problem is

$$C_{t+1} = \frac{\bar{C}(r-\delta)}{1+r} + \frac{1+\delta}{1+r}C_t - \varepsilon_{t+1}$$

16.3 First-Order Linear Difference Equations

A first-order linear difference equation is of the form

$$y_t = \lambda y_{t-1} + \gamma x_t \tag{16.24}$$

where x_t can be a known function of time such as

$$x_t = \alpha_0 + \alpha_1 t \tag{16.25}$$

or be a deterministic or stochastic sequence of variables, that is $x_t \in \{x_j\}_{j=-\infty}^{\infty}$. Because of γx_t on the RHS the equation is called non-homogeneous. The homogeneous part reflects the internal dynamics of the system when no external force is applied to it, whereas x_t is an external force that could be in the form of a stochastic shock to the system or a policy or any other exogenous factor affecting y. The solution of a nonhomogeneous difference equation consists of the sum of the solutions of the homogeneous and the nonhomogeneous parts. We start with the solution of the homogeneous part and an analysis of its behavior.

16.3.1 Solution of First-Order Linear Homogeneous Difference Equations

The first-order linear homogeneous equation

$$y_t = \lambda y_{t-1} \tag{16.26}$$

has a straightforward solution. Observe that

$$\frac{y_t}{y_{t-1}} = \lambda$$

Therefore,

$$\begin{aligned} y_t &= \frac{y_t}{y_{t-1}} \frac{y_{t-1}}{y_{t-2}} \cdots \frac{y_1}{y_0} y_0 \\ &= \lambda\, \lambda\, \ldots\, \lambda\, y_0 \\ &= y_0 \lambda^t \end{aligned} \tag{16.27}$$

In the above solution, we implicitly assumed that we know that the value of y at time $t = 0$ is y_0. Lacking such knowledge, we simply can write the solution as

$$y_t = A\lambda^t \tag{16.28}$$

We can verify that a function of the form $y_t = A\lambda^t$ is a solution of (16.26) because

$$y_t = A\lambda^t = \lambda A\lambda^{t-1} = \lambda y_{t-1}$$

In accordance with our previous chapter terminology we shall call $y_t = A\lambda^t$ the *complementary solution*. If, in addition, we have an initial condition that specifies the value of y for some t, say, y_0 for $t = 0$, then we can find a specific solution for the equation.

Example 16.7 The solution of the difference equation

$$y_t = \frac{1}{2} y_{t-1}$$

is

$$y_t = A\left(\frac{1}{2}\right)^t$$

We can verify that the above is indeed the solution by checking that

$$\frac{1}{2} A\left(\frac{1}{2}\right)^{t-1} = A\left(\frac{1}{2}\right)^t = y_t$$

Now suppose that the initial condition is

$$y_t = y_0 \qquad \text{for} \qquad t = 0.$$

Then the specific solution will be

$$y_t = y_0\left(\frac{1}{2}\right)^t$$

Example 16.8 Given the difference equation

$$y_t = 3.5\, y_{t-1}$$

and the initial condition

$$y_0 = 2.5$$

the solution is

$$y_t = 2.5(3.5)^t$$

The nature of a complementary solution will prove important in analyzing the behavior of a difference equation. If $|\lambda| > 1$, y_t keeps growing, whereas for $|\lambda| < 1$ it will approach zero as $t \to \infty$. On the other hand, when $\lambda > 0$, y_t does not change sign and if, say, $y_0 > 0$, then $y_t > 0, \ \forall t$, but if $\lambda < 0$, y_t switches back and forth from negative to positive and positive to negative values. The cases of $\lambda = 1$ and $\lambda = -1$ result in

$$y_t = \begin{cases} A & \text{if} \quad \lambda = 1 \\ (-1)^t A & \text{if} \quad \lambda = -1 \end{cases}$$

These possibilities are illustrated in Fig. 16.1. An intuitively appealing way to verify the solution of a difference equation is by tracing the trajectory of y_t using both the original equation and its solution, and comparing them. In addition, such an exercise will allow us to explore the nature of the solution. We can perform the necessary computation using Matlab or Excel. The code is written for the problem

$$y_t = 0.5y_{t-1},$$
$$y_0 = 2.5$$

but you may want to experiment with different equations and initial values.

Matlab code

```
% Specify the number of periods y is to be computed
n = 15;
% Initialize two arrays to hold the computed
% values of y one from the original equation and one
% from the solution
y1 = zeroes(n,1);
y2 = zeroes(n,1);
% specify the equation and the initial condition
lambda = 0.5;
y1(1) = 2.5;
% Note that Matlab starts its index of an
% array with 1
for i=2:n
        y1(i) = lambda.* y1(i-1);
end
for i=1:n
       t = i;
       y2(i) = y1(1).*(lambda.^(i-1));
end
```

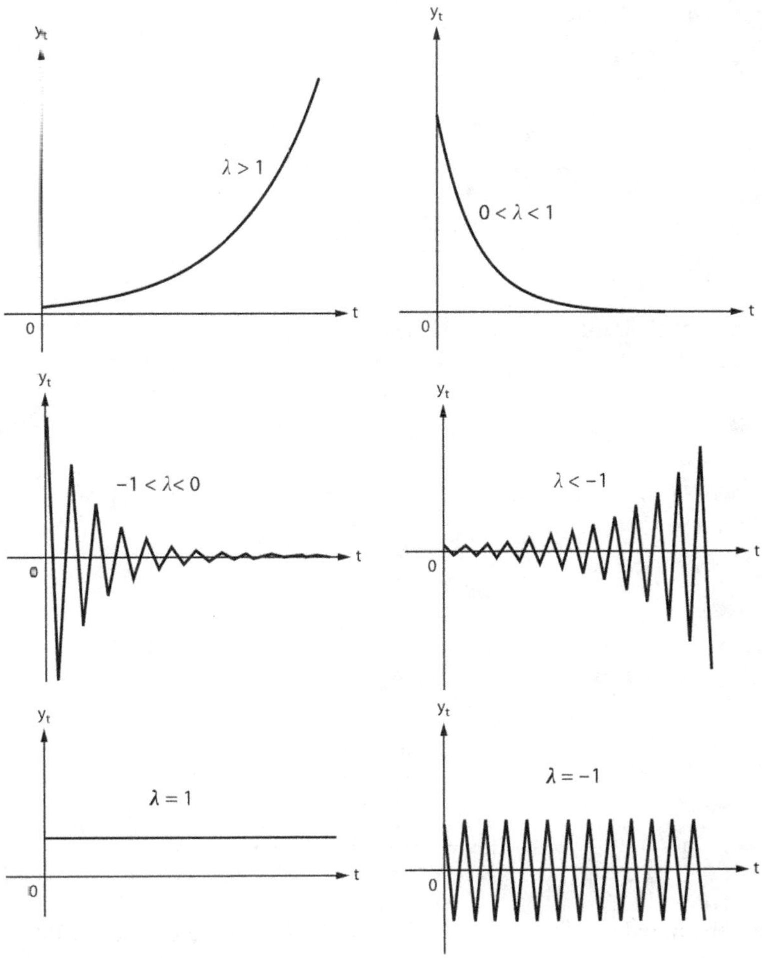

Fig. 16.1 Behavior of a first-order homogeneous difference equation

```
y1
y2
plot(t,y1)
plot(t,y2)
```

16.3.2 Solution of First-Order Nonhomogeneous Equations

In solving a nonhomogeneous equation, we can follow the same procedure as we did for homogeneous equation. Let

$$y_t = \lambda y_{t-1} + \gamma x_t \tag{16.29}$$

By successive substitutions, we have

$$\begin{aligned} y_t &= \lambda^2 y_{t-2} + \gamma x_t + \lambda\gamma x_{t-1} \\ &= \lambda^3 y_{t-3} + \gamma x_t + \lambda\gamma x_{t-1} + \lambda^2\gamma x_{t-2} \\ &= \ldots \\ &= A\lambda^t + \gamma \sum_{i=0}^{t-1} \lambda^i x_{t-i} \end{aligned} \tag{16.30}$$

Let us now consider different possibilities for x_t.
1. *Nonhomogeneous part is a constant.* We have

$$y_t = \lambda y_{t-1} + \alpha \tag{16.31}$$

In other words, x_t takes a constant value for all periods equal to x_0, and $\alpha = \gamma x_0$. The solution is

$$\begin{aligned} y_t &= A\lambda^t + \gamma \sum_{j=0}^{t-1} \lambda^j x_0 \\ &= A\lambda^t + \sum_{j=0}^{t-1} \lambda^j \alpha \\ &= A\lambda^t + \alpha \frac{1-\lambda^t}{1-\lambda} \\ &= \left(A - \frac{\alpha}{1-\lambda}\right)\lambda^t + \frac{\alpha}{1-\lambda} \end{aligned} \tag{16.32}$$

If we add the initial condition $y_t = y_0$, for $t = 0$, the specific solution will be

$$y_t = \left(y_0 - \frac{\alpha}{1-\lambda}\right)\lambda^t + \frac{\alpha}{1-\lambda} \tag{16.33}$$

If the equation is stable, that is, if $|\lambda| < 1$, then

$$\lim_{t\to\infty} y_t = \frac{\alpha}{1-\lambda} \tag{16.34}$$

which in economics is referred to as intertemporal equilibrium. That is, if $y_t = \alpha/(1-\lambda)$, there is no reason for it to change

Example 16.9 Let

$$y_t = 0.75 y_{t-1} + 4, \qquad y_0 = 18$$

The solution is

$$y_t = \left(18 - \frac{4}{1-0.75}\right)(0.75^t) + \frac{4}{1-0.75} = 2(0.75^t) + 16$$

Example 16.10 (Monthly Payments of a Loan). We are already familiar with the formula to calculate the monthly payment D of a loan of A dollars borrowed for n years at the rate r percent. It is

$$D = A\frac{(1+r/12)^{12n}(r/12)}{(1+r/12)^{12n}-1}$$

For example, the monthly payment of a $200,000 loan borrowed for 15 years at 6% is

$$200000\frac{(1+0.06/12)^{180}(0.06/12)}{(1+0.06/12)^{180}-1} = 1687.71$$

You can check this with the PMT function in Excel. Note that the payment is made at the end of the month; therefore, the command line is

```
PMT(6%/12,15*12,200000,,0)
```

We can derive the formula using difference equations. Let y_0 be the amount borrowed and y_t the amount owed at the end of the month t, $t = 1, \ldots, 12n$. Then

$$y_t = y_{t-1}\,(1+r/12) - D$$

Solving this equation we have

$$y_t = \left(y_0 - \frac{D}{r/12}\right)(1+r/12)^t + \frac{D}{r/12}$$

Noting that $y_0 = A$ and $y_{12n} = 0$, we have

$$0 = \left(A - \frac{D}{r/12}\right)(1+r/12)^{12n} + \frac{D}{r/12}$$

which upon solving for D gives the monthly payment formula.

Example 16.11 (A Model of Real Business Cycles). In this model the main driving forces of the economy are technological change and population growth. The model assumes a Cobb-Douglas production function, where output (Y) depends on capital (K), labor (L), and technology (A).

$$Y_t = K_t^{\alpha}(A_tL_t)^{1-\alpha} \tag{16.35}$$

What is not consumed is saved and capital accumulation is the result of saving, and it takes one period for saving to turn into capital. Unrealistically, however, it is assumed that capital lasts only one period, that is, it is totally used up in the process of production.

$$K_t = Y_{t-1} - C_{t-1} = sY_{t-1} \tag{16.36}$$

Population (N) grows at the rate η and labor participation rate ($\bar{l}$) is constant. Therefore,

$$L_t = \bar{l}\bar{N}e^{\eta t}$$

which implies

$$\ln L_t = \ln \bar{l} + \ln \bar{N} + \eta t \tag{16.37}$$

There are two main sources of technological growth: a constant accumulation of knowhow, improvements, and innovations, γt, and a random component of innovations, u_t. Thus,

$$A_t = \bar{A}e^{\gamma t + u_t}$$

or

$$\ln A_t = \ln \bar{A} + \gamma t + u_t \tag{16.38}$$

We will deal with this model in three stages of increasing complexity. First, let us, for the time being, set $u_t = 0, \ \forall t$. In other words, technological progress has a constant rate of growth γ. Taking the logarithm of the production function and substituting for $\ln L_t$ and $\ln A_t$, we have

$$\begin{aligned}
\ln Y_t &= \alpha \ln K_t + (1-\alpha)\ln A_t + (1-\alpha)\ln L_t \\
&= \alpha \ln s + \alpha \ln Y_{t-1} + (1-\alpha)(\bar{A} + \gamma t) + (1-\alpha)(\ln \bar{l} + \bar{N} + \eta t) \\
&= B + \alpha \ln Y_{t-1} + (1-\alpha)(\eta + \gamma)t
\end{aligned} \tag{16.39}$$

where

$$B = \alpha \ln s + (1-\alpha)(\bar{A} + \ln \bar{l} + \bar{N})$$

The model is now reduced to a first-order difference equation in $\ln Y_t$. We shall discuss the solution of this model below.

2. *The nonhomogeneous part is a linear trend.* We have

$$y_t = \lambda y_{t-1} + \alpha + \beta t \tag{16.40}$$

that is,

$$\gamma x_t = \alpha + \beta t \tag{16.41}$$

Substituting (16.41) in (16.30), the solution is

$$y_t = A\lambda^t + \alpha \sum_{j=0}^{t-1} \lambda^j + \beta \sum_{j=0}^{t-1} \lambda^j (t-j) \tag{16.42}$$

But

$$\begin{aligned}
\sum_{j=0}^{t-1} \lambda^j (t-j) &= t\sum_{j=0}^{t-1} \lambda^j - \sum_{j=0}^{t-1} j\lambda^j \\
&= t\frac{1-\lambda^t}{1-\lambda} - \frac{1}{1-\lambda}t\lambda^t - \lambda\frac{1-\lambda^t}{(1-\lambda)^2} + \frac{1}{1-\lambda}t\lambda^t + \frac{1}{1-\lambda}t\lambda^t \\
&= \frac{1}{1-\lambda}t - \frac{\lambda}{(1-\lambda)^2} + \frac{\lambda}{(1-\lambda)^2}\lambda^t
\end{aligned}$$

Therefore, the solution is

$$y_t = A\lambda^t + \frac{\alpha}{1-\lambda} - \frac{\alpha}{1-\lambda}\lambda^t + \beta\left[\frac{1}{1-\lambda}t - \frac{\lambda}{(1-\lambda)^2} + \frac{\lambda}{(1-\lambda)^2}\lambda^t\right]$$

Using the initial condition $y_t = y_0$, for $t = 0$, we have

$$y_t = \left(y_0 - \frac{\alpha}{1-\lambda} + \frac{\beta\lambda}{(1-\lambda)^2}\right)\lambda^t + \frac{\alpha}{1-\lambda} - \beta\frac{\lambda}{(1-\lambda)^2} + \beta\frac{1}{1-\lambda}t \tag{16.43}$$

We can now solve the equation in Example 16.10. Letting the output at time $t = 0$ be Y_0, the solution to the equation (16.39) is

$$\ln Y_t = \left(\ln Y_0 - \frac{B}{1-\alpha} + \frac{\alpha(\eta+\gamma)}{1-\alpha}\right)\alpha^t + \frac{B}{1-\alpha} - \frac{\alpha(\eta+\gamma)}{1-\alpha} + (\eta+\gamma)t$$

Suppose the economy starts at the point

$$\ln Y_0 = \frac{B}{1-\alpha} - \frac{\alpha(\eta+\gamma)}{1-\alpha}$$

Then the long-run rate of growth of Y_t, in the absence of any stochastic technological shock, would be $\eta + \gamma$, that is, the rate of growth of population plus the long-term growth rate of technology.

What the solution in (16.43) tells us is that y_t has a trend. Let us assume that in (16.40) $|\lambda| < 1$ and the first term in (16.43) vanishes as t increases. Then y_t converges to a linear trend. Many economic variables behave in this way. In such cases, it would be desirable to estimate the trend and study the deviation of y from this trend. In (16.40) we can define:

$$\beta^* = \frac{\beta}{1-\lambda} \qquad \alpha^* = \frac{\alpha}{1-\lambda} - \frac{\lambda\beta}{(1-\lambda)^2} \tag{16.44}$$

Therefore,

$$\beta = (1-\lambda)\beta^* \qquad \alpha = (1-\lambda)\alpha^* + \lambda\beta^* \tag{16.45}$$

Substituting (16.45) in (16.40) and rearranging terms, we have

$$y_t - \alpha^* - \beta^* t = \lambda[y_{t-1} - \alpha^* - \beta^*(t-1)] \tag{16.46}$$

Define

$$z_t = y_t - \alpha^* - \beta^* t \tag{16.47}$$

(16.46) can now be written as a first-order homogeneous difference equation

$$z_t = \lambda z_{t-1} \tag{16.48}$$

Such a transformation is particularly interesting for economic analysis because, instead of analyzing the variable of interest, say, GDP, we can analyze its fluctuations around a deterministic trend.

Example 16.12 (A Model of Real Business Cycles, continued) Going back to (16.39),

$$\ln Y_t = B + \alpha \ln Y_{t-1} + (1-\alpha)(\eta + \gamma)t$$

Define

$$z_t = \ln Y_t - (\eta + \gamma)t$$

We have

$$z_t = B + \alpha z_{t-1} \tag{16.49}$$

By subtracting the growth rate from $\ln Y_t$ we are able to study the fluctuations of the output around its long-run trend. We still have a first-order difference equation and the solution is left to the reader.

3. *The nonhomogeneous part is an exogenous deterministic or stochastic variable.* The equation and its solution are

$$y_t = \lambda y_{t-1} + \gamma x_t \tag{16.50}$$

and

$$y_t = A\lambda^t + \gamma \sum_{i=0}^{t-1} \lambda^i x_{t-i} \tag{16.51}$$

From an economic point of view the stability of this equation is of great importance. If $\lambda > 1$, then the impact of past values of x_t become more and more important as time passes. For example, if y is the GDP and x, government expenditures, then $\lambda > 1$ implies that one dollar of government expenditures in 1945 had a greater impact on the GDP in 2011 than a dollar spent in 2010 or 2011. Similarly, if x_t represents innovation in the production process, then an innovation in 1900 would have a greater effect on the well-being of Americans in 2011 than a similar innovation in 1995. This is hard to accept. It seems more reasonable to assume that the opposite is true, and the effect of a shock or an exogenous variable dies down as time passes.

Example 16.13 (A Model of Real Business Cycles, continued). We now go back to our original assumption that technological progress has a deterministic trend and a stochastic shock:

$$A_t = \bar{A}e^{\gamma t + u_t}$$

or in logarithmic form

$$\ln A_t = \ln \bar{A} + \gamma t + u_t \tag{16.52}$$

In this case the final difference equation becomes

$$z_t = B + \alpha z_{t-1} + (1-\alpha)u_t \tag{16.53}$$

with solution

$$z_t = \left(z_0 - \frac{B}{1-\alpha}\right)\alpha^t + \frac{B}{1-\alpha} + (1-\alpha)\sum_{j=0}^{t-1} \alpha^j u_{t-j} \tag{16.54}$$

Because $0 < \alpha < 1$, the model is stable and as time goes by the effects of shocks to the aggregate production function subside. Thus, the model has a deterministic trend, but fluctuations around the long-run trend are caused by innovations.

16.3.3 Exercises

E.16.5 Solve the following first order difference equations.

$$
\begin{array}{lll}
i. & y_t = 0.5y_{t-1} + 6 & \\
ii. & y_t = 1.75y_{t-1} + 0.3t & \\
iii. & y_t = 1.3y_{t-1} + 5 - 0.6t & \\
iv. & y_t = 0.85y_{t-1} + 0.4x_t & \\
v. & y_t = 0.6y_{t-1}, \quad y_0 = 2.6 & \\
vi. & y_t = 0.75y_{t-1} + 2, & y_0 = 3 \\
vii. & y_t = 2y_{t-1} + 5 + 0.5t, & y_0 = 7.5 \\
viii. & y_t = 1.5y_{t-1} + x_t, & y_0 = 2
\end{array}
$$

E.16.6 Solve the first order difference equation resulting from Hall's model,

$$C_t = \lambda C_{t-1} + \varepsilon_t$$

and discuss the implications of different values of λ for the behavior of consumption.

E.16.7 Determine if the solutions to equations in E.16.5 are stable.

E.16.8 Write a program that calculates the monthly payment of a loan given the amount borrowed, the interest rate, and the life of the loan in month.

E.16.9 Write a program that given the information about a loan, generates a monthly report informing the borrower how much is owed after the last payment and how much she has paid in interest so far.

E.16.10 Write a program that, given the information about the loan, at any instance can tell the borrower how much is owed. In other words, how much the customer needs to pay to pay off the loan right away? Assume that interest is accrued on a daily basis.

16.4 Second-Order Linear Difference Equations

In studying second-order difference equations, we will follow the same procedure as we did for the first-order equations. That is, we first deal with homogeneous equations and then nonhomogeneous equations. In finding the solution of the latter, the lag operator will prove very useful. We will take a short detour to cover this subject. Of course, the subject is useful in its own right, because the lag operator finds many uses in econometrics.

16.4.1 Solution of Second-Order Linear Homogeneous Difference Equations

A second-order linear difference equation has the general form

$$y_t = b_1 y_{t-1} + b_2 y_{t-2} \tag{16.55}$$

In view of the solution of the first-order equations, we may try a solution of the form $y = A\lambda^t$ on (16.55):

$$A\lambda^t = b_1 A\lambda^{t-1} + b_2 A\lambda^{t-2} \tag{16.56}$$

Dividing through by $A\lambda^{t-2} \neq 0$, we get the *characteristic equation* or *auxiliary equation*

$$\lambda^2 - b_1\lambda - b_2 = 0$$

which has the solutions

$$\lambda_1 = \frac{b_1 + \sqrt{b_1^2 + 4b_2}}{2}, \qquad \lambda_2 = \frac{b_1 - \sqrt{b_1^2 + 4b_2}}{2}$$

As a result, we have two solutions,

$$y_t = A_1 \left(\frac{b_1 + \sqrt{b_1^2 + 4b_2}}{2} \right)^t$$

and

$$y_t = A_2 \left(\frac{b_1 - \sqrt{b_1^2 + 4b_2}}{2} \right)^t$$

We need not choose between the two but take a linear combination of them as the solution to (16.55):

$$y_t = A_1 \left(\frac{b_1 + \sqrt{b_1^2 + 4b_2}}{2} \right)^t + A_2 \left(\frac{b_1 - \sqrt{b_1^2 + 4b_2}}{2} \right)^t \tag{16.57}$$

To see that this is a solution that satisfies the original equation, consider

$$\begin{aligned}\lambda_1^t &= \left(\frac{b_1+\sqrt{b_1^2+4b_2}}{2}\right)^2 \lambda_1^{t-2}\\ &= \left(b_1\frac{b_1+\sqrt{b_1^2+4b_2}}{2}+b_2\right)\lambda_1^{t-2}\\ &= b_1\left(\frac{b_1+\sqrt{b_1^2+4b_2}}{2}\right)^{t-1}+b_2\left(\frac{b_1+\sqrt{b_1^2+4b_2}}{2}\right)^{t-2}\end{aligned} \tag{16.58}$$

Similarly,

$$\lambda_2^t = b_1\left(\frac{b_1-\sqrt{b_1^2+4b_2}}{2}\right)^{t-1}+b_2\left(\frac{b_1-\sqrt{b_1^2+4b_2}}{2}\right)^{t-2} \tag{16.59}$$

Substituting (16.58) and (16.59) in (16.57), shows that it is indeed the solution to (16.55).

Example 16.14 Find the solution of the difference equation

$$y_t = \frac{7}{2}y_{t-1} - \frac{3}{2}y_{t-2}$$

Solving the characteristic equation

$$b^2 - \frac{7}{2}b + \frac{3}{2} = 0$$

we have

$$b_1 = 3 \qquad \text{and} \qquad b_2 = \frac{1}{2}$$

and

$$y_t = A_1 3^t + A_2\left(\frac{1}{2}\right)^t$$

Because the solution of second-order equation has two free parameters, A_1 and A_2, we need two initial conditions to turn our solution into a specific solution.

Example 16.15 In Example 16.14, let

$$y_0 = 1, \qquad y_1 = \frac{3}{2}$$

Then

$$\begin{aligned} A_1 + A_2 &= 1 \\ 3A_1 + \frac{1}{2}A_2 &= \frac{3}{2} \end{aligned}$$

Thus, the specific solution is

$$y_t = \frac{2}{5}3^t + \frac{3}{5}\left(\frac{1}{2}\right)^t$$

Example 16.16 Let

$$y_t = 3y_{t-1} + 28y_{t-2}$$

with initial conditions

$$y_1 = 23, \quad y_2 = 293$$

The complementary function is

$$y_t = A_1(-4)^t + A_2 7^t$$

and the specific solution

$$y_t = 3(-4)^t + 5(7^t)$$

16.4.2 Behavior of the Solution of Second-Order Equation

We noted that the behavior of the solution of the first-order equation depends on the sign and magnitude of λ. The same is true for the case of the second-order equation except that here we have two magnitudes, λ_1 and λ_2, and they are solutions to a quadratic equation. They may be real and distinct roots, repeated roots, or conjugate complex roots.

1. *Real Distinct Roots*. In this case both λ_1 and λ_2 are real and $\lambda_1 \neq \lambda_2$. Examples 16.14 and 16.15 depict this case. The solution is stable iff

$$\max\{|\lambda_1|, |\lambda_2|\} < 1 \tag{16.60}$$

because in that case

$$\lim_{t\to\infty} y_t = \lim_{t\to\infty} A_1\lambda_1^t + \lim_{t\to\infty} A_2\lambda_2^t = 0 \tag{16.61}$$

Equations in Examples 16.14 and 16.15 are both unstable, but the following is an example of a stable equation

Example 16.17 The equation

$$y_t = 0.4\, y_{t-1} + 0.21\, y_{t-2}$$

is stable because its solution is

$$y_t = A_1 0.7^t + A_2(-0.3)^t$$

2. *Repeated Real Roots*. In this case, both λ_1 and λ_2 are real, but because

$$b_1^2 + 4b_2 = 0$$

we have

$$\lambda_1 = \lambda_2 = \lambda = \frac{b_1}{2} = \sqrt{-b_2}$$

If we follow our procedure above, we get

$$y_t = A_1\lambda^t + A_2\lambda^t = (A_1 + A_2)\lambda^t \tag{16.62}$$

Thus, something is lost and the solution has been reduced to that of a first-order equation. When we have a repeated root, the solution becomes

$$y_t = A_1\lambda^t + A_2 t\lambda^t \tag{16.63}$$

To verify that indeed this is a solution, we write

$$A_1\lambda^t + A_2 t\lambda^t = b_1 A_1\lambda^{t-1} + b_1 A_2(t-1)\lambda^{t-1} + b_2 A_1\lambda^{t-2} + b_2 A_2(t-2)\lambda^{t-2}$$

Recalling that $b_1 = 2\lambda$ and $b_2 = -\lambda^2$ we have

$$\begin{aligned} A_1\lambda^t + A_2 t\lambda^t &= 2A_1\lambda^t + 2A_2(t-1)\lambda^t - A_1\lambda^t - A_2(t-2)\lambda^t \\ &= A_1\lambda^t + A_2 t\lambda^t \end{aligned}$$

Example 16.18 Let

$$y_t = 4y_{t-1} - 4y_{t-2}$$

Check that the solution is

$$y_t = A_1 2^t + A_2 t 2^t$$

Example 16.19 Let

$$y_t = 5y_{t-1} - 6.25y_{t-2}$$

Check that the solution is

$$y_t = A_1 2.5^t + A_2 t 2.5^t$$

Example 16.20 Given

$$y_t = 0.6\, y_{t-1} - 0.09\, y_{t-2}$$

and

$$y_0 = 2, \qquad y_1 = 2.1$$

the solution is

$$y_t = 2(0.3^t) + 5t(0.3^t)$$

For a difference equation with repeated real roots, the criterion for stability is the same as for the case of distinct roots except that we have only one λ. That is,

$$|\lambda| < 1$$

3. *Conjugate complex roots*. In this case, we have

$$b_1^2 + 4b_2 < 0$$

and the solution to the characteristic equation is of the form $\lambda = h_1 \pm h_2 i$, that is,

$$\lambda_1 = \frac{b_1}{2} + i\frac{\sqrt{-(b_1^2 + 4b_2)}}{2}, \qquad \lambda_2 = \frac{b_1}{2} - i\frac{\sqrt{-(b_1^2 + 4b_2)}}{2}$$

Recall that complex numbers can be written as

$$\lambda_1 = \rho(\cos\theta + i\sin\theta), \qquad \lambda_2 = \rho(\cos\theta - i\sin\theta)$$

where

$$\rho = \sqrt{h_1^2 + h_2^2} = \sqrt{-b_2}$$

$$\theta = \tan^{-1}\frac{h_2}{h_1} = \tan^{-1}\sqrt{-\left(1 + \frac{4b_2}{b_1^2}\right)}$$

Thus,

$$\begin{aligned}y_t &= A_1[\rho(\cos\theta + i\sin\theta)]^t + A_2[\rho(\cos\theta - i\sin\theta)]^t \\ &= A_1[\rho^t(\cos\theta t + i\sin\theta t)] + A_2[\rho^t(\cos\theta t - i\sin\theta t)] \\ &= \rho^t[(A_1 + A_2)\cos\theta t + (A_1 - A_2)i\sin\theta t]\end{aligned} \tag{16.64}$$

where the second equality is based on De Moivre's theorem (Chap. 2). Because y_t is a real variable and the RHS of (16.64) involves complex variables, it necessitates that A_1 and A_2 be complex numbers. Moreover, they are complex conjugates (see Exercise E.16.11). Note that

$$\begin{aligned}\cos\theta t &= \cos(\theta t + 2k\pi) \\ \sin\theta t &= \sin(\theta t + 2k\pi)\end{aligned}$$

Therefore, we can write (16.64) as

$$y_t = \rho^t[(A_1 + A_2)\cos(\mathrm{mod}(\theta t, 2\pi)) + (A_1 - A_2)i\sin(\mathrm{mod}(\theta t, 2\pi))]$$

where $\mathrm{mod}(\theta t, 2\pi)$ is the remainder of the division of θt by 2π. Let

$$\begin{aligned}A_1 &= a(\cos B + i\sin B) \\ A_2 &= a(\cos B - i\sin B)\end{aligned} \tag{16.65}$$

where a and B are real constants

$$\begin{aligned}B &= \tan^{-1}\left(-\frac{(A_1 - A_2)i}{A_1 + A_2}\right) \\ a &= \frac{A_1 + A_2}{2\cos B}\end{aligned}$$

Because $A_1 + A_2 = 2a\cos B$ and $A_1 - A_2 = 2ai\sin B$, combining (16.65) and (16.64), we have

$$\begin{aligned}y_t &= 2a\rho^t(\cos B\cos\theta t - \sin B\sin\theta t) \\ &= C\rho^t\cos(\theta t + B)\end{aligned}$$

where $C = 2a$.

Trigonometric functions are cyclical and oscillating. If $\rho < 1$ then we have damped oscillation; if $\rho > 1$ the model has explosive oscillations; and when $\rho = 1$ we observe constant oscillation.

Example 16.21 Consider the equation

$$y_t = 2y_{t-1} - 5y_{t-2}$$

We have

$$\lambda_1 = 1 + 2i \qquad \lambda_2 = 1 - 2i$$

Therefore,

$$\rho = \sqrt{5}\theta = \tan^{-1}(2) \approx 0.352\pi$$

Thus,

$$y_t = 5^{t/2}[(A_1 + A_2)\cos(0.352\pi t) + i(A_1 - A_2)\sin(0.352\pi t)]$$

As before, we need values of y_t at two points to find a specific solution. Let

$$y_0 = 2, \qquad y_1 = 4$$

Then

$$A_1 + A_2 = 2, \qquad (A_1 - A_2)i = 1$$

Therefore,

$$A_1 = 1 - 0.5i, \qquad A_2 = 1 + 0.5i$$

and

$$y_t = 5^{t/2}[2\cos(0.352\pi t) + \sin(0.352\pi t)]$$

Alternatively,

$$A_1 + A_2 = 2\cos B \qquad (A_1 - A_2)i = -2\sin B$$

Therefore,

$$B = \tan^{-1}\left(-\frac{(A_1 - A_2)i}{A_1 + A_2}\right) = \tan^{-1}\left(-\frac{1}{2}\right) \approx -0.148\pi$$

and

$$C = 2a = \frac{A_1 + A_2}{\cos B} = \frac{2}{0.894} = 2.236$$

Thus,

$$y_t = 2.236(5^{t/2})\cos(0.352\pi t - 0.148\pi)$$

The reader should convince herself that indeed the three versions of the equation in the above example depict the same trajectory for y_t. This can be done in Matlab. Note that here the solution is explosive because $\rho > 1$.

Matlab code

```
% Define three arrays
y = zeros(15,1);
y1 = zeros(15,1);
y2 = zeros(15,1);
% Specify the equation and the initial conditions
b1 = 2;
b2 = -5;
y(1) = 2;
y(2) = 4;
h1 = b1./2;
h2 = ((-(b1.^{2+4*b2)).^0.5)./2;
rho = (-b2).^0.5;
theta = atan((-(1+4*b2./(b1.^2))).^0.5);
A1A2 = y(1);
A2A1= (y(2)-rho.*y(1)*cos(theta))./(rho.*sin(theta));
B = atan(-A2A1./A1A2);
C = A1A2./cos(B);
for i=1:13
      y(i+2) = b1*y(i+1)+b2*y(i);
end
for i=1:15
      t=i-1;
      y1(i)=(rho.^t).*(A1A2.*cos(mod(theta.*t,2*pi))
           + A2A1.*sin(mod(theta.*t,2*pi)));
      y2(i)=C.*(rho.^t).*cos(mod((theta.*t+B),2*pi));
end
y
y1
y2
```

Example 16.22 Solve the following difference equation

$$y_t = 1.2\,y_{t-1} - y_{t-2}$$

with the initial conditions

$$y_0 = 3,\ \ y_1 = 2.5$$

We have

$$\rho = 1, \qquad \theta \approx 0.295\pi$$

Thus,

$$y_t = 3\cos(0.295t\pi) + 0.875\sin(0.295t\pi)$$

Writing the solution in terms of a cosine only is left to the reader (see E.16.13). Note that the equation has constant fluctuations because $\rho = 1$.

16.4.3 Computer Solution of Difference Equations

The function `rsolve` in Maple can be used to solve difference equations. We shall illustrate its use with a few examples. Consider the first order difference equation

$$y_t = 2y_{t-1} + 7$$

To solve this equation in Maple you input

$$rsolve\ (y(t) = 2\,y(t-1) + 7, y(t))$$

You get

$$y(0)\ 2^t + 7\,2^t - 7$$

Which should be read as

$$y_t = (y_0 + 7)2^t - 7$$

The reader should check to see that this is indeed the solution. We could also specify an initial condition, for example let $y_0 = 5$. We have:

$$rsolve\ (\{y(t) = 2\,y(t-1) + 2, y(0) = 5\},\ \{y(t)\})$$

We get

$$\{y(t) = 12\,2^t - 7\}$$

The coefficients need not be numerical, we can input:

$$rsolve\ (\{y(t) = \mathrm{a}\cdot \mathrm{y}(t-1) + \mathrm{b}, y(0) = \mathrm{B}\},\ \{y(t)\})$$

Make sure that you input a*y(t–1), you get:

$$\left\{y(t) = Ba^t - \frac{b}{-1+a} + \frac{b\,a^t}{-1+a}\right\}$$

Higher order equations can be solved in a similar manner. For example, inputting

$$rsolve\ (\{y(t) = 3.5\ \mathrm{y}(t-1) - 1.5\ \mathrm{y}(t-2)\},\ \{y(t)\})$$

We get the solution:

$$y(t) = -\left(\frac{1}{5}y(0) - \frac{2}{5}y(1)\right)3^t - \frac{1}{2}\left(-\frac{12}{5}y(0) + \frac{4}{5}y(1)\right)\left(\frac{1}{2}\right)^t$$

16.4.4 The Lag Operator

The lag operator is a very useful device in the study of discrete dynamic economic and econometric models. Like any other linear operator, it can be treated like a variable or number. Thus, instead of working with long expressions, we can perform all of the necessary operations and transformations on a polynomial of lag operators.

Definition 16.2 Let $\{x_t\}_{t=-\infty}^{\infty}$ be a sequence of real numbers.[2] Then the lag operator L is such that

$$L\,x_t = x_{t-1} \tag{16.66}$$

It follows that

$$L^2 x_t = L x_{t-1} = x_{t-2}$$

and in general

$$L^n x_t = x_{t-n} \tag{16.67}$$

Thus, the first-order nonhomogeneous difference equation

$$y_t = \lambda y_{t-1} + a \tag{16.68}$$

can be rewritten as

$$y_t = \lambda L y_t + a \tag{16.69}$$

or, rearranging

$$(1 - \lambda L) y_t = a \tag{16.70}$$

Let us assume that $\lambda < 1$. Then dividing through, we have

$$y_t = \frac{a}{1 - \lambda L} = a \sum_{j=0}^{\infty} \lambda^j \tag{16.71}$$

[2] Not all series go as far back as the time of the Big Bang. We have used the notation for generality. If a series starts at time $t = t_0$, then we can simply write $x_t = 0,\ \forall t < t_0$.

Because the lag operator can be treated as a variable, we can use it as an argument in a function. In particular we can have lag polynomials.

Example 16.23

$$y_t = \mu + a_1 y_{t-1} + a_2 y_{t-2} + \ldots + a_n y_{t-n}$$

can be written as

$$(1 - a_1 L - a_2 L^2 - \ldots - a_n L^n) y_t = A(L) y_t = \mu$$

where $A(L)$ is the shorthand for the lag polynomial. Lag polynomials have roots and can be factored out

Example 16.24

$$A(L) = 1 + L - 6L^2 = (1 - 2L)(1 + 3L)$$

Again, as polynomials they can be added, subtracted, multiplied, and divided.

Example 16.25 Let

$$\begin{aligned} A(L) &= 1 - \phi L \\ B(L) &= 1 + \theta L + \theta^2 L^2 \end{aligned}$$

Then

$$A(L) + B(L) = 1 + (\theta - \phi)L + \theta^2 L^2$$

$$A(L) - B(L) = -(\phi + \theta)L - \theta^2 L^2$$

$$A(L)B(L) = 1 + (\theta - \phi)L + \theta(\theta - \phi)L^2 - \phi\theta^2 L^3$$

$$\frac{B(L)}{A(L)} = \frac{1 + \theta L + \theta^2 L^2}{1 - \phi L}$$

Lag polynomials that can be written as the ratio of two polynomials are called *rational lag polynomials*. The following are examples of such polynomials:

$$\sum_{j=0}^{\infty} (\lambda L)^j = \frac{1}{1 - \lambda L}, \qquad |\lambda| < 1$$

$$\sum_{j=0}^{n-1} (\lambda L)^j = \frac{1-(\lambda L)^n}{1-\lambda L}$$

$$\sum_{j=0}^{\infty} (j+1)(\phi L)^j = \frac{1}{(1-\phi L)^2}, \qquad |\lambda| < 1$$

$$1 + 2\sum_{j=1}^{\infty} (\phi L)^j = \frac{1+\phi L}{1-\phi L}, \qquad |\phi| < 1$$

$$1 + \frac{\phi-\theta}{\phi}\sum_{j=1}^{\infty} (\phi L)^j = \frac{1-\theta L}{1-\phi L}, \qquad |\phi| < 1$$

Lag operators not only make the description of a difference equation easier but will also make the solution quite straightforward. Consider the following difference equation:

$$y_t = \lambda y_{t-1} + \gamma x_t \tag{16.72}$$

with initial conditions

$$y_t = y_0, \quad \text{for} \quad t = 0$$

Rewrite the equation as

$$(1-\lambda L)y_t = \gamma x_t \tag{16.73}$$

Multiply both sides by

$$\frac{1-(\lambda L)^t}{1-\lambda L} \tag{16.74}$$

We get

$$\begin{aligned} y_t &= y_0\lambda^t + \gamma\frac{1-(\lambda L)^t}{1-\lambda L}x_t \\ &= y_0\lambda^t + \gamma\sum_{j=0}^{t-1}\lambda^j x_t \end{aligned} \tag{16.75}$$

which is the solution we found by successive substitutions. We will see that a similar shortcut can be applied to the case of a second-order nonhomogeneous equation with boundary conditions.

16.4.5 Solution of Second-Order Nonhomogeneous Difference Equations

A nonhomogeneous second-order equation is of the form

$$y_t = b_1 y_{t-1} + b_2 y_{t-2} + \gamma x_t \tag{16.76}$$

Using lag operators, we can rewrite the equation as

$$y_t = (b_1 L + b_2 L^2) y_t + \gamma x_t \tag{16.77}$$

or

$$(1 - b_1 L - b_2 L^2) y_t = \gamma x_t \tag{16.78}$$

Let λ_1 and λ_2 be the solutions of the characteristics equation:

$$1 - b_1 L - b_2 L^2 = (1 - \lambda_1 L)(1 - \lambda_2 L) \tag{16.79}$$

Furthermore, assume that $\max\{|\lambda_1|\,|\lambda_2|\} < 1$; in other words, the model is stable. Dividing both side of (16.78) by (16.79), we have

$$y_t = \frac{1}{(1 - \lambda_1 L)(1 - \lambda_2 L)} \gamma x_t = \frac{1}{\lambda_1 - \lambda_2} \left(\frac{\lambda_1}{1 - \lambda_1 L} - \frac{\lambda_2}{1 - \lambda_2 L} \right) \gamma x_t \tag{16.80}$$

where the second equality is based on the decomposition of proper rational fractions. The particular solution is

$$y_t = \frac{\lambda_1 \gamma}{\lambda_1 - \lambda_2} \sum_{j=0}^{\infty} \lambda_1^j x_{t-j} + \frac{\lambda_2 \gamma}{\lambda_2 - \lambda_1} \sum_{j=0}^{\infty} \lambda_2^j x_{t-j} \tag{16.81}$$

and the general solution

$$y_t = A_1 \lambda_1^t + A_2 \lambda_2^t + \frac{\lambda_1 \gamma}{\lambda_1 - \lambda_2} \sum_{j=0}^{\infty} \lambda_1^j x_{t-j} + \frac{\lambda_2 \gamma}{\lambda_2 - \lambda_1} \sum_{j=0}^{\infty} \lambda_2^j x_{t-j} \tag{16.82}$$

Of course, from a practical point of view, writing x_t for an economic variable and then letting $t \to -\infty$ (that is, a couple years after the Big Bang) stretches credulity. We may want to think of a time to start in the reasonable past and assume the value of x_t for all periods prior to that date to be zero.

Example 16.26 Solve the following equation:

$$y_t = 1.2\, y_{t-1} - 0.35\, y_{t-2} + 0.9\, x_t$$

Because $\lambda_1 = 0.7$ and $\lambda_2 = 0.5$, the solution is

$$y_t = A_1(0.7)^t + A_2(0.5)^t + 3.15\sum_{j=0}^{\infty} 0.7^j x_{t-j} - 2.25\sum_{j=0}^{\infty} 0.5^j x_{t-j}$$

Now suppose we have the following initial conditions:

$$y_t = \begin{cases} y_0 & t = 0 \\ y_1 & t = 1 \end{cases}$$

Then we can determine the specific solution

$$y_0 = A_1 + A_2 + \frac{\lambda_1\gamma}{\lambda_1 - \lambda_2}\sum_{j=0}^{\infty}\lambda_1^j x_{-j} + \frac{\lambda_2\gamma}{\lambda_2 - \lambda_1}\sum_{j=0}^{\infty}\lambda_2^j x_{-j} \tag{16.83}$$

and

$$y_1 = A_1\lambda_1 + A_2\lambda_2 + \frac{\lambda_1\gamma}{\lambda_1 - \lambda_2}\sum_{j=0}^{\infty}\lambda_1^j x_{1-j} + \frac{\lambda_2\gamma}{\lambda_2 - \lambda_1}\sum_{j=0}^{\infty}\lambda_2^j x_{1-j} \tag{16.84}$$

Solving for A_1 and A_2, we get

$$\begin{aligned} A_1 &= \frac{1}{\lambda_1 - \lambda_2}\left(y_1 - \lambda_2 y_0 - \gamma x_1 - \gamma\lambda_1\sum_{j=0}^{\infty}\lambda_1^j x_{-j}\right) \\ A_2 &= \frac{1}{\lambda_2 - \lambda_1}\left(y_1 - \lambda_1 y_0 - \gamma x_1 - \gamma\lambda_2\sum_{j=0}^{\infty}\lambda_2^j x_{-j}\right) \end{aligned} \tag{16.85}$$

Therefore the complete solution is

$$\begin{aligned} y_t = {} & \frac{y_1 - \lambda_2 y_0 - \gamma x_1 - \gamma\lambda_1\sum_{j=0}^{\infty}\lambda_1^j x_{-j}}{\lambda_1 - \lambda_2}\lambda_1^t \\ & + \frac{y_1 - \lambda_2 y_0 - \gamma x_1 - \gamma\lambda_1\sum_{j=0}^{\infty}\lambda_1^j x_{-j}}{\lambda_1 - \lambda_2}\lambda_2^t \\ & + \frac{\lambda_1\gamma}{\lambda_1 - \lambda_2}\sum_{j=0}^{\infty}\lambda_1^j x_{t-j} + \frac{\lambda_2\gamma}{\lambda_2 - \lambda_1}\sum_{j=0}^{\infty}\lambda_2^j x_{t-j} \end{aligned} \tag{16.86}$$

Simplifying, we have

$$y_t = \left(\frac{y_1 - \lambda_2 y_0}{\lambda_1 - \lambda_2}\right)\lambda_1^t + \left(\frac{y_1 - \lambda_1 y_0}{\lambda_2 - \lambda_1}\right)\lambda_2^t$$
$$+\frac{\lambda_1\gamma}{\lambda_1 - \lambda_2}\sum_{j=0}^{t-2}\lambda_1^j x_{t-j} + \frac{\lambda_2\gamma}{\lambda_2 - \lambda_1}\sum_{j=0}^{t-2}\lambda_2^j x_{t-j} \tag{16.87}$$

Example 16.27 In Example 16.26, let

$$y_0 = 1, \qquad y_1 = 2$$

Then we have $A_1 = 7.5$ and $A_2 = -6.5$. And the specific solution is

$$y_t = 7.5(0.7)^t - 6.5(0.5)^t + 3.15\sum_{j=0}^{t-2}0.7^j x_{t-j} - 2.25\sum_{j=0}^{t-2}0.5^j x_{t-j}$$

In solving the problem of second-order nonhomogeneous equations we assumed that $\max\{|\lambda_1|, |\lambda_2|\} < 1$ and the equation is stable. Indeed, the solution we obtained is more general and applies to both stable and unstable cases. To see this, let us consider the following problem:

$$y_t - b_1 y_{t-1} + b_2 y_{t-2} = \alpha + \gamma x_t \tag{16.88}$$

$$y_t = \begin{cases} y_0 & t = 0 \\ y_1 & t = 1 \end{cases}$$

Let λ_1 and λ_2 be solutions to the equation $z^2 - b_1 z - b_2 = 0$. We have

$$(1 - \lambda_1 L)(1 - \lambda_2 L)y_t = \alpha + \gamma x_t \tag{16.89}$$

Multiply both sides of the equation by

$$\frac{\lambda_1}{\lambda_1 - \lambda_2}(1 - \lambda_2 L)\left[1 - (\lambda_1 L)^{t-1}\right] + \frac{\lambda_2}{\lambda_2 - \lambda_1}(1 - \lambda_1 L)\left[1 - (\lambda_2 L)^{t-1}\right] \tag{16.90}$$

and divide both sides by $(1 - \lambda_1 L)(1 - \lambda_2 L)$. On the LHS, we get

$$y_t - \left(\frac{y_1 - \lambda_2 y_0}{\lambda_1 - \lambda_2}\right)\lambda_1^t - \left(\frac{y_1 - \lambda_1 y_0}{\lambda_2 - \lambda_1}\right)\lambda_2^t \tag{16.91}$$

On the RHS, we get

$$
\begin{aligned}
&\frac{\lambda_1}{\lambda_1-\lambda_2}\frac{(1-\lambda_2 L)[(1-(\lambda_1 L)^{t-1}]}{(1-\lambda_1 L)(1-\lambda_2 L)}(\alpha+\gamma x_t)\\
&+\frac{\lambda_2}{\lambda_2-\lambda_1}\frac{(1-\lambda_1 L)[1-(\lambda_2 L)^{t-1}]}{(1-\lambda_1 L)(1-\lambda_2 L))}(\alpha+\gamma x_t)
\end{aligned}
\tag{16.92}
$$

Simplifying, we have

$$
\begin{aligned}
&\frac{\alpha}{\lambda_1-\lambda_2}\left[\frac{\lambda_1}{1-\lambda_1}-\frac{\lambda_2}{1-\lambda_2}\right]-\frac{\alpha\lambda_1^t}{\lambda_1-\lambda_2}-\frac{\alpha\lambda_2^t}{\lambda_2-\lambda_1}\\
&+\frac{\lambda_1\gamma}{\lambda_1-\lambda_2}\sum_{j=0}^{t-2}\lambda_1^j x_{t-j}+\frac{\lambda_2\gamma}{\lambda_2-\lambda_1}\sum_{j=0}^{t-2}\lambda_2^j x_{t-j}
\end{aligned}
\tag{16.93}
$$

Thus, the complete equation is

$$
\begin{aligned}
y_t=&\frac{1}{\lambda_1-\lambda_2}\left(y_1-\lambda_2 y_0-\frac{\alpha}{1-\lambda_1}\right)\lambda_1^t\\
&+\frac{1}{\lambda_2-\lambda_1}\left(y_1-\lambda_1 y_0-\frac{\alpha}{1-\lambda_2}\right)\lambda_2^t\\
&+\frac{\alpha}{\lambda_1-\lambda_2}\left[\frac{\lambda_1}{1-\lambda_1}-\frac{\lambda_2}{1-\lambda_2}\right]\\
&+\frac{\lambda_1\gamma}{\lambda_1-\lambda_2}\sum_{j=0}^{t-2}\lambda_1^j x_{t-j}+\frac{\lambda_2\gamma}{\lambda_2-\lambda_1}\sum_{j=0}^{t-2}\lambda_2^j x_{t-j}
\end{aligned}
\tag{16.94}
$$

Example 16.28 (The Real Business Cycle Model continued).

The last time we visited this model we were still dealing with a first-order difference equation. But now let u_t, the random technological innovation, follow a first-order Markov process

$$u_t=\rho u_{t-1}+\nu_t, \qquad |\rho|<1$$

where ν_t is white noise. Recall that

$$z_t=B+\alpha z_{t-1}+(1-\alpha)u_t \tag{16.95}$$

Therefore,

$$\rho z_{t-1}=\rho B+\rho\alpha z_{t-2}+\rho(1-\alpha)u_{t-1} \tag{16.96}$$

Subtracting (16.96) from (16.95), we get

$$z_t = B(1-\rho) + (\rho+\alpha)z_{t-1} - \rho\alpha z_{t-2} + (1-\alpha)v_t \tag{16.97}$$

or

$$(1 - (\rho+\alpha)L + \rho\alpha L^2)z_t = B(1-\rho) + (1-\alpha)v_t \tag{16.98}$$

Now we have a second-order difference equation whose solution is left to the reader (see Exercise E.16.15). Note that the roots of the lagged polynomial are ρ and α, both of which are real and less than one. Therefore, the homogeneous part will not generate sinusoidal fluctuations: rather it will gradually diminish. All the fluctuations around the trend will come from stochastic innovations.

Example 16.29 (A Keynesian Business Cycle Model). Samuelson suggested the following model to explain fluctuations in the economy. It is based on the interaction of multiplier and acceleration principles:

$$\begin{aligned} Y_t &= C_t + I_t + G \\ C_t &= \beta Y_{t-1} \\ I_t &= \gamma(C_t - C_{t-1}) \end{aligned}$$

Combining the equations,

$$Y_t = \beta(1+\gamma)Y_{t-1} - \beta\gamma Y_{t-2} + G$$

The particular solution is

$$Y_p = \frac{1}{1-\beta}G$$

and the roots of the characteristic equation are

$$\lambda = \frac{\beta(1+\gamma) \pm \sqrt{\beta^2(1+\gamma)^2 - 4\beta\gamma}}{2}$$

Thus, the model will produce oscillations if

$$\beta^2(1+\gamma)^2 - 4\beta\gamma < 0$$

or equivalently

$$\beta < \frac{4\gamma}{(1+\gamma)^2}$$

the nature of fluctuations depends on the magnitude of $\beta\gamma$. Thus, the amplitude of the function will

$$\begin{array}{c} \text{increase} \\ \text{be constant} \\ \text{decrease} \end{array} \quad \text{as} \quad \beta\gamma \begin{cases} > 1 \\ = 1 \\ < 1 \end{cases} \quad \Rightarrow \quad \beta \begin{cases} > 1/\gamma \\ = 1/\gamma \\ < 1/\gamma \end{cases}$$

John Hicks proposed a similar model. In his model the fluctuations are increasing, but Hicks's model exhibits fluctuations within an upper and a lower bound. The upper bound is full employment output, which is fixed at any point in time, but it is growing over time. The lower limit is provided by the fact that investment has a lower bound. If no new investment is undertaken, the change in capital stock is equal to depreciation.

16.4.6 Exercises

E.16.11 Solve the following second order difference equations.

$$\begin{array}{ll} i. & y_t = -2y_{t-1} + 3y_{t-2} \\ ii. & y_t = 0.8y_{t-1} - 0.12y_{t-2} \\ iii. & y_t = 0.4y_{t-1} - 0.77y_{t-2} \\ iv. & y_t = 3y_{t-1} - 2.25y_{t-2} \\ v. & y_t = 1.2y_{t-1} - 0.36y_{t-2} \\ vi. & y_t = 2y_{t-1} - 2y_{t-2} \\ vii. & y_t = 3y_{t-1} - 2.61y_{t-2} \\ viii. & y_t = 4y_{t-1} - 4.25y_{t-2} \end{array}$$

E.16.12 Show that when λ_1 and λ_2 are complex conjugates, so are A_1 and A_2.

E.16.13 Write the solution of the equation in Example 16.22 in the form

$$y_t = C\rho^t \cos(\theta t + B)$$

E.16.14 Let

$$\begin{aligned} A(L) &= 1 - aL \\ B(L) &= 1 - b_1L - b_2L^2 \\ C(L) &= 1 - (b_1 + c)L + b_1cL^2 \\ D(L) &= 1 - 2aL + a^2L^2 \end{aligned}$$

Find

$$\begin{array}{llll} i. & A(L)+B(L) & ii. & B(L)+C(L) \\ iii. & A(L)-B(L) & iv. & B(L)C(L) \\ v. & A(L)D(L) & vi. & B(L)D(L) \\ vii. & \dfrac{A(L)}{D(L)} & viii. & \dfrac{B(L)}{C(L)} \end{array}$$

E.16.15 Solve the second-order difference equation of the real business cycle model in Example 16.28.

E.16.16 Write a Matlab program that solves a second-order homogeneous difference equation with numerical coefficients. [*Hint*: First write three functions for the cases of real distinct roots, repeated real roots, and complex roots. Then write a function that solves the characteristic equation and determines the type of solution. Finally tie all four pieces together in a main program.]

16.5 *n*-th-Order Difference Equations

Empirical models that use monthly or weekly time series data on occasions involve a large number of lags resulting in a difference equation of order 12 or more. But it is quite rare to encounter a difference equation of more than order two in theoretical work. Even if we encounter them, there is little interest in solving them. Such models are better analyzed by numerical simulation. Here we present a Matlab routine for the simulation of a difference equation of order 12 with a linear trend and an exogenously determined variable x. The routine is written in a way that can be adapted to any equation. Suppose we want to calculate the trajectory of the following equation:

$$y_{t+1} = c_1 y_{t-1} + c_2 y_{t-2} + \ldots + c_{12} y_{t-12} + a + bt + hx_t \tag{16.99}$$

Further assume that we have set up an Excel file that contains the values of x_t, $t = 13, \ldots, T$, where T is the last period for which we would like to find y_t. The data in this file called `x.xls` are arranged as shown below.

File Containing x_t

A
x_{13}
x_{14}
$\vdots$
x_T

A second file contains the initial values, y_j, $j = 1, \ldots, 12$, and the coefficients, c_j's. This file is called `yc.xls` and the data are arranged as shown below. Note the order of y_j's and c_j's that is chosen for the convenience of programming. Moreover, remember that Matlab's index starts with one and does not allow indexing from zero. Therefore, t starts from 1. If your data are such that t should start from zero, you need to adjust the code by defining $t = i{-}1$. Now we can have our Matlab code.

File Containing y_j's and c_j's

A	B
y_1	c_{12}
y_2	c_{11}
$\vdots$	$\vdots$
y_{12}	c_1

Matlab code

```
% Specify T, n, and parameters a, b, and h
T = 30;
n = 12;
a = 1.5;
b = 0.03;
h = 1.1;
% Define the arrays to hold y's, x's and c's
y = zeros(T,1);
x = zeros(T-n,1);
c = zeros(n,1);
% Read from Excel files
x = xlsread('x.xls');
A = xlsread('yc.xls');
for i=1:n
        y(i) = A(i,1);
        c(i) = A(i,2);
end
for i=n+1:T
        t = i;
        S = 0;
        for j=1:n
              S = S + c(j)*y(j+i-n-1);
        end
        y(i) = S + a + b*t + h*x(i-n);
end
y
```

16.5.1 Exercises

E.16.17 Using the code in the book or your own program, calculate the trajectory of the following equations for $T = 30$ periods.

$$y_t = 0.8y_{t-1} + 0.64y_{t-2} + 0.512y_{t-3} + 10 + 0.06t$$
$$y_0 = 100, \quad y_1 = 110, \quad y_2 = 125$$
$$y_t = 0.9y_{t-1} - 0.5y_{t-2} + 0.09y_{t-3} + 0.06t$$
$$y_0 = 100, \quad y_1 = 110, \quad y_2 = 125$$
$$y_t = 1.1y_{t-1} - 0.65y_{t-2} + 0.17y_{t-3} - 0.04y_{t-4}$$
$$y_0 = 65, \quad y_1 = 70, \quad y_2 = 74, \quad y_3 = 79$$

E.16.18 In E.16.17 change the initial conditions one at a time to observe the effect of the starting points on the trajectory.

Chapter 17
Dynamic Systems

17.1 Systems of Differential Equations

The rationale behind dynamic models in general and in economic analysis in particular was discussed in Chaps. 14, 15, and 16. The reason for a system of difference or differential equations follows the same logic, except that here more than one dynamic process is at work and, therefore, we need more than one equation to describe these processes. Analysis of dynamic systems is a vast field of inquiry and could be the subject of a multivolume book. Here we confine ourselves to two topics that are deemed most useful for economic analysis: the solution of linear systems of homogeneous differential equations with constant coefficients and qualitative analysis of a system of differential equations using *phase portrait*.

The solution of linear systems of differential equations is greatly facilitated if you are already familiar with Jordan canonical form, which involves the diagonalization of an arbitrary matrix (we learned of such an operation regarding positive definite matrices in Chap. 7), and the exponential of a matrix, that is, exp(**A**). Thus, we will take up these two topics as a prelude to the subjects of this chapter. It is recommended that the reader reread the topics of eigenvalues and eigenvectors in Chap. 7 as they are used extensively in this chapter.

A system of first-order differential equations is of the form

$$\begin{aligned}\dot{y}_1 &= f^1(y_1,\ldots,y_n)\\ \dot{y}_2 &= f^2(y_1,\ldots,y_n)\\ &\cdots\\ \dot{y}_n &= f^n(y_1,\ldots,y_n)\end{aligned} \tag{17.1}$$

where $\dot{y}_i = dy_i/dt,\ i = 1,\ldots,n$. The system is called linear if functions $f^1,\ldots,f^n$ are all linear. In that case we can write the system as

$$\begin{bmatrix}\dot{y}_1\\ \dot{y}_2\\ \vdots\\ \dot{y}_n\end{bmatrix} = \begin{bmatrix}a_{11} & a_{12} & \ldots & a_{1n}\\ a_{21} & a_{22} & \ldots & a_{2n}\\ \vdots & \vdots & \ddots & \vdots\\ a_{n1} & a_{n2} & \ldots & a_{nn}\end{bmatrix}\begin{bmatrix}y_1\\ y_2\\ \vdots\\ y_n\end{bmatrix} + \begin{bmatrix}h^1(t)\\ h^2(t)\\ \vdots\\ h^n(t)\end{bmatrix} \tag{17.2}$$

K. Dadkhah, *Foundations of Mathematical and Computational Economics,* 2nd ed.,
DOI 10.1007/978-3-642-13748-8_17, © Springer-Verlag Berlin Heidelberg 2011

If $h^i,\ i = 1, \ldots, n$ are equal to zero, then the system is called homogeneous.[1] In compact matrix notation such a system can be written as

$$\dot{\mathbf{y}} = \mathbf{A}\mathbf{y} \tag{17.3}$$

Example 17.1 In Chap. 15 we encountered the two-equation model

$$\frac{dq}{dt} = (\delta + \phi)q + p\theta_1 K - p\theta_0$$
$$\frac{dK}{dt} = \alpha\frac{q}{p} - \phi K$$

where q and p are, respectively, the price of capital goods and consumer goods, whereas K is the capital stock, ϕ the depreciation rate, and δ is the rate of interest.

Example 17.2 In Chap. 14 we discussed the Ramsey problem and derived the following differential equations:

$$\frac{dc}{dt} = c\sigma(c)[f'(k) - \theta - n]$$
$$\frac{dk}{dt} = f(k) - c - nk$$

where c and k are, respectively, per capita consumption and the capital-labor ratio, θ, rate of time preference, and n, rate of growth of labor.

An important concept regarding systems of differential equations is that of the *equilibrium (critical) point*. In an equilibrium point the system comes to rest, and there is no tendency for its components to change. Such a point is characterized by $\dot{\mathbf{y}} = \mathbf{0}$.

Example 17.3 The system in Example 17.2 attains its equilibrium when

$$c^*\sigma(c^*)[f'(k^*) - \theta - n] = 0$$

and

$$f(k^*) - c^* - nk^* = 0$$

because $c^*\sigma(c^*) \neq 0$, the equilibrium point c^*, k^* is characterized by

$$f'(k^*) = \theta + n$$
$$f(k^*) = c^* + nk^*$$

---

[1] Whereas the theory developed in this chapter is quite general and pertains to a system of two and more differential equations, for the ease and clarity of exposition, with a few exceptions, we will confine ourselves to systems of two equations.

Example 17.4 In Chap. 14 we also discussed the derivation of Tobin's q as a dynamic optimization problem. We can summarize the model in a system of differential equations:

$$\dot{K} = I(q) - \delta K$$
$$\dot{q} = (r + \delta)q - \Pi'(K)$$

where K is the capital stock of the firm, I, investment, δ, the depreciation rate, q, the ratio the market value of additional units of capital to its replacement cost (price of capital was set at 1), r, interest rate, and Π' the marginal profit of capital.

In this chapter we will concentrate on the solution of first-order linear homogeneous equations. Such models have a much wider range of applications than it may seem at first glance. In addition to many phenomena that can be modeled using first-order linear models, any linear differential equation of higher order can be transformed into a system of first-order differential equations. For example, a second-, third-, ..., n-th-order linear differential equation can be transformed into a system of two, three, ..., n first-order linear differential equations.

Equally important, when dealing with nonlinear differential equations, on many occasions, we are interested in the behavior of the system near the equilibrium or critical point. When the neighborhood (around the equilibrium) in which we are interested is small, we can linearize the system. Before going any further we will elaborate on these two points, equivalence of a second-order linear differential equation and a system of two first-order linear equations, and the linearization of a system of nonlinear differential equations.

17.1.1 Equivalence of a Second-Order Linear Differential Equation and a System of Two First-Order Linear Equations

An n-th order linear homogeneous differential equation can be transformed into a system of n first-order linear homogeneous equations and vice versa. Whereas the relationship holds for any finite n, we shall illustrate it for the case of $n = 2$. Consider the equation

$$\frac{d^2y}{dt^2} + a_1\frac{dy}{dt} + a_2y = 0 \tag{17.4}$$

Let $dy/dt = x$, then $d^2y/dt^2 = dx/dt$, and we can write

$$\begin{aligned} \frac{dx}{dt} + a_1x + a_2y = 0 \\ \frac{dy}{dt} - x = 0 \end{aligned} \tag{17.5}$$

or in matrix form

$$\begin{bmatrix} dx/dt \\ dy/dt \end{bmatrix} = \begin{bmatrix} -a_1 & -a_2 \\ 1 & 0 \end{bmatrix} \begin{bmatrix} x \\ y \end{bmatrix} \tag{17.6}$$

Alternatively, suppose we have a system of equations of the form

$$\begin{aligned} \frac{dx}{dt} &= a_{11}x + a_{12}y \\ \frac{dy}{dt} &= a_{21}x + a_{22}y \end{aligned} \tag{17.7}$$

Take the derivative of the second equation,

$$\frac{d^2y}{dt^2} = a_{21}\frac{dx}{dt} + a_{22}\frac{dy}{dt} \tag{17.8}$$

Substituting for dx/dt from the first equation and for x from the second equation,

$$\begin{aligned} \frac{d^2y}{dt^2} &= a_{21}(a_{11}x + a_{12}y) + a_{22}\frac{dy}{dt} \\ &= a_{11}\left(\frac{dy}{dt} - a_{22}y\right) + a_{21}a_{12}y + a_{22}\frac{dy}{dt} \end{aligned} \tag{17.9}$$

Rearranging

$$\frac{d^2y}{dt^2} - (a_{11} + a_{22})\frac{dy}{dt} + (a_{21}a_{12} - a_{11}a_{22})y = 0 \tag{17.10}$$

which is a second-order differential equation.

Example 17.5 The angular momentum of a pendulum moving in a medium with friction is given by the second-order differential equation

$$mL^2\frac{d^2\theta}{dt^2} + cL\frac{d\theta}{dt} + mgL\sin\theta = 0 \tag{17.11}$$

where m is the mass of the pendulum attached to a solid, weightless rod (see Fig. 17.1), $g = 9.8$ is acceleration due to gravity, and θ is taken to be positive in the counterclockwise direction. The term $cLd\theta/dt$ measures the friction and c depends on the viscosity of the medium in which the pendulum moves. For example, if the pendulum is moving in the air, then c is taken to be zero and the equation reduces to

$$mL^2\frac{d^2\theta}{dt^2} + mgL\sin\theta = 0 \tag{17.12}$$

Fig. 17.1 Pendulum

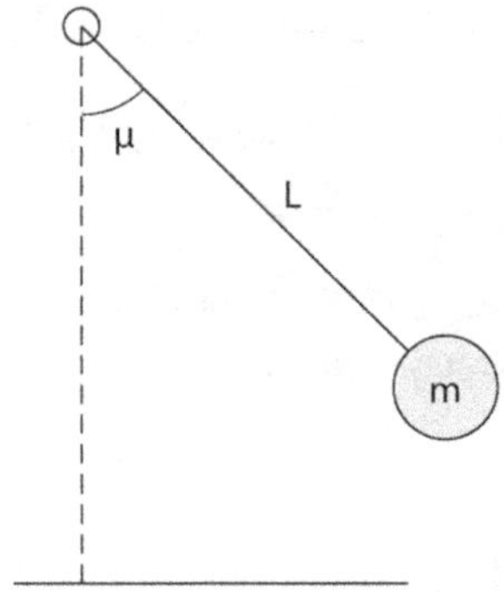

We can simplify (17.11) by dividing both sides of the equation by mL^2 :

$$\frac{d^2\theta}{dt^2} + \frac{c}{mL}\frac{d\theta}{dt} + \frac{g}{L}\sin\theta = 0 \tag{17.13}$$

and in obvious change of notation,

$$\frac{d^2\theta}{dt^2} + \gamma\frac{d\theta}{dt} + \omega^2\sin\theta = 0 \tag{17.14}$$

(17.14) is a second-order differential equation, but we can transform it to a system of first-order equations. Let

$$y_1 = \theta, \qquad y_2 = \frac{d\theta}{dt} \tag{17.15}$$

Then

$$\frac{d^2\theta}{dt^2} = \frac{dy_2}{dt} \tag{17.16}$$

Combining (17.14), (17.15), and (17.16), we can write our system

$$\begin{aligned} \dot{y}_1 &= y_2 \\ \dot{y}_2 &= -\omega^2\sin y_1 - \gamma y_2 \end{aligned} \tag{17.17}$$

Example 17.6 In Chap. 15 we showed that the system of equations

$$\begin{aligned} \frac{dq}{dt} &= (\delta+\phi)q + p\theta_1 K - p\theta_0 \\ \frac{dK}{dt} &= \alpha\frac{q}{p} - \phi K \end{aligned} \tag{17.18}$$

can be transformed into the second-order equation

$$\frac{d^2K}{dt^2} - \delta\frac{dK}{dt} - (\phi^2 + \delta\phi + \alpha\theta_1)K = -\alpha\theta_0 \tag{17.19}$$

We can transform the second-order equation into a system of two first-order equations. Let

$$y_1 = K, \qquad y_2 = \frac{dK}{dt} \tag{17.20}$$

Then (17.19) can be written as

$$\frac{dy_2}{dt} = \delta y_2 + (\phi^2 + \delta\phi + \alpha\theta_1)y_1 - \alpha\theta_0 \tag{17.21}$$

and transformed into a system of two first order equations:

$$\begin{aligned} \dot{y}_1 &= y_2 \\ \dot{y}_2 &= (\phi^2 + \delta\phi + \alpha\theta_1)y_1 + \delta y_2 - \alpha\theta_0 \end{aligned} \tag{17.22}$$

Although the system in (17.22) is quite different from the equation in (17.19), the solutions are identical. Furthermore, while the solution of the latter involves only K, because dq/dt depends on K we can solve for q as well.

17.1.2 Linearizing Nonlinear Systems of Differential Equations

A nonlinear system can be written as

$$\begin{aligned} \dot{y}_1 &= f(y_1, y_2) \\ \dot{y}_2 &= g(y_1, y_2) \end{aligned} \tag{17.23}$$

Let the point (y_1^*, y_2^*) be the equilibrium or critical point, that is,

$$\begin{aligned} f(y_1^*, y_2^*) &= 0 \\ g(y_1^*, y_2^*) &= 0 \end{aligned} \tag{17.24}$$

Using Taylor expansion near the equilibrium point, we can write

$$\begin{aligned} f(y_1, y_2) &= f(y_1^*, y_2^*) + f_1(y_1^*, y_2^*)(y_1 - y_1^*) + f_2(y_1^*, y_2^*)(y_2 - y_2^*) + R_1 \\ g(y_1, y_2) &= g(y_1^*, y_2^*) + g_1(y_1^*, y_2^*)(y_1 - y_1^*) + g_2(y_1^*, y_2^*)(y_2 - y_2^*) + R_2 \end{aligned}$$

where $f_i = \partial f/\partial y_i$ and $g_i = \partial g/\partial y_i$, $i = 1, 2$, and R_1 and R_2 are Lagrange remainders. We can approximate (17.23) with the linear system

$$\begin{aligned}\frac{d}{dt}(y_1 - y_1^*) &= f_1(y_1^*, y_2^*)(y_1 - y_1^*) + f_2(y_1^*, y_2^*)(y_2 - y_2^*)\\ \frac{d}{dt}(y_2 - y_2^*) &= g_1(y_1^*, y_2^*)(y_1 - y_1^*) + g_2(y_1^*, y_2^*)(y_2 - y_2^*)\end{aligned} \tag{17.25}$$

Letting $z_1 = y_1 - y_1^*$ and $z_2 = y_2 - y_2^*$, we can rewrite the system in the approximate linear form as

$$\begin{bmatrix} \dot{z}_1 \\ \dot{z}_2 \end{bmatrix} = \begin{bmatrix} f_1(y_1^*, y_2^*) & f_2(y_1^*, y_2^*) \\ g_1(y_1^*, y_2^*) & g_2(y_1^*, y_2^*) \end{bmatrix} \begin{bmatrix} z_1 \\ z_2 \end{bmatrix} \tag{17.26}$$

If the functions f and g are twice differentiable, and we confine ourselves to very small deviations from the equilibrium point, then the remainders tend to zero, and the linearized system is a very good approximation. Thus, as long as we are interested in analyzing the behavior of the model at or near the equilibrium, the system is almost linear and (17.26) is all we need.

Example 17.7 Consider the following system of nonlinear differential equations:

$$\begin{aligned}\dot{x} &= x + a_1 x^2 + a_2 xy\\ \dot{y} &= y + b_1 y^2 + b_2 xy\end{aligned}$$

The equilibrium point of the system is (0, 0). Using Taylor expansion, we can linearize the model as

$$\begin{bmatrix} \dot{x} \\ \dot{y} \end{bmatrix} = \begin{bmatrix} 1 & 0 \\ 0 & 1 \end{bmatrix} \begin{bmatrix} x \\ y \end{bmatrix}$$

Example 17.8 Consider the system of equations in (17.17) for the angular momentum of a pendulum. The equilibrium point of this system is (0, 0). Let us linearize the model near this point. We have

$$\begin{aligned}f(y_1, y_2) &= y_2\\ g(y_1, y_2) &= -\omega^2 \sin y_1 - \gamma y_2\end{aligned}$$

Therefore,

$$\begin{aligned}f_1(0,0) &= 0, & f_2(0,0) &= 1\\ g_1(0,0) &= -\omega^2, & g_2(0,0) &= -\gamma\end{aligned}$$

and the linearized system is

$$\begin{aligned}\dot{y}_1 &= y_2\\ \dot{y}_2 &= -\omega^2 y_1 - \gamma y_2\end{aligned}$$

Example 17.9 To linearize the economic model in Example 17.2, note that the Taylor expansion of *dc/dt* near the equilibrium point (c^*, k^*) is

$$\begin{aligned}&\left(\sigma(c^*) + c^*\frac{d\sigma}{dc}\right)[f'(k^*) - \theta - n](c - c^*) + f''(k^*)c^*\sigma(c^*)(k - k^*)\\ &\qquad = f''(k^*)c^*\sigma(c^*)(k - k^*)\end{aligned}$$

The Taylor expansion of *dk/dt* is

$$([f'(k^*) - n](k - k^*) - (c - c^*) = \theta(k - k^*) - (c - c^*)$$

Denoting $f''(k^*)c^*\sigma(c^*)$ by $-\beta$, we can write the linearized system as

$$\begin{bmatrix} dc/dt \\ dk/dt \end{bmatrix} = \begin{bmatrix} 0 & -\beta \\ -1 & 0 \end{bmatrix}\begin{bmatrix} c - c^* \\ k - k^* \end{bmatrix}$$

17.2 The Jordan Canonical Form

Solution of systems of linear differential equations and estimation of some econometric models are considerably facilitated if we can diagonalize certain matrices or factorize them into two or three matrices. In Chap. 7 we discussed the diagonalization of a symmetric matrix. Here we extend the analysis to the case of an arbitrary matrix. Because our main objective is application, we shall not give a proof of Jordan form; rather we illustrate the proposition with many examples.

17.2.1 Diagonalization of a Matrix with Distinct Real Eigenvalues

Let **A** be an $n \times n$ matrix, **P** a matrix whose columns are eigenvectors[2] of **A**, and **D** a diagonal matrix whose diagonal elements are eigenvalues of **A**. That is,

[2]To obtain the Jordan canonical form one can use either normalized eigenvectors—that is, vectors of unit length—or eigenvectors of any length. In the text both normalized and nonnormalized eigenvectors are used. Normalized eigenvectors, however, are required for the radial decomposition of a positive definite matrix, discussed in Chap. 7. Note that the `eig` function in Matlab returns normalized eigenvectors.

$$\mathbf{D} = \begin{bmatrix} \lambda_1 & & & \\ & \ddots & & \\ & & \lambda_j & \\ & & & \ddots \end{bmatrix} \tag{17.27}$$

and

$$\mathbf{P} = [\mathbf{x}_1\ \mathbf{x}_2\ \ldots\ \mathbf{x}_n] \tag{17.28}$$

Then we have

$$\begin{aligned} \mathbf{P}^{-1}\mathbf{A}\mathbf{P} &= \mathbf{D} \\ \mathbf{P}\mathbf{D}\mathbf{P}^{-1} &= \mathbf{A} \end{aligned} \tag{17.29}$$

Example 17.10 Consider the matrix

$$\mathbf{A} = \begin{bmatrix} 3\ 6 \\ 4\ 1 \end{bmatrix}$$

whose eigenvalues are 7 and -3 and its eigenvectors are

$$\mathbf{x}_1 = \begin{bmatrix} 3 \\ 2 \end{bmatrix}, \quad \text{and} \quad \mathbf{x}_1 = \begin{bmatrix} 1 \\ -1 \end{bmatrix}$$

Then

$$\mathbf{P} = \begin{bmatrix} 3 & 1 \\ 2 & -1 \end{bmatrix}, \quad \mathbf{D} = \begin{bmatrix} 7 & 0 \\ 0 & -3 \end{bmatrix}$$

and

$$\mathbf{P}^{-1} = \begin{bmatrix} 0.2 & 0.2 \\ 0.4 & -0.6 \end{bmatrix}$$

It can be checked that the relationships in (17.29) hold. The reader is urged to try the same procedure with other matrices and using a Matlab program, verify the validity of relationships in (17.29).

17.2.2 Block Diagonal Form of a Matrix with Complex Eigenvalues

Let $\mathbf{A}$ be an $n \times n$ matrix whose eigenvalues are complex. Its eigenvectors will be of the form

$$\mathbf{x}_j = \mathbf{u}_j + i\mathbf{v}_j \qquad j = 1, \ldots, n.$$

Let us form the matrix $\mathbf{P}$ as

$$\mathbf{P} = [\mathbf{v}_1\ \mathbf{u}_1\ \mathbf{v}_2\ \mathbf{u}_2 \ldots \mathbf{v}_n\ \mathbf{u}_n]$$

Note that $\mathbf{v}'$s precede $\mathbf{u}'$s. Then

$$\mathbf{P}^{-1}\mathbf{A}\mathbf{P} = \begin{bmatrix} a_1 & -b_1 & & & & \\ b_1 & a_1 & & & & \\ & & \ddots & & & \\ & & & a_j & -b_j & \\ & & & b_j & a_j & \\ & & & & & \ddots \end{bmatrix}$$

That is, $\mathbf{P}^{-1}\mathbf{A}\mathbf{P}$ is a block diagonal matrix with off-diagonal elements all equal to zero. Note that if we define $\mathbf{P}$ as

$$\mathbf{P} = [\mathbf{u}_1\ \mathbf{v}_1\ \mathbf{u}_2\ \mathbf{v}_2 \ldots \mathbf{u}_n\ \mathbf{v}_n]$$

that is, if the places of $\mathbf{u}'$s and $\mathbf{v}'$s are interchanged, the diagonal block of the $\mathbf{P}^{-1}\mathbf{A}\mathbf{P}$ will be

$$\begin{matrix} a_1 & b_1 \\ -b_1 & a_1 \end{matrix}$$

Example 17.11 Consider the matrix

$$\mathbf{A} = \begin{bmatrix} 1 & 9 \\ -1 & 1 \end{bmatrix}$$

The eigenvalues of $\mathbf{A}$ are $1 \pm 3i$, and the eigenvectors are

$$\begin{bmatrix} 3 \\ 0 \end{bmatrix}, \begin{bmatrix} 0 \\ 1 \end{bmatrix}$$

Thus, we can write

$$\mathbf{P} = \begin{bmatrix} 0 & 3 \\ 1 & 0 \end{bmatrix}$$

$$\mathbf{P}^{-1}\mathbf{A}\mathbf{P} = \begin{bmatrix} 0 & 1 \\ 1/3 & 0 \end{bmatrix} \begin{bmatrix} 1 & 9 \\ -1 & 1 \end{bmatrix} \begin{bmatrix} 0 & 3 \\ 1 & 0 \end{bmatrix} = \begin{bmatrix} 1 & -3 \\ 3 & 1 \end{bmatrix}$$

and

$$\mathbf{P}\mathbf{D}\mathbf{P}^{-1} = \begin{bmatrix} 0 & 3 \\ 1 & 0 \end{bmatrix} \begin{bmatrix} 1 & -3 \\ 3 & 1 \end{bmatrix} \begin{bmatrix} 0 & 1 \\ 1/3 & 0 \end{bmatrix} = \begin{bmatrix} 1 & 9 \\ -1 & 1 \end{bmatrix}$$

17.2.3 An Alternative Form for a Matrix with Complex Roots

We can decompose a matrix with complex eigenvalues into three matrices in an alternative way, which indeed simplifies its applications in systems of linear differential equations. Here instead of restricting eigenvectors to have real elements, we allow them to be complex. As a result, $\mathbf{P}$, $\mathbf{P}^{-1}$, and $\mathbf{D}$ will all be complex.

Example 17.12 For the matrix in Example 17.11 we have

$$\mathbf{P} = \begin{bmatrix} 3 & 3 \\ i & -i \end{bmatrix}, \qquad \mathbf{P}^{-1} = \begin{bmatrix} 1/6 & -0.5i \\ 1/6 & 0.5i \end{bmatrix}$$

and

$$\mathbf{D} = \begin{bmatrix} 1+3i & 0 \\ 0 & 1-3i \end{bmatrix}$$

Calculation shows that

$$\mathbf{P}^{-1}\mathbf{A}\mathbf{P} = \mathbf{D}$$

and

$$\mathbf{P}\mathbf{D}\mathbf{P}^{-1} = \mathbf{A}$$

Example 17.13 For the matrix,

$$\mathbf{B} = \begin{bmatrix} 6.0 & 2.8 \\ -3.6 & 1.1 \end{bmatrix}$$

$$\mathbf{P} = \begin{bmatrix} -0.5104 - 0.4207i & -0.5104 + 0.4207i \\ 0.7500 & 0.7500 \end{bmatrix}$$

$$\mathbf{P}^{-1} = \begin{bmatrix} 1.1885i & 0.6667 + 0.8089i \\ -1.1885i & 0.6667 - 0.8089i \end{bmatrix}$$

and

$$\mathbf{D} = \begin{bmatrix} 3.5500 + 2.0193i & 0 \\ 0 & 3.5500 - 2.0193i \end{bmatrix}$$

Again we can ascertain that

$$\mathbf{PDP}^{-1} = \mathbf{B}$$

and

$$\mathbf{P}^{-1}\mathbf{BP} = \mathbf{D}$$

17.2.4 Decomposition of a Matrix with Repeated Roots

Let $\mathbf{A}$ be an $n \times n$ matrix with eigenvalues $\lambda_1, \ldots, \lambda_n$ repeated with different multiplicities and with $\mathbf{v}_1, \ldots \mathbf{v}_n$ its generalized eigenvectors.[3] Let

$$\mathbf{P} = [\mathbf{v}_1 \; \mathbf{v}_2 \; \ldots \mathbf{v}_n] \tag{17.30}$$

Then

$$\mathbf{P}^{-1}\mathbf{AP} = \mathbf{D} + \mathbf{N} \tag{17.31}$$

where again $\mathbf{D}$ is a diagonal matrix whose diagonal elements are eigenvalues of $\mathbf{A}$, and $\mathbf{N}$ is a nilpotent matrix of order k, where k is the highest multiplicity among its eigenvalues. $\mathbf{N}$ is of the form

$$\mathbf{N} = \begin{bmatrix} 0 & 1 & 0 & \ldots & 0 \\ 0 & 0 & 1 & \ldots & 0 \\ & & & \ddots & \\ 0 & 0 & 0 & \ldots & 1 \\ 0 & 0 & 0 & \ldots & 0 \end{bmatrix} \tag{17.32}$$

Because $\mathbf{N}$ is a nilpotent of order k, we have

$$\mathbf{N}^{k-1} \neq \mathbf{0} \qquad \text{and} \qquad \mathbf{N}^k = \mathbf{0}$$

Alternatively, we can write $\mathbf{A}$ as

$$\mathbf{A} = \mathbf{S} + \mathbf{N}^* \tag{17.33}$$

where $\mathbf{S}$ can be diagonalized with diagonal elements equal to the eigenvalues of $\mathbf{A}$:

$$\mathbf{P}^{-1}\mathbf{SP} = \mathbf{D} \qquad \text{or} \qquad \mathbf{S} = \mathbf{PDP}^{-1} \tag{17.34}$$

and $\mathbf{N}^*$ is a nilpotent matrix such that

$$\mathbf{P}^{-1}\mathbf{N}^*\mathbf{P} = \mathbf{N} \qquad \text{or} \qquad \mathbf{N}^* = \mathbf{PNP}^{-1} \tag{17.35}$$

[3] The generalized eigenvector is explained in Chap. 7.

Example 17.14 Consider the matrix

$$\mathbf{A} = \begin{bmatrix} 2 & 1 & 4 \\ 0 & 2 & 1 \\ 0 & 0 & 2 \end{bmatrix}$$

which has $\lambda = 2$ as its eigenvalue with multiplicity 3. We have

$$\mathbf{P} = \begin{bmatrix} 1 & 0 & 0 \\ 0 & 1 & -4 \\ 0 & 0 & 1 \end{bmatrix}, \qquad \mathbf{P}^{-1} = \begin{bmatrix} 1 & 0 & 0 \\ 0 & 1 & 4 \\ 0 & 0 & 1 \end{bmatrix}$$

$$\mathbf{P}^{-1}\mathbf{A}\mathbf{P} = \begin{bmatrix} 2 & 1 & 0 \\ 0 & 2 & 1 \\ 0 & 0 & 2 \end{bmatrix} = \begin{bmatrix} 2 & 0 & 0 \\ 0 & 2 & 0 \\ 0 & 0 & 2 \end{bmatrix} + \begin{bmatrix} 0 & 1 & 0 \\ 0 & 0 & 1 \\ 0 & 0 & 0 \end{bmatrix} = \mathbf{D} + \mathbf{N}$$

The same would be true in the case of repeated complex eigenvalues, except that in such cases we have

$$\mathbf{P}^{-1}\mathbf{A}\mathbf{P} = \begin{bmatrix} a & -b & 1 & 0 & & & \\ b & a & 0 & 1 & & & \\ & & a & -b & 1 & 0 & \\ & & b & a & 0 & 1 & \\ & & & & & & \ddots \end{bmatrix}$$

$$= \begin{bmatrix} a & -b & & & \\ b & a & & & \\ & & a & -b & \\ & & b & a & \\ & & & & \ddots \end{bmatrix} + \begin{bmatrix} & & 1 & 0 & & & \\ & & 0 & 1 & & & \\ & & & & 1 & 0 & \\ & & & & 0 & 1 & \\ & & & & & & \ddots \end{bmatrix}$$

$$= \mathbf{D} + \mathbf{N}$$

Example 17.15 For the matrix

$$\mathbf{A} = \begin{bmatrix} 0 & -1 & 0 & 0 \\ 1 & 0 & 0 & 0 \\ 0 & 0 & 0 & -1 \\ 2 & 0 & 1 & 0 \end{bmatrix}$$

The characteristic polynomial is

$$\lambda^4 + 2\lambda^2 + 1 = (\lambda^2 + 1)^2$$

and the eigenvalues are $\lambda = \pm i$ with multiplicity two. Calculating the generalized eigenvalues (for details, see Chap. 7) and forming the **P** matrix, we have

$$\mathbf{P} = \begin{bmatrix} 0 & 0 & 0 & 1 \\ 0 & 0 & -1 & 0 \\ 1 & 0 & 0 & 0 \\ 0 & 1 & -1 & 0 \end{bmatrix}$$

Thus,

$$\mathbf{P}^{-1}\mathbf{AP} = \begin{bmatrix} 0 & 0 & 1 & 0 \\ 0 & -1 & 0 & 1 \\ 0 & -1 & 0 & 0 \\ 1 & 0 & 0 & 0 \end{bmatrix} \begin{bmatrix} 0 & -1 & 0 & 0 \\ 1 & 0 & 0 & 0 \\ 0 & 0 & 0 & -1 \\ 2 & 0 & 1 & 0 \end{bmatrix} \begin{bmatrix} 0 & 0 & 0 & 1 \\ 0 & 0 & -1 & 0 \\ 1 & 0 & 0 & 0 \\ 0 & 1 & -1 & 0 \end{bmatrix}$$

$$= \begin{bmatrix} 0 & -1 & 0 & 0 \\ 1 & 0 & 0 & 0 \\ 0 & 0 & 0 & -1 \\ 0 & 0 & 1 & 0 \end{bmatrix} + \begin{bmatrix} 0 & 0 & 1 & 0 \\ 0 & 0 & 0 & 1 \\ 0 & 0 & 0 & 0 \\ 0 & 0 & 0 & 0 \end{bmatrix}$$

17.2.5 Exercises

E.17.1 Find the Jordan canonical form of the following matrices.

$$\begin{bmatrix} 1 & 1 \\ 0 & -1 \end{bmatrix} \quad \begin{bmatrix} 3 & -1 \\ 1 & -2 \end{bmatrix} \quad \begin{bmatrix} 3 & 4 \\ 2 & 7 \end{bmatrix} \quad \begin{bmatrix} 1 & -5 \\ -4 & 0 \end{bmatrix}$$

$$\begin{bmatrix} 2 & -3 \\ -5 & 4 \end{bmatrix} \quad \begin{bmatrix} 2.5 & 1.5 \\ -1.5 & 3.0 \end{bmatrix} \quad \begin{bmatrix} 12 & -11 \\ 10 & 21 \end{bmatrix} \quad \begin{bmatrix} 5 & 3 \\ -2 & 9 \end{bmatrix}$$

$$\begin{bmatrix} 1 & -1 \\ 0 & 1 \end{bmatrix} \quad \begin{bmatrix} 1 & -1 \\ 1 & 4 \end{bmatrix}$$

E.17.2 Find the Jordan canonical form of the following matrices.

$$\begin{bmatrix} -2 & 1 & 0 \\ 0 & -2 & 1 \\ 0 & 0 & -2 \end{bmatrix} \quad \begin{bmatrix} 2 & 3 & -2 \\ 0 & 5 & 4 \\ 1 & 0 & -1 \end{bmatrix} \quad \begin{bmatrix} 1 & 2 & 3 \\ -2 & 3 & 2 \\ -3 & -1 & 1 \end{bmatrix}$$

E.17.3 Use Matlab to check your results for E.17.1 and E.17.2. Recall that Matlab normalizes the eigenvectors. You need to normalize your eigenvectors or reverse the normalizing process of Matlab to be able to check your results.

17.3 Exponential of a Matrix

In some applications, particularly for the solution of a system of linear differential equations, the exponential of a matrix plays a crucial role. We are interested in finding

$$\exp(\mathbf{A}t) \tag{17.36}$$

where $\mathbf{A}$ is an $n \times n$ matrix and t is a variable. The Taylor series of (17.47) near the point $t = 0$ is

$$\exp(\mathbf{A}t) = \sum_{j=0}^{\infty} \frac{1}{j!} \mathbf{A}^j t^j \tag{17.37}$$

If $\mathbf{A}$ is a diagonal matrix with diagonal elements $a_k, k = 1, \ldots, n$, then because

$$\sum_{j=0}^{\infty} \frac{1}{j!} \mathbf{A}^j t^j = \begin{bmatrix} \sum_{j=0}^{\infty} \frac{1}{j!} a_1^j t^j & 0 & \ldots & 0 \\ 0 & \sum_{j=0}^{\infty} \frac{1}{j!} a_2^j t^j & \ldots & 0 \\ \vdots & \vdots & \ddots & \vdots \\ 0 & 0 & \ldots & \sum_{j=0}^{\infty} \frac{1}{j!} a_n^j t^j \end{bmatrix} \tag{17.38}$$

$\exp(\mathbf{A}t)$ will also be diagonal with diagonal elements of the form

$$\sum_{j=0}^{\infty} \frac{1}{j!} a_k^j t^j = e^{a_k t} \qquad k = 1, \ldots, n \tag{17.39}$$

Example 17.16 Let

$$\mathbf{A} = \begin{bmatrix} 7 & 0 \\ 0 & -4 \end{bmatrix}$$

Then

$$\exp(\mathbf{A}t) = \begin{bmatrix} e^{7t} & 0 \\ 0 & e^{-4t} \end{bmatrix}$$

If $\mathbf{A}$ is not diagonal, then we can consider its Jordan canonical form which we recall, depending on the nature of eigenvalues, can take three different forms (Sect. 17.2). We discuss each case in turn.

17.3.1 Real Distinct Roots

Suppose we have a nondiagonal matrix $\mathbf{A}$ with real distinct eigenvalues, and we are interested in finding $\exp(\mathbf{A}t)$. If we can transform $\mathbf{A}$ into a diagonal matrix, we are home free. Because the matrix has real and distinct eigenvalues, we know that we can write its Jordan canonical form as

$$\mathbf{A} = \mathbf{PDP}^{-1} \tag{17.40}$$

where $\mathbf{D}$ is diagonal with its diagonal elements equal to eigenvalues of $\mathbf{A}$. Substituting (17.40) in (17.37), we can write

$$\exp(\mathbf{A}t) = \sum_{j=0}^{\infty} \frac{1}{j!} (\mathbf{PDP}^{-1})^j t^j \tag{17.41}$$

Now note that

$$(\mathbf{PDP}^{-1})^j = \underbrace{\mathbf{PDP}^{-1}\mathbf{PDP}^{-1} \ldots \mathbf{PDP}^{-1}}_{j} = \mathbf{PD}^j\mathbf{P}^{-1} \tag{17.42}$$

Thus,

$$\exp(\mathbf{A}t) = \sum_{j=0}^{\infty} \frac{1}{j!} \mathbf{PD}^j\mathbf{P}^{-1} t^j = \mathbf{P}\exp(\mathbf{D}t)\mathbf{P}^{-1} \tag{17.43}$$

That is,

$$\exp(\mathbf{A}t) = \mathbf{P} \begin{bmatrix} e^{\lambda_1 t} & 0 & \dots & 0 \\ 0 & e^{\lambda_2 t} & \dots & 0 \\ \vdots & \vdots & \ddots & \vdots \\ 0 & 0 & \dots & e^{\lambda_n t} \end{bmatrix} \mathbf{P}^{-1} \tag{17.44}$$

where λ_j's are eigenvalues of $\mathbf{A}$.

Example 17.17 Let $\mathbf{A}$ be the same as in Example 17.10. Then

$$\begin{aligned} \exp(\mathbf{A}t) &= \mathbf{P}\exp(\mathbf{D}t)\mathbf{P}^{-1} \\ &= \begin{bmatrix} 3 & 1 \\ 2 & -1 \end{bmatrix} \begin{bmatrix} e^{7t} & 0 \\ 0 & e^{-3t} \end{bmatrix} \begin{bmatrix} 0.2 & 0.2 \\ 0.4 & -0.6 \end{bmatrix} \\ &= \begin{bmatrix} 0.6e^{7t} + 0.4e^{-3t} & 0.6e^{7t} - 0.6e^{-3t} \\ 0.4e^{7t} - 0.4e^{-3t} & 0.4e^{7t} + 0.6e^{-3t} \end{bmatrix} \end{aligned}$$

17.3.2 Complex Roots

We are interested in finding $\exp(\mathbf{A}t)$ where $\mathbf{A}$ is a matrix with complex eigenvalues, that is, the k-th eigenvalue is of the form $\lambda_k = a_k \pm b_k i$. We can write $\mathbf{A}$ as

$$\mathbf{A} = \mathbf{PDP}^{-1}, \qquad \mathbf{D} = \mathbf{P}^{-1}\mathbf{AP} \tag{17.45}$$

where $\mathbf{D}$ is a block diagonal matrix and each block is of the form

$$\begin{matrix} a_k & -b_k \\ b_k & a_k \end{matrix}$$

Now

$$\exp(\mathbf{A}t) = \sum_j^{\infty} \frac{1}{j!} \left(\mathbf{PDP}^{-1}\right)^j t^j = \mathbf{P} \left(\sum_j^{\infty} \frac{1}{j!} \mathbf{D}^j t^j \right) \mathbf{P}^{-1} \tag{17.46}$$

Recall that D is a block diagonal matrix with blocks of the form

$$\begin{bmatrix} a & -b \\ b & a \end{bmatrix} = \begin{bmatrix} \Re(\lambda) & -\Im(\lambda) \\ \Im(\lambda) & \Re(\lambda) \end{bmatrix} \tag{17.47}$$

where $\Re(\lambda)$ and $\Im(\lambda)$ are, respectively, the real and imaginary parts of λ.

Using induction, we can show that

$$\begin{bmatrix} a & -b \\ b & a \end{bmatrix}^j = \begin{bmatrix} \Re(\lambda^j) & -\Im(\lambda^j) \\ \Im(\lambda^j) & \Re(\lambda^j) \end{bmatrix} \tag{17.48}$$

Therefore, for a 2×2 matrix, we have

$$\begin{aligned} \exp(\mathbf{A}t) &= \mathbf{P}\left(\sum_{j=0}^{\infty} \frac{1}{j!} \begin{bmatrix} \Re(\lambda^j) & -\Im(\lambda^j) \\ \Im(\lambda^j) & \Re(\lambda^j) \end{bmatrix} t^j \right) \mathbf{P}^{-1} \\ &= \mathbf{P} \sum_{j=0}^{\infty} \begin{bmatrix} \Re\left(\dfrac{\lambda^j t^j}{j!}\right) & -\Im\left(\dfrac{\lambda^j t^j}{j!}\right) \\ \Im\left(\dfrac{\lambda^j t^j}{j!}\right) & \Re\left(\dfrac{\lambda^j t^j}{j!}\right) \end{bmatrix} \mathbf{P}^{-1} \\ &= \mathbf{P} \begin{bmatrix} \Re(e^{\lambda t}) & -\Im(e^{\lambda t}) \\ \Im(e^{\lambda t}) & \Re(e^{\lambda t}) \end{bmatrix} \mathbf{P}^{-1} \end{aligned} \tag{17.49}$$

But,

$$e^{\lambda t} = e^{(a+bi)t} = e^{at}(\cos bt + i \sin bt)$$

Thus, (17.49) can be written as

$$\exp(\mathbf{A}t) = \mathbf{P}e^{at} \begin{bmatrix} \cos bt & -\sin bt \\ \sin bt & \cos bt \end{bmatrix} \mathbf{P}^{-1} \tag{17.50}$$

Example 17.18 Consider the matrix in Example 17.11:

$$\mathbf{A} = \begin{bmatrix} 1 & 9 \\ -1 & 1 \end{bmatrix}$$

whose eigenvalues were $1 \pm 3i$. Using the first method, we had

$$\mathbf{P} = \begin{bmatrix} 0 & 3 \\ 1 & 0 \end{bmatrix} \qquad \mathbf{D} = \begin{bmatrix} 1 & -3 \\ 3 & 1 \end{bmatrix}$$

Thus,

$$\exp(\mathbf{A}t) = \begin{bmatrix} 0 & 3 \\ 1 & 0 \end{bmatrix} e^t \begin{bmatrix} \cos 3t & -\sin 3t \\ \sin 3t & \cos 3t \end{bmatrix} \begin{bmatrix} 0 & 1 \\ 1/3 & 0 \end{bmatrix} = e^t \begin{bmatrix} \cos 3t & 3\sin 3t \\ -1/3 \sin 3t & \cos 3t \end{bmatrix}$$

Recall that we learned of two methods of diagonalizing a matrix with complex roots. Using the second method, we arrive at an identical result:

$$\mathbf{D} = \begin{bmatrix} 1+3i & 0 \\ 0 & 1-3i \end{bmatrix}$$

Therefore,

$$\exp(\mathbf{D}t) = \begin{bmatrix} e^{(1+3i)t} & 0 \\ 0 & e^{(1-3i)t} \end{bmatrix} = e^t \begin{bmatrix} \cos 3t + i \sin 3t & 0 \\ 0 & \cos 3t - i \sin 3t \end{bmatrix}$$

and

$$\exp(\mathbf{A}t) = \begin{bmatrix} 3 & 3 \\ i & -i \end{bmatrix} e^t \begin{bmatrix} \cos 3t + i \sin 3t & 0 \\ 0 & \cos 3t - i \sin 3t \end{bmatrix} \begin{bmatrix} 1/6 & -0.5i \\ 1/6 & 0.5i \end{bmatrix}$$

$$= e^t \begin{bmatrix} \cos 3t & 3 \sin 3t \\ -\frac{1}{3} \sin 3t & \cos 3t \end{bmatrix}$$

17.3.3 Repeated Roots

Recall that in the case of repeated roots the **P** matrix consists of generalized eigenvectors and we have

$$\mathbf{A} = \mathbf{P}(\mathbf{D} + \mathbf{N})\mathbf{P}^{-1} \tag{17.51}$$

where **D** is diagonal and **N** is as shown in (15.42). Thus,

$$\begin{aligned} \exp(\mathbf{A}t) &= \mathbf{P}\left(\sum_{k=0}^{\infty} \frac{1}{j!}(\mathbf{D}+\mathbf{N})^j t^j\right)\mathbf{P}^{-1} \\ &= \mathbf{P}\exp[(\mathbf{D}+\mathbf{N})t]\mathbf{P}^{-1} \\ &= \mathbf{P}\exp(\mathbf{D}t)\exp(\mathbf{N}t)\mathbf{P}^{-1} \end{aligned} \tag{17.52}$$

We have already discussed the form of $\exp(\mathbf{D}t)$ and, therefore, we concentrate on the term $\exp(\mathbf{N}t)$. Write

$$\exp(\mathbf{N}t) = \sum_{j=0}^{\infty} \frac{1}{j!}\mathbf{N}^j t^j \tag{17.53}$$

You can check that because $\mathbf{N}$ is nilpotent,

$$\mathbf{N}^j = \mathbf{0}, \qquad \forall j \geq k$$

where k is the highest multiplicity of any repeated eigenvalue of the matrix $\mathbf{N}$. Thus,

$$\begin{aligned} \exp(\mathbf{N}t) &= \sum_{j=0}^{k-1} \frac{1}{j!}\mathbf{N}^j t^j \\ &= \mathbf{I} + \mathbf{N}t + \ldots + \frac{1}{(k-1)!}\mathbf{N}^{k-1}t^{k-1} \end{aligned} \tag{17.54}$$

Substituting (17.54) in (17.52) we have

$$\exp(\mathbf{A}t) = \mathbf{P}\exp(\mathbf{D}t)\left(\mathbf{I} + \mathbf{N}t + \ldots + \frac{1}{(k-1)!}\mathbf{N}^{k-1}t^{k-1}\right)\mathbf{P}^{-1} \tag{17.55}$$

Example 17.19 Let us find $\exp(\mathbf{A}t)$ where

$$\mathbf{A} = \begin{bmatrix} 2 & 1 \\ 0 & 2 \end{bmatrix}$$

Then $\lambda_1 = \lambda_2 = 2$, and

$$\mathbf{P} = \mathbf{P}^{-1} = \begin{bmatrix} 1 & 0 \\ 0 & 1 \end{bmatrix}$$

Thus,

$$\begin{aligned} \exp(\mathbf{A}t) &= \begin{bmatrix} 1 & 0 \\ 0 & 1 \end{bmatrix}\begin{bmatrix} e^{2t} & 0 \\ 0 & e^{2t} \end{bmatrix}\left\{\begin{bmatrix} 1 & 0 \\ 0 & 1 \end{bmatrix} + \begin{bmatrix} 0 & t \\ 0 & 0 \end{bmatrix}\right\}\begin{bmatrix} 1 & 0 \\ 0 & 1 \end{bmatrix} \\ &= \begin{bmatrix} e^{2t} & 0 \\ 0 & e^{2t} \end{bmatrix}\begin{bmatrix} 1 & t \\ 0 & 1 \end{bmatrix} = \begin{bmatrix} e^{2t} & te^{2t} \\ 0 & e^{2t} \end{bmatrix} \end{aligned}$$

Example 17.20 In Example 17.14, we had

$$\mathbf{A} = \begin{bmatrix} 2 & 1 & 4 \\ 0 & 2 & 1 \\ 0 & 0 & 2 \end{bmatrix}, \quad \mathbf{P} = \begin{bmatrix} 1 & 0 & 0 \\ 0 & 1 & -4 \\ 0 & 0 & 1 \end{bmatrix}, \quad \mathbf{P}^{-1} = \begin{bmatrix} 1 & 0 & 0 \\ 0 & 1 & 4 \\ 0 & 0 & 1 \end{bmatrix}$$

and matrix **A** had eigenvalue $\lambda = 2$ with multiplicity of three. Using (17.55), we can write exp(**A**t) as

$$\mathbf{P}\begin{bmatrix} e^{2t} & 0 & 0 \\ 0 & e^{2t} & 0 \\ 0 & 0 & e^{2t} \end{bmatrix}\left\{\begin{bmatrix} 1 & 0 & 0 \\ 0 & 1 & 0 \\ 0 & 0 & 1 \end{bmatrix} + \begin{bmatrix} 0 & t & 0 \\ 0 & 0 & t \\ 0 & 0 & 0 \end{bmatrix} + \begin{bmatrix} 0 & 0 & t^2 \\ 0 & 0 & 0 \\ 0 & 0 & 0 \end{bmatrix}\right\}\mathbf{P}^{-1}$$
$$= \begin{bmatrix} 1 & 0 & 0 \\ 0 & 1 & -4 \\ 0 & 0 & 1 \end{bmatrix}\begin{bmatrix} e^{2t} & 0 & 0 \\ 0 & e^{2t} & 0 \\ 0 & 0 & e^{2t} \end{bmatrix}\begin{bmatrix} 1 & t & t^2 \\ 0 & 1 & t \\ 0 & 0 & 1 \end{bmatrix}\begin{bmatrix} 1 & 0 & 0 \\ 0 & 1 & 4 \\ 0 & 0 & 1 \end{bmatrix}$$
$$= \begin{bmatrix} e^{2t} & te^{2t} & 4te^{2t} + t^2e^{2t} \\ 0 & te^{2t} & te^{2t} \\ 0 & 0 & e^{2t} \end{bmatrix}$$

17.3.4 Exercises

E.17.4 Find the exponential of all matrices in E.17.1 and E.17.2.

17.4 Solution of Systems of Linear Differential Equations

A system of linear homogeneous equations can be written as

$$\dot{\mathbf{y}} = \mathbf{A}\mathbf{y} \tag{17.56}$$

where

$$\mathbf{y}(t) = \begin{bmatrix} y_1 \\ \vdots \\ y_n \end{bmatrix}, \qquad \dot{\mathbf{y}} = \begin{bmatrix} dy_1/dt \\ \vdots \\ dy_n/dt \end{bmatrix}$$

and **A** is a square matrix of order n. We can also have a set of initial conditions that will turn (17.56) into an initial value problem:

$$\dot{\mathbf{y}} = \mathbf{A}\mathbf{y}, \qquad \mathbf{y}(0) = \mathbf{y_0} \tag{17.57}$$

The solution to (17.57) is

$$\mathbf{y}(t) = \exp(\mathbf{A}t)\mathbf{y_0} \tag{17.58}$$

Furthermore, this solution is unique. More formally we have the fundamental theorem of linear systems.

Theorem 17.1 *The unique solution to the initial value problem in* (17.57) *where* $\mathbf{A}$ *is an* $n \times n$ *matrix and* $\mathbf{y}(\mathbf{0}) \in \Re^n$ *is given by* (17.58).

Proof First, letting $t = 0$ results in $\mathbf{y}(0) = \mathbf{y_0}$, satisfying the initial conditions. Second, note that

$$\dot{\mathbf{y}} = \mathbf{A}\exp(\mathbf{A}t)\mathbf{y_0}$$

Substituting the solution and its derivative in (17.57), we have an identity, verifying that (17.58) is indeed the solution:

$$\dot{\mathbf{y}} = \mathbf{A}\exp(\mathbf{A}t)\mathbf{y_0} = A\mathbf{y}$$

To prove that this solution is unique, first we show that if $\mathbf{y}(t)$ is a solution, then $\mathbf{z}(t) = \exp(-\mathbf{A}t)\mathbf{y}(t)$ is a constant because

$$\begin{aligned}\dot{\mathbf{z}} &= -\mathbf{A}\exp(\mathbf{A}t)\mathbf{y}(t) + \exp(-\mathbf{A}t)\dot{\mathbf{y}}(t)\\ &= -\mathbf{A}\exp(\mathbf{A}t)\mathbf{y}(t) + \exp(-\mathbf{A}t)\mathbf{A}\mathbf{y}(t)\\ &= \mathbf{0}\end{aligned}$$

The second equality is based on the fact that if $\mathbf{y}(t)$ is a solution, then $\dot{\mathbf{y}}(t) = \mathbf{A}\mathbf{y}(t)$. In addition, because the solution has to satisfy the initial condition, then $\mathbf{z} = \mathbf{y}_0$. Thus, any solution will be equal to

$$\exp(\mathbf{A}t)\mathbf{z} = \exp(\mathbf{A}t)\mathbf{y}_0$$

and therefore, the solution is unique. Given our preparatory work in the previous sections, we should have no difficulty in solving systems of linear differential equations. We have the general form of the solution and in each case the specific form of the solution depends on $\exp(\mathbf{A}t)$, which in turn hinges on the nature of the eigenvalues of matrix $\mathbf{A}$. We take up each case in turn and illustrate with examples.

17.4.1 Decoupled Systems

A decoupled system is of the form

$$\begin{aligned}\dot{y}_1 &= a_1 y_1\\ \dot{y}_2 &= a_2 y_2\end{aligned} \tag{17.59}$$

or in matrix form

$$\begin{bmatrix} \dot{y}_1 \\ \dot{y}_2 \end{bmatrix} = \begin{bmatrix} a_1 & 0 \\ 0 & a_2 \end{bmatrix} \begin{bmatrix} y_1 \\ y_2 \end{bmatrix} \tag{17.60}$$

More compactly

$$\dot{\mathbf{y}} = \mathbf{A}\mathbf{y} \tag{17.61}$$

$\mathbf{A}$ is diagonal and the diagonal elements are the eigenvalues of the matrix. The solution is

$$\mathbf{y} = \exp(\mathbf{A}t)\mathbf{c} = \begin{bmatrix} e^{a_1 t} & 0 \\ 0 & e^{a_2 t} \end{bmatrix} \begin{bmatrix} c_1 \\ c_2 \end{bmatrix} = \begin{bmatrix} c_1 e^{a_1 t} \\ c_2 e^{a_2 t} \end{bmatrix} \tag{17.62}$$

Example 17.21 Solve the following system of equations:

$$\begin{bmatrix} \dot{y}_1 \\ \dot{y}_2 \end{bmatrix} = \begin{bmatrix} 3 & 0 \\ 0 & -2 \end{bmatrix} \begin{bmatrix} y_1 \\ y_2 \end{bmatrix}$$

The solution is

$$\begin{bmatrix} y_1 \\ y_2 \end{bmatrix} = \begin{bmatrix} c_1 e^{3t} \\ c_2 e^{-2t} \end{bmatrix}$$

If initial conditions are given, then we can also determine $\mathbf{c}$.

Example 17.22 In Example 17.20 let

$$\mathbf{y}(0) = \begin{bmatrix} 5 \\ 2 \end{bmatrix}$$

Then the solution is

$$\begin{bmatrix} y_1 \\ y_2 \end{bmatrix} = \begin{bmatrix} 5e^{3t} \\ 2e^{-2t} \end{bmatrix}$$

Figure 17.2 shows the time path of variables y_1 and y_2. Note that the equilibrium (critical) point of the system is $y_1 = 0,\ y_2 = 0$. As t increases y_1 moves away from equilibrium, but y_2 approaches it. If both coefficients a_1 and a_2 were negative, both functions would approach equilibrium. A system whose components approach the equilibrium point as t increases is called stable. In other words, when a system is stable, a deviation from equilibrium causes the system to return to its equilibrium. It is clear that the system in Example 17.21 is unstable. But if both diagonal elements of matrix $\mathbf{A}$ were negative, the system would have been stable. Finally note that the coefficients a_1 and a_2 are also the eigenvalues of $\mathbf{A}$. We shall return to this issue in the next few subsections and finally will generalize the condition for the stability of a system.

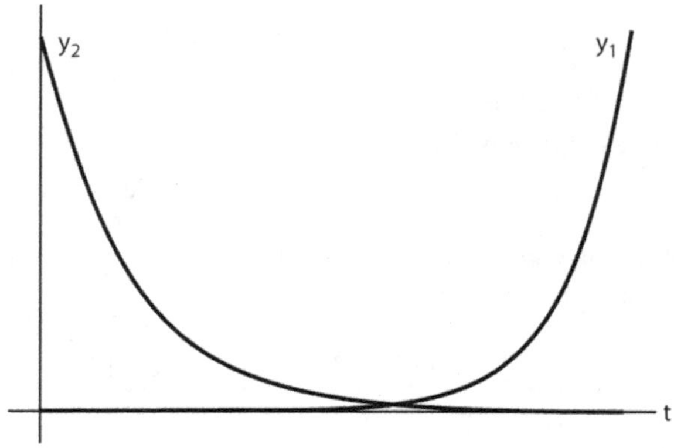

Fig. 17.2 Time path of variables in a decoupled system

17.4.2 Systems with Real and Distinct Roots

Consider the system of equations

$$\begin{aligned} \dot{y}_1 &= a_{11}y_1 + a_{12}y_2, \qquad & y_1(0) &= y_{10} \\ \dot{y}_2 &= a_{21}y_1 + a_{22}y_2, \qquad & y_2(0) &= y_{20} \end{aligned} \tag{17.63}$$

Here we are dealing with the same system as in (17.60) except that the matrix

$$\mathbf{A} = \begin{bmatrix} a_{11} & a_{12} \\ a_{21} & a_{22} \end{bmatrix}$$

is not diagonal, but it has real and distinct roots. Based on what we know about the exponential of such a matrix, the solution is

$$\begin{aligned} \mathbf{y}(t) &= \exp(\mathbf{A}t)\mathbf{y_0} \\ &= \mathbf{P}\exp(\mathbf{D}t)\mathbf{P}^{-1}\mathbf{y_0} \\ &= \mathbf{P}\begin{bmatrix} e^{\lambda_1 t} & 0 \\ 0 & e^{\lambda_2 t} \end{bmatrix}\mathbf{P}^{-1}\mathbf{y_0} \end{aligned} \tag{17.64}$$

where λ_1 and λ_2 are eigenvalues of the matrix $\mathbf{A}$.

Example 17.23 Consider the dynamic system

$$\begin{aligned} \dot{y}_1 &= 3y_1 + 6y_2 \qquad & y_1(0) &= 1 \\ \dot{y}_2 &= 4y_1 + y_2 \qquad & y_2(0) &= 2 \end{aligned}$$

The matrix

$$\mathbf{A} = \begin{bmatrix} 3 & 6 \\ 4 & 1 \end{bmatrix}$$

is not diagonal but has real and distinct eigenvalues $\lambda_1 = 7$ and $\lambda_2 = -3$. Using the result obtained in Example 17.17, the solution to the system is

$$\begin{bmatrix} y_1 \\ y_2 \end{bmatrix} = \begin{bmatrix} 0.6e^{7t} + 0.4e^{-3t} & 0.6e^{7t} - 0.6e^{-3t} \\ 0.4e^{7t} - 0.4e^{-3t} & 0.4e^{7t} + 0.6e^{-3t} \end{bmatrix} \begin{bmatrix} 1 \\ 2 \end{bmatrix} = \begin{bmatrix} 1.8e^{7t} - 0.8e^{-3t} \\ 1.2e^{7t} + 0.8e^{-3t} \end{bmatrix}$$

The equilibrium point of the system is $y_1 = 0,\ y_2 = 0$. It is clear that because one of the eigenvalues is positive, the system is not stable. Consider that

$$\lim_{t\to\infty} y_1 = \lim_{t\to\infty} \left(1.8e^{7t} - 0.8e^{-3t}\right) = \lim_{t\to\infty} 1.8e^{7t} \to \infty$$

The same is true for y_2. Now it should be clear that the stability of a system with real roots requires that both eigenvalues be negative. This conclusion will be generalized further in the following subsection.

17.4.3 Systems with Complex Roots

For a system with complex roots $\lambda = a \pm bi$, the solution will take the form

$$\mathbf{y}(t) = \mathbf{P}e^{at} \begin{bmatrix} \cos bt & -\sin bt \\ \sin bt & \cos bt \end{bmatrix} \mathbf{P}^{-1}\mathbf{y}(0) \tag{17.65}$$

Example 17.24 The solution to the system of equations

$$\begin{aligned} \dot{y}_1 &= y_1 + 9y_2 & \qquad y_1(0) &= 3 \\ \dot{y}_2 &= -y_1 + y_2 & \qquad y_2(0) &= 1 \end{aligned}$$

Is

$$\begin{bmatrix} y_1 \\ y_2 \end{bmatrix} = e^t \begin{bmatrix} \cos 3t & 3\sin 3t \\ -1/3 \sin 3t & \cos 3t \end{bmatrix} \begin{bmatrix} 3 \\ 1 \end{bmatrix} = \begin{bmatrix} 3e^t(\cos 3t + \sin 3t) \\ e^t(\cos 3t - \sin 3t) \end{bmatrix}$$

Figure 17.3 shows the time path of the branches of the system. The equilibrium point of the system is attained at $y_1 = 0,\ y_2 = 0,$ but although each branch crosses the horizontal line periodically, the two do not coincide. Moreover, amplitudes of the functions increase as t increases. But were the real part of the eigenvalues negative, then amplitudes would be decreasing and gradually both functions would converge

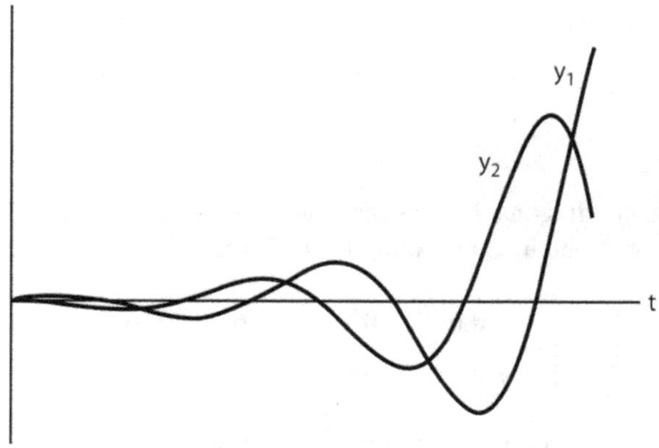

Fig. 17.3 Time path of a system with complex roots

to zero. Thus, the condition for the stability of the system is that the real part of the eigenvalues should be negative.

We can now combine all we have said so far and generalize the condition for the stability of a system.

Proposition 17.2 A system is stable if the real parts of its eigenvalues are negative. This proposition also applies to the case of repeated roots discussed below.

17.4.4 Systems with Repeated Roots

The solution of a system with repeated roots is

$$\mathbf{y}(t) = \mathbf{P}\exp(\mathbf{D}t)\left(\mathbf{I} + \mathbf{N}t + \ldots + \frac{1}{(n-1)!}\mathbf{N}^{k-1}t^{k-1}\right)\mathbf{P}^{-1}\mathbf{y}(0) \tag{17.66}$$

where k is the highest multiplicity of the eigenvalues of $\mathbf{A}$

Example 17.25 Consider the system of equations

$$\begin{aligned} \dot{y}_1 &= 2y_1 + y_2 \qquad & y_1(0) = 4 \\ \dot{y}_2 &= \ y_2 \qquad & y_2(0) = 7 \end{aligned}$$

Applying the general formula in Example (17.2), we can write the solution as

$$\begin{bmatrix} y_1 \\ y_2 \end{bmatrix} = \begin{bmatrix} e^{2t} & te^{2t} \\ 0 & e^{2t} \end{bmatrix} \begin{bmatrix} 4 \\ 7 \end{bmatrix} = \begin{bmatrix} 4e^{2t} + 7te^{2t} \\ 7e^{2t} \end{bmatrix}$$

Example 17.26 Solve the system of equations

$$\begin{aligned} \dot{y}_1 &= 2y_1 + y_2 + 4y_3 & y_1(0) &= 1 \\ \dot{y}_2 &= 2y_2 + y_3 & y_2(0) &= 1 \\ \dot{y}_3 &= 2y_3 & y_3(0) &= 1 \end{aligned}$$

Using the results obtained in Example 17.20, we can write the solution as

$$\begin{bmatrix} y_1 \\ y_2 \\ y_3 \end{bmatrix} = \begin{bmatrix} e^{2t} & te^{2t} & 4te^{2t} + t^2e^{2t} \\ 0 & te^{2t} & te^{2t} \\ 0 & 0 & e^{2t} \end{bmatrix} \begin{bmatrix} 1 \\ 1 \\ 1 \end{bmatrix}$$

or

$$\begin{aligned} y_1 &= e^{2t} + 5te^{2t} + t^2e^{2t} \\ y_2 &= 2te^{2t} \\ y_3 &= e^{2t} \end{aligned}$$

17.4.5 Exercises

E.17.5 Solve the following systems of equations

i. $\dot{y}_1 = y_1 + y_2$, $\dot{y}_2 = -y_2$

ii. $\dot{y}_1 = 3y_1 - y_2$, $\dot{y}_2 = y_1 - y_2$

iii. $\dot{y}_1 = 3y_1 + 4y_2$, $\dot{y}_2 = 2y_1 + 7y_2$

iv. $\dot{y}_1 = y_1 - 5y_2$, $\dot{y}_2 = -4y_1$

v. $\dot{y}_1 = 2y_1 - 3y_2$, $\dot{y}_2 = -5y_1 + 4y_2$

vi. $\dot{y}_1 = 12y_1 - 11y_2$, $\dot{y}_2 = 10y_1 + 21y_2$

vii. $\dot{y}_1 = 5y_1 + 3y_2$, $\dot{y}_2 = -2y_1 + 9y_2$

viii. $\dot{y}_1 = y_1 - y_2$, $\dot{y}_2 = y_2$

ix. $\dot{y}_1 = y_1 - y_2$, $\dot{y}_2 = y_1 + 4y_2$

17.5 Numerical Analysis of Systems of Differential Equations

The numerical techniques for solving differential equations, which we discussed in Chap. 15, are applicable to systems of differential equations without modification. Consider Euler's method. Here we have

$$\begin{bmatrix} \dot{y}_1 \\ \dot{y}_2 \end{bmatrix} = \begin{bmatrix} f(y_1, y_2) \\ g(y_1, y_2) \end{bmatrix} \tag{17.67}$$

Using the Taylor formula, we can write

$$\begin{aligned}\begin{bmatrix} y_1(t_{n+1}) \\ y_2(t_{n+1}) \end{bmatrix} &= \begin{bmatrix} y_1(t_n) \\ y_2(t_n) \end{bmatrix} + \begin{bmatrix} \dot{y}_1(t_n)h_n \\ \dot{y}_2(t_n)h_n \end{bmatrix} \\ &= \begin{bmatrix} y_1(t_n) \\ y_2(t_n) \end{bmatrix} + \begin{bmatrix} f(y_1(t_n), y_2(t_n))h_n \\ g(y_1(t_n), y_2(t_n))h_n \end{bmatrix}\end{aligned} \tag{17.68}$$

where the step h_n is the interval $t_{n+1} - t_n$ and determines the accuracy of approximation.

Example 17.27 Consider the following system:

$$\begin{bmatrix} \dot{y}_1 \\ \dot{y}_2 \end{bmatrix} = \begin{bmatrix} y_2 \\ -\alpha \sin y_1 - \beta y_2 \end{bmatrix}$$

Starting from the point of $y_1\,(t_0) = y_{10}$ and $y_2\,(t_0) = y_{20}$, we can simulate the model as

$$\begin{bmatrix} y_1(t_{n+1}) \\ y_2(t_{n+1}) \end{bmatrix} = \begin{bmatrix} y_1(t_n) \\ y_2(t_n) \end{bmatrix} + \begin{bmatrix} y_2(t_n)h_n \\ -[\alpha \sin y_1(t_n) + \beta y_2(t_n)]h_n \end{bmatrix}$$

In the same fashion other methods such as Runge-Kutta can be extended to systems of differential equations. Matlab has a number of functions for simulating systems of differential equations. We shall illustrate one of these functions with the help of the example above. First, we need to specify the system in a function file.

Matlab code

```
function dy = pendulum(t,y)
% Define a vector of zeros
dy = zeros(2,1);
% Define the system
dy(1) = y(2);
dy(2) = -0.5*sin(y(1))- 3*y(2);
```

Now we can simulate the model using one of several solvers in Matlab.

Matlab code

```
[T, Y] =ode15s(@pendulum, [0 20], [1, 1])
% [0 20] is the interval of solution, and [1, 1] are
% the initial conditions.
```

There are other functions available in Matlab and there are a number of refinements one can make to the procedures. We leave these issues to the reader and the `help` function of Matlab.

17.5.1 Exercises

E.17.6 Write a program to simulate the systems in E.17.5 based on (17.68).

E.17.7 Use one of Matlab's `ode15s` to solve the systems in E.17.5.

17.6 The Phase Portrait

There are many instances where we cannot explicitly solve a system of differential equations because either the system is nonlinear or because the specific forms of the functions are not specified. The latter instance happens with many economic models, such as those described in Examples 17.1 and 17.2. The reason is that economists are generally reticent about specifying the exact form of many functions, such as the utility and production functions. But we may still be interested in analyzing the behavior of the model. In particular, we may want to know if the equilibrium is stable or not, or under what initial conditions we may return to equilibrium and under what conditions the system may diverge forever. A tool for this kind of analysis is the *phase portrait*. In the solutions of the systems of differential equations obtained in Sect. 17.4, both y_1 and y_2 were functions of t. We may try to eliminate t and write y_2 as a function of y_1.

Example 17.28 In Example 17.21 the solutions were

$$y_1(t) = c_1 e^{3t}, \qquad y_2(t) = c_2 e^{-2t}$$

Solving for y_2 in terms of y_1 we have

$$y_2 = c_2 \sqrt[3]{\left(\frac{c_1}{y_1}\right)^2}$$

The same argument holds, in principle, for all linear and nonlinear systems of two differential equations because the solutions will be of the form $y_1 = F_1(t)$ and $y_2 = F_2(t)$. Although we may not be able to solve explicitly for one in terms of the other, we can write

$$y_2 = F_2(F_1^{-1}(y_1))$$

where F_1^{-1} is the inverse function of F_1 although we may not be able to write it in closed form.

Practically, given the initial conditions for any value of t, we can obtain a unique pair of y_1, y_2 and graph the point on the y_1, y_2 plane. A plane whose axes are y_1 and y_2 is called the *phase plane*, and the geometric representation of pairs of y_1 and y_2 together with the direction of their movement as t increases is called a phase portrait. The phase portrait of the decoupled system in Example 17.21 is depicted in Fig. 17.4. As can be seen the system is generally unstable. Only if the initial position

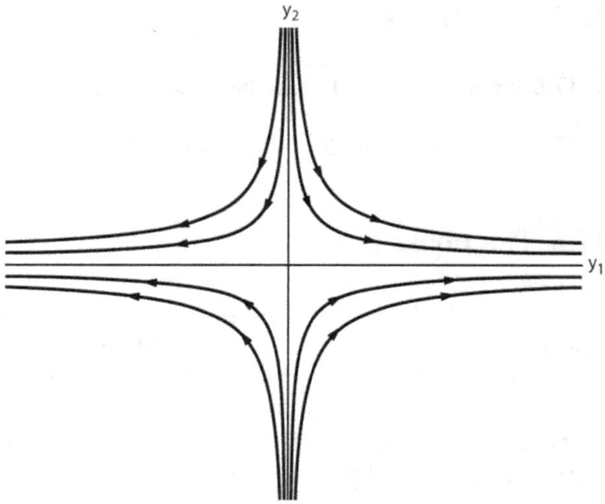

Fig. 17.4 Phase portrait of a decoupled system

of the system is on the vertical axis, that is, if we start from a point where $y_1 = 0$, then we return to the equilibrium point; else we move away from it. Figure 17.5 shows the phase portrait of the system in Example 17.24. Here we have two complex eigenvalues and the nature of the relationship between y_1 and y_2 is different from that of the previous example. But, again the system is unstable because the real part of the eigenvalues is positive. Thus, a small deviation from equilibrium starts a motion away from equilibrium.

Example 17.29 In contrast, the system

$$\dot{y}_1 = -y_1$$
$$\dot{y}_2 = -2y_2$$

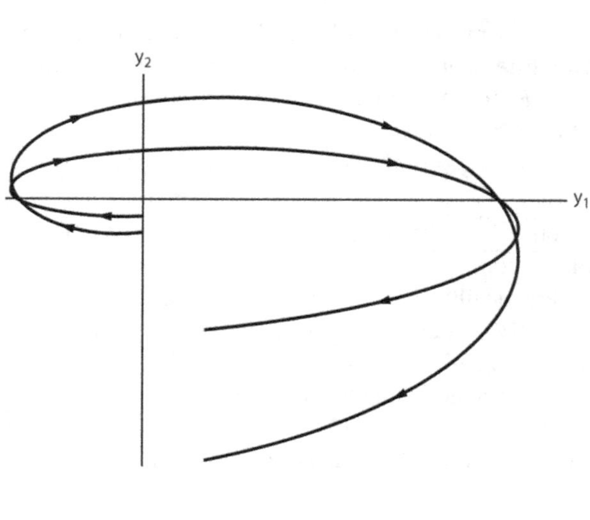

Fig. 17.5 Phase portrait of a system with complex roots

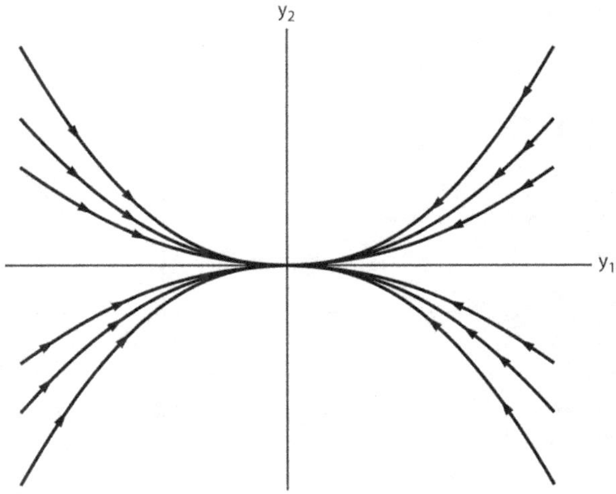

Fig. 17.6 Phase portrait of a stable system with real roots

whose phase portrait is shown in Fig. 17.6, is stable. Note that here we can write

$$y_2 = \frac{c_2}{c_1^2} y_1^2$$

where c_1 and c_2 are the initial values of y_1 and y_2, respectively.

To extend the use of a phase portrait to the stability analysis of nonlinear models, consider the system of equations

$$\begin{aligned} \dot{y}_1 &= f(y_1, y_2) \\ \dot{y}_2 &= g(y_1, y_2) \end{aligned}$$

For all combinations of y_1 and y_2 for which $f(y_1, y_2) > 0$, $\dot{y}_1 > 0$ and, therefore, for those combinations y_1 is increasing. The reverse is true for all combinations for which $f(y_1, y_2) < 0$. The same statements hold for the function g and y_2, that is, y_2 is increasing when $g(y_1, y_2) > 0$ and decreasing when $g(y_1, y_2) < 0$. Thus, based on the knowledge of the functions f and g, we can determine the direction of the movement of the system at any point in the phase plane. The direction of such movements is shown by horizontal and vertical arrows.

Example 17.30 Consider the following system:

$$\begin{aligned} \dot{y}_1 &= -y_1 \\ \dot{y}_2 &= y_1 - y_2 \end{aligned}$$

It follows that whenever y_1 is positive, it is decreasing and whenever y_2 is less than y_1, then y_2 is increasing. The movement of the system in different regions is shown in Fig. 17.7. As can be seen, the system is stable, because in every region it is directed toward the equilibrium. This is not true for an unstable system. For

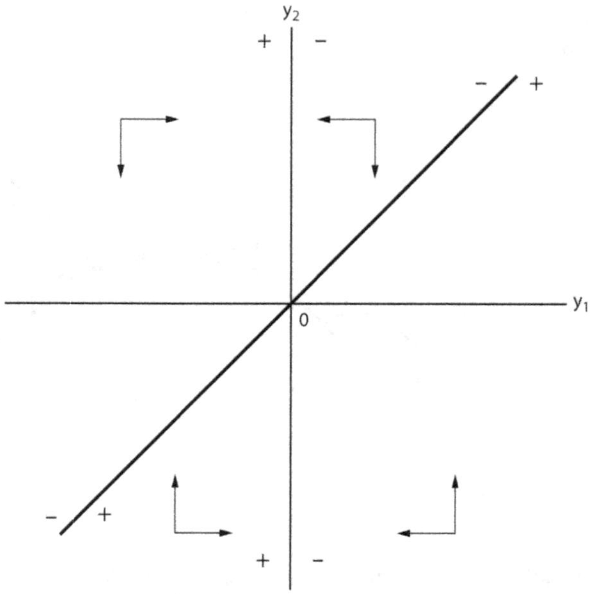

Fig. 17.7 Stability analysis of the model in example

example, if we replace the first equation with $\dot{y}_1 = y_1$, the system will not be stable in all regions. The reader may want to follow the procedure of Fig. 17.7 and ascertain that indeed the altered system is not stable everywhere.

Example 17.31 The phase portrait of the Ramsey model is shown in Fig. 17.8. As can be seen, the model is not stable everywhere. There are three equilibrium points, $k = k^*, c = c^*$ (point *E*), $k = c = 0$, and the point to the right of $k = k^*$ where

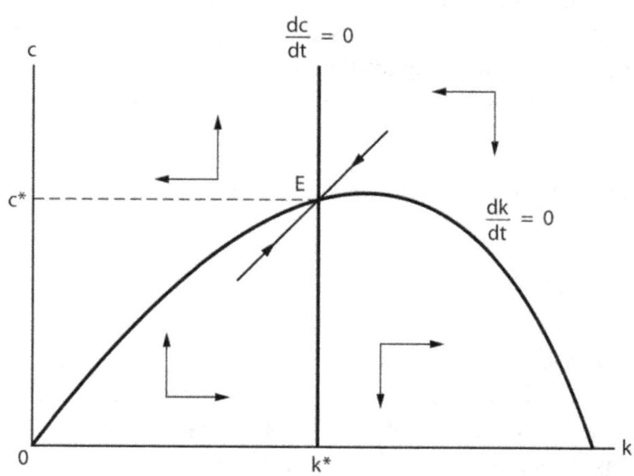

Fig. 17.8 Stability analysis of the Ramsey model

the curve $dk/dt = 0$ intersects the horizontal axis. Among the three points only E is stable. There are only two regions in which the movements of both k and c are toward the equilibrium point E. In these regions we can find a path (shown with the arrows passing through E) that converges to the equilibrium. All other paths will either diverge or end up in an unstable equilibrium.

17.6.1 Exercises

E.17.8 Draw the phase portraits of the systems in E.17.5, and analyze their stability.

Solutions to Selected Problems

Chapter 2

E.2.3 51076
E.2.5 2, 1.5
E.2.6 For 12% interest rate it is equal to \$1,673,155.37
E.2.9 110,037,600
E.2.13 For the interest rate of 8%

$$i.\, 0.0219\%, \quad ii.\, 8.3278\%, \quad iii.\, 8.3287\%$$

E.2.14 $x = 0, y = 0$, and $x = 2, y = 4$
E.2.16 $i. -1, \quad ii.\, 0.6124, \quad iii.\, 0.4957$
E.2.21

$$i. \quad \frac{1}{\tan\theta + \dfrac{1}{\tan\theta}} = \frac{1}{\dfrac{\sin\theta}{\cos\theta} + \dfrac{\cos\theta}{\sin\theta}} = \frac{\sin\theta\cos\theta}{\sin^2\theta + \cos^2\theta} = \sin\theta\cos\theta$$

E.2.22

$$i. \quad z_1 = \sqrt{2}\left(\cos\frac{\pi}{4} + i\sin\frac{\pi}{4}\right) = \sqrt{2}\exp(i\pi/4)$$

$$ii. \quad z_2 = \sqrt{2}\left(\cos\frac{\pi}{4} - i\sin\frac{\pi}{4}\right) = \sqrt{2}\exp(-i\pi/4)$$

Chapter 3

E.3.10

$$\begin{array}{lll} i. & x_1 = 1.1250 + 0.9922\mathrm{i}, & x_2 = 1.1250 - 0.9922\mathrm{i} \\ ii. & x_1 = x_2 = -3 & \\ iii. & x_1 = 2 + i, & x_2 = 2 - i \end{array}$$

K. Dadkhah, *Foundations of Mathematical and Computational Economics,* 2nd ed.,
DOI 10.1007/978-3-642-13748-8, © Springer-Verlag Berlin Heidelberg 2011

E.3.13

$$
\begin{aligned}
&x_1 = 3.678,\ \ x_2 = 0.161 + 0.8887i,\ \ x_3 = 0.161 - 0.8887i\\
&x_1 = -3.8554,\ \ x_2 = -1.1803,\ \ x_3 = 1.5178 + 1.1968i,\ \ x_4 = 1.5178 - 1.1968i\\
&x_1 = 1.4726,\ \ x_2 = -0.8133 + 2.5405i,\ \ x_3 = -0.8133 - 2.5405i\\
&x_4 = -0.4231 + 1.7234i,\ \ x_5 = -0.4231 - 1.7234i
\end{aligned}
$$

Chapter 4

E.4.1

$$
\begin{aligned}
&E(X) = 12.18, \quad E(X^2) = 4124.45,\\
&E(X^3) = 208848.75, \quad E(X^4) = 51959603.8\\
&E(Y) = 1.791, \quad E(Y^2) = 3.2363,\\
&E(Y^3) = 5.899, \quad E(Y^4) = 10.8441
\end{aligned}
$$

E.4.5

$$
\begin{aligned}
&iii.\ \ E(Z|X=0) = -1.9421, \quad E(Z|X=1) = 0.1874\\
&iv.\ \ E(Z) = E(Z|X=0)P(Z|X=0) + E(Z|X=1)P(Z|X=0)\\
&\qquad\quad = -1.9421 \times 0.52 + 0.1874 \times 0.48 = -0.92
\end{aligned}
$$

Chapter 5

E.5.2 Rounded to 2 decimal places, the norms are

$$
\|\mathbf{a}\| = 12.69, \quad \|\mathbf{b}\| = 15.17, \quad \|\mathbf{c}\| = 1.41, \quad \|\mathbf{d}\| = 12.08, \quad \|\mathbf{e}\| = 25.78
$$

E.5.3 Rounded to 2 decimal places, the distances are

$$
\begin{aligned}
&d(\mathbf{a},\mathbf{b}) = 3.32, \qquad d(\mathbf{a},\mathbf{d}) = 16.58, \qquad d(\mathbf{b},\mathbf{c}) = 15.75\\
&d(\mathbf{c},\mathbf{e}) = 26.59, \qquad d(\mathbf{d},\mathbf{e})| = 15.59
\end{aligned}
$$

E.5.4

$$
\begin{aligned}
&\theta_{\mathbf{ab}} = 0.051\pi, \qquad \theta_{\mathbf{ad}} = 0.467\pi, \qquad \theta_{\mathbf{bc}} = 0.622\pi\\
&\theta_{\mathbf{ce}} = 0.783\pi, \qquad \theta_{\mathbf{de}} = 0.146\pi
\end{aligned}
$$

E.5.10 Denoting the orthogonal vectors by $\mathbf{w}_i$, $i = 1, 2, 3$, **we have:**

$$
i.\ \ \mathbf{w}_1 = \begin{bmatrix} 3 \\ 2 \\ 5 \end{bmatrix} \quad \mathbf{w}_2 = \begin{bmatrix} 1.1579 \\ 0.1053 \\ 0.7368 \end{bmatrix} \quad \mathbf{w}_3 = \begin{bmatrix} -2.8246 \\ -3.4388 \\ -4.5436 \end{bmatrix}
$$

Chapter 6

E.6.1

$$ii.\ \mathbf{B}(\mathbf{D}+\mathbf{G}') = \begin{bmatrix} 87 & 105 & 57 \\ -52 & -8 & 4 \end{bmatrix}, \quad vi.\ \mathbf{B}'\mathbf{C}' = \begin{bmatrix} 15 & -9 \\ -8 & -22 \end{bmatrix}$$

E.6.5 For the third matrix, we have

$$\Delta = 60, \quad tr = 15, \quad \rho = 3$$

and the inverses is

$$\frac{1}{60}\begin{bmatrix} 53 & -29 & -51 \\ 14 & -2 & -18 \\ -5 & -18 & 15 \end{bmatrix} = \begin{bmatrix} 0.8833 & -0.4833 & -0.85 \\ 0.2333 & -0.0333 & -0.3 \\ -0.0833 & -0.3 & 0.25 \end{bmatrix}$$

E.6.9

$$\begin{array}{ll} i. & x = 1, \quad y = 2, \\ iv. & x_1 = 15, \quad x_2 = -20, \quad x_3 = 12, \\ v. & x_1 = x_2 = x_3 = 0 \end{array}$$

Chapter 7

E.7.1 $\mathbf{C}$ is positive definite, because

$$3 > 0, \quad \begin{vmatrix} 3 & -6 \\ -6 & 18 \end{vmatrix} = 18 > 0, \quad \begin{vmatrix} 3 & -6 & -2 \\ -6 & 18 & 5 \\ -2 & 5 & 14 \end{vmatrix} = 225 > 0$$

E.7.3 Matrix $\mathbf{X}'\mathbf{X}$ is positive semidefinite because for any vector $\mathbf{z}$, we can write:

$$\mathbf{z}'\mathbf{X}'\mathbf{X}\mathbf{z} = \mathbf{y}'\mathbf{y} = \sum_{i=1}^{n} y_i^2 \geq 0$$

E.7.14

$$\mathbf{X}'\mathbf{e} = \mathbf{X}'[\mathbf{y} - \mathbf{X}(\mathbf{X}'\mathbf{X})^{-1}\mathbf{X}'\mathbf{y}] = \mathbf{X}'\mathbf{y} - \mathbf{X}'\mathbf{y} = \mathbf{0}$$

$$\mathbf{y}'\mathbf{y} = (\hat{\mathbf{y}}' + \mathbf{e}')(\hat{\mathbf{y}} + \mathbf{e}) = \hat{\mathbf{y}}'\hat{\mathbf{y}} + \mathbf{e}'\mathbf{e} + 2\hat{\mathbf{y}}'\mathbf{e}$$

but

$$\hat{\mathbf{y}}'\mathbf{e} = \mathbf{y}'\mathbf{X}(\mathbf{X}'\mathbf{X})^{-1}\mathbf{X}'\mathbf{e} = \mathbf{0}$$

E.7.16 Eigenvalues and eigenvectors of the last matrix are:

$$\lambda_1 = 1.1816, \qquad \lambda_1 = 9.1665, \qquad \lambda_1 = 119.6519$$

$$\mathbf{v_1} = \begin{bmatrix} -0.9344 \\ -0.3042 \\ 0.1856 \end{bmatrix}, \quad \mathbf{v_2} = \begin{bmatrix} 0.3293 \\ -0.9360 \\ 0.1240 \end{bmatrix}, \quad \mathbf{v_3} = \begin{bmatrix} 0.1360 \\ 0.1770 \\ 0.9748 \end{bmatrix}$$

E.7.17

$$\pi^* = \begin{bmatrix} 0.8743 & 0.0946 & 0.0311 \end{bmatrix}'$$

E.7.19 Since $\mathbf{Ax} = \lambda\mathbf{x}$, we have

$$\mathbf{A^{-1}Ax} = \mathbf{x} = \lambda\mathbf{A^{-1}x} \quad \Rightarrow \quad \mathbf{A^{-1}x} = \frac{1}{\lambda}\mathbf{x}$$

Chapter 8

E.8.4

i. 117, *ii.* doesn't exist, *iii.* 4.828314, *iv.* 0.000227, *v.* 0

E.8.6

$$i. \ -\frac{2}{x^3}, \qquad ii. \ \frac{1}{\sqrt{2x}}, \qquad iii. \ \frac{2x+1}{2\sqrt{x^3}}, \qquad iv. \ \frac{3}{2}x^{-1/2} + \frac{1}{3}x^{-2/3} - x^{-2}$$

E.8.8 *i.* Monotonically increasing because for $x > 0$

$$\frac{dy}{dx} = \frac{1}{x} > 0$$

iv. Monotonically decreasing because

$$\frac{dy}{dx} = -9x^2 - 2 < 0$$

E.8.11

$$i. \ \frac{6}{x^4}, \qquad ii. \ -\frac{1}{\sqrt{(2x)^3}}, \qquad iii. \ -\frac{2x+3}{4\sqrt{x^5}}, \qquad iv. \ -\frac{3}{4}x^{-3/2} - \frac{2}{9}x^{-5/3} + 2x^{-3}$$

Chapter 9

E.9.1

$$i.\ \frac{\partial u}{\partial x} = 2x + 2y, \quad ii.\ \frac{\partial z}{\partial x} = \frac{y^2 - x^2}{(x^2 + y^2)^2}, \quad iii.\ \frac{\partial z}{\partial x} = \sin^2 y$$

$$\frac{\partial u}{\partial y} = 2x + 3y^2, \quad \frac{\partial z}{\partial y} = -\frac{2xy}{(x^2 + y^2)^2}, \quad \frac{\partial z}{\partial y} = x \sin 2y$$

E.9.2

$$i.\ \frac{\partial^2 u}{\partial x^2} = 2, \quad ii.\ \frac{\partial^2 z}{\partial x^2} = \frac{2x^3 - 6xy^2}{(x^2 + y^2)^3}, \quad iii.\ \frac{\partial^2 z}{\partial x^2} = 0$$

$$\frac{\partial^2 u}{\partial y^2} = 6y, \quad \frac{\partial^2 z}{\partial y^2} = \frac{-2y^3 + 6x^2 y}{(x^2 + y^2)^3}, \quad \frac{\partial^2 z}{\partial y^2} = 2x \cos 2y$$

$$\frac{\partial^2 u}{\partial x \partial y} = 2 \quad \frac{\partial^2 z}{\partial x \partial y} = \frac{-2x^3 - 6xy^2}{(x^2 + y^2)^3} \quad \frac{\partial^2 z}{\partial x \partial y} = \sin 2y$$

E.9.5 For the CES production function, we have

$$f_K = \alpha K^{\rho-1}(\alpha K^\rho + \beta L^\rho)^{\frac{1-\rho}{\rho}} \qquad f_L = \beta L^{\rho-1}(\alpha K^\rho + \beta L^\rho)^{\frac{1-\rho}{\rho}}$$

$$\sigma = \frac{d(K/L)}{d(f_L/f_K} \cdot \frac{f_L/f_K}{K/L} = \frac{1}{1-\rho}$$

E.9.9

$$i.\ \frac{dy}{dt} = \frac{3(\gamma_1 + \gamma_2)}{(\gamma_2 t + 3)^2}, \quad iii.\ \frac{du}{dt} = t^{t^2}\left((1 + 2\ln t - \frac{1}{t^2}\right)$$

E.9.10 *ii.*, *iv.* and *v.* are not homogeneous. *i.* is homogeneous of degree 3, and *iii.* of degree *a* - *b*.

E.9.13

$$i.\ \frac{dx_2}{dx_1} = -\frac{x_2 - \beta}{x_1 - \alpha}, \qquad ii.\ \frac{dx_2}{dx_1} = -\frac{\alpha x_2}{\beta x_1}$$

Chapter 10

E.10.1

$$ii.\ \sqrt{2x} = \sqrt{2x_0} + \frac{1}{\sqrt{2x_0}}(x - x_0) - \frac{1}{2\sqrt{(2x_0)^3}}(x - x_0)^2 + \frac{1}{2\sqrt{(2x)^5}}(x - x_0)^3$$

$$iii.\ e^{3x} = e^{3x_0}[1 + 3(x - x_0) + \frac{9}{2}(x - x_0)^2 + \frac{9}{2}(x - x_0)^3]$$

E.10.2

$$\begin{aligned} &i.\ \cos(2x) = -1 + 0 + 2\left(x - \tfrac{\pi}{2}\right)^2 + 0, \\ &ii.\ \sin(2x) = 0 - \left(x - \tfrac{\pi}{2}\right) + 0 + \tfrac{8}{6}\left(x - \tfrac{\pi}{2}\right)^3 \end{aligned}$$

E.8.15

$$E(e^y) = e^{\hat{y}}, \qquad \mathrm{var}(e^y) = \hat{\sigma}^2 e^{2\hat{y}}$$

Chapter 11

E.11.1

$$\begin{aligned} &i.\ \int \frac{dx}{x\ln x} = \int \frac{d\ln x}{\ln x} = \ln|\ln x| + C \\ &ii.\ \int (x^2 + x + \sqrt{x})dx = \frac{1}{3}x^3 + \frac{1}{2}x^2 + \frac{2}{3}x^{3/2} + C \\ &iii.\ \int 6x^5 dx = x^6 + C \\ &iv.\ \int \frac{dx}{\cos^3 3x} = \frac{1}{3}\int \frac{d3x}{\cos^3 3x} = \frac{1}{3}\int \sec^2 3x d3x = \frac{1}{3}\tan 3x + C \\ &v.\ \int \frac{dx}{2x-4} = \frac{1}{2}\int \frac{d(2x-4)}{2x-4} = \frac{1}{2}\ln|2x-4| + C \end{aligned}$$

E.11.2

$$\begin{aligned} &i.\ \int_0^2 x^3 dx = \left.\frac{x^4}{4}\right|_0^2 = 4 \\ &ii.\ \int_0^1 2e^x dx = 2e^x|_0^1 = 2e - 2 \\ &iii.\ \int_{\pi/2}^{\pi} \cos x dx = \sin x|_{\pi/2}^{\pi} = 0 - 1 = -1 \\ &iv.\ \int_{-\pi/2}^{0} \sin x dx = (-\cos x)|_{-\pi/2}^{0} = -1 \\ &v.\ \int_0^3 \frac{dx}{1+x} = \ln(1+x)|_0^3 = \ln 4 = 1.3862944 \end{aligned}$$

E.11.8

$$\begin{aligned} i.\ \Gamma(\alpha+1) &= \int_0^\infty e^{-t}t^\alpha dt \\ &= (-e^{-t}t^\alpha)|_0^\infty - \int_0^\infty (-e^{-t})\alpha t^{\alpha-1} dt \\ &= \alpha \int_0^\infty (-e^{-t})t^{\alpha-1} dt \\ &= \alpha\Gamma(\alpha) \end{aligned}$$

E.11.10

$$ii.\ \frac{1}{2x-1}, \qquad iii.\ 5, \qquad v.\ \int_{-y}^{y} \frac{x}{y}dx + 2y\ln y$$

Chapter 12

E.12.1

$$\begin{array}{llll} i. & x=1, \quad y=4, & \min & \\ ii. & x_1=5.6458, & y_1=-23.3468, & \min \\ & x_2=0.3543, & y_2=1.3468, & \max \\ viii. & x=0, \quad y=1, & \max & \\ ix. & x=0, \quad y=0, & \min & \end{array}$$

E.12.4

$$\begin{array}{lllll} i. & x=2, & y=5, & z=-3 & \min \\ v. & x=\dfrac{1}{3}, & y=\dfrac{1}{2}, & z=\dfrac{1}{432} & \max \end{array}$$

Chapter 13

E.13.1

$$\begin{array}{lllll} i. & x=\dfrac{16}{3}, & y=\dfrac{20}{3}, & z=\dfrac{29}{3}, & \lambda=-\dfrac{20}{3} \\ ii. & x=7.5, & y=7.5, & z=4.0298, & \lambda=-\dfrac{2}{15} \end{array}$$

E.13.2

$$x_1=46.12546, \; x_2=41.51292, \; x_3=37.36162, \quad z=10.1175, \; \lambda=0.02168$$

E.13.11

$$\begin{array}{lllll} i. & x=7, & y=5, & \mu=-7, & \lambda=-2 \\ ii. & x=6, & y=6, & \mu=-6, & \lambda=0 \end{array}$$

E .13.12

$$x=5, \quad y=5, \quad z=1, \quad \mu_1=-\frac{5}{2}, \quad \mu_2=-\frac{25}{2}, \quad \lambda=-10$$

Chapter 14

E.14.1

$$\begin{array}{llll} i. & -\dfrac{d}{dt}(t+2y')=0 & \Rightarrow & y''=-\dfrac{1}{2} \\ ii. & 2y+4y'-\dfrac{d}{dt}(4y+6y')=0 & \Rightarrow & 6y''-2y=0 \end{array}$$

$$iii.\quad -2(t-y)=0 \qquad \Rightarrow \quad y=t$$
$$iv.\quad 2y+2e^t-\frac{d}{dt}(2y')=0 \qquad \Rightarrow \quad y''-y=e^t$$

E.14.2

$$i.\ f_{y'y'}=2,\ \ \text{min}, \qquad ii.\ f_{y'y'}=6,\ \ \text{min}$$
$$iii.\ f_{y'y'}=0,\ \ \text{min}, \qquad iv.\ f_{y'y'}=2,\ \ \text{min}$$

E.14.4 The least cost paths are: ABFIJ and ADEIJ.

E.14.5

$$x\frac{1-\delta^k}{1-\delta}=\delta E[v_{n-k+1}(y)] \quad \Rightarrow \quad x=\frac{\delta(1-\delta)}{1-\delta^k}E[v_{n-k+1}(y)]$$

E.14.6

$$i.\ H=yu-2y^2-5u^2+\lambda(y+3u)$$
$$u-4y+\lambda=-\lambda'$$
$$y-10u+3\lambda=0$$
$$y'=y+3u$$

$$ii.\ H=(3y+u^2)^{1/2}+2\lambda u$$
$$\frac{3}{2\sqrt{y+u^2}}=-\lambda'$$
$$\frac{u}{\sqrt{y+u^2}}+2\lambda=0$$
$$y'=2u$$

Chapter 15

E.15.1

$$i.\ \ y=Ae^{-4t}, \qquad vi.\ y=Ae^{t^2},$$
$$vii.\ y=Ae^{-t}+\frac{1}{2}e^t, \qquad xii.\ \frac{t}{y}=C$$

E.15.2

$$i.\ \ y=3e^{-0.4t}$$
$$v.\ \ y=\ln(t+12.1825),$$
$$vi.\ y=\left(y_0+\frac{1}{5}\right)e^{2t}-\frac{2}{5}\sin t-\frac{1}{5}\cos t$$

E.15.5

$$i.\ \ y=A_1e^{-4t}+A_2e^{-3t}$$
$$vi.\ \ y=e^{-0.5t}(B_1\cos 2.4t+B_2\sin 2.4t)$$
$$=e^{-0.5t}(B_1\cos 0.76\pi t+B_2\sin 0.76\pi t)$$

E.15.6

$$
\begin{aligned}
&i.\quad y = A_1 e^{-4t} + A_2 e^{-3t} - \frac{5}{12}\\
&vi.\quad y = e^{-0.5t}(B_1 \cos 2.4t + B_2 \sin 2.4t) + \frac{3}{4} + \frac{1}{2}t
\end{aligned}
$$

E.15.7

$$
\begin{aligned}
&i.\quad y = \frac{9}{7}e^{-4t} + \frac{5}{7}e^{-3t}\\
&v.\quad y = e^{-0.5t}(\cos 2.4t + 0.625 \sin 2.4t)
\end{aligned}
$$

Chapter 16

E.16.2

$$
L = E_t \sum_{j=t}^{T} \frac{U(C_j)}{(1+\delta)^{j-t}} + \lambda \left(\sum_{j=t}^{T} \frac{C_j - W_j}{(1+r)^{j-t}} - A_t \right)
$$

$$
\frac{\partial L}{\partial C_j} = E_t \frac{U'(C_j)}{(1+\delta)^{j-t}} + \frac{\lambda}{(1+r)^{j-t}} = 0
$$

$$
\text{for } j = t+1 : \quad E_t \frac{U'(C_j)}{1+\delta} + \frac{\lambda}{1+r} = 0
$$

$$
\text{for } j = t : \quad U'(C_t) + \lambda = 0
$$

$$
\therefore \quad E_t U'(C_{t+1}) \frac{1+r}{1+\delta} = U'(C_t), \quad E_t U'(C_{t-1}) = \frac{1+\delta}{1+r} U'(C_t)
$$

E.16.5

$$
\begin{aligned}
&i.\quad y_t = (A-12)0.5^t + 12\\
&ii.\quad y_t = (A+0.93)1.75^t - 0.4t - 3.11\\
&vii.\quad y_t = 14.5(2)^t - 0.5t - 6\\
&viii.\quad y_t = 2(1.5)^t + \sum_{i=0}^{t-1} 1.5^i x_{t-i}
\end{aligned}
$$

E.16.11

$$
\begin{aligned}
&i.\quad y_t = A_1(-3)^t + A_2\\
&ii.\quad y_t = A_1(0.6)^t + A_2(0.2)^t\\
&iii.\quad y_t = 1.71^t[(A_1 + A_2)\cos(1.45t) + (A_1 - A_2)i \sin(1.45t)]\\
&iv.\ y_t = A_1(1.5)^t + A_2 t(1.5)^t\\
&v.\ y_t = A_1(0.6)^t + A_2 t(0.6)^t
\end{aligned}
$$

Chapter 17

E.17.5

$$i.\quad \begin{bmatrix} y_1 \\ y_2 \end{bmatrix} = \begin{bmatrix} 1 & -1 \\ 0 & 2 \end{bmatrix} \begin{bmatrix} e^t & 0 \\ 0 & e^{-t} \end{bmatrix} \begin{bmatrix} 1 & 0.5 \\ 0 & 0.5 \end{bmatrix} \begin{bmatrix} y_{10} \\ y_{20} \end{bmatrix}$$

$$= \begin{bmatrix} y_{10}e^t + \frac{1}{2}y_{20}e^t - \frac{1}{2}y_{20}e^{-t} \\ y_{20}e^{-t} \end{bmatrix}$$

$$v.\quad \begin{bmatrix} y_1 \\ y_2 \end{bmatrix} = \begin{bmatrix} 1 & -3 \\ 1 & 5 \end{bmatrix} \begin{bmatrix} e^{7t} & 0 \\ 0 & e^{-t} \end{bmatrix} \begin{bmatrix} 5 & 3 \\ -1 & 1 \end{bmatrix} \begin{bmatrix} y_{10} \\ y_{20} \end{bmatrix}$$

$$= \begin{bmatrix} (5y_{10} + 3y_{20})e^{-t} + 3(y_{10} - y_{20})e^{7t} \\ (5y_{10} + 3y_{20})e^{-t} - 5(y_{10} - y_{20})e^{7t} \end{bmatrix}$$

$$vii.\quad \begin{bmatrix} y_1 \\ y_2 \end{bmatrix} = \begin{bmatrix} -i & i \\ 1 & 1 \end{bmatrix} \begin{bmatrix} e^7(\cos 2t + i \sin 2t) & 0 \\ 0 & e^t(\cos 2t - i \sin 2t) \end{bmatrix}$$

$$\begin{bmatrix} 0.5i & 0.5 \\ 0.5i & 0.5 \end{bmatrix} \begin{bmatrix} y_{10} \\ y_{20} \end{bmatrix}$$

$$= \begin{bmatrix} e^t(y_{10} \cos 2t + y_{20} \sin 2t) \\ e^t(y_{20} \cos 2t - y_{10} \sin 2t) \end{bmatrix}$$

Index